EVOLUTION AND ECOLOGY OF MACAQUE SOCIETIES

EVOLUTION AND ECOLOGY OF MACAQUE SOCIETIES

Edited by

JOHN E. FA

*The International Training Centre for Breeding and
Conservation of Endangered Species, Jersey
Wildlife Preservation Trust, Les Augrès Manor,
Jersey, Channel Islands*

and

DONALD G. LINDBURG

*Center for Reproduction of Endangered Species,
Zoological Society of San Diego, USA*

CAMBRIDGE
UNIVERSITY PRESS

Published by the Press Syndicate of the University of Cambridge
The Pitt Building, Trumpington Street, Cambridge CB2 1RP
40 West 20th Street, New York, NY 10011-4211, USA
10 Stamford Road, Oakleigh, Melbourne 3166, Australia

First published 1996

Printed in Great Britain at the University Press, Cambridge

A catalogue record for this book is available from the British Library

Library of Congress cataloguing in publication data

Evolution and ecology of macaque societies / edited by John E. Fa and Donald G. Lindburg.
p.   cm.
Includes index.
ISBN 0 521 41680 9 (hc)
1. Macaques – Behavior.   2. Macaques – Ecology.   3. Macaques – Evolution.
4. Animal societies.   I. Fa, John E.   II. Lindburg, Donald G., 1932–
QL737.P93E96 1996
599.8′ – dc20   95-13515 CIP

ISBN 0 521 41680 9 hardback

VN

Contents

Contributors

Burton, F. D.
Department of Anthropology, Scarborough College, University of Toronto, Scarborough, Ontario, Canada M1C 1A4

Caldecott, J. O.
Independent consultant in conservation and biodiversity management, 79, Windsor Road, Cambridge, CB4 3JL, UK

Chan, L.
Department of Anthropology, Scarborough College, University of Toronto, Scarborough, Ontario, Canada M1C 1A4

de Ruiter, J. R.
Department of Comparative Physiology, Ethology and SocioEcology Group, University of Utrecht, De UithofPadualaan 14, University of Utrecht, Utrecht 3058 TB, The Netherlands
Present address: Institute of Zoology, Zoological Society of London, Regent's Park, London NW1 4RY, UK

Fa, J. E.
Jersey Wildlife Preservation Trust, Les Augrès Manor, Jersey, Channel Islands

Fedigan, L. M.
Department of Anthropology, University of Alberta, Edmonton, Alberta, Canada T6G 2H4

Feistner, A. T. C.
Jersey Wildlife Preservation Trust, Les Augrès Manor, Jersey, Channel Islands

Froehlich, J. W.
Department of Anthropology, University of New Mexico, Albuquerque, NM 87131-1086, USA

Gadsby, E. L.
Housing Estate P.O. Box 107, Calabar, Cross River State, Nigeria, or c/o Flora and Fauna International (FFPS), Gt Eastern House, Tenison Road,

Cambridge CB1 2BU, UK

Gonder, M. K.
Department of Anthropology, University Station, University of Alabama, Birmingham, AL 35294, USA

Griffin, L.
Department of Anthropology, University of Alberta, Edmonton, Alberta, Canada T6G 2H4

Harvey, N. C.
Center for Reproduction of Endangered Species, Zoological Society of San Diego, P.O. Box 551, San Diego, CA 92112-0551, USA

Harya Putra, D. K.
Lab. Fisiologi, Fakultas Peternakan Unud, Universitas Udayana, Denpasar, Bali, Indonesia

Hauser, M. D.
Department of Anthropology & Psychology, Program in Neuroscience, Harvard University, Cambridge, MA 02138, USA

Hill, D. A.
The Center for African Area Studies, Kyoto University, Shimoadachi-cho, Sakyo-ku, Kyoto 606, Japan
Present address: School of Biological Sciences, University of Sussex, Falmer Brighton, BN1 9QG, UK

Hoelzer, G. A.
Genetics Laboratory, Department of Anthropology, Columbia University, New York, NY 10027, USA
Present address: Department of Biology and Department of Environmental and Resource Sciences, University of Nevada Reno, Reno, NV 89557, USA

Iwamoto, T.
Department of Biology, Faculty of Education, Miyazaki University, 1-1 Gakuen-Kibanbadai-Nishi, Miyazaki 889-21, Japan

Jiang Haisheng
South China Institute of Endangered Animals, 105 Xingang Road West, Guangzhou, Peoples' Republic of China

Kohlhaas, A.
Environmental Population and Organismic Biology, University of Colorado at Boulder, Boulder, CO 80309-0334, USA
Present address: Department of Biological Sciences, California State University, Stanislaus, Turlock, CA 95382, USA

Kuester, J.
Abteilung Funktionelle Morphologie, Ruhr-Universität, Bochum, Postfach 102140, D-44780 Bochum, Germany

Lind, R.
Jersey Wildlife Preservation Trust, Les Augrès Manor, Jersey, Channel Islands

Lindburg, D. G.
Center for Reproduction of Endangered Species, Zoological Society of San Diego, P.O. Box 551, San Diego, CA 92112-0551, USA

Liu Zhenhe
South China Institute of Endangered Animals, 105 Xingang Road West, Guangzhou, Peoples' Republic of China

Maruhashi, T.
Department of Human and Cultural Science, Musashi University, Toyotama-kami 1-26, Nerimaku, Tokyo 176, Japan

Maryanski, A.
Department of Sociology, University of California, Riverside, CA 92521, USA

Mehlman, P. T.
Laboratory Animal Breeders & Services, P.O. Box 557, Yemassee, SC 29945, USA

Melnick, D. J.
Department of Anthropology, Columbia University, New York, NY 10027, USA

Ménard, N.
URA 373 CNRS, Université de Rennes, Station Biologique de Paimpont, 35380 Paimpont, Rennes, France

Nakagawa, N.
Shion Junior College, 6-11-1, Omika, Hitachi, Ibaraki 319-12, Japan

Oi, T.
Forestry and Forest Products Research Institute, Tohoku Research Center, 72, Nabeyashiki, Shimokuriyagawa, Morioka, Iwate 020-01, Japan

Okayasu, N.
Laboratory of Human Evolution Studies, Faculty of Science, Kyoto University, Sakyo-ku, Kyoto 606, Japan

Paul, A.
Institut Für Anthropologie, Universität Göttingen, Buegerstrasse 50, D-37073 Göttingen, Germany

Qu Wenyuan
Department of Biology, Henan Normal University, Sinxiang, Henan, People's Republic of China

Rhine, R. J.
Department of Psychology, University of California, Riverside, CA 92521, USA

Scheffrahn, W.
Institute of Anthropology, University of Zurich, Winterthurer Strasse 190, Zurich CH 8057, Switzerland

Southwick, C. H.
Environmental Population and Organismic Biology, University of Colorado at Boulder, Boulder, CO 80309-0334, USA

Soumah, A. G.
Primate Research Institute, Kyoto University, Kanrin, Inuyama, Aichi 484, Japan

Sprague, D. S.
Laboratory of Human Evolution Studies, Faculty of Science, Kyoto University, Sakyo, Kyoto 606, Japan
Present address: Center for African Area Studies, Kyoto University, Sakyo, Kyoto 606, Japan

Supriatna, J.
Department of Anthropology, University of New Mexico, Albuquerque, NM 87131, USA, and Universitas Indonesia, Depok 16424, Indonesia

Suzuki, S.
Laboratory of Human Evolution Studies, Faculty of Science, Kyoto University, Sakyo, Kyoto 606, Japan

Takasaki, H.
The Center for African Area Studies, Kyoto University, Shimoadachi-cho, Sakyo, Kyoto 606, Japan

Tsukahara, T.
Department of Anthropology, Faculty of Science, University of Tokyo, Hongu, Bunkyo-ku, Tokyo 113, Japan

Vallet, D.
URA 373 CNRS, Université de Rennes, Station Biologique de Paimpont, 35380 Paimpont, Rennes, France

van Amerongen, A.
Department of Comparative Physiology, Ethology and SocioEcology Group, University of Utrecht, De UithofPadualaan 14, University of Utrecht, The Netherlands

van Hooff, J. A. R. A. M.
Ethology and SocioEcology Group, University of Utrecht, De UithofPadualaan 14, University of Utrecht, The Netherlands

van Noordwijk, M. A.
Department of Comparative Physiology, Ethology and SocioEcology Group, University of Utrecht, De UithofPadualaan 14, University of Utrecht, The Netherlands
Present address: Department of Biological Anthropology and Anatomy, Duke University, The Wheeler Building, 3705-B Erwin Road, Durham, NC 27705, USA

van Schaik, C. P.
Department of Biological Anthropology and Anatomy, Duke University, The Wheeler Building, 3705-B Erwin Road, Durham, NC 27705, USA

Wheatley, B. P.
Department of Anthropology, University Station, University of Alabama, Birmingham, AL 35294-3350, USA

Yokota, N.
Oita Junior College, 3-3-8, Chiyo-Machi, Oita, Oita 870-91, Japan

Zeller, A.
Department of Anthropology, University of Waterloo, Waterloo, Ontario, Canada N2L 3G1

Zhang Yongzu
Institute of Geography, Academia Sinica, Beijing, People's Republic of China

Zhao Qi-Kun
Kunming Institute of Zoology, Chinese Academy of Sciences, Kunming, Yunnam 650223, People's Republic of China

Note: The names of Chinese contributors have, by request, been given in Chinese style, with the surname first, in the chapter titles and List of contributors.

General introduction

J. E. FA AND D. G. LINDBURG

Macaque studies – new insights and future perspectives

The genus *Macaca* is probably one of the most widespread primate groups in the world. Its 19 extant species are found in northwestern Africa and on a number of island and continental areas in southern and eastern Asia. The radiation of this group is considered to have taken place relatively recently, around 5 million years ago, and yet the number of species that has emerged is unequalled by any other group of primates. Such a variety of forms, species that differ in morphology and ecology, represent a most alluring group of animals for the study of problems associated with species radiation and adaptation to different environments – from semi-arid, warm and cold temperate to tropical forest habitats (to say nothing of environments now modified by man).

What we have proposed to do in this book is to give readers a look at 'variation in a most variable' primate group by offering examples of evolutionary biology and ecology that appear from studying the macaques. This book represents an effort to consolidate information now available on the wide variety of macaque species that are currently being studied. As with its counterpart, the 1980 book *The macaques: studies in ecology, behaviour and evolution*, edited by one of us, we have also tried to include reports from all recent field investigations of little-known macaques. Once again, this objective has only been partly realised since there are still glaring gaps in our knowledge of some taxa. Nonetheless, the great satisfaction in the present book is that it includes a greater number of investigations than the first as a result of the growing interest in the field. This is patently reflected in the number of contributors and the size of the book. Whereas the 1980 book comprised 13 chapters, the present book has double this number. Equally, the number of authors has gone up from 23 to 53 in line with a notable increment in field and laboratory studies of the group. While neither book has been able to embrace all the individuals working on macaques (constraints are imposed as much by the editors' awareness of field studies as by fieldworkers' readiness for publication), each represents a significant proportion of contemporary research.

The present book is divided into four main sections: Biogeography and evolution; Population biology, ecology and conservation; Mating and social systems; and Communication. Chapters are set to lead the reader from general overviews of topics, such as Hoelzer and Melnick's new insights into the evolutionary relationships of the macaques, genetic consequences of macaque social organisations as well as the intricacies of vocal and non-vocal communication, to more specific analyses and findings of a species' ecology, reproduction and social behaviour. All contributions, however, aim to present much more than new data. They cover widely their respective subject areas and in themselves are significant summaries of the 'state of the art'.

The first section of the book, on biogeography and evolution, starts with a new framework of macaque evolution. This chapter builds upon past schemas based on palaeontology, morphology and genetics but expands on this with new molecular genetic evidence. The following two chapters (Scheffrahn *et al.*; Froehlich and Supriatna) give clear examples of speciation processes in two main groups of macaques: the wide ranging long-tailed macaques and the restricted but differentiated Sulawesi macaques. Both studies employ modern genetic techniques to decipher speciation patterns.

The book's second part focuses on ecological topics. Various chapters concentrate on intra-specific comparisons of macaques living in differing habitats. This is a crucial topic for continuous study, since the plasticity shown by these primates can allow us to understand their ability to cope with habitats ranging from snowy elevations to rain forests. In this light, Chinese rhesus macaques are compared between tropical and temperate environments (Southwick *et al.*), Barbary macaque population dynamics are observed in forest and scrub in Algeria (Ménard and Vallet) and Japanese macaques (Nakagawa *et al.*; Maruhashi and Takasaki) are examined within a variety of habitats on the Japanese islands. Another chapter in this section concentrates on group dynamics of *Macaca nigrescens* in Sulawesi (Kohlhaas and Southwick) and three others take count of the influence of human intervention on the behaviour and ecology of long-tailed macaques in Bali (Wheatley *et al.*), Barbary macaques in Gibraltar (Fa and Lind) and Tibetan macaques in China (Zhao). Throughout these three papers, the conservation of the species in these human-affected environments is thoroughly discussed.

Our third section in the book presents novel approaches to deciphering reproductive and social systems. Although it has been long recognised that macaques differed significantly in mating patterns, some authors arguing for an ecological influence on this, information collected until now has been merely observational. Using the DNA fingerprinting techniques now available, we are able to probe deeper into mate choice and reproductive isolation among the different species (Burton and Chan; Lindburg and Harvey; Oi), quantify lifetime reproductive success for males and females (Paul and Kuester) and understand the mechanisms involved in the structuring of macaque societies (Hill and Okayasu; Melnick and Hoelzer; Sprague *et al.*). The importance of long-term studies is clearly epitomised by Rhine and Maryanski's 21 year study of social history and by Fedigan and Griffin's analysis of reproductive seasonality.

The last section of the book presents three chapters on communication mechanisms among macaques. Mehlman's observations reveal fascinating ways in which Barbary macaques use branch-shaking behaviour to communicate. Zeller's study of facial expression opens up a rich new field of investigation into kin recognition. The last chapter in the book, by Hauser, aptly gathers our current understanding of vocal communication in the macaques.

We are most grateful to the contributors to this volume for their interest throughout the venture (or adventure) of editing their work. We are also greatly indebted to Alan Crowden, Tracey Sanderson and Harriet Stewart for their unfailing help throughout the production of this book. Tracey and Harriet endured the last moments of gestation of the book (the latter similarly engaged in her own gestation of twins) and without them, we are certain this would have been a lesser volume. The unfailing support and constant help of Monique Williamson, the senior editor's wife, is more than gratefully acknowledged. Monique gave of her time in organising and editing papers, typing material, sometimes even translating, and all in all encouraging the completion of the book.

Part I
Biogeography and evolution

1

Evolutionary relationships of the macaques

G. A. HOELZER AND D. J. MELNICK

Introduction

The evolutionary diversification of the genus *Macaca* has yielded 19 extant species (Fooden, 1980) occupying ranges in north-western Africa and across southern and eastern Asia. In addition to the living species, numerous extinct forms have been identified from the fossil record of North Africa, Europe and Asia (Szalay & Delson, 1979; Delson, 1980). Interestingly, this radiation, one of the most extensive in primate evolutionary history, has occurred only over the past five million years (Delson, 1980). By contrast, although the hominid lineage of primates originated at about the same time as the lineage of the macaques, and in the same part of the world, it is currently represented by only a single species (*Homo sapiens*). Therefore, an accurate phylogeny of the extant taxa of macaques is critical, not only to our understanding of the evolutionary history of this genus, but to our general understanding of primate radiations. In addition, a reliable phylogeny will provide a strong foundation for studies of the evolution of macaque behaviour and ecology using the comparative method (Brooks & McLennan, 1991; Harvey & Pagel, 1991).

The genus *Macaca* is placed within the tribe Papionini, which also includes the genera *Papio* (baboons), *Mandrillus* (mandrills and drills), *Theropithicus* (geladas) and *Cercocebus* (mangabeys). Although phylogenetic relationships among these genera are being re-evaluated (e.g. Disotell, Honeycutt & Rovulo, 1992) the *Macaca* clearly appears to constitute a monophyletic group (i.e. a group that contains all descendants of the most recent common ancestor of the constituent taxa).

Species of *Macaca* have been variously separated into several species groups (see Fa (1989) for a review of this literature). However, more recently, a consensus has been reached on the existence of at least three species groups (Fooden, 1976, 1980; Delson, 1980), although the relationships among species within groups are less clear. The species groups are the *silenus* group, the *fascicularis* group and the *sinica* group (Table 1.1). The taxonomic position of two species, *M. sylvanus* and *M. arctoides*, is still debated. *Macaca sylvanus*, the lone remaining African representative of the genus, is generally considered to be the most primitive extant macaque species. It is not evident, however, whether it belongs to a monophyletic group that includes other extant macaques, or whether it should be in its own monospecific group. Similarly, *M. arctoides*, morphologically quite distinct, is sometimes included in a group with other macaques and sometimes accorded status as a monospecific group. Finally, because to date no strong evidence has been brought to bear on the branching order of a phylogenetic tree connecting the different species groups, we will confine our review to data pertaining only to the evolutionary relationships of species within species groups.

Sources of phylogenetic data

Until recently, macaque systematics was based solely on morphological characters, such as the male and female reproductive organs (Fooden, 1976, 1980; Fooden & Lanyon, 1989), or features of the dentition and cranium (Delson, 1980). However, the analysis of molecular data is substantially enhancing our ability to reconstruct accurately phylogenetic relationships (Hillis & Moritz, 1990). This chapter focuses on the contributions of molecular data to our understanding of macaque phylogeny. It is not our intention to suggest that molecular data are qualitatively superior to morphological data for phylogenetic reconstruction. Indeed, the best estimations of phylogeny should consider both morphological and molecular data. Nevertheless, it is the recent development of molecular techniques that has made possible the measurement of molecular diversity and the reinterpretation of macaque phylogeny.

To date, sources of molecular data for the reconstruction of macaque phylogeny have included allozymes (Nozawa *et al.*, 1982; Cronin, Cann & Sarich, 1980; Kawamoto, Takenaka & Brotoisworo, 1982; Melnick & Kidd, 1985; summarised in Fooden & Lanyon, 1989) and mitochondrial DNA (mtDNA) restriction fragments, restriction sites and nucleotide sequences (George, 1982; Harihara *et al.* 1988; Hayasaka *et al.*, 1986, 1988*b*; Hayasaka, Gojobari & Horai, 1988*a*; Hoelzer, Hoelzer & Melnick 1992; Zhang & Shi, 1989; Melnick & Hoelzer, 1992; Melnick *et al.*, 1993). There are advantages and disadvantages to each of these sources.

Table 1.1. *The classification of extant macaque* (Macaca)
species based on Delson (*1980*) *and Fooden* (*1976*)

Taxa	
Delson	Fooden
sylvanus group	*silenus–sylvanus* group
M. *sylvanus*	M. *sylvanus*
silenus group	M. *silenus*
M. *silenus*	M. *nemestrina*
M. *nemestrina*	M. *tonkeana*
M. *tonkeana*	M. *maura*
M. *maura*	M. *ochreata*
M. *ochreata*	M. *brunnescens*
M. *brunnescens*	M. *hecki*
M. *hecki*	M. *nigrescens*
M. *nigrescens*	M. *nigra*
M. *nigra*	
fascicularis group	*fascicularis* group
M. *mulatta*	M. *mulatta*
M. *cyclopis*	M. *cyclopis*
M. *fuscata*	M. *fuscata*
M. *fascicularis*	M. *fascicularis*
	arctoides group
	M. *arctoides*
sinica group	*sinica* group
M. *arctoides*	M. *radiata*
M. *radiata*	M. *sinica*
M. *sinica*	M. *assamensis*
M. *assamensis*	M. *thibetana*
M. *thibetana*	

Allozymes

Allozymes, like morphology, reflect variation in the nuclear genome, which includes most of the genetic information defining a macaque individual. Thus, the genetic history of the nuclear genome essentially defines the historical relationships of the taxa under study. Unfortunately, several factors limit the usefulness of allozyme data (Murphy *et al.*, 1990). First, allozymes may evolve too slowly to provide a sufficient number of phylogenetically informative characters when closely related species are compared. Second, if divergence is too great, then too few alleles will be shared. Similarly, populations that exhibit reduced genetic variability due to past population 'bottle-necks' and genetic drift, such as may be the case with many of the insular macaque species, may also share few allozyme alleles. Either way the lack of shared genetic variability results in a lack of phylogenetic information (Fooden & Lanyon, 1989). A third constraint on the use of allozymes is a practical one. The number of useful allozyme loci known and available for study is relatively small (<40), thus limiting the number of phylogenetic characters that can be measured. Finally, because allozymes represent a phenotypic expression of the genome, it is often unclear to what extent natural selection has affected the evolution of the protein (see Karl & Avise, 1992). If selection was an important factor, then convergent or parallel evolution could interfere with the reconstruction of the phylogenetic tree.

Mitochondrial DNA

Along with the nuclear genome, a small circular molecule of DNA is found in the mitochondria of eukaryotes (see reviews in Brown, 1985; Melnick, Hoelzer & Honeycutt, 1992). The analysis of mtDNA sequence variation overcomes many of the limitations inherent in allozyme data. Mitochondrial DNA evolves at a substantially faster rate than nuclear DNA (Brown, George & Wilson, 1979), so recently separated taxa can be readily distinguished. Nevertheless, the occurrence of both conserved and variable characters in mtDNA sequences allows similarity to be assessed over a relatively wide range of divergences (Hillis & Davis, 1988). In addition, the number of potential characters in a mtDNA sequence is very high (e.g. restriction sites and nucleotide base sequences). Finally, although much of the mitochondrial genome is phenotypically expressed, many neucleotide substitutions can occur without affecting the gene products, thus reducing the opportunities for adaptive convergent evolution.

The principal drawback in using mtDNA in phylogenetic research is that the distribution of mtDNA diversity among taxa may differ from the distribution of nuclear DNA variation (Melnick & Hoelzer, 1992; Melnick *et al.*, 1993). This is possible because the mode of inheritance is different for the two genomes. The haploid mitochondrial genome is inherited strictly from the mother (Hutchinson *et al.*, 1974), and there is generally no recombination between mtDNA molecules from different parents (Hayashi, Tagashira & Yoshida, 1985). Therefore, individuals within the same matriline share the same mtDNA clone or haplotype. Furthermore, this monomorphism may extend to both the social group and the local population (see chapter 19, this volume). Because female macaques almost never leave their natal group (Lindburg, 1969), the population genetics of nuclear and mitochondrial genomes can be markedly different in macaques.

However, the same features that can cause mtDNA 'gene trees' to differ from the true 'species trees', in some instances, also simplify the reconstruction of evolutionary relationships. Unlike nuclear genomes, mtDNA haplotypes exhibit a history of dichotomous branching; the form of historical branching assumed and displayed by cladistic analysis. In addition, the geographical locations of individuals bearing particular mtDNA haplotypes can be mapped onto the phylogenetic tree, thus allowing one to trace the biogeographical dispersal history of the DNA lineages (Avise *et al.*, 1987). In the particular case of macaque evolution, numerous species seem to have arisen as a result of geographical isolation on islands following changes in sea level. Subsequent sea level fluctuations could have re-established nuclear gene flow between populations for brief periods, potentially obscuring the proper branching order of macaque relationships. Mitochondrial DNA would have most likely been unaffected by the intermittent contact of divergent populations because female macaques are highly philopatric, whereas male macaques usually disperse from their natal social group (Pusey & Packer, 1987).

Future sources of data

The ideal type of molecular data to use in reconstructing phylogenetic trees would consist of several unlinked nuclear DNA sequences that evolved at a rate yielding a moderate level of divergence between taxa. Sequences that evolve relatively fast would be used to compare closely related taxa, whereas slowly evolving sequences would be used to compare more distantly related taxa. Because these sequences come from the nuclear genome, the phylogenetic information obtained would best reflect the true species' relationships, but would probably lack the biogeographical information contained in the mtDNA sequences. Therefore, a research programme involving a detailed and simultaneous survey of the nuclear and mitochondrial genomes would provide a more comprehensive set of data with which to reconstruct macaque phylogeny, and the biogeographical context of the evolutionary events it reflects.

Unfortunately, data from nuclear sequences are not readily available at this time. However, the degree to which sequencing is becoming automated, and the rate at which sequence data is being produced, suggests that sufficient nuclear genetic information is likely to become available over the next decade.

Phylogenetic analysis

There are two common quantitative approaches used for reconstructing phylogenetic relationships: phenetics and cladistics (Felsenstein, 1982, 1988; Swofford & Olson, 1990). Many different cladistic and phenetic methods have been proposed, and the field of phylogenetic systematics is developing rapidly at this time (see Harvey & Pagel, 1991; Miyamoto & Cracraft, 1991; Penny, Hendy & Steel, 1992). All of these methods produce phylogenetic estimates based on various assumptions of the evolutionary processes that led to diversification of the group. Therefore, the strongest evidence for phylogenetic relationships comes from the congruent results of different approaches and independent data sets. In the absence of congruent results, we must choose between alternative phylogenies by assessing which method employed the most reasonable assumptions for that group.

The phenetic approach requires the estimation of a distance or similarity statistic, such as the widely used parameter for genetic distance, d. The taxa are then linked on branches of the tree based on levels of similarity observed between pairs of taxa (Felsenstein, 1984). Algorithms commonly used to build trees with phenetic distances include UPGMA (Sneath & Sokal, 1973), Fitch–Margoliash (Fitch & Margoliash, 1967), and Neighbor–Joining (Saitou & Nei, 1987).

In constrast, the cladistic approach to reconstructing phylogenetic trees (Hennig, 1966) minimises the number of changes in discrete character states by using some form of maximum parsimony (Sober, 1989; Swofford & Olson, 1990). Clusters of in-group taxa are then defined by their shared, derived characters. Consequently, it is possible for a cladogram to link two very different taxa based on the derived character states that they share, relative to an ancestral condition, even though one or both are more similar overall to other taxa excluded from this clade.

Most of the trees presented in this chapter have been rooted to an outgroup (usually *M. nemestrina*). That is, a taxon known to be excluded from the group under consideration is included in the analysis (Swofford & Olson, 1990). Inclusion of the out-group polarises the character states by defining the primitive condition of the group's common ancestor.

Morphological variation among macaques has not been subjected to rigorous phylogenetic analysis. Therefore, the phylogenetic trees based on morphology that are presented in this chapter were constructed by traditional, non-quantitative methods. These trees are essentially the results of intuitive assessments of similarity and disparity among species by a recognised expert (e.g. Fooden, 1969; Delson 1980).

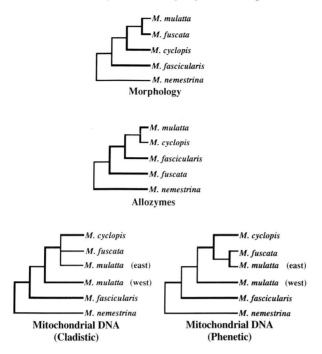

Fig. 1.1. Phylogenetic hypotheses for the *fascicularis* group: the tree derived from morphological data is based on a phenetic analysis by Delson (1980); the allozyme tree is based on a phenetic analysis by Fooden & Lanyon (1989); and the results of both phenetic and cladistic analyses of mitochondrial DNA data are based on data from Melnick *et al.* (1993; see also Hayasaka *et al.*, 1988a,b).

The *fascicularis* group

While there appears to be consensus about the members of this species group, the exact relationships between the member species have not been conclusively determined (Fig. 1.1). *Macaca fascicularis* is placed as a sister group to the other species based on morphology and mtDNA, whereas the allozyme data place *M. fuscata* as the most primitive member of the group. *Macaca mulatta* and *M. cyclopis* are judged to be recently derived and closely related species in all the data sets presented here. However, there is a conflict as to whether *M. cyclopis* or *M. fuscata* is most closely related to *M. mulatta*.

The mtDNA tree appears to resolve this conflict by including samples from numerous geographically separated individuals from each species. The different mtDNA haplotypes found within a single species generally clustered together on a branch of the phylogenetic tree that did not include

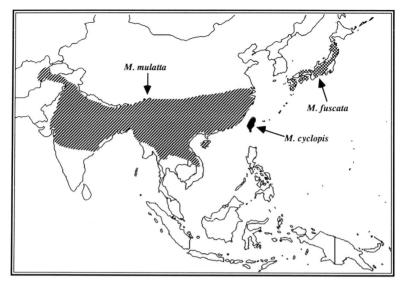

Fig. 1.2. The geographical ranges of three species in the *fascicularis* group: *Macaca mulatta*, *M. cyclopis* and *M. fuscata* (Fooden, 1980, 1982).

any haplotypes from other species, indicating a monophyletic origin for the mtDNA genome of that species. However, a major division was found between eastern and western *M. mulatta* populations (Melnick *et al.*, 1993). The mtDNA haplotypes from these populations are paraphyletic; the eastern *M. mulatta* haplotype is more closely related to the *M. fuscata* haplotype than to the western *M. mulatta* haplotype (Fig. 1.1). A phenetic analysis indicated that *M. cyclopis* may also be most closely linked to the eastern *M. mulatta* haplotype.

These results are supported by a consideration of the history of glacial events in the region, along with the present geographical ranges of the two island species (*M. fuscata* and *M. cyclopis*) and the widespread mainland species (*M. mulatta*; Fig 1.2). The division between the eastern and the western haplotypes of *M. mulatta* occurs along the Bramaputra river valley, which was the site of a major glacial intrusion about 180 000 years ago. This also corresponds geographically with the east–west separation between the two sub-species of *M. assamensis* (see below). It seems likely that the glacial intrusion presented an obstacle to migration between the temporarily isolated populations of *M. mulatta*, leading to the geographical separation of mtDNA haplotypes that we see today (Melnick *et al.*, 1993). At the same time, the glacial advance caused sea levels to lower, connecting the islands of Japan and Taiwan to eastern mainland China. *Macaca fuscata*, the Japanese macaque, and *M. cyclopis*, the Formosan macaque, are endemic to these islands. Therefore, it is likely that both *M. fuscata* and *M. cyclopis* evolved from an eastern *M. mulatta* ancestor when they became isolated on the islands of Japan and Taiwan, respectively, by the rise in sea level that followed a major glacial event.

The paraphyletic distribution of the two *M. mulatta* populations on the mtDNA tree would appear to disqualify *M. mulatta* as a true species, but remember the distinction between a mtDNA gene tree and a species tree. It is not necessary for the mtDNA haplotypes within a species to cluster together in a phylogenetic analysis. In fact, it is the unusual topological features of a mtDNA tree, such as intra-specific paraphyly, that yield the most interesting hypotheses regarding the evolutionary history of the taxa being analysed.

The *sinica* group

The evolutionary relationships of the species included in the *sinica* group remain problematical. The trees generated by morphological and allozyme data are similar (Fig. 1.3). Both clearly link *M. radiata* and *M. sinica* as derived sister taxa within the group. The principal distinction between the two trees lies in the placement of *M. arctoides* as either the sister species to the rest of the group (allozymes; also see the mtDNA tree), or as a member of a clade that includes *M. assamensis* and *M. thibetana* (morphology).

The use of several mtDNA haplotypes from each species distorts the topology of the *sinica* group tree even more strongly than it did the *fascicularis* group tree. Again, the distortion is fruitful ground for forming a hypothesis regarding the historical biogeography of the group.

The departure from intra-specific monophyly is seen at its most extreme in the case of *M. assamensis*. Indeed, the positions of the two distinct *M. assamensis* haplotypes do not define a clade at all, let alone an exclusive clade, indicating a polyphyletic origin for the mtDNA of *M. assamensis*. This species is composed of eastern and western sub-species, *M. assamensis assamensis* and *M. assamensis pelops*, respectively. The geographic division between these sub-species is located at approximately the same position as the division between eastern and western *M. mulatta* haplotypes (Fig. 1.4; Hoelzer *et al.*, 1992; Melnick *et al.*, 1993). The western sub-species is morphologically intermediate between the eastern sub-species and *M. radiata* in some respects (e.g. tail length), and the mtDNA data strongly link it with *M. radiata*. The mtDNA of the eastern sub-species is more similar to both *M. sinica* and *M. thibetana* than it is to the mtDNA of the western sub-species. On the basis of this evidence, and biogeographical

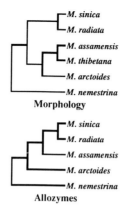

Morphology

Allozymes

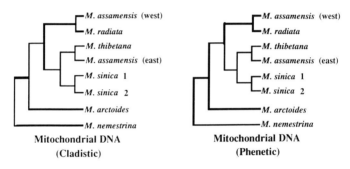

Mitochondrial DNA
(Cladistic)

Mitochondrial DNA
(Phenetic)

Fig. 1.3. Phylogenetic hypotheses for the *sinica* group. The tree derived from morphological data is based on a phenetic analysis by Delson (1980); the allozyme tree is based on a phenetic analysis by Fooden & Lanyon (1989); and the results of both phenetic and cladistic analyses of mitochrondial DNA data is based on Hoelzer *et al.* (1992).

data suggesting that *M. radiata* was once quite widespread (Fooden, 1988), Hoelzer *et al.* (1992) have suggested that the western sub-species of *M. assamensis* was a remnant population of *M. radiata* that became isolated as the *M. radiata* species range contracted. Subsequent hybridisation with *M. assamensis* males that migrated from the East could have swamped the nuclear gene pool of this small, isolated population yielding a chimaeric population that primarily resembled *M. assamensis* in its morphology and nuclear genome, but retained the *M. radiata* mtDNA genome.

The mtDNA data also link the two distinct mtDNA haplotypes of *M. sinica* to the eastern *M. assamensis* sub-species and *M. thibetana*. A 'bootstrap' analysis of these data provides support for this clade (82 of 100 'bootstrap' replicates; Hoelzer *et al.*, 1992). However, the 'bootstrap' did not support linking the two *M. sinica* haplotypes as sister taxa within this

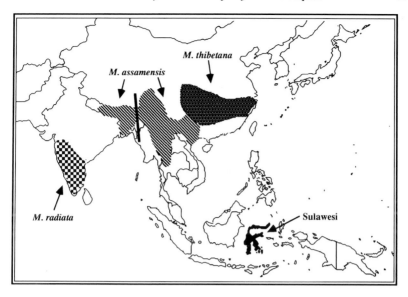

Fig. 1.4. The geographic ranges of three species in the *sinica* group: *Macaca radiata*, *M. assamensis* and *M. thibetana* (Fooden, 1980, 1982). The division between the eastern and western sub-species of *M. assamensis* is indicated by a solid line. In addition, the island of Sulawesi, the home of seven endemic macaque species, is indicated.

clade. Rather, the base of this group formed a trichotomy that linked the two *M. sinica* haplotypes with a clade containing the eastern *M. assamensis* sub-species and *M. thibetana*. These results challenge the *sinica* group trees based on morphology and allozymes, which have the *assamensis–thibetana* lineage branching off before the *sinica–radiata* clade.

The phylogenetic position of *M. arctoides*, which differed in the morphological and allozyme studies, was not clearly determined by the mtDNA data. Statistical analysis of the mtDNA tree, along with investigation of trees requiring only a few more character state changes, indicated that the position of *M. arctoides* is unstable (Hoelzer *et al.*, 1992). Although the most parsimonious tree placed *M. arctoides* as a sister taxon to the rest of the *sinica* group, it was classified as a member of the *sinica–assamensis–thibetana* clade in trees only slightly longer.

Several major questions remain concerning the evolutionary relationships of *sinica* group species. Is *M. arctoides* a member of this group? If it is a member, is it a primitive sister taxon or part of an internal clade? Is *M. sinica* ancestral to the *assamensis–thibetana* lineage, or is it a recently derived sister species to *M. radiata*? More data from all sources will be required to answer these questions.

The *silenus–sylvanus* group

The data on morphology (Fooden, 1976) and mtDNA (Williams, 1990) form a consensus regarding the general relationships of species in the *silenus–sylvanus* group (Fig. 1.5), although the analysis of allozymes is still ambiguous with respect to this group (Fooden & Lanyon, 1989). *Macaca sylvanus* is ancestral to the group; however, an analysis of the entire genus with a non-macaque out-group may show this species to be ancestral to the entire Asian clade of macaques, thus removing it from this species group (*sensu* Delson, 1980). *Macaca silenus* is the next most primitive species in this group, followed by *M. nemestrina*. A clade including the species complex endemic to the island of Sulawesi (see Fig. 1.4 for the location of Sulawesi) is derived from a common ancestor with *M. nemestrina*. Although we discuss the seven forms of Sulawesi macaques as distinct species, it should be noted that the taxonomic status of these species is questionable (Groves, 1980). It is clear that some hybridisation occurs between some pairs of neighbouring species.

Fooden (1969) proposed that macaques originally colonised Sulawesi during the Pleistocene period by waif dispersal from Borneo. This contention was later challenged by Groves (1980), who suggested that macaques may have arrived in the Pliocene. This view is supported by the affinity between much of the Sulawesi mammal fauna and the Pliocene Siva–Malayan fauna (Groves, 1976). These authors also disagreed about which of the extant species is most primitive; Fooden (1969) suggested that *M. tonkeana* is the most ancient taxon, whereas Groves (1980) placed *M. maura* in that position.

In general, the relationships among the Sulawesi macaques remain ambiguous. Different morphological data sets have yielded contradictory trees; and trees based on both allozyme and mtDNA data are different still (Fig. 1.6). The morphological analyses have agreed on two pairs of sister species, *M. ochreata* with *M. brunnescens*, and *M. nigra* with *M. nigrescens*, but these pairs are not confirmed by either source of molecular data. Again, different mtDNA haplotypes from a single species appear on separate branches of the mtDNA tree. One *M. tonkeana* haplotype is associated with *M. brunnescens*, while another is associated with the rest of the Sulawesi macaques. This supports Fooden's (1969) hypothesis that *M. tonkeana* is ancestral to the group. The greatest agreement among the alternative phylogenies is a close relationship between *M. nigra* and *M. hecki*. Nevertheless, given the inconsistent relationships exhibited by this set of phylogenetic trees, one must conclude that there is currently no consensus on the phylogenetic relationships of the Sulawesi macaques.

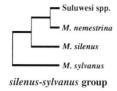

silenus-sylvanus group

Fig. 1.5. Consensus phylogenetic hypothesis for the *silenus–sylvanus* group based on morphology (Delson, 1980) and mitochondrial DNA (D. J. Melnick, unpublished data). The seven macaques endemic to the island of Sulawesi are thought to be a monophyletic group, and this clade is represented by a single branch on this tree. Note that there is no out-group indicated on this tree because there have been no cladistic or phenetic studies of this group that used an out-group comparison.

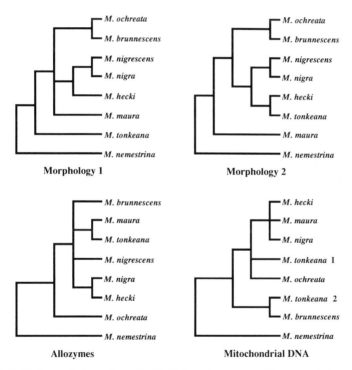

Fig. 1.6. Phylogenetic hypotheses for the Sulawesi macaques. Two morphology-based trees are shown (Morphology 1 – Fooden, 1969; Morphology 2 – Albrecht, 1977, 1978; Groves, 1980) to indicate the lack of consensus on the relationships of the Sulawesi macaques. The allozyme tree is from Fooden & Lanyon (1989), and the mitochondrial DNA tree is from D. J. Melnick (unpublished data). Only cladistic analyses are shown for the allozyme and mitochondrial DNA data.

Conclusions

The genus *Macaca* is divided into three species groups (or as many as five if *M. sylvanus* and/or *M. arctoides* are placed in their own groups) based on morphology. The historical branching pattern that gave rise to these groups requires futher study (but see Fooden & Lanyon, 1989), but a large amount of data has now been gathered that addresses the relationships of species within species groups. Much of these data come from the study of the variation in allozyme alleles and mtDNA sequences.

The mtDNA data have been especially fruitful in generating phylogenetic hypotheses. The clonal mode of inheritance of the mitochondrial genome, which is distinct from the mode of inheritance for nuclear DNA, allows one to trace the relationships of conspecific mtDNA haplotypes to phylogenetic branches that are deeper than the origin of the species. This can lead to the appearance of intra-specific paraphyly within the phylogeny, as evidenced in the mtDNA trees for each of the macaque species groups. The occurrence of intra-specific paraphyly usually implies that the species that branch off within the clade defined by the paraphyletic species are daughter species of the diverse species. This is suggested for the paraphyletic species *M. mulatta* and, possibly, for *M. tonkeana*. However, in the case of *M. assamensis*, the transformation of a small population through extensive inter-specific hybridisation was hypothesised. In addition, the geographical distribution of different haplotypes often suggests particular biogeographical patterns of dispersal and speciation.

The unanswered questions regarding the placement of *M. sylvanus* and *M. arctoides* in their own groups, and the evolutionary relationships of the Sulawesi macaques, must be addressed by collecting more data from all sources. It is possible that the Sulawesi species may also be so closely related that the topology of the mtDNA tree would bear little resemblance to the true species tree. Nevertheless, a well-supported mtDNA tree may yield a wealth of information when compared with a consensus tree based on nuclear genetic variation and morphology. More extensive data sets are also needed so that all macaque species, along with a species of baboon (genus *Papio*) as an out-group, can be included in the same analyses. Only in this way can we determine the number and branching order of monophyletic clades within the *Macaca*.

References

Albrecht, G. H. (1977). Methodological approaches to morphological variation in Primate populations: the Celebesian macaques. *Yearbook of Physical Anthropology*, **20**, 290–308.

Albrecht, G. H. (1978). The craniofacial morphology of the Sulawesi macaques. Multivariate approaches to biological problems. *Contributions to Primatology*, vol. 13. Basel: S. Karger.

Avise, J. C., Arnold, A., Ball, R. M., Bermingham, E., Lamb, T., Neigel, J. E., Reeb, C. A. & Saunders, N. C. (1987). Intraspecific phylogeography: the mitochondrial DNA bridge between population genetics and systematics. *Annual Review of Ecology and Systematics*, **18**, 489–522.

Brooks, D. R. & McLennan, D. A. (1991). *Phylogeny, ecology, and behavior*. Chicago: University of Chicago Press.

Brown, W. M. (1985). The mitochondrial genome of animals. In *Molecular evolutionary genetics*, ed. R. J. MacIntyre, pp. 95–130. New York: Plenum Press.

Brown, W. M., George, M. & Wilson, A. C. (1979). Rapid evolution of animal mitochondrial DNA. *Proceedings of the National Academy of Sciences, USA*, **71**.

Cronin, J. E., Cann, R. & Sarich, V. (1980). Molecular evolution and systematics of the genus *Macaca*. In *The macaques: studies in ecology, behavior, and evolution*, ed. D. G. Lindburg, pp. 31–51. New York: Van Nostrand Reinhold.

Delson, E. (1980). Fossil macaques, phyletic relationships and a scenario of development. In *The macaques: Studies in ecology, behavior, and evolution*, ed. D. G. Lindburg, pp. 10–30. New York: Van Nostrand Reinhold.

Disotell, T. R., Honeycutt, R. L. & Rovulo, M. (1992). Mitochondrial DNA phylogeny of the Old-World monkey tribe Papionini. *Molecular Biology and Evolution*, **9**, 1–13.

Fa, J. E. (1989). The genus *Macaca*: a review of taxonomy and evolution. *Mammal Reviews*, **19**, 45–81.

Felsenstein, J. (1982). Numerical methods for inferring evolutionary trees. *Quarterly Reviews of Biology*, **57**, 379–404.

Felsenstein, J. (1984). Distance methods for inferring phylogenies: a justification. *Evolution*, **38**, 16–24.

Felsenstein, J. (1988). Phylogenies and quantitative methods. *Annual Review of Ecology and Systematics*, **19**, 445–71.

Fitch, W. M. & Margoliash, E. (1967). Construction of phylogenetic trees. *Science*, **155**, 279–84.

Fooden, J. (1969). Taxonomy and evolution of the monkeys of Celebes (Primates, Cercopithecidae). *Bibliotheca Primatologica*, No. 10. Basel: S. Karger.

Fooden, J. (1976). Provisional classification and key to the living species of macaques (Primates: *Macaca*). *Folia Primatologica*, **25**, 225–36.

Fooden, J. (1980). Classification and distribution of living macaques. In *The macaques: studies in ecology, behavior, and evolution*, ed. D. G. Lindburg, pp. 1–9. New York: Van Nostrand Reinhold.

Fooden, J. (1982). Ecogeographic segregation of macaque species *Primates*, **6**, 574–9.

Fooden, J. (1988). Taxonomy and evolution of the *sinica* group of macaques: 6. Interspecific comparisons and synthesis. *Fieldiana: Zoology, New Series*, **45**, 1–44.

Fooden, J. & Lanyon, S. M. (1989). Blook protein allele frequencies and phylogenetic relationships in *Macaca*: a review. *American Journal of Primatotogy*, **17**, 209–41.

George, M., Jr (1982). Mitochondrial DNA evolution in Old World monkeys. PhD thesis, University of California, Berkeley.

Groves, C. P. (1976). The origin of the mammalian fauna of Sulawesi. *Zeitschrift für Saugetierkunde*, **41**, 201–16.

Groves, C. P. (1980). Speciation in *Macaca*: The view from Sulawesi. In *The macaques: studies in ecology, behavior, and evolution*. ed. D. G. Lindburg, pp. 84–124. New York: Van Nostrand Reinhold.

Harihara, S., Saitou, N., Hirai, M., Aoto, N., Tero, K., Cho, F., Honjo, S. & Omoto, K. (1988). Differentiation of mitochondrial DNA types in *Macaca fascicularis*. *Primates*, **29**, 117–27.

Harvey, P. H. & Pagel, M. D. (1991). *The comparative method in evolutionary biology*. Oxford Series in Ecology and Evolution. Oxford: University Press.

Hayasaka, K., Gojobori, T. & Horai, S. (1988a). Molecular phylogeny and evolution of primate mitochondrial DNA. *Molecular Biology and Evolution*, **5**, 626–44.

Hayasaka, K., Horai, S., Gojobori, T., Shotake, T., Nozawa, K. & Matsunaga, E. (1988b). Phylogenetic relationships among Japanese, rhesus, Formosan, and crab-eating monkeys, inferred from restriction-enzyme analysis of mitochondrial DNAs. *Molecular Biology and Evolution*, **5**, 270–81.

Hayasaka, K., Horai, S., Shotake, T., Nozawa, K. & Matsunga, E. (1986). Mitochondrial DNA polymorphism in Japanese monkeys, *Macaca fuscata*. *Japanese Journal of Genetics*. **61**, 345–59.

Hayashi, J. -I., Tagashira, Y. & Yoshida, M. C. (1985). absence of extensive recombination between inter- and intra-species mitochondrial DNA in mammalian cells. *Experimental Cell Research*, **160**, 287–95.

Hennig, W. (1966). *Phylogenetic Systematics*. Urbana, IL: University of Illinois Press.

Hillis, D. M. & Davis, S. K. (1988). Ribosomal DNA: intra-specific polymorphism, concerted evolution, and phylogeny reconstruction. *Systematic Zoology*, **32**, 63–6.

Hillis, D. M. & Moritz, C. (1990). *Molecular systematics*. Sunderland, MA: Sinauer Press.

Hoelzer, G. A., Hoelzer, M. A. & Melnick, D. J. (1992). The evolutionary history of the *sinica* group of macaque monkeys as revealed by mtDNA restriction site analysis. *Molecular Phylogenetics and Evolution*, **1**, 215–22.

Hutchinson, C. A., Newbold, C. E., Potter, S. S. & Edgell, M. H. (1974). Maternal inheritance of mammalian mitochondrial DNA. *Nature*, **251**, 536–8.

Karl, S. A. & Avise, J. C. (1992). Balancing selection at allozyme loci in oysters: implications from nuclear RFLPs. *Science*, **256**, 100–2.

Kawamoto, Y., Takenaka, O. & Brotoisworo, E. (1982). Preliminary report on genetic variations within and between species of Sulawesi macaques. *Kyoto University Overseas Research Report of Studies of Asian Non-human Primates*, **2**, 23–37.

Lindburg, D. G. (1969). Rhesus monkeys: mating season mobility of adult males. *Science*, **166**, 1176–8.

Melnick, D. J. & Hoelzer, G. A. (1992). Differences in male and female macaque dispersal lead to contrasting distributions of nuclear and mitochondrial DNA variation. *International Journal of Primatology*, **13**, 1–15.

Melnick D. J., Hoelzer, G. A. & Honeycutt, R. L. (1992). Mitochondrial DNA: its uses in anthropological research. In *Molecular applications in biological anthropology*, ed. E. J. Devor, pp. 179–233.

Melnick, D. J., Hoelzer, G. A., Absher, R. & Ashley, M. V. (1993). MtDNA diversity in rhesus monkeys reveals overestimates of divergence time and paraphyly with neighbouring species. *Molecular Biology and Evolution*, **10**, 282–95.

Melnick, D. J. & Kidd, K. K. (1985). Genetic and evolutionary relationships among

Asian macaques. *International Journal of Primatology*, **6**, 123–60.

Miyamoto, N. M. & Cracraft, J. (1991). *Phylogenetic analysis of DNA sequences*. Oxford: Oxford University Press.

Murphy, R. W., Sites, Jr J. W., Buth, D. G. & Haufler, C. H. (1990). Proteins: isozyme electrophoresis. In *Molecular systematics*, ed. D. M. Hillis & C. Moritz, pp. 45–126. Sunderland, MA: Sinauer Press.

Nozawa, K., Shotake, T., Kawamoto, Y. & Tanabe, Y. (1982). Population genetics of Japanese monkeys. II. Blood protein and polymorphisms and population structure. *Primates*, **23** 252–71.

Penny, D., Hendy, M. D. & Steel, M. A. (1992). Progress with methods for constructing evolutionary trees. *TREE* 7, 73–9.

Pusey, A. E. & Packer, C. (1987). Dispersal and philopatry. In *Primate societies*, ed. B. B. Smuts, D. L. Cheney, R. M. Seyfarth, R. W. Wrangham & T. T. Strusaker, pp. 250–66. Chicago: University of Chicago Press.

Saitou, N. & Nei, M. (1987). The neighbour-joining method: a new method for reconstructing phylogenetic trees. *Molecular Biology and Evolution*, **4**, 406–25.

Sneath, P. H. A. & Sokal, R. R. (1973). *Numerical taxonomy*. San Francisco, CA: W. H. Freeman.

Sober, E. (1989). *Reconstructing the past: parsimony, evolution and inference*. Cambridge, MA: MIT Press.

Swofford, D. L. & Olson, G. J. (1990). Phylogeny reconstruction. In *Molecular Systematics*, ed. D. M. Hillis & C. Moritz, pp. 411–501. Sunderland, MA: Sinauer Press.

Szalay, F. S. & Delson, E. (1979). *Evolutionary history of the primates*. New York, NY: Academic Press.

Williams, A. K. (1990). The evolution of mitochondrial DNA in the *silenus–sylvanus* species group of macaques. PhD thesis, Columbia University.

Zhang, Y. & Shi, X. (1989). Mitochondrial DNA polymorphism in five species of the genus *Macaca*. *Chinese Journal of Genetics*, **16**, 326–38.

2

Genetic relatedness within and between populations of *Macaca fascicularis* on Sumatra and off-shore islands

W. SCHEFFRAHN, J. R. DE RUITER, AND
J. A. R. A. M. VAN HOOFF

Introduction

Perspectives in primate genetics

The principal aim of primate genetics is the study of the evolution of primate populations. Evolutionary genetics is based on the genetic variability: (1) between different primate taxa, i.e. genetic inter-specific variability; and (2) within primate populations, i.e. genetic polymorphism and intra-specific variability. The variability at loci of genetic markers (blood proteins), nuclear DNA, and mitochondrial DNA can be used to outline both the evolution of genetic markers and the species carrying these markers, and to analyse the causal forces that govern the process of evolution of primate populations.

In contrast to studies on the evolution of species and their genetic markers, research on population genetics in free-ranging primates is still fairly scarce; it started developing in the late 1970s. From this time primate geneticists have been increasingly concentrating on the study of micro-evolutionary processes within extant primate populations. Recently, there has been a series of field studies on ecology, behaviour and population genetics (some completed, some still in progress) that determine the composition and structure of gene pools, particularly in populations of the genera of *Cercopithecus*, *Papio* and *Macaca* (e.g. Nozawa *et al.*, 1975, 1982; Shotake, Nozawa & Tanebe, 1977; Kawamoto & Ischak, 1981; Kawamoto, Nozowa & Ischak, 1981; Shotake, 1981; Turner, 1981; Kawamoto, Shotake & Nozawa, 1982*a*; Kawamoto, Takenaka & Brotoisworo, 1982*b*; Takenaka & Brotoisworo, 1982; Dracopoli *et al.*, 1983; Kawamoto, Ischak & Supriatna, 1984; Melnick, Pearl & Richard, 1984; Kawamoto *et al.*, 1985; Melnick, 1987, 1988; Camperio-Ciani *et al.*, 1989; Rogers, 1989; Pope, 1992; de Ruiter *et al.*, 1992; Morin *et al.*, 1993; de Ruiter & van

Hooff, 1993; Scheffrahn *et al.*, 1993; de Ruiter, van Hooff & Scheffrahn, 1994*a*; de Jong, de Ruiter & Haring, 1994; de Ruiter *et al.*, 1994**b**).

Several authors (e.g. Kawamoto *et al.*, 1982*b*; Melnick *et al.*, 1984; Rogers, 1989) analysed genetic variability between social groups within a local population, and found a relatively large genetic differentiation between adjacent groups. This observation is comparable with the earlier concepts of Wahlund (1928) and Wright (1931) based on gene diversity within human populations. It also appears to support the theory that social constraints on gene flow result in drift and heterogeneity among kin groups, and thereby accelerate evolutionary change, a change fostered either through inter-demic selection (Wade, 1982; Wilson, 1983) or through reproductive isolation of chromosomal variants (White, 1968; Bush 1975; Wilson *et al.*, 1975).

In a recent detailed analysis, Pope (1992) also found a large genetic differentiation of social groups, but at the same time she found a relatively low differentiation between larger units at the level of local populations. This pattern has been reported in other studies as well (for primates, Dracopoli *et al.*, 1983; Kawamoto *et al.*, 1984; Melnick *et al.*, 1984; and for black-tailed prairie dogs; Chesser, 1983). It indicates that in these species variability exists largely at the level of adjacent social groups. Melnick *et al.* (1984) concluded that rates of gene flow were probably too high for inter-demic selection and proposed that clumping of genetic characteristics in sub-populations consisting of relatives might result in a founder effect and subsequent genetic drift. This could lead to speciation if part of such a population became isolated (Melnick *et al.*, 1984). One may well wonder which level of genetic differentiation is relevant with respect to the processes of population differentiation and speciation that presumably take place after the colonisation or isolation of areas. The present chapter analyses both the genetic structure within a single population and the divergence between different populations, thereby contributing to the discussion about the importance of genetic structure and isolation of sub-populations for evolutionary change.

Genetic study of Indonesian populations of Macaca fascicularis

Wild populations of *M. fascicularis* have been shown to possess a high degree of genetic variation, as expressed in polymorphism and heterozygosity (Kawamoto & Ischak, 1981; Kawamoto *et al.*, 1981, 1982*a,b*, 1984). Kawamoto and co-authors have concentrated mainly on the questions of genetic variability between locations on the same island and between populations of different islands that are situated on the Sunda shelf, and which were connected relatively recently in evolutionary time (see below). We collected blood samples from long-tailed macaques at 13 different sites on Sumatra and its off-shore islands. This allowed us to compare specifically the effects of three different principles underlying genetic differentiation in the same study. We attempted to assess the relative contribution of these three principles to differentiation and investigated possible interactions between these mechanisms. We distinguished:

1. Socially mediated differentiation apparent in the genetic differentiation of adjacent social groups. The relationships between social causes and population genetic effects have been analysed in computer data and genetic paternity tests (de Ruiter *et al.*, 1992, 1994*b*; de Jong *et al.*, 1994).
2. Isolation by distance. Because individual migration (gene flow) has a limited range, greater differentiation is expected to occur over larger distances, where the habitat is continuous (Malécot, 1950).
3. Isolation in time. Depending on population size and variability of the founder population, random genetic drift causes a certain degree of genetic differentiation. There is a theoretical relationship between genetic divergence and isolation time (Li & Nei, 1975; Nozawa *et al.*, 1977; Smith & Coss 1984).

For the third aspect, we collected samples on islands of different sizes that have been isolated for different evolutionary periods of time. The variation in isolation time was due to differences in the depth of the ocean basins connecting these islands with the mainland of Sumatra during geohistoric fluctuations in sea level.

Material and methods

Sampling sites, sample sizes and blood collection

Blood samples were collected from 234 long-tailed macaques (*M. fascicularis*, Raffles 1821) during a combined behaviour and genetic field study between 1984 and 1989 (in summer 1985 W. Scheffrahn participated in this work at Ketambe station). In addition, we collected samples from four individuals of the Mentawai macaque *M. nemestrina pagensis* (which were used as an out-group in the cluster analysis). Group size in long-tailed macaques varies between 5 and 60 individuals (van Schaik & van Noordwijk, 1985; de Ruiter *et al.*, 1992) and sample sizes are indicated in Table 2.1, below.

Figure 2.1 shows the locations where samples were collected. All sites on the Sumatran mainland (sites 1–9) represent social groups or parts of

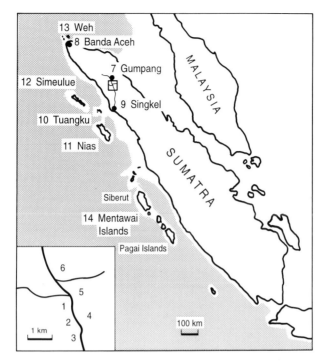

Fig. 2.1. Map of Sumatra, indicating the sampled populations. The numbers correspond to those in Tables 2.1 and 2.4–2.6 and in Fig. 2.2. Dotted area indicates sea depth of more than 200 m.

groups. Sites 1–6 comprise the Ketambe adjacent social groups that live in a 4 km stretch along the Alas river. As summarised in Table 2.1, site 7 was located 25 km upstream to the north, site 9 was located 150 km downstream to the south, and site 8 was located 350 km to the north-west of Ketambe. We also sampled blood from long-tailed macaques on four islands. For two of these off-shore islands, sites 11 and 12, all samples were obtained from pets kept for non-economic reasons. On the other two islands, sites 10 and 13, blood samples were taken from individuals of a wild social group, supplemented at the latter site with samples from four pets. These captive animals must have originated from various social groups on the islands, as they were morphologically distinct from the Sumatran macaques (see below).

The Mentawai macaques were kept in captivity after being confiscated from illegal traders and, judging by their morphology, probably originated from Siberut island (Whitten & Whitten, 1982). We obtained samples from four individuals (one of which was the offspring of two individuals also represented in the sample).

Table 2.1. *Study populations, locations and number of individuals*

Site	No. of samples	No. of pets	No. of loci	Distance to Ketambe (km)	Island size (km²)	Isolation time (years)
M. *fascicularis*						
Ketambe 1	59		6	Same area		
Ketambe 2	25		6	Same area		
Ketambe 3	8		6	Same area		
Ketambe 4	20		6	Same area		
Ketambe 5	24		6	Same area		
Ketambe 6	20		6	Same area		
Sumatra 7	8		6	25		
Sumatra 8	13		6	350		
Sumatra 9	10		6	150		
Island 10	12		4		300	10 000
Island 11	11	11	4		4000	> 100 000
Island 12	12	12	4		2000	> 100 000
Island 13	12	4	4		250	> 100 000
Total	234					
M. *nemistrina*						
Island 14	4	4	6			
Total	238					

The islands where we studied long-tailed macaques were of different sizes and had different isolation histories (Table 2.1). Island site 10 was about 300 km² in size, island 11 was 4000 km², island 12 was 2000 km², and island 13, was 250 km². The isolation history of the islands can to some extent be inferred from their geology. Only island site 10 lies on the Sunda shelf, separated from the mainland by a shallow channel with a current maximum depth of about 65 m (according to a survey of the west coast, provided by the Awy National Marine Navigation Authority, 1987). Therefore, this island must have been connected to mainland Sumatra during the last glaciation and the associated decrease in sea level, until about 10 000 years ago (Eudey, 1980; Jongsma, 1970; Whitten *et al.*, 1984). The other islands, 11–14, are in deep seas and all of them are thought to have been isolated from the mainland for at least 100 000, and perhaps up to a million, years (Eudey, 1980; Whitten *et al.*, 1984). Both sites 11 and 14, but not sites. 12 and 13, might have been connected to Sumatra by a narrow land bridge, but probably not during the last glacial period, which ended 10 000 years ago. The occurrence of endemic species (Heany, 1986; Whitten *et al.*, 1984) is a further indication that isolation has been relatively effective. Moreover, the macaques on all these deep sea islands are morphologically differentiated.

It is not precisely known when and how these islands became populated.

It is highly unlikely that transfer by swimming or even floating on a tree trunk could have occurred because distances from Sumatra are too large. Human transfer of macaques in early times is a hypothetical possibility, but on all these deep sea islands the coats of macaques are very dark, almost black. It has been suggested that this is an adaptive change related to the island conditions, a dark coat being better than a brown one for thermo-dynamic reasons. However, a brown coat is better protection against feline predators, so the shift in coat colour could have occurred because there are no feline predators on these islands (Sugardjito *et al.*, 1984; van Schaik & van Noordwijk, 1985). On shelf-island site 10 and two other shelf-islands that we checked, which were located south-west of site 13, macaques had a light (brown) coat colour. The fact that only the macaques on deep sea islands (11–14) but not on shelf-islands had a dark coat, suggests that the difference in coat colour reflects a difference in isolation history of the primate populations, and that these primate populations may have been isolated since the islands became separate from the mainland.

To sum up, we obtained samples from six adjacent social groups at Ketambe (sites 1–6). Since the river proved not to be a geographical barrier for migrating males who could swim across, the social structure was the mechanism for genetic differentiation among these groups. We obtained samples at three sites (7–9) at distances 25 km, 150 km and 350 km from Ketambe respectively. All these sites were presumably isolated from the Ketambe groups by distance only, since macaques inhabit not only virgin rain forest but also disturbed habitat and the species is quite common throughout the region. Finally, we obtained samples from four islands (sites 10–13) for which isolation had been complete for different lengths of evolutionary time. These consisted of one small island with a short isolation time (10 000 years), and two large and one small island with long isolation times (probably at least several 100 000 years).

Genetic markers

The animals were trapped and anaesthetised, then blood samples were taken and they were examined (30 min each) before being released (de Ruiter, 1992). Blood was taken from the femoral vein and collected in EDTA-treated tubes (EDTA is ethylenediaminetetra-acetic acid). After centrifugation, the plasma and blood cells were stored in liquid nitrogen. The cell fraction was mixed with an isotonic glycerol solution to prevent cell lysis (Rowe, Davis & Moore-Jankowski, 1972). The haemolysate for the typing of cellular enzymes was prepared shortly before use. Electrophoresis was carried out for 29 blood proteins between 1988 and 1990. The proteins investigated were albumin (ALB), α-anti-trypsin (PI), vitamin-D-binding protein (DBP) or group-specific component (GC), transferrin (TF), properdin factor B (BF), complement factor 3 (C3), serum amylases (AMY 1 and 2), haemoglobin-beta (HB), adenosine deaminase (ADA), adenylate kinase (AK 1), 6-phosphogluconate dehydrogenase (6-PGD), galactose-1-phosphate uridyl transferase (GALT), glyoxalase (GLO), phosphoglucomutases (PGM 1, 2 and 3), carbonic anhydrases (CA 1 and 2), acid phosphatase (ACP), esterases (ESD), isocitrate dehydrogenase (IDH), glucose-6-phosphate dehydrogenase (G6PD), diaphorases (DIA 1 and 2), phosphohexose isomerase (PHI), malate dehydrogenase (MDH) and lactate dehydrogenases (LDH A and B). Electrophoresis was performed on agarose gels or by isoelectric focusing (IEF) with ampholines or immobilines (Scheffrahn, 1992). In the present analysis, we include only the blood protein markers that proved to be reproducible and could be shown to follow Mendelian rules; these were seven highly variable proteins (de Ruiter *et al.*, 1992). Five of these highly variable markers were investigated at all sites. For the Sumatran mainland sites (locations 1–9), we could use two extra marker systems to increase resolution (Table 2.2). The resolution obtained was at least of the same order of magnitude as in the studies by Kawamoto *et al.* (1982b) and Melnick *et al.* (1984).

Statistics

Calculations were performed with the BIOSYS-1 computer program (Swofford & Selander, 1989). We computed allele frequencies and performed a chi-squared test for deviation from Hardy–Weinberg equilibrium with Levene's (1949) correction for small sample size. Mean heterozygosity (unbiased and direct-count estimate) and mean number of alleles per locus were also calculated for each troop, and for the whole Sumatran population. The genetic similarity and genetic distance (D) between populations were calculated (1) with five markers (PI, DBP, TF, AMY *2 and BF) for the total of 14 sites, and (2) with the two additional markers (IDH, 6-PGD) at the Sumatran mainland sites 1–9 and reference site 14. We used Rogers' (1972) formulas, as modified according to Wright (1978) and Nei (1978). We prefer Rogers' modified genetic distance and similarity because it meets the requirements for the triangle inequality (Farris, 1972). In addition, we calculated Nei's genetic distance. F-statistics (F_{ST}) were computed with Weir & Cockerham's (1984) method that corrects for small sample sizes. Standard errors were computed by way of 'bootstrapping' over loci (Haring, 1993).

Table 2.2. *Blood proteins, number of alleles and methods applied*

Blood protein	Abbreviation	No. of alleles	Method
Isocitrate dehydrogenase	IDH	3	AGE
			AGE
6-Phosphogluconate dehydrogenase	6-PGD	2	AGE
			AGE
Protease inhibitors (α-1 anti-trypsin)	PI	4	IEF imm. mod.
Vitamin-D-binding protein	DBP (GC)	7	IEF imm. mod.
Transferrin	TF	6	AGE
Properdin factor B	BF	4	AGE
Amylase 2	AMY 2	5	AGE mod.

AGE, agarose gel electrophoresis; IEF imm., isoelectric focusing with immobilines; mod, modified.

Results

Genetic variation

Of the 29 systems examined, 15 revealed some kind of genetically determined variability. These were: the plasma proteins PI, DBP (GC), TF, C3 and BF; the haemoglobins; the plasma enzyme AMY; and the red blood cell enzymes DIA, IDH, GALT, 6-PGD, CA1, CA2, PGM1 and PGM2. This means a degree of polymorphism (P) of $P = 0.517$. Allele frequencies of the seven markers used in this analysis are shown in Table 2.3. At the locus IDH we found two alleles. The 6-PGD is one of the less variable blood protein systems; yet in most groups we found two alleles, although in two groups the system was monomorphic with the allele 6-PGD*1 ($= 6$-PGD*A). PI also showed two alleles in most groups studied, but at island sites 11–13 a third allele, and at Ketambe site 3 a fourth allele was found. DBP (GC) is one of the most variable blood protein systems in primates (Constans *et al.*, 1987). With an improved IEF technique using immobilines (Scheffrahn 1992), we were able to detect seven alleles at the DBP locus in our 14 populations, and these alleles had a well-balanced distribution. Therefore, the DBP represented the most suitable marker system for our main aim, which was to describe genetic variation and relatedness between the different *Macaca fascicularis* sites. In the TF systems we found six alleles, three of which (TF*1, TF*2 and TF*3) were found in most Ketambe groups (1, 3–6) and these showed a balanced distribution throughout the area.

Table 2.3. *Blood protein markers, alleles, frequencies and populations*

Locus							Population							
	1	2	3	4	5	6	7	8	9	10	11	12	13	14
IDH														
N	59	24	8	19	23	19	7	12	9	0	0	0	0	0
1	0.712	0.875	0.875	0.737	0.652	0.632	0.786	0.500	0.889	0.000	0.000	0.000	0.000	0.000
2	0.288	0.125	0.125	0.263	0.348	0.368	0.214	0.500	0.111	0.000	0.000	0.000	0.000	0.000
6-PGD														
N	58	25	8	18	23	19	7	12	9	0	0	0	0	0
1	0.819	0.880	0.938	0.861	0.783	0.947	1.000	0.792	1.000	0.000	0.000	0.000	0.000	0.000
2	0.181	0.120	0.063	0.139	0.217	0.053	0.000	0.208	0.000	0.000	0.000	0.000	0.000	0.000
PI														
N	58	25	8	20	24	19	8	13	0	12	7	9	11	4
1	0.776	0.800	0.750	0.875	0.854	0.737	0.688	0.731	0.833	0.792	0.571	0.389	0.136	0.000
2	0.224	0.200	0.188	0.125	0.146	0.263	0.313	0.269	0.167	0.208	0.357	0.556	0.818	1.000
3	0.000	0.000	0.000	0.000	0.000	0.000	0.000	0.000	0.000	0.000	0.071	0.056	0.045	0.000
4	0.000	0.000	0.063	0.000	0.000	0.000	0.000	0.000	0.000	0.000	0.000	0.000	0.000	0.000
DBP														
N	59	25	8	20	24	20	8	13	10	12	11	12	12	4
1	0.110	0.320	0.000	0.050	0.063	0.100	0.000	0.000	0.150	0.000	0.000	0.000	0.083	1.000
2	0.542	0.340	0.438	0.125	0.313	0.150	0.625	0.769	0.450	0.375	0.000	0.042	0.750	0.000
3	0.051	0.020	0.000	0.075	0.083	0.050	0.125	0.000	0.150	0.000	0.000	0.000	0.000	0.000
4	0.093	0.240	0.125	0.400	0.208	0.476	0.125	0.192	0.250	0.625	0.545	0.958	0.167	0.000
5	0.203	0.080	0.438	0.175	0.271	0.200	0.125	0.000	0.000	0.000	0.000	0.000	0.000	0.000
6	0.000	0.000	0.000	0.175	0.063	0.025	0.000	0.000	0.000	0.000	0.000	0.000	0.000	0.000
7	0.000	0.000	0.000	0.000	0.000	0.000	0.000	0.038	0.000	0.000	0.455	0.000	0.000	0.000

TF

N	59	25	7	20	24	19	8	13	10	12	11	12	12	4
1	0.127	0.000	0.071	0.175	0.083	0.026	0.188	0.192	0.150	0.208	0.045	0.042	1.000	0.000
2	0.678	0.640	0.786	0.475	0.708	0.947	0.750	0.731	0.800	0.792	0.955	0.917	0.000	0.375
3	0.195	0.360	0.143	0.350	0.208	0.026	0.000	0.000	0.000	0.000	0.000	0.042	0.000	0.000
4	0.000	0.000	0.000	0.000	0.000	0.000	0.000	0.000	0.050	0.000	0.000	0.000	0.000	0.000
5	0.000	0.000	0.000	0.000	0.000	0.000	0.000	0.000	0.000	0.000	0.000	0.000	0.000	0.625
6	0.000	0.000	0.000	0.000	0.000	0.000	0.000	0.000	0.000	0.000	0.000	0.000	0.000	0.000
7	0.000	0.000	0.000	0.000	0.000	0.000	0.063	0.077	0.000	0.000	0.000	0.000	0.000	0.000

BF

N	59	25	8	19	24	19	8	13	9	12	9	12	12	4
1	0.000	0.000	0.063	0.000	0.083	0.053	0.125	0.308	0.111	0.333	0.222	0.042	0.250	0.000
2	0.881	0.860	0.813	0.684	0.625	0.763	0.750	0.615	0.278	0.583	0.556	0.917	0.708	0.000
3	0.051	0.000	0.000	0.105	0.042	0.026	0.000	0.077	0.167	0.083	0.111	0.000	0.042	0.000
4	0.068	0.140	0.125	0.211	0.250	0.158	0.125	0.000	0.444	0.000	0.111	0.042	0.000	1.000

AMY

N	58	25	8	19	24	20	8	13	10	12	10	11	12	4
1	0.862	0.900	0.813	0.895	0.792	0.825	0.750	0.808	0.850	0.542	0.900	1.000	0.083	0.750
2	0.069	0.100	0.125	0.053	0.167	0.165	0.250	0.115	0.150	0.417	0.000	0.000	0.000	0.000
3	0.000	0.000	0.000	0.000	0.000	0.000	0.000	0.077	0.000	0.000	0.000	0.000	0.000	0.000
4	0.069	0.000	0.063	0.053	0.042	0.000	0.000	0.000	0.000	0.042	0.100	0.000	0.458	0.250
6	0.000	0.000	0.000	0.000	0.000	0.000	0.000	0.000	0.000	0.000	0.000	0.000	0.458	0.000

N is number of samples analysed and the loci are 1, House Group; 2, Antana; 3, Gunung Mesjid; 4, Kanan; 5, Kiri; 6, Gurah B; 7, Gumpang; 8, Banda Aceh; 9, Singkel; 10, Tuangku; 11, Nias; 12, Simeulue; 13, Weh; 14, Mentawai.

TF*1 and TF*2 were found at 11 of the 13 *fascicularis* sites. Although the properdin factor B system has not often been studied in non-human primates, appears generally to have a high degree of inter-specific and intra-specific variability. We found four alleles in this system; these alleles showed the most even allelic distribution of all the loci investigated. At the AMY locus we demonstrated five alleles, two of which were common at all the various sites except for site 12, where only allele AMY*1 was found. In summary, at these seven loci we were able to demonstrate 31 alleles in our *M. fascicularis* populations, with an average of 4.4 alleles per locus. The average number of alleles varies somewhat between sites and when we compare the five loci investigated at all sites, we found a somewhat lower number of alleles at the island sites (10–13, Table 2.4).

The mean expected heterozygosity for the Sumatran mainland was 0.388 ranging from 0.342 (site 2) to 0.459 (site 5), see Table 2.4. This variation is probably due partly to a difference in sample size (Table 2.1) and partly to the stochastic processes of genetic drift. The approximate Hardy–Weinberg equilibrium indicates that there is no strong selection at the investigated loci. The heterozygosity values are not directly comparable to average heterozygosity values obtained in other studies on Sumatran *M. fascicularis* because we did not include monomorphic loci and the protein systems we used were the more variable ones. However, if one multiples average heterozygosity by average degree of polymorphism (*P*) one obtains an estimate of total heterozygosity of 0.201 for this study. This figure is still much higher than the heterozygosity of 0.035 found by Kawamoto *et al.* (1981*b*, 1984). The reason for this must be the choice of the more variable loci to analyse (reflected in a high *P* value) and, moreover, that these loci had a higher number of alleles. Comparison of the average heterzygosity values (five loci) of the mainland populations with those of the island populations reveals that the variability is not lower in the shelf-island population (site 10), but is lower on one of the deep sea islands (site 12).

The fixation of alleles, the uniqueness of variants and a peculiar distribution of alleles within the site (Table 2.3), all produce the characteristic picture of the genetic relationship between the populations studied. The Mentawai macaques (14) are sharply differentiated by the occurrence of only one allele at three loci: PI*2, DBP*1, BF*3. We found a unique allele in transferrin (TF*5) not present in our *M. fascicularis* populations. The *M. fascicularis* samples from the island sites 11, 12 and 13 are characterised by their atypical allele distributions, including an allele PI*3 that was not found at any other site. Furthermore, we found fixations at the AMY and the TF locus at sites 12 and 13, respectively.

Table 2.4. *Mean number of alleles over loci (direct count and Hardy–Weinberg, HW, expected values ± SEM)*

Population	Mean no. of alleles per locus	Mean heterozygosity	
		Direct count	HW expected
1, Ketambe – House group	2.9	0.375 ± 0.056	0.381 ± 0.057
2, Ketambe – Antara	2.4	0.339 ± 0.069	0.342 ± 0.075
3, Ketambe – Gunung	2.7	0.329 ± 0.048	0.356 ± 0.061
4, Ketambe – Kanan	3.0	0.492 ± 0.108	0.424 ± 0.084
5, Ketambe – Kiri	3.1	0.452 ± 0.075	0.459 ± 0.066
6, Ketambe – Gurah B	3.0	0.367 ± 0.100	0.356 ± 0.082
7, Sumatra – Gumpang	2.4	0.413 ± 0.100	0.383 ± 0.070
8, Sumatra – Banda Aceh	2.6	0.415 ± 0.064	0.426 ± 0.030
9, Sumatra – Singkel	2.6	0.403 ± 0.112	0.368 ± 0.101
10, Island – Tuangku	2.4	0.583 ± 0.070	0.459 ± 0.049
11, Island – Nias	2.6	0.296 ± 0.114	0.407 ± 0.112
12, Island – Simeulue	2.4	0.178 ± 0.098	0.196 ± 0.098
13, Island – Weh	2.6	0.336 ± 0.160	0.359 ± 0.100
14, Islands – Mentawai	1.4	0.036 ± 0.036	0.199 ± 0.095

Populations 1–9 and 14 with seven loci, populations 10–13 with five loci.

Genetic similarity and distance between populations

Table 2.5 shows how Rogers' genetic similarity indices roughly decrease on Sumatra from the north (site 8) down to the south, across populations 7 (0.852), 1–6 (0.804) and 9 (0.778). The genetic distance between these populations increases. Table 2.6 gives these figures for 14 sites (five loci). When the related species on the Mentawai Islands were taken as an out-group, Rogers' similarity values produced the results shown in Fig. 2.2a,b; these reveal several clusters. The lowest level in the tree contains the six Ketambe groups and, at a higher level, a second cluster contains sites 7, 8, 9 and 10 are represented in a cluster.

The island sites are included in Fig. 2.2b. Island sites 11 and 12 are in a cluster, and island site 13 stands on its own at a higher level still. Island site 13 contains the most differentiated population; the fixation of one transferrin allele (TF*1) contributes substantially to this unique position. A significant point is that island site 10, unlike sites on other islands, falls in the same group as the sites 7–9 on mainland Sumatra. Apparently, the relatively short isolation time of about 10 000 years did not result in appreciable genetic differentiation. Trees based on genetic distance values (Nei, 1978; Rogers, 1972) produced similar clusters.

Table 2.5. *Matrix of genetic similarity (above diagonal: Rogers, 1972) and genetic distance (below diagonal: Nei, 1978) (seven loci) for sites 1–9 and 14*

Population	1	2	3	4	5	6	7	8	9	14
1	—	0.890	0.890	0.865	0.896	0.846	0.868	0.847	0.803	0.489
2	0.019	—	0.886	0.870	0.852	0.839	0.826	0.771	0.831	0.514
3	0.003	0.017	—	0.830	0.869	0.865	0.880	0.793	0.837	0.497
4	0.043	0.024	0.037	—	0.889	0.846	0.795	0.765	0.797	0.494
5	0.014	0.025	0.004	0.012	—	0.870	0.835	0.835	0.824	0.475
6	0.048	0.046	0.026	0.041	0.018	—	0.856	0.812	0.815	0.491
7	0.008	0.035	0.000	0.065	0.021	0.032	—	0.852	0.850	0.508
8	0.037	0.097	0.064	0.113	0.049	0.074	0.014	—	0.778	0.454
9	0.079	0.065	0.049	0.068	0.041	0.061	0.026	0.084	—	0.539
14	0.618	0.536	0.621	0.620	0.656	0.591	0.584	0.685	0.478	—

For loci on Sumatra, see Table 2.4.

Discussion

Socially mediated differentiation

We found strong genetic differentiation of adjacent social groups (sites 1–6). In most studies, the number of groups and number of loci investigated were too limited to compute a meaningful F_{ST} value (Weir & Cockerham, 1984) as a measure of genetic differentiation. In most primate populations, where genetic differentiation of four or more adjacent social groups has been studied with at least four variable loci, comparable results have been obtained. Rogers' (1989) data on baboons (five groups, 4 loci) gave an F_{ST} of 0.053 ± 0.026; Kawamoto *et al.*'s (1984) data on Sumbawa (five groups; four loci) gave $F_{ST} = 0.022 \pm 0.012$ and for the six ketambe macaque groups (11 loci included in our study) F_{ST} was 0.045 ± 0.012. This socially based differentiation was first pointed out by Melnick *et al.* (1984). They used a slightly different F_{ST} calculation method that yields comparable results (Nei & Chesser, 1983), and found an F_{ST} of 0.04. Pope (1992), using the same method, found F_{ST} values of 0.14 and 0.23 in two populations; she emphasised the role of harem structure with respect to the pattern of variability. In a computer simulation (de Jong *et al.*, 1994) in which we modelled the Ketambe macaque population, we demonstrated that this genetic differentiation is due to strong genetic drift operating at the social level; this genetic differentiation results from sex-dependent migration and differential reproduction in males. In Sumatran long-tailed macaques, the linear arrangement of many populations, including those in Ketambe, along rivers and coast lines may enhance group differentiation by the

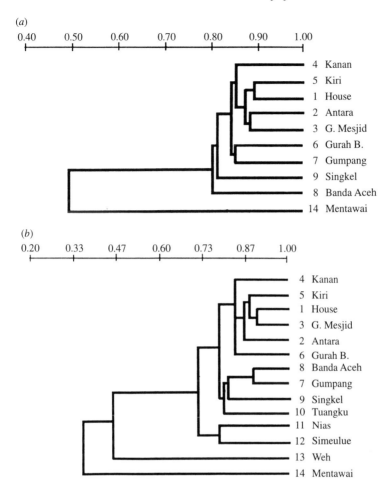

Fig. 2.2. Genetic similarity (Rogers, 1972) between populations on Sumatra: (*a*) seven loci and (*b*) five loci.

'stepping stone' configuration of the social groups (Crow & Kimura, 1970, de Jong *et al.*, 1994).

It should be appreciated that one differentiation cause may very well have an influence on other differentiation mechanisms. For instance, the pattern resulting from socially mediated differentiation is a coarse pattern containing groups of individuals that resemble one another but which are distinct from other groups. The isolation of one or a number of such groups must have consequences different from that of isolation if genetic variability were more evenly distributed over a population.

Table 2.6. *Distance and similarity values (five loci) for sites 1–14*

Population	1	2	3	4	5	6	7	8	9	10	11	12	13	14
1	—	0.891	0.902	0.825	0.871	0.826	0.866	0.834	0.795	0.746	0.723	0.705	0.491	0.375
2	0.021	—	0.852	0.849	0.854	0.836	0.799	0.772	0.790	0.730	0.713	0.706	0.433	0.387
3	0.000	0.030	—	0.805	0.889	0.862	0.862	0.814	0.787	0.745	0.740	0.706	0.454	0.352
4	0.067	0.035	0.057	—	0.875	0.823	0.751	0.732	0.774	0.727	0.731	0.679	0.421	0.371
5	0.025	0.026	0.000	0.021	—	0.852	0.836	0.798	0.842	0.775	0.739	0.661	0.437	0.368
6	0.066	0.051	0.025	0.061	0.027	—	0.840	0.795	0.803	0.799	0.820	0.790	0.414	0.363
7	0.002	0.053	0.000	0.105	0.021	0.045	—	0.892	0.811	0.804	0.738	0.687	0.542	0.358
8	0.045	0.099	0.049	0.152	0.073	0.104	0.000	—	0.809	0.827	0.750	0.659	0.562	0.363
9	0.099	0.101	0.086	0.100	0.034	0.073	0.047	0.064	—	0.778	0.731	0.633	0.415	0.398
10	0.137	0.141	0.107	0.137	0.089	0.056	0.057	0.065	0.097	—	0.743	0.676	0.484	0.324
11	0.179	0.154	0.140	0.139	0.129	0.049	0.149	0.161	0.129	0.111	—	0.787	0.413	0.401
12	0.207	0.166	0.183	0.163	0.198	0.066	0.185	0.227	0.247	0.141	0.061	—	0.409	0.361
13	0.683	0.893	0.798	0.936	0.939	0.992	0.574	0.578	0.976	0.717	1.045	0.882	—	1.166
14	0.990	0.893	1.122	1.056	1.080	0.951	1.046	1.023	0.830	1.190	0.818	0.893	1.166	—

Population loci as Table 2.4. Above diagonal, Rogers' (1972) genetic similarity; below diagonal, Nei's (1978) genetic distance.

Isolation by distance

The finding that the groups on the mainland of Sumatra were more different genetically when they were further apart is in accordance with the isolation by distance principle, and was also reported by Kawamoto *et al.* (1982*b*) for Sulawesi. In this study, the geographically most distant populations of *M. nigra* in the north and *M. maura* in the south had the highest genetic distance (*D*), and were linked by intermediate groups in a clinal gradient. In our population and on Sulawesi, the isolation by distance principle appears to lead to clinal variation.

Isolation in time and island size

Monkeys on the deep sea islands (sites 11–13) may have somewhat lower heterozygosity values compared with mainland populations. Such a result would be expected on the basis of the Goodman–Ishimoto island theorem, which predicts a reduction in genetic variability in relation to island size (Goodman *et al.*, 1965; Ishimoto, 1973). Genetic divergence between the populations at the various island sites shows a clear pattern. The three deep sea island sites 11–13 have become substantially differentiated, as is evident from the cluster analysis. The smallest island (site 13) of these stands on its own. In contrast, the small *shelf*-island (site 10) lies within the cluster of the Sumatran mainland (Fig. 2.2). The genetic divergence and loss of variability on deep sea islands could be due to the founder effect, to genetic drift, or to both. Assuming a comparable history of the primate populations at the time of isolation one can draw several conclusions: first, that differentiation apparently results from long isolation time; and secondly, that this is particularly true for small populations. Genetic drift over long periods, particularly in small populations, explains this result. On shelf-island site 10, which is about the same size as island site 13, the relatively short isolation time has not caused great differentiation. This implies that the founder effect has not been an important principle of genetic differentiation under the existing conditions. If the founder effect were an important principle, one would expect genetic differentiation on a relatively small island, immediately after isolation.

Social structure, speciation, genetic drift and founder effect

From these observations, it is clear that genetic divergence per se is not a reliable indicator of isolation time. One should, therefore, be very cautious about classifying these populations into sub-species or even species on the basis of their genetic divergence alone. Several authors (Crockett-Wilson & Wilson, 1977; van Schaik & van Noordwijk, 1985) pointed out that differences between an island population and a population on the mainland may result from natural selection, and may in particular represent adaptions to the absence of feline predators. For example, on the island of Simeulue (site 12) group size is considerably smaller than on Sumatra. Nevertheless, dominant males on Simeulue have a specific call, a signal which may be used for group spacing. These differences in social behaviour, as well as the difference in coat colour, may be interpreted as adaptations to the absence of feline predators (van Schaik & van Noordwijk, 1985). If the behaviour differences reported by van Schaik are not apparent after a relatively short isolation time of a population, as may be the case, these differences as well as morphological differences must be taken into account in decisions concerning systematics.

Interestingly, differentiation of adjacent social groups was large relative to the differentiation attributable to limited gene flow at larger distances (Fig. 2.2(a) sites 1–6 and sites 7–9, respectively). This may in part explain the absence of a founder effect on shelf-island site 10. As noted before by Melnick (1987), in a population of strongly differentiated social groups, a founder effect may only be expected if a very small group of individuals, such as one social group or even a matriline, becomes isolated. Several social groups as founders would already contain a large part of the regional variation. Melnick argued that such small groups of founders may be responsible for the differentiation of the macaques. He suggested that there is evidence of this in the geographical distribution of differentiated macaque populations and of macaque species; these occur in areas with islands and ecological instability during evolutionary history. Subsequent colonisation with repeated founder effects, as proposed by Melnick, would, however, give a pattern where each population further down the line contains part of the previous population in the linear arrangement. The divergence in different directions of *M. fascicularis* populations in a linear arrangement, as reported by Kawamoto et al. (1984), indicates that these populations stem from residual populations that have become more differentiated through genetic drift, as Fooden & Lanyon (1989) have suggested. Our survey stresses the relative importance of genetic drift over a long time compared to the more immediate effect in island populations of long-tailed macaques.

Pope (1992) also noted the important influence social structure may have in processes of genetic differentiation. She argued that the new combinations of genes that arise because of the social structure may have facilitated adaptations to new areas. Our macaques also show this socially mediated genetic structure, but this structure does not seem to have led to a strong founder effect and subsequent speciation, even in populations which have been isolated for more than 100 000 years. Nevertheless, the populations on the deep sea islands may be recognised as different species on the basis of their different social structure (see above). Other principles, however, must be involved in speciation. The genetic structure of these populations does not explain the rapid speciation suggested in social mammals. The problem of the relationship between genetic differentiation and speciation must be addressed in order to resolve the question of the role of socio-genetic structure in speciation. From the present data, it is evident that there are no indications that, in primates, spatial isolation plays the significant role in speciation that has often been assumed. The genus *Presbytis*, for example, which occupies a habitat similar to that of *M. fascicularis* and faces very similar ecological constraints (e.g. Slabbekoorn, 1993), is represented on Sumatra by 3 species and 19 sub-species.

Summary

The genetic relatedness between 13 sites of long-tailed macaques (*M. fascicularis*) ($n = 234$) on Sumatra and its off-shore islands was analysed on the basis of allele frequencies of seven multi-allelic blood protein markers (IDH, 6-PGD, PI, DBP or GC, TF, BF, AMY). We compared genetic variation, average heterozygosity, genetic similarity and genetic distance, and came to the following conclusions.

1. Differentiation of adjacent social groups caused by social structure is relatively large compared to differentiation caused by distance.
2. For sites at various distances from our main study populations, differentiation from the main population appeared to be a function of distance.
3. Isolation of a sub-population did not lead to genetic differentiation, indicating the absence of a strong founder effect.
4. Long isolation times led to substantial differentiation, presumably as a result of genetic drift.
5. The effect of isolation was (as would be expected) much stronger on a very small island.

Acknowledgements

We are very grateful to the Indonesian Institute of Sciences (LIPI) and its Biological Division (Puslitbang Biologi) and the Indonesian Conservation

Service (PHPA) for permission to carry out this research. Dr Jito Sugardjito, Ir Harris Surono and Ir Chalic Deri, Drs Suharto Djojosudharmo and Drs Nazwar Surono were very helpful with respect to local matters. We thank Jeff Rogers for allowing us to cite data from his PhD thesis, and Robert Haring for assistance with computation of F-statistics. The authors are grateful to Gerdien de Jong, Hans Slabbekoorn and Chris Pryce for valuable comments on the manuscript, and to J. Schneller for assisting with the BIOSYS-1 computer program. We thank W.S.'s son, Raymund Scheffrahn, for excellent field assistance in the summer of 1985, and Irma Voogd, Gerda Greuter and Andrea Camperio-Ciani for their contributions to the work. We are very grateful to J.R. de R.'s wife, Margo de Ruiter for participating in all the research, including the fieldwork on isolated islands. The fieldwork was financed by WOTRO, the Netherlands Foundation for the Advancement of Tropical Research. A subsidy by the Harry Frank Guggenheim Foundation allowed J.R. de R. to take part in the genetic analysis. Important financial contributions were made by the Schultz Stiftung, the Nederlandse Stichting voor Antropobiologie, and Lucie Burger's Stichting voor Vergelijkend Gedragsonderzoek.

References

Bush, G. L. (1975). Modes of animal speciation. *Annual Review of Ecology and Systematics*, **6**, 339–64.

Camperio-Ciani, A., Stanyon, R., Scheffrahn, W. & Sampurno, B. (1989). Evidence of gene flow between Sulawesi macaques. *American Journal of Primatology*, **17**, 257–70.

Chesser, R. K. (1983). Genetic variation within and among populations of the black tailed prairy dog. *Evolution*, **372**, 320–31.

Constans, J., Gouaillard, C., Bouissou, C. & Dugoujon, J. M. (1987). Polymorphism of the vitamin D binding protein (DBP) among primates: an evolutionary analysis. *American Journal of Physical Anthropology*, **73**, 365–77.

Crockett-Wilson, C. & Wilson, W. L. (1977). Behavioral and morphological variations among primate populations in Sumatra. *Yearbook of Physical Anthropology*, **20**, 207–33.

Crow, J. F. & Kimura, M. (1970). *An introduction to population genetics theory*. New York, Harper & Row.

de Jong G., de Ruiter, J. R. & Haring R. (1994). Genetic structure of a population with social structure and migration. In *Conservation genetics*, ed. V. Loeschcke, J. Tomiuk & S. K. Jain, pp. 147–64. Basel: Birkhauser Verlag.

de Ruiter, J.R. (1992). Capturing wild long-tailed macaques. *Folia Primatologica*, **59**, 89–104.

de Ruiter, J. R., Haring, R. N. F., de Jong, G., Scheffrahn, W. & van Hooff, J. A. R. A. M. (1994*b*). The influence of social structure on genetic variability and relatedness in populations with stable social groups: a computer simulation study based on paternity testing and population genetic analysis of

Macaca fascicularis. In *Behaviour and genes in natural populations of long-tailed macaques* (Macaca fascicularis), ed. J. R. de Ruiter, pp. 107–30. Privately published.

de Ruiter, J. R., Scheffrahn, W., Trommelen, G. J. J. M., Uitterlinden, A. G., Martin, R. D. & van Hooff, J. A. R. A. M. (1992). Male social rank and reproductive success in wild long-tailed macaques. In *Paternity in primates: tests and theories,* ed. R. D. Martin, A. Dixson & J.Wickings, pp. 175–91. Basel: S. Karger.

de Ruiter, J. R. & van Hooff, J. A. R. A. M. (1993). Male dominance rank and reproductive success in primate groups. *Primates,* **34**, 513–23.

de Ruiter, J. R., van Hooff, J. A. R. A. M. & Scheffrahn, W. (1994a). Social and genetic aspects of paternity in wild long-tailed macaques (*Macaca fascicularis*). *Behaviour,* **129**, 1203–24.

Dracopoli, N. C., Brett, F. L., Turner, T. R., Jolly, C. L. (1983). Patterns of genetic variability in the serum proteins of the Kenyan vervet monkey (*Cercopithecus aethiops*). *American Journal of Anthropology,* **61**, 39–49.

Eudey, A. A. (1980). Pleistocene glacial phenomena and the evolution of Asian macaques. In *The macaques: studies in ecology, behavior and evolution,* ed. D. G. Lindburg, pp. 52–84. New York: Van Nostrand Reinhold.

Farris, J. S. (1972). Estimating phylogenetic trees from distance matrices. *American Naturalist,* **106**, 645–68.

Fooden, J. & Lanyon, S. M. (1989). Blood-protein allele frequencies and phylogenetic relationships in *Macaca:* a review. *American Journal of Primatology,* **17**, 209–41.

Goodman, M., Kulkarni, A., Poulik, E. & Reklys, E. (1965). Species and geographic differences in the transferrin polymorphism of macaques. *Science,* **147**, 884.

Haring, R. (1993). The influence of social structure on relatedness and qualities of various relatedness estimators. MSc thesis, University of Utrecht.

Heany, L. R. (1986). Biogeography of mammals in SE Asia: estimates of rates of colonization, extinction and speciation. *Biological Journal of the Linnaean Society,* **28**, 127–65.

Ishimoto, G. (1973). Blood protein variations in Asian macaques. III. Characteristics of the macaque blood protein polymorphism. *Journal of the Anthropological Society of Nippon,* **18**, 1–13.

Jongsma, D. (1970). Eustatic sea level changes in the Arafura Sea. *Nature,* **228**, 150–51.

Kawamoto, Y. (1982). A reexamination of electromorphs of plasma transferrin in the Indonesian crab-eating macaque (*Macaca fascicularis*). *Kyoto University Overseas Research Report of Studies on Asian Non-Human Primates,* **2**, 65–73.

Kawamoto, Y. & Ischak, T. M. (1981). Genetic differentiation of the Indonesian crab-eating macaque (*Macaca fascicularis*) I. Preliminary report on blood protein polymorphism. *Primates,* **22**, 237–52.

Kawamoto, Y., Ischak, T. M. & Supriatna, J. (1984). Genetic variations within and between troops of crab-eating macaques (*Macaca fascicularis*) on Sumatra, Java, Bali, Lombok and Sumbawah, Indonesia. *Primates,* **25**, 31–59.

Kawamoto, Y., Nozowa, K. & Ischak, T. M. (1981). Genetic variability and differentiation of local populations in the Indonesian crab-eating macaque (*Macaca fascicularis*). *Kyoto University Overseas Report of Studies on Indonesian Macaques,* **1**, 15–39.

Kawamoto, Y., Shotake, T. & Nozawa, K. (1982a). Genetic differentiation among three genera of the family Cercopithecidae. *Primates*, **23**, 272–86.

Kawamoto, Y., Takenaka, O. & Brotoisworo, E. (1982*b*). Preliminary report on genetic variations within and between species of Sulawesi macaques. *Kyoto University Overseas Research Report of Studies on Asian Non-human Primates*, **2**, 23–37.

Kawamoto, Y., Takenaka, O., Suryobroto, B. & Brotoisworo, E. (1985). Genetic differentiation of Sulawesi macaques. *Kyoto University Overseas Research Report of Studies on Asian Non-human Primates*, **4**, 41–61.

Levene, H. 1949. On a matching problem arising in genetics. *Annals of Mathematics and Statistics*, **20**, 91–4.

Li, W. H. & Nei, M. (1975). Drift variances of heterozygosity and genetic distance in transient states. *Genetic Research*, **25**, 229–48.

Malécot, G. (1950) Quelques schémas probabilistes sur la variabilité des populations naturelles. *Annales d'Université de Lyons Scientifiques*, A, **13**, 37–60.

Melnick, D. J. 1987. The genetic consequences of primate social organization: a review of macaques, baboons and vervet monkeys. *Genetica*, **73**, 117–35.

Melnick, D. J. (1988). The genetic structure of a primate species: rhesus macaques and other cercopithecine monkeys. *International Journal of Primatology*, **9**, 195–231.

Melnick, D. J. & Kidd, K. K. (1983). The genetic consequences of social group fission in a wild population of rhesus monkeys (*Macaca mulatta*). *Behavioural Ecology and Sociobiology*, **12**, 229–36.

Melnick, D. J., Pearl, M. C. & Richard A. F. (1984). Male migration and inbreeding avoidence in wild rhesus monkeys. *American Journal of Primatology*, **7**, 229–43.

Morin P., Wallis, J, Moore, J.J., Chakraborty, R. & Woodruff, D. (1993). Non-invasive sampling and DNA amplification for paternity exclusion, community structure, and phylogeography in wild chimpanzees. *Primates*, **34**, 347–56.

Nei, M. (1978). Estimation of average heterozygosity and genetic distance from a small number of individuals. *Genetics*, **89**, 583–90.

Nei, M. & Chesser, R. K. (1983). Estimation of fixation indices and gene diversities, *Annals of Human Genetics*, **47**, 253–9.

Nozawa, K., Shotake, T., Kawamoto, Y. & Tanabe, Y. (1982). Population genetics of Japanese monkeys. II. Blood protein polymorphism and population structure. *Primates*, **23**, 252–71.

Nozawa, K., Shotake, T., Ohkura, Y., Kitajima, M. & Tanabe, Y. (1975). Genetic variation within and between troops of *Macaca fuscata fuscata*. *Contemporary Primatology*, **5**, 75–89.

Nozawa, K., Shotake, T., Ohkura, Y. & Tanabe, Y. (1977). Genetic variations within and between species of Asian macaques. *Japanese Journal of Genetics*, **52**, 15–30.

Pope, T. R. (1990). The reproductive consequences of male cooperation in the red howler monkey: paternity exclusion in multi and single male troops using genetic markers. *Behavioural Ecology & Sociobiology*, **27**, 439–46.

Pope, T. R. (1992). The influence of dispersal patterns and mating system on genetic differentiation within and between populations of the red howler monkey (*Alouatta seniculus*). *Evolution*, **46**, 1112–18.

Rogers, J. S. (1972). Measures of genetic similarity and genetic distance. *Studies in Genetics*, University of Texas Publishers, **7213**, 145–53.

Rogers, J. A. (1989). Genetic structure and microevolution in a population of Tanzanian yellow baboon (*Papio hamadryas cynocephalus*). PhD thesis, Yale University.

Rowe, A. W., Davis, J. H. & Moore-Jankowski, J. (1972). Preservation of red cells from the nonhuman primates. In *Primates in Medicine*, vol. 7, pp. 117–30. Basel: S. Karger.

Scheffrahn, W. (1992). Elektrophoretische Methoden. In *Anthropologie. Handbuch der vergleichen den biologie des menschen*, vols. 1 and 2, ed. R. Knubmann, H. W. Jurgens, I. Schwidetzky & G. Ziegelmayer, pp. 371–421. Stuttgart: G. Fischer.

Scheffrahn, W., Ménard, N., Vallet, D. & Gaci, B. (1993). Ecology, demography and population genetics of Barbary macaques in Algeria. *Primates*, **34**, 381–94.

Shotake, T. (1981). Population genetical study of natural hybridization between *Papio anubis* and *P. hamadryas*. *Primates*, **22**, 285–308.

Shotake, T., Nozawa, K. & Tanabe, Y. (1977). Blood protein variations in baboons, I. Gene exchange and genetic distance between *Papio anubis, Papio hamadryas* and their hybrid. *Japanese Journal of Genetics*, **52**, 223–37.

Slabbekoorn, H. W. (1993). Energetics of patch use. Foraging strategies of two sympatric primates (*Macaca fascicularis* and *Presbytis thomasi*). MSc thesis, University of Utrecht.

Smith, D. G. & Coss, R. G. (1984). Callibrating the molecular clock; estimates of ground squirrel divergence made using fossil and geological time markers. *Molecular Biology of Evolution* **1**, 249–59.

Sugardjito, J., van Noordwijk, M. A., van Schaik, C. P. & Tatang, M. S. (1984). The Simeulue monkeys (*Macaca fascicularis fusca*, Miller 1903). Report to IUCN.WWF primate specialist group, Washington.

Swofford, D. J. & Selander, R. B. (1989). *BIOSYS-1*. Champaign, IL: Natural History Survey.

Takenaka, O. & Brotoisworo, E. (1982). Preliminary report on Sulawesi macaques – their distribution and inter-specific differences. *Kyoto University Overseas Research Report of Studies on Asian Non-Human Primates*, **2**, 11–22.

Turner, T. R. (1981). Blood protein variation in a population of Ethiopian vervet monkeys (*Cercopithecus aethiops aethiops*). *American Journal of Physical Anthropology*, **55**, 225–32.

van Schaik, C. P. & van Noordwijk, M. A. (1985). Evolutionary effect of absence of fields on the social organization of the macaques on the island of Simeulue (*Macaca fascicularis fusca*, Miller 1903). *Folia Primatologica*, **44**, 138–47.

Wade, M. J. (1982). Group selection: migration and the differentiation of small populations. *Evolution*, **36**, 949–61.

Wahlund, S. (1928). Zusammensetzung von Populationen und Korrelationserscheinungen vom Standpunkt der Vererbungslehre aus betrachtet. *Hereditas*, **11**, 65–106.

Weir, B. A. & Cockerham, C. C. (1984). Estimating F-statistics for the analysis of population structure. *Evolution*, **38**, 1358–70.

White, J. D. (1968). Models of speciation. *Science*, **159**, 1065–70.

Whitten, A. J., Damanik, S. J., Anwar, J. & Hisyam, N. (1984). *The ecology of Sumatra*. Jogyakarta: Gajah Mada University Press.

Whitten, A. J. & Whitten, J. E. J. (1982). Preliminary observations of the Mentawai macaque on Siberut Island, Indonesia. *International Journal of Primatology*, **3**, 445–59.

Wilson, D. S. (1983). The group selection controversy: history and current status. *Annual Review of Ecology and Systematics*, **14**, 159–87.

Wilson, A.C., Bush, G. L., Case, S. M. & King, M. C. (1975). Social structure of mammalian populations and rate of chromosomal evolution. *Proceedings of the National Academy of Sciences, USA*, **72**, 5061–5.

Wright, S. (1931). Evolution in Mendelian populations. *Genetics* **16**, 97–159.

Wright, S. (1978). *Evolution and the genetics of populations*, vol. 4, *Variability within and among natural populations*. Chicago: University of Chicago Press.

3

Secondary intergradation between *Macaca maurus* and *M. tonkeana* in South Sulawesi, and the species status of *M. togeanus*

J. W. FROEHLICH AND J. SUPRIATNA

Introduction

Hybrid zones between phenotypically distinct taxa have recently received considerable attention from evolutionary biologists (Harrison, 1990) because they are natural laboratories for investigating not only the contentious issue of species identification, but also the theoretical predictions of alternative species definitions. Inclusive mate recognition systems (Paterson, 1985; Masters *et al.*, 1987), exclusive premating reinforcement of postmating isolation (Dobzhansky, 1940; Raubenheimer & Crowe, 1987; Butlin, 1989; Coyne & Orr, 1989), and other mechanisms that maintain morphogenetic cohesion within and between species (Templeton, 1989) can all be studied in the different stages and circumstances of hybridisation (see Hewitt, 1989). Despite serious reservations to the contrary (Endler, 1977, 1982), primary, ecogeographical clines may be distinguishable from secondary intergradation (Mayr, 1942) by historical and ecological patterns inferred from variation in hybrid zones (Fleischer & Rothstein, 1988; Harrison, 1990). In particular, these patterns of variation can be used to test the alternative prediction of stable tension zones between parapatric species and those of full introgression between formerly allopatric, polymorphic sub-species (Barton & Hewitt, 1989).

For some of these issues of species definition and process, the socio-ecological characteristics of the primates provide useful insights (Marks, 1987; Godfrey & Marks, 1991), but there have been few detailed studies of primate hybridisation. Not one primate example is used in any of the extensive reviews cited in the previous paragraph. Overlooked is the intensive, two decade investigation of what we consider a narrow tension zone between hamadryas and anubis baboons (Kummer, Goetz & Angst, 1970; Nagel, 1973; Sugawara, 1988), although Shotake (1981) argued for widespread, presumably differential introgression and Phillips-Conroy, Jolly & Brett (1991) contended that the recognition criterion of a single species is satisfied. Introgressive hybridisation, with assortative mating against rare phenotypes, has also been predicted for a single 'ring species' of spider monkeys, formerly considered two separate species (Froehlich, Supriatna & Froehlich, 1991). Hybrids have occasionally been observed in gibbons (Brockelman & Gittins, 1984), documented in other savanna baboon sub-species (Samuels & Altmann, 1986; Hayes, Freedman & Oxnard, 1990), and described only under exceptional circumstances in guenons (Struhsaker, Butynski & Lwanga, 1988).

Although frequently observed in captivity among macaque species (Bernstein & Gordon, 1980), fertile hybrids have also only rarely been reported in nature (Fooden, 1964; Bernstein, 1966; Southwick & Southwick, 1983). This leads to the conclusion that the possible is probable only when habitat disturbance disrupts normal behavioural differences, such as is predicted by the concept of recognition species in the absence of significant postmating isolation. The long debated and remarkable species diversity of Sulawesi macaques provides a laboratory for further investigation of macaque speciation mechanisms. While Fooden (1969) distinguished seven species, Thorington & Groves (1970) contended that there might be only one. Of particular interest are the alternative interpretations for the marginally sympatric populations of *Macaca tonkeana* and *maurus*.

From brief observations made in 1975 at Maroangin (Fig. 3.1), Groves (1980) proposed that these two species were isolated by distinct pelage differences, suggesting reinforcement by reproductive character displacement in the light coloured *M. maurus*. Starting a decade later, our contradictory observations at Maroangin found mixed social groups, with breeding between male *tonkeana* and female *maurus*, and many young individuals indeterminate by their intermediate phenotypes (Supriatna, 1991). Similarly Ciani *et al.* (1989) in a study of a few hybrid pets, documented steep clines between neighbouring taxa, although they concluded that these may represent 'hybrid sinks', with only limited, differential introgression.

The divergence of *M. tonkeana* is further complicated by the evolution of an insular sub-species in the Togian Islands (Fig. 3.1) with distinctively different coloration. Sody (1949) gave this population species status based on pelage and size comparison of a 1939 museum collection from Malenge Island. Although Fooden (1969) further documented cranial and tail length differences, based largely on biased samples that were more than 75% from the Togian Islands (see Albrecht, 1978), Fooden did not accept even separate sub-species status. Regardless of this issue, the inclusion of the Togian

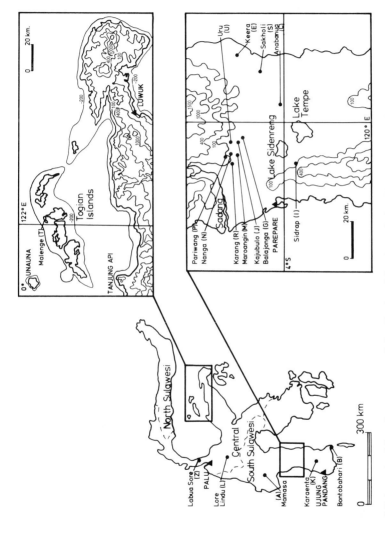

Fig. 3.1. Map of Sulawesi with enlargements of the Togian Islands and the Tempe Depression, showing sample locations and, in parentheses, the letters used symbolically in subsequent figures.

sample in the present study, with larger comparative samples of mainland animals, provides a broad perspective on the conspecific status of *M. tonkeana* and *maurus*, and the processes that contributed to their divergence.

Methods

Sampling procedures

Since previous studies of species diversity in Sulawesi lacked adequate sampling across parapatric borders (Ciani *et al.*, 1989), we designed our investigation of the *tonkeana/maurus* boundary to examine the width, phenotypic concordance, and other features of a predicted secondary hybrid zone. Wild-caught animals were studied over a distance of more than 450 km from Central Sulawesi to the southern tip of South Sulawesi. The range of variation was anchored by 'pure species' samples of *M. tonkeana* from Lore Lindu and *M. maurus* at Karaenta and Bontobahari (Fig. 3.1); all three areas are nature reserves. At Lore Lindu National Park, the animals were captured outside the western boundary at an elevation of about 1000 m. Karaenta is on a forested small limestone mountain at 150 m elevation. The essentially sea level Bontobahari sample was actually trapped in the adjacent village of Lemo-lemo, where monkey sleeping trees are still protectively owned by the boat building villagers.

Putative hybrids comprised 60% of the wild-caught sample of 105 monkeys; they represented western and eastern transect-like opportunistic samples across the species boundary in the Tempe Depression, which was periodically flooded during the Quaternary period (Whitten, Mustafa & Henderson, 1987). Because it is presently constrained by agricultural clearing and a steep escarpment, the western hybrid zone in the Maroangin area was only about 10 km wide, in the foothills of the mountainous range of *M. tonkeana* (Fig. 3.1). In this hilly area of small cattle pastures and gallery forests (Fig. 3.2), we captured animals over a distance of about 15 km, at elevations from 50 m to 150 m (e.g. Pariwang). The second 'transect' was about 30 km further east; sampling from relic forests in low hills along the eastern margin of the Tempe Depression, it represented a hybrid zone width of more than 35 km.

For portions of the eastern 'transect', especially around Anabanua, we opportunistically sampled locally caught pet monkeys. In all, 20 reasonably well-documented pets were used to augment our wild-caught animals and to add small, 'pure species' samples from Sidrap and Mamasa. For morphological comparisons, we also examined the skulls and pelts of specimens at the American Museum of Natural History, the United States

Fig. 3.2. View of the Maroangin study area, looking north towards the escarpment, with the Pariwang locality immediately below the peak at the left.

National Museum, and the Museum Zoologicum Bogoriense in Indonesia. These provided a sample of 35 animals from Malenge Island in the Togian archipelago, plus small numbers from the base of the narrow isthmus joining North Sulawesi (Labua Sore), and supplements to the wild-caught samples from Lore Lindu and Uru. From all these sources, data on 177 specimens were analysed in various parts of this investigation.

The wild-caught macaques were provisioned and habituated to traps for several weeks before capture. Individual or group traps were built primarily to the designs of villagers, out of local materials. The trapped animals were tranquillised with Ketamine HCl using a blow-gun dart. Each animal was ear tagged, tattooed and data processed before being released to its social group or relocated away from crop raiding. Pet animals were treated similarly.

Data collection and analysis

For each animal we took five standardised photographs that would later be used to categorise pelage patterns in conjunction with museum pelts. As in previous studies (Albrecht, 1978; Ciani *et al.*, 1989), we scored the colour and distribution of hair patterns on ordinal scales of three to four categories. Seventeen features were separately recorded from adult and sub-adult animals of both sexes. In a sample mainly of pets, Hamada *et al.* (1988) found no sex or age difference in body colour, except for greying in some older individuals (cf. Fooden, 1969), which we did not include. Due to the potentially confounding effects of diet and housing, we also did not score pets for body colour.

The 17 pelage traits included hair colour (from light brown to black) at nine locations on the trunk and limbs, the shape and colour of the crown and contrasting, bushy cheek hair, the extent and contrasting colour of the rump patch, and the shape and extent of the gluteal fields. Only for the latter characteristic feature of *M. maurus* (Fig. 3.4) was the greatest expression given the lowest ordinal score, so that high scores always reflected the most *tonkeana* expression in hybrids.

The ordinal pelage data were reduced to two dimensions by principal component factor analysis, and transformed scores for all specimens obtained. Since these two orthogonal variables separated the three principal samples of *M. maurus*, *tonkeana* and *togeanus*, their factor loadings defined these taxonomic distinctions and the intermediate patterns of all hybrid samples. As a further clarification of pelage patterning in hybrids, we used a three factor, oblique rotation to dissociate the coadapted taxonomic aggregate of pelage colour, facial hair and perineal patterns.

By taking comparable measurements directly on prominent, skeletal landmarks of wild-caught animals, supplemented and corroborated by measuring dental casts from these specimens, we were able to enlarge and compare our field samples with museum cranial specimens. Many of the latter also included data on body, foot and tail lengths, taken when the specimens were collected. Thirteen dimensions were used in this analysis, including measurements of the cranium, face and anterior dentition. In order to pool both sexes for adequate samples (e.g. of the 21 adult *M. maurus* captured, only four were female), we eliminated sexual dimorphism by standardising the overall data separately for each sex before combining them (see Froehlich *et al.*, 1991). Tail length was also eliminated in the pooling process; due to the complications of sexual swelling, it usually could not be measured in females. The procedures produced a sample of 92 adults for multivariate analysis.

The metric data were reduced to two dimensions by canonical discriminant analysis. Since this method uses *a priori* sample designations, unlike factor analysis, we followed a procedure of first using 59 specimens of pure or nearly pure species comprising only five groups to define the relevant discriminant space, before the scores of remaining hybrids were then calculated and plotted in that space. This provided an unbiased (i.e. non-discriminating) display of size and shape in hybrid samples relative to parental species. The interpretation of these relationships was then facilitated by calculating discriminant loadings, or correlations between the original data and the discriminant scores for the entire sample of 92.

Dermatoglyphic prints were taken on both hands and both feet using powdered graphite and wide, transparent tape. We also scored the undamaged volar surfaces of museum pelts by direct observation, since adequate replication was impossible. A three digit, ordinal scale was designed for scoring macaque variation on all of the volar pads of both palms and soles, defining bilateral asymmetry, shape and complexity of the ridge patterns, similar to the system described for howling monkeys (Froehlich & Thorington, 1982).

Since discriminant analysis has been well established in previous dermatoglyphic studies of both human (Froehlich & Giles, 1981; Froehlich, 1987) and non-human primates (Froehlich & Froehlich, 1986, 1987), we reduced the present data set by the same procedures as the metric analysis. Samples of pure or nearly pure species of both sexes and all ages comprised an initial canonical analysis on 119 specimens in six groups, before unbiased scores were interpolated for the remaining hybrid samples.

Whole blood (typically 5 ml/kg body weight) was drawn from the femoral vein of tranquillised monkeys, heparinised, and separated immediately by low speed centrifugation. The white cells were pelleted, washed and stored in liquid nitrogen. To date, preliminary analyses only include the polymorphic serum protein transferrin in 111 individuals and three restriction enzyme probes (*Bgl*I, *Hind*I and *Hinc*I) that showed systematic mitochondrial DNA polymorphism in a selected subset of 22 monkeys from across the hybrid zone (Supriatna, 1991). Since mitochondrial haplotypes were consistently different for the two pure species samples, the results are only reported here as representing one matrilineage or the other in the hybrids.

Finally, the three morphological data sets were analysed for concordance in the hybrid zone by rank-order (Spearman) correlations of individual hybrid scores on the multivariate vectors. In this analysis, we included the pure *M. maurus* samples from Bontobahari and Karaenta in order to anchor the regression line within one geographical population. By not including the Lore Lindu *tonkeana* sample used to define polarity in the reduction analyses, however, we did not bias either the direction of the

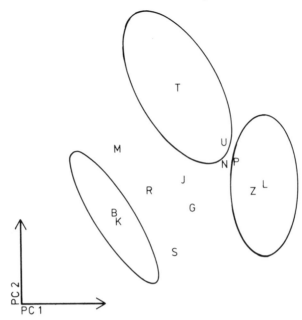

Fig. 3.3. Plot of sample centroids (group means) on the first two principal components (PC) for 17 pelage traits. The three 'pure species' samples are also indicated by 95% confidence ellipses, with Karaenta and Bontobahari combined for this purpose. Letter symbols are identified in Fig. 3.1.

regression line in the hybrids or the degree of association between their individual scores in, otherwise, uncorrelated phenotypic traits.

Results

Pelage patterns

The three 'pure species' samples are completely separated in Fig. 3.3, where the 95% confidence ellipses are plotted on the first two principal components for pelage traits. Indeed, with the samples available there was absolutely no overlap between the Togian and mainland populations of *M. tonkeana*. In descriptive terms, the Togian monkeys have bicoloured, chocolate brown coats, with a darker head, especially along the midline, lighter limbs and ventrum, and show a distinctive lateral stripe in sub-adults. With their long, buff coloured cheek whiskers and extended rump patches, the Togian animals are clearly distinguishable from typical, all black *M. tonkeana*, even with the latter's sharply contrasting light grey cheeks and rump patches. With large, circular gluteal fields, reduced rump patches above the knee, and a lack of crested crown hair (Fig 3.4), *M. maurus* samples are also

Fig. 3.4. *Macaca maurus* male from the Bontobahari locality, showing diagnostic
perineal and facial features.

separable from their *M. tonkeana* neighbours. A full list of these pelage
traits and their loadings, or relative contributions to the two axes, is
presented in Table 3.1.

Although the confidence ellipses for *M. maurus* and *tonkeana* are far
apart in Fig. 3.3, the multivariate gap between them is filled by a continuous
distribution of hybrid centroids, much as these hybrid samples bridge the
geographical space between the two taxa. Moreover, the relative hybrid
distribution in intermediate pelage patterns closely follows their expected
clinal positions along our sampling transects (see Fig. 3.1), especially when
projected onto the first principal component, which utilises 67% of the
variance in the data. Although some sample sizes are small, the spread of
these hybrid populations on the second axis might also suggest additional
variability from the recombination of pelage genes, notably at Sakholi (S)
in the middle of the eastern transect.

The distribution of pelage phenotypes across the hybrid zone is further
clarified in Fig. 3.5, where an oblique rotation of the principal components
is plotted for all samples except those from Togian. By relaxing the
orthogonal criterion, this rotation decouples the different aspects of pelage
phenotypes, even though they are taxonomically correlated (inter-factor

Table 3.1. *Principal component factor loadings for pelage patterns in 137 adult and sub-adult specimens of* Macaca maurus, tonkeana *and* togeanus

	Principal components	
Pelage traits[a]	1	2
Crown crest	0.70	0.54
Crown colour	0.72	−0.02
Cheek colour	0.73	0.09
Cheek whiskers	0.74	0.37
Shoulder colour	0.93	−0.03
Back colour	0.92	−0.10
Abdomen colour	0.86	−0.20
Chest colour	0.85	−0.32
Upper arm colour	0.85	−0.43
Forearm colour	0.91	−0.27
Hand colour	0.92	−0.15
Thigh colour	0.92	−0.25
Foot colour	0.91	−0.26
Rump patch size	0.79	0.49
Rump patch colour	0.73	−0.21
Gluteal field shape	0.74	0.59
Gluteal field size	0.66	0.60
Total proportion of variance	0.67	0.12

[a]All variables were scaled from low scores in *maurus* to highest in *tonkeana*. Therefore, the prominent *maurus* feature of gluteal fields has high scores for small, elliptical patterns.

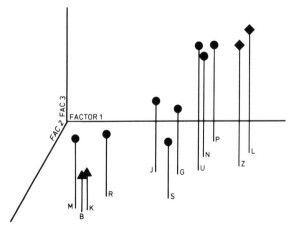

Fig. 3.5. Plot of sample centroids in the approximate space defined by an oblique rotation of pelage principal components. Letter symbols are identified in Fig. 3.1, and FAC represents 'factor'.

Table 3.2. *Oblique reference structure for rotated pelage factors of 137 animals, including the Togian sample not plotted in Fig. 3.5*

Pelage traits	Partial factor loadings		
	1	2	3
Crown crest	−0.05	0.11	0.60
Crown colour	0.53	−0.29	0.30
Cheek colour	0.04	0.61	−0.06
Cheek whiskers	−0.09	0.50	0.25
Shoulder colour	0.54	−0.04	0.21
Back colour	0.55	0.00	0.13
Abdomen colour	0.66	−0.14	0.09
Chest colour	0.70	−0.07	−0.05
Upper arm colour	0.72	0.05	−0.21
Forearm colour	0.65	0.03	−0.04
Hand colour	0.54	0.11	0.02
Thigh colour	0.64	0.03	−0.02
Foot colour	0.58	0.16	−0.11
Rump patch size	0.01	0.18	0.54
Rump patch colour	0.39	0.30	−0.17
Gluteal field shape	0.00	−0.01	0.71
Gluteal field size	−0.02	−0.08	0.74
Total proportion of variance	0.64	0.45	0.45

$r = 0.61$ to 0.70) and the graph is a necessary approximation without these correlations. The oblique reference structure (i.e. partial loadings) in Table 3.2 shows that the first axis is essentially coat colour, especially on the ventrum, limbs and crown, while the second highlights the contrasting bushy cheeks and rump, and the third residually combines the erect crown hair, extended rump patch, and reduced gluteal fields of *M. tonkeana*. Although the hybrids still bridge the gap between the pure species poles in the graph, they appear to fall into three distinct clusters of *maurus*-like, *tonkeana*-like and intermediate.

Body size and shape

For the morphometric data, a plot of the first two canonical vectors (Fig. 3.6) shows a distribution closely similar to the pelage patterns, albeit with some overlap by the 95% confidence ellipses. Nevertheless, the approximate Wilks' F values for both functions are highly significant ($p < 0.0001$). The first axis in the discriminant loadings (Table 3.3), with 56% of the variance, is a body size vector, emphasising the relatively short heads and narrow muzzles of *M. maurus* (Fig. 3.4) in contrast to the larger bodied

Table 3.3. *Discriminant loadings for metric data of* Macaca maurus, tonkeana *and* togeanus *based on 92 adult animals, including 33 hybrids not used in the initial canonical analysis*

	Canonical vector	
Body measurements	1	2
Prosthion–inion length	0.70	−0.46
Nasion–inion length	0.50	−0.46
Prosthion–nasion length	0.41	−0.03
Bizygomatic breadth	0.63	−0.35
Bifrontal breadth	0.34	0.23
Bimalar breadth	0.76	0.05
Canine root breadth	0.56	−0.07
Upper bicanine breadth	0.75	−0.35
Upper bi-incisor breadth	0.75	0.00
Lower bicanine breadth	0.63	−0.20
Lower bi-P_3 breadth	0.63	−0.38
Crown–rump length	0.71	0.09
Foot length	0.38	0.10
Proportion of initial variance	0.56	0.31

P_3 is the 'first' premolar.

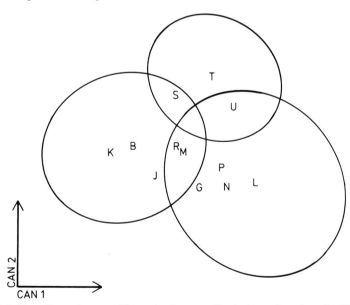

Fig. 3.6. Plot of sample centroids on the first two discriminant functions for 13 body measurements, with 95% confidence ellipses for the three 'pure species' samples (Karaenta and Bontobahari pooled). Letter symbols are identified in Fig. 3.1; CAN1 and CAN2 are canonical vectors 1 and 2.

M. tonkeana. However, on the second canonical axis, accounting for 31% of the variance, the Togian animals are distinguished by interesting shape differences. While they are virtually identical in body length and brow ridges (bifrontal), their heads are much shorter, more rounded, and their canines are relatively smaller than in the mainland populations.

These size and shape comparisons can be seen in the individual measurements for the three major samples in Table 3.4 (*M. maurus* females at Bontobahari could not be included because we only captured one). Although the *M. maurus* male data are clearly smaller in all measurements, as already demonstrated on the first canonical axis, for clarity the table shows only the significant differences between the Togian and mainland samples of *M. tonkeana*. Other than substantiating the significance of the shape differences described above, these data indicate that the tails of Togian animals are remarkably longer (>30%) than typical *M. tonkeana*, indeed longer than any other Sulawesi macaque (Fooden, 1969). The degree of sexual dimorphism for body length in Togian is also greater, with males exceeding females by 16.5% compared with 13.5% in mainland *M. tonkeana*. Again, this is closer to the presumed ancestral (Melnick & Kidd, 1985; Fooden & Lanyon, 1989) condition in *M. nemestrina* (21%) found at the equator (Albrecht, 1980). By contrast, in our sample and the larger one of Watanabe *et al.* (1987), *M. maurus* males at Karaenta are only about 11% larger than females.

The intermediate, bridging distribution of hybrid samples in Fig. 3.6 is quite similar to the pelage relationships in Fig. 3.3. Again, most hybrid samples are positioned along a line between the centroids of the two "pure" species populations, and their relative positions generally correspond. However, the Maroangin (M) and Karang (R) samples, thought by Groves (1980) to be pure *M. maurus* or pure *M. tonkeana*, respectively, and closely clustered with more southern samples in the oblique pelage display, are now midway between the two taxa. The Sakholi (S) animals from the eastern transect (and the Uru (U) sample, previously called typical *M. tonkeana*) also show slightly divergent positions suggestive of some changes in shape due to recombination.

Dermatoglyphic patterns

The dermatoglyphic data reduction in Fig. 3.7 differs from pelage and metric analyses with respect to relationships among the three major samples. Remarkably, the first axis maximises the separation between the

Table 3.4. *Body measurements (mm) in adult male and female Macaca tonkeana from Lore Lindu, M. togeanus, and male M. maurus from Bontobahari*

	Males						Females			
	maurus		*tonkeana*		*togeanus*		*tonkeana*		*togeanus*	
	N	Mean	N	Mean	N	Mean	N	Mean	N	Mean
Prosthion-inion length	11	139.2	8	158.5**	15	147.3	8	133.1**	9	124.8
Nasion-inion length	11	96.2	8	101.5*	11	95.9	8	93.0**	9	87.4
Prosthion-nasion length	11	65.9	8	75.4	11	71.7	8	56.3	9	56.3
Bizygomatic breadth	11	92.0	8	105.9**	15	97.2	8	86.3**	9	80.1
Bifrontal breadth	11	75.4	8	82.7	15	79.7	8	67.9	9	68.4
Bimalar breadth	11	77.5	8	90.1	15	86.5	8	75.4	9	73.6
Canine root breadth	11	37.8	8	42.0*	15	39.0	8	33.9	9	32.6
Upper bicanine breadth	11	39.8	8	46.6***	11	41.1	8	34.0**	9	31.6
Upper bi-incisor	11	21.2	8	24.9*	15	23.6	8	22.1	9	21.7
Lower bicanine breadth	11	31.4	7	34.3	11	32.7	8	26.6*	9	25.0
Lower bi-P_3 breadth	11	29.1	7	34.8**	11	30.6	8	29.6	9	26.9
Crown-rump length	11	501.6	7	586.9	14	584.6	8	517.2	9	502.0
Tail length	11	36.0	7	45.1	14	59.2***				
/foot length	11	163.5	7	170.7	14	173.9	8	151.2	9	149.3

*$p < 0.05$; **$p < 0.01$; ***$p < 0.001$. N, indicates number of individuals measured; for P_3 see Table 3.3.
In females, tail length was obliterated by sexual swelling; too few M. maurus were sampled to be included. Significance of *t*-test comparisons is only indicated for the larger measurements between the same sex of M. tonkeana and togeanus.

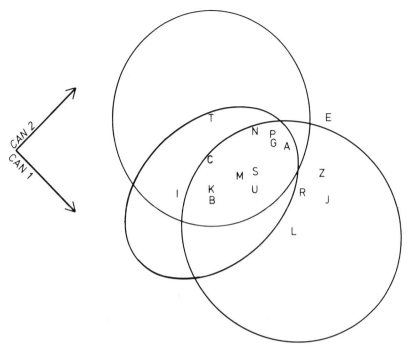

Fig. 3.7. Plot of sample centroids on the first two discriminant functions for dermatoglyphic data, with 95% confidence ellipses for the three 'pure species' samples (Karaenta and Bontobahari pooled). Letter symbols are identified in Figs. 3.1 and 3.6.

Togian and mainland samples of *M. tonkeana*, with about the same degree of overlap in the confidence ellipses as along the second vector of the metric analysis. This highly significant ($p < 0.0001$), initial discrimination is based upon 43% of the variance in the data, while the second axis separating the *M. maurus* samples, with their joint centroid well within the other two confidence ellipses, is based on only 21% of the variance and a lower, but still significant ($p < 0.007$), F value.

Consequently, the hybrid samples have high variance and are less clearly distributed along an intermediate vector, but primarily on the second axis. This is the reason we have rotated the axes in the dermatoglyphic display by 45°, and for easier comparisons with the previous metric and pelage figures. Nevertheless, the dermatoglyphic data reduction includes a greater number of hybrid samples because of the inclusion of immatures and pets, both of which groups were born with their volar patterns. In some cases the relative positions of the hybrid means are again reversed, and almost all are within the confidence ellipse of *M. tonkeana* from Lore Lindu, but the overall

Table 3.5. *Discriminant loadings for dermatoglyphic data of* Macaca maurus, tonkeana *and* togeanus *based on 177 monkeys, including 58 hybrids not used in the initial canonical analysis*

	Canonical variates	
Dermatoglyphic traits	1	2
Hands		
R Inter-digital III	−0.35	−0.43
L Inter-digital III	−0.32	−0.43
R Inter-digital IV	−0.34	0.00
L Inter-digital IV	−0.18	−0.15
R Thenar	0.29	−0.02
L Thenar	0.33	0.08
Feet		
R Inter-digital III	−0.18	−0.37
L Inter-digital III	−0.16	−0.21
R Inter-digital IV	−0.06	−0.18
L Inter-digital IV	−0.04	−0.24
R Distal hypothenar	−0.50	−0.34
L Distal hypothenar	−0.38	−0.25
R Accessory hypothenar	−0.02	−0.26
L Accessory hypothenar	−0.16	−0.05
R Calcar	0.23	−0.07
L Calcar	−0.04	−0.17
R Thenar	0.34	−0.49
L Thenar	0.31	−0.43
R Parathenar	0.39	−0.14
L Parathenar	0.34	−0.08
R Mainline II	0.02	0.20
L Mainline II	−0.08	0.14
Proportion of initial variance	0.43	0.21

clinal pattern is similar to the pelage and metric analyses. This is particularly the case along the eastern, 35 km 'transect', where the geometrical relationships of the three samples (C, S and E) in discriminant space are congruent with their geographical positions (cf. Figs. 3.1 and 3.7).

Discriminant loadings of the first axis, which distinguishes the highly differentiated Togian sample, and to a lesser degree *M. maurus* from *M. tonkeana*, represent a complex combination of trait complexity and simplicity on the various volar pads (Table 3.5). With low scores on this vector, Togian animals have complex patterning on the third and fourth inter-digitals of the hand, and on the distal hypothenar of the foot, combined with low complexity of the thenar of both appendages and in the

Table 3.6. *Transferrin allele frequencies and lineage affiliation of mitochondrial DNA haplotypes in samples of 'pure species' and hybrids of* Macaca maurus *and* tonkeana

	Transferrin		Mitochondrial DNA	
	N	G allele	N	Lineage
Pure *M. tonkeana*				
Lore Lindu	11	0.54	3	*tonkeana*
West hybrid zone				
Pariwang	19	0.84	2	*tonkeana*
Nanga	11	0.86	1	*tonkeana*
Uru/Kajubulo	3	1.00	1	*tonkeana*
Karang	3	1.00	2	*maurus*
Maroangin	12	0.96	5	*maurus*
Balajonga	4	0.62	1	*maurus*
East hybrid zone				
Keera	5	0.60	0	?
Sakholi	9	0.61	2	*maurus*
Anabanua	3	0.67	0	?
Pure *M. maurus*				
Sidrap	2	0.75	0	?
Karaenta	15	0.63	2	*maurus*
Bontobahari	14	0.86	3	*maurus*

N, indicates number of individuals sampled.

pedal parathenar. An assortment of lesser, sometimes unilateral, distinctions can also be discerned in the table. Similarly, the *M. maurus* sample is further separated on the lower end of the second axis with comparable or even greater complexity on the left inter-digital pad of the hand, and a high degree of patterning in the third and fourth inter-digitals, and on the thenar of the foot. Most of these second vector differences are similar to the univariate comparisons previously reported, primarily for pets (Suryobroto & Iwamoto, 1985).

Blood data and trait concordance

The dermatoglyphic cline of intermediate samples in the eastern hybrid zone is mirrored by an apparent cline in transferrin gene frequencies (Table 3.6), although only the Sakholi sample has been haplotyped for *M. maurus* mitochondrial DNA to establish the direction of gene flow. Except for the intermediate frequency of the Balajonga sample, however, all of the western hybrid samples have exceptionally high transferrin G frequencies near fixation, suggesting localised genetic drift, presumably from historical population 'bottle-necking'. So far all of the western samples also show either *M. maurus* or *tonkeana* mitochondrial haplotypes in a distribution consistent with their 'transect' positions. The implication of this pattern of relationships is that only males are crossing and back-crossing the hybrid zone, with the philopatric females maintaining their haplotype lineages.

The nuclear genes carried back and forth by these males show noteworthy concordance among the three independent phenotypic data sets that they control polygenically. In all multivariate data reductions, the clinal positions of the hybrid sample means are remarkably consistent between the distributions of the 'pure species' samples, even though recombination is clearly at work in some of the exceptions. Since hybrid individuals are by definition (Harrison, 1990) combinations of the two parental gene pools, we predicted that concordance might also be seen in individual scores for the different data set reductions, as each animal has similar proportions of each gene pool for each of the intrinsically uncorrelated traits. Table 3.7 shows that the individual rank-order correlations among principal component and discriminant scores range between 0.54 and 0.69, all at significance levels of $P < 0.0001$. More importantly, these values mean that upwards of 50% of the variance in one data set is accounted for by another, even though metric, pelage and dermatoglyphic phenotypes are uncorrelated within populations.

The table of correlations among individual vector scores also shows that the first pelage principal component accounts for almost all of the variance of the obliquely rotated factors. Each of the latter two is also correlated with the second dermatoglyphic and first metric vectors. Despite the fact that oblique pelage factors tended to cluster many hybrid samples near the *M. maurus* or *tonkeana* poles, the clinal, rank-order concordance among individual phenotypic scores in the different traits is still quite significant.

Discussion

Sub-species introgression

Concordant clines in three independent sets of morphological data, augmented in some cases by allozyme or mitochondrial haplotype data, provide strong support for concluding that hybridisation has originated by historical, or secondary intergradation, rather than as a clinal response to distance or current ecogeographical factors (Hewitt & Barton, 1980;

Table 3.7. *Spearman correlation coefficients for individual hybrid specimen scores on canonical vectors from dermatoglyphic and metric data, and factor analyses of pelage patterns, including oblique rotations* (Ob1, 2 and 3)

	Met 1	PC 1	Ob 1	Ob 2	Ob 3
Discriminant scores					
Dermatoglyph 2	0.59	0.54	0.51	0.49	0.52
Metric 1 (met 1)		0.59	0.66	0.54	0.70
Pelage factor scores					
Principal component 1 (PC 1)			0.97	0.78	0.88
Oblique factor 1 (Ob 1)				0.69	0.80
Oblique factor 2 (Ob 2)					0.69

All values are significant at the $p < 0.0001$ level. The data set for these correlations included samples of pure *M. maurus* to anchor the regression line, but none of the Lore Lindu animals used to define *M. tonkeana* polarity in the canonical analyses.

Harrison 1990). Identical responses to the environment in several independent characters are improbable (Barton & Hewitt, 1989). While body size in macaques might well respond to gradients (cf. Albrecht, 1980; Fooden, 1982) along the steeply contoured western 'transect' (Fig. 3.1), this is not likely over the much wider, more uniform eastern transect; nor does it explain similar changes in pelage and dermatoglyphs across both transects, and in transferrin gene frequencies in the east.

Other than being allopatrically differentiated in two parental taxa and correlated in individuals by virtue of similar proportions of the two genomes for all traits, the only alternative explanation for such coincident clines would be very widespread coadaptation (i.e. 'linkage disequilibrium') in the two genomes (Harrison, 1990) leading to the prediction of reduced fitness for recombinant hybrids in a tension zone (Barton & Hewitt, 1989). For such complexly polygenic, yet highly heritable traits as dermatoglyphics, this high degree of coadaptation is unlikely, since it would involve a very large portion of the genome (Froehlich, 1976).

There is, however, some suggestion in our data of smaller, coadapted gene complexes in the oblique rotation of pelage features. In Fig. 3.5, most hybrid samples were clustered near either parental pole, while only three were intermediate. It might be noted in passing that this pattern implies a very steep, stepped cline, which is another characteristic of secondary intergradation (Endler, 1977). It also explains some of the previous confusion in distinguishing the two species in the hybrid zone (e.g. Groves, 1980). The various pelage features do appear to resist recombination; but does this necessarily imply reduced hybrid fitness (cf. Barton & Hewitt, 1989), instead of something like developmental genetic inertia (cf. Lerner, 1954) in correlated set of traits? The fact that some hybrid samples, characterised as nearly pure species in their pelage patterns, are recombined with metric or dermatoglyphic patterns of the opposite taxon would seem to belie the reduced hybrid fitness prediction. For example, the Karang (R) sample is *maurus*-like in its pelage, intermediate in size and close to *M. tonkeana* in its dermatoglyphic patterns. Alternatively, the *tonkeana*-like Nanga (N) have *M. maurus* dermatoglyphs.

Another characteristic of tension zones with reduced hybrid fitness is that they are narrow relative to the dispersal range of the animals (Barton & Hewitt, 1989). In the case of the anubis and hamadryas baboon hybrid zone this criterion may be met (Kummer *et al.*, 1970), since daily travel distances of *Papio hamadryas* range from 6.5 to 13.2 km and the hybrid zone along the Awash River has a width of only 25 km (Kummer, 1968, 1971), although its position has been shifting (Phillips-Conroy & Jolly, 1986), probably leaving cross-species genes in its wake (Shotake, 1981; Harrison, 1990). In the case of our eastern transect, however, the width of the hybrid zone exceeds 35 km, while the day range of *M. maurus* and the Maroangin hybrids is no greater than 1.5 km (Supriatna, 1991) comparable to the closely related *M. nemestrina* (Caldecott, 1986; Oi, 1990). Therefore, it would appear that the width of this hybrid zone is relatively large for primates, suggesting that full introgression has been taking place without significant selection against hybrids.

Full introgression is also indicated by our preliminary data for an independent marker of parental type (see Harrison, 1990). Hybrid samples characterised by *M. maurus* mitochondrial haplotypes show a mixture of pelage, metric and dermatoglyphic features, demonstrating nuclear gene flow from *M. tonkeana* populations. In the eastern transect, transferrin alleles also appear to introgress from north to south. Since macaques are almost always female philopatric (Melnick & Pearl, 1987), these data are consistent with the prediction that male immigration is providing this gene flow. Shared vocalisations and affiliative facial expression (Petit & Thierry, 1992) would presumably facilitate immigration. Perhaps the much larger bodied, more sexually dimorphic *tonkeana* migrants are even at a competitive advantage, in the absence of strongly positive assortative mating.

The distribution of morphological features and of transferrin allele frequencies near fixation almost throughout the western hybrid transect also suggests that back-crossing has occurred at a time subsequent to the 'bottle-necking' event that presumably concentrated the *maurus* allele. Not only are hybrid samples with *M. tonkeana* mitochondrial haplotypes characterised by intermediate morphological features, but they also share the very high transferrin frequencies. The small Kajubulo (J) sample is a most interesting example with *maurus* size, intermediate pelage, and yet *tonkeana* dermato-glyphs in combination with *tonkeana* mitochondrial DNA. The robust Pariwang (P) and Nanga (N) samples make even stronger cases for introgressive back-crossing.

It would appear, therefore, that the hybrid zone represents secondary contact between allopatric taxa, wherein there is no apparent reduction in hybrid fitness nor significant barrier to full, reciprocal introgression, other than that provided by distance and geography in the past, and human disturbance in the present. Two very different and easily diagnosable populations (Cracraft, 1989) are, nevertheless, completely inter-fertile and, therefore, comprise only one recognition species (Paterson, 1985). Since *Macaca maurus* Schinz, 1825 has priority, we conclude that mainland *tonkeana* should be subsumed as a second sub-species. This close relationship is supported by broad comparisons of blood protein data (Fooden & Lanyon, 1989), but it is inconsistent with the taxonomy of Groves (1980), who regarded *M. hecki* as a sub-species of *tonkeana*.

The reconstructed geological history of Sulawesi suggests alternative periods of geographical contiguity and isolation between the two *M. maurus* sub-species. World sea level was about 10 m higher around 4000 years ago; 120 000 years ago it was elevated by nearly 20 m (Bard, Hamelin & Fairbanks, 1990). In the Tempe Depression, transgression during the Holocene period is documented by mangrove pollen just east of Lake Tempe and, incredibly, by the oral traditions of local people recalling a time when direct sailing was possible between the two coasts (Whitten *et al.*, 1987). Even today during minor flooding the two lakes are connected (Fig. 2.1); major floods reach almost to the east coast. During the Pleistocene, periods of inundation would have alternated with intervals much drier than today (Whitten *et al.*, 1987); presumably this cycle not only produced alternating periods of vicariance and potential hybridisation between the two sub-species, but also small population 'bottle-necks' (i.e. refugia; see Froehlich *et al.*, 1991), one of which locally almost fixed the transferrin allele.

Species status of the Togian taxon

Similar, but even more complex, biogeographical patterns of intermittent isolation must have occurred between the Central Sulawesi peninsula and the Togian Islands (Fig. 3.1), although there is little direct evidence. The most detailed bathymetric contours available (Hamilton, 1978) show the 200 m shelves only moderately separated near Poat Island, between the mainland and the Togian archipelago, making it probable that the latter was an extension of the peninsula during lower sea levels in the Pleistocene (Morley & Flemley, 1987; Hutchison, 1989; Bard *et al.*, 1990). The remarkable combination of atolls, barrier reefs and fringing reefs, unique for Indonesia (Whitten *et al.*, 1987), also implies a complex recent history for Togian. Moreover, the isthmus that joins the tip of the peninsula would have been much narrower during high sea levels. Given that it is traversed by an active series of thrust faults (Hamilton, 1979), it is conceivable that the elevations shown in Fig. 3.1 were somewhat lower in the past (Hutchison, 1989), with inter-glacial sea transgression potentially isolating the tip.

With this background, we must address the taxonomic status of the Togian monkeys. Are they a third sub-species of *M. maurus*, or do they deserve separate species status as proposed by Sody (1949)? The third alternative, that they are not even sub-specifically distinct from *M. tonkeana* (Fooden, 1969) is clearly falsified by their diagnosable differences implicit in the pelage, morphometric and dermatoglyphic separations in multivariate space, as well as being unlikely from their peripheral isolate biogeography. However without further field work in the isthmus near Luwuk and genetic studies of living animals, we cannot resolve this issue; but we are inclined to predict that *M. togeanus* is a valid species.

The differences of *M. togeanus* from mainland samples in facial and cranial shape, without diminished body size and with longer tails and greater sexual dimorphism, would seem to falsify a simple interpretation of island dwarfing in large mammals (Sondaar, 1977; Case, 1978; Heaney, 1978; Lomolino, 1985; Roth, 1990). By contrast, the sub-species of the pig-tailed macaque (*M. nemestrina pagensis*) from the Mentawai Islands is about 16% smaller in both body and head lengths than mainland populations at comparable latitude (Albrecht, 1980), even though these islands are well within the 200 m bathymetric contour (Hamilton, 1978), making Pleistocene land bridges more likely than for Togian.

The wide separation of the Togian animals in dermatoglyphic patterning, based on 48% of the variance, implies major changes in polygenic frequencies. They are distinct from both sub-species of *M. maurus*, which show more clinal overlap. This divergence in the polygenic patterning of volar ridges is consistent with an interpretation of peripheral isolate speciation by a small founding population. Such a prediction may be testable with genetic studies of living animals, but it is also consistent with the major alterations in metric proportions, which show a clinal pattern on the mainland. Alternatively, the relatively long tails and high sexual dimorphism of Togian monkeys may be relics of ancestral species in Sulawesi. Either interpretation implies isolation.

The pelage patterns of Togian monkeys are 100% distinct in our factor analyses and easily diagnosable. It is noteworthy, however, that they share many of the same features with *tonkeana*, such as the crown tuft, cheek whiskers and extended rump patch, but with distinctively different colours. On the mainland, both the size and the colour of these features may be clinally distributed. With their characteristic long buff whiskers and sharply bicoloured bodies, *togeanus* specimens are more distinctive than all black *tonkeana*, even with the latter's contrasting cheeks and rump.

All of these pelage and some of the morphometric patterns are consistent with an interpretation of selection for a different mate recognition system in *M. togeanus*, possibly implying reinforcement in an intermittent tension zone between the two taxa. Given the complex biogeography of the region, where the narrow isthmus, rather than the 200 m shelf, may have been an allopatric isolating barrier, we further predict that this tension zone may still exist in the vicinity of Luwuk and that *M. togeanus* still ranges over the distal peninsula to the north east.

Except for the fact that a modest number of about 250 monkeys still exist on Malenge Island, we know nothing about the distribution or appearance of these animals elsewhere to test our predictions. Apparently typical *tonkeana* have been surveyed in the Lombuyan Game Reserve west of the isthmus, but there is no report for the sole and pitifully small (198 ha) Pati-pati Game Reserve on the distal peninsula (Watanabe & Brotoisworo, 1989). Indeed, it is conceivable that this critical population is now extirpated and we may never confirm its species status. The conservation status of the Melenge Island population was tenuous when it was recently censused. Certainly, there are presently no adequate nature reserves in the whole region to assure that extinction is not the Togian taxon's destiny in the future.

Conclusions

1. Full, reciprocal introgression over a wide hybrid zone demonstrates that *M. maurus* and *tonkeana* are conspecific and should be considered as polymorphic sub-species of *M. maurus*.
2. Comparably diagnosable pelage patterns, distinct alterations in body shape, and greater dermatoglyphic differences suggest that *M. togeanus* may be a valid species.

3. We also predict that this taxon may range over the undocumented, distal eastern peninsula of Sulawesi, north-east of Luwuk, with a hybrid tension zone in parapatry with *M. maurus tonkeana* that may permit confirmation of *M. togeanus* species status.
4. Non-existent protection and the tenuous status of the only known population of *M. togeanus* on Malenge Island lend urgency to the confirmation of these predictions and the formulation of adequate conservation initiatives.

Recent fieldwork (NGS grant 5288-94) supports local oral traditions that the monkeys of Malenge Island were artifically transported there about 1920 from near Tanjung Api (see Fig. 3.1); they occur nowhere else in the Togian archipelago. Furthermore, even more genetically and physically distinctive populations are widespread on the central peninsula of Sulawesi, perhaps even existing sympatrically with typical *M. tonkeana*.

Acknowledgements

This research was supported by an intramural grant to J. Froehlich, a WWF–US grant to J. Supriatna, an USPHS/NIH Grant RR–00169 to the California Primate Research Center and N. Lerche, an USPHS/NIH Grant RR–04391–01 to C. Southwick, a NSF Grant BNS–8909–775 to D. Melnick, and NGS grants to J. Erwin. Additional help with allozyme analysis was generously provided by W. H. Stone and S. Manis of Trinity University. The authors thank Universities Indonesia (UI) and Lembaga Ilmu Pengetahuan Indonesia (LIPI) for their sponsorship, numerous students for their help in the field, and S. Somadikarta, J. Sugardjito, C. Darsono, G. Musser, R. Thorinton, Pak Boeadi, and J. Fooden for invaluable assistance. Finally, we acknowledge the invaluable assistance of the curators and support staff of the Field Museum of Natural History, the American Museum of Natural History, the Museum Zoologicum Bogoriense, and the Smithsonian National Museum of Natural History.

References

Albrecht, G. H. (1978). The craniofacial morphology of the Sulawesi macaques: multivariate approaches to biological problems. *Contributions to Primatology*, vol. 13. Basel: S. Karger.

Albrecht, G. H. (1980). Latitudinal, taxonomic, sexual, and insular determinants of size variation in pig-tail macaques, *Macaca nemestrina*. *International Journal of Primatology*, 1, 141–52.

Bard, E., Hamelin, B. & Fairbanks, R. G. (1990). U-Th ages obtained by mass spectrometry in corals from Barbados: sea level during the past 130,000 years. *Nature*, **346**, 456–8.

Barton, N. H. & Hewitt, G. M. (1989). Adaptation, speciation and hybrid zones. *Nature*, **341**, 497–503.

Bernstein, I. S. (1966). Naturally occurring primate hybrid. *Science*, **154**, 1559–60.

Bernstein, I. S. & Gordon, T. P. (1980). Mixed taxa introductions, hybrids and macaque systematics. In *The macaques: studies in ecology, behavior and evolution*, ed. D. G. Lindburg, pp. 125–47. New York: Van Nostrand Reinhold.

Brockelman, W. Y. & Gittins, S. P. (1984). Natural hybridization in the *Hylobates lar species* group: implications for speciation in gibbons. In *The lesser apes: evolutionary and behavioural biology*, ed. H. Preuschoft, D. J. Chivers, W. Y. Brockelman & N. Creel, pp. 498–532. Edinburgh: Edinburgh University Press.

Butlin, R. (1989). Reinforcement of premating isolation. In *Speciation and its consequences*, ed. D. Otte & J. A. Endler, pp. 158–79. Sunderland, MA: Sinauer Press.

Caldecott, J. O. (1986). An ecological and behavioural study of the pig-tailed macaque. *Contributions to Primatology*, vol. 21. Basel: S. Karger.

Case, T. J. (1978). A general explanation for insular body size trends in terrestrial vertebrates. *Ecology*, **59**, 1–18.

Ciani, A. C., Stanyon, R., Scheffrahn, W. & Sampurno, B. (1989). Evidence of gene flow between Sulawesi macaques. *American Journal of Primatology*, **17**, 257–70.

Coyne, J. A. & Orr, H. A. (1989). Patterns of speciation in *Drosophila*. *Evolution*, **43**, 362–81.

Cracraft, J. (1989). Speciation and its ontology: the empirical consequences of alternative species concepts for understanding patterns and processes of differentiation. In *Speciation and its consequences*, ed. D. Otte & J. A. Endler, pp. 28–59. Sunderland, MA: Sinauer Press.

Dobzhansky, T. (1940). Speciation as a stage in evolutionary divergence. *American Naturalist*, **74**, 312–21.

Endler, J. A. (1977), *Geographic variation, speciation, and clines.* Princeton: Princeton University Press.

Endler, J. A. (1982). Problems in distinguishing historical from ecological factors in biogeography. *American Zoologist*, **22**, 441–52.

Fleischer, R. C. & Rothstein, S. I. (1988). Known secondary contact and rapid gene flow among subspecies and dialects in the brown-headed cowbird. *Evolution*, **42**, 1146–58.

Fooden, J. (1964). Rhesus and crab-eating macaques: intergradation in Thailand. *Science*, **143**, 363–5.

Fooden, J. (1969). *Taxonomy and evolution of the monkeys of Celebes* (*Primates: Cercopithecidae*). Bibliotheca Primatologica, no. 10. Basel: S. Karger.

Fooden, J. (1982). Ecogeographic segregation of macaque species. *Primates*, **23**, 574–9.

Fooden, J. & Lanyon, S. M. (1989). Blood-protein allele frequencies and phylogenetic relationships in *Macaca*: a review. *American Journal of Primatology*, **17**, 209–41.

Froehlich, J. W. (1976). The quantitative genetics of fingerprints. In *The measures of man: methodologies in biological anthropology*, ed. E. Giles & J. S. Friedlaender, pp. 260–320. Cambridge, MA: Peabody Museum Press.

Froehlich, J. W. (1987). Fingerprints as phylogenetic markers in the Solomon Islands. In *The Solomon Islands Project: a long-term study of health, human biology, and culture change*, ed. J. S. Friedlaender, pp. 174–214. Oxford: Clarendon Press.

Froehlich, J. W. & Froehlich, P. H. (1986). Dermatoglyphics and subspecific systematics of mantled howler monkeys (*Alouatta palliata*). In *Current perspectives in primate biology*, ed. D. M. Taub & F. A. King, pp. 107–21. New York: Van Nostrand Reinhold.

Froehlich, J. W. & Froehlich, P. H. (1987). The status of Panama's endemic howling monkeys. *Primate Conservation*, **8**, 58–62.

Froehlich, J. W. & Giles, E. (1981). A multivariate approach to fingerprint variation in Papua New Guinea: implications for prehistory. *American Journal of Physical Anthropology*, **54**, 73–91.

Froehlich, J. W. & Thorington, R. W., Jr (1982). The genetic structure and socio-ecology of howler monkeys (*Alouatta palliata*) on Barro Colorada Island. In *The ecology of a tropical forest: seasonal rhythms and long-term changes*, ed. E. G. Leigh, Jr, A. S. Rand & D. M. Windsor, pp. 291–305. Washington, DC: Smithsonian Institution Press.

Froehlich, J. W., Supriatna, J. & Froehlich, P. H. (1991). Morphometric analyses of *Ateles*: systematic and biogeographic implications. *American Journal of Primatology*, **25**, 1–22.

Godfrey, L. & Marks, J. (1991). The nature and origins of primate species. *Yearbook of Physical Anthropology*, **34**, 39–68.

Groves, C. P. (1980). Speciation in *Macaca*: The view from Sulawesi. In *The macaques: studies in ecology, behavior and evolution*, ed. D. G. Lindburg, pp. 84–124. New York: Van Nostrand Reinhold.

Hamada, Y., Watanabe, T., Takenaka, O., Suryobroto, B. and Kawamoto, Y. (1988). Morphological studies on the Sulawesi macaques. I. Phyletic analysis of body color. *Primates*, **29**, 65–80.

Hamilton, W. (1978). *Tectonic map of the Indonesian region*. Miscellaneous Investigations Series Map I–875-D. Washington, DC: United States Geological Survey.

Hamilton, W. (1979). *Tectonics of the Indonesian region*. United States Geological Survey Professional Paper 1078. Washington, DC: United States Government Printing Office.

Harrison, R. G. (1990). Hybrid zones: windows on evolutionary process. In *Oxford surveys in evolutionary biology*, ed. D. Futuyma & J. Antonovics, pp. 69–128. Oxford: Oxford University Press.

Hayes, V. J., Freedman, L. & Oxnard, C. E. (1990). The taxonomy of savannah baboons: an odontomorphometric analysis. *American Journal of Primatology*, **22**, 171–90.

Heaney, L. R. (1978). Island area and body size of insular mammals: evidence from the tri-colored squirrel. (*Callosciurus prevosti*) of Southeast Asia. *Evolution*, **32**, 29–44.

Hewitt, G. M. (1989). The subdivision of species by hybrid zones. In *Speciation and its consequences*, ed. D. Otte & J. A. Endler, pp. 85–110. Sunderland, MA: Sinauer Press.

Hewitt, G. M. & Barton, N. H. (1980). The structure and maintenance of hybrid zones as exemplified by *Podisma pedetris*. In *Insect cytogenetics*, ed. R. L. Blackman, G. M. Hewitt & M. Ashburner, pp. 149–69. Oxford: Blackwell Scientific Publications.

Hutchison, C. S. (1989). *Geological evolution of South-east Asia.* Oxford: Clarendon Press.

Kummer, H. (1968). *Social organization of hamadryas baboons.* Chicago: University of Chicago Press.

Kummer, H. (1971). *Primate societies: group techniques of ecological adaptation.* Chicago: Aldine-Atherton.

Kummer, H., Goetz, W. & Angst, W. (1970). Cross-species modifications of social behavior in baboons. In *Old World monkeys: evolution, systematics, and behavior,* ed. J. R. Napier & P. H. Napier, pp. 352–63. New York: Academic Press.

Lerner, I. M. (1954). *Genetic Homeostasis.* Edinburgh: Oliver & Boyd.

Lomolino, M. V. (1985). Body size of mammals on islands: the island rule reexamined. *American Naturalist,* **125,** 310–16.

Marks, J. (1987). Social and ecological aspects of primate cytogenetics. In *The evolution of human behaviour: Primate models,* ed. W. G. Kinzey, pp. 139–50. Albany: State University of New York Press.

Masters, J. C., Rayner, R. J., McKay, I. J., Potts, A. D., Nails, D., Ferguson, J. W., Weissenbacher, B. K., Alsopp, M. & Anderson, M. L. (1987). The concept of species: recognition versus isolation. *South African Journal of Science,* **83,** 534–7.

Mayr, E. (1942). *Systematics and the origin of species.* New York: Columbia University Press.

Melnick, D. J. & Kidd, K. K. (1985). Genetic and evolutionary relationships among Asian macaques. *International Journal of Primatology,* **6,** 123–59.

Melnick, D. J. & Pearl, M. C. (1987). Cercopithecines in multimale groups: genetic diversity and population structure. In *Primate societies,* ed. B. B. Smuts, D. L. Cheney, R. M. Seyfarth, R. W. Wrangham & T. Struhsaker, pp. 121–34. Chicago: University of Chicago Press.

Morley, R. J. & Flemley, J. R. (1987). Late Cainozoic vegetational and environmental changes in the Malay archipelago. In *Biogeographical evolution of the Malay archipelago,* ed. T. C. Whitmore, pp. 50–9. Oxford: Clarendon Press.

Nagel, U. (1973). A comparison of anubis baboons, hamadryas baboons and their hybrids at a species border in Ethiopia. *Folia Primatologica,* **19,** 104–65.

Oi, T. (1990). Population organization of wild pig-tailed macaques (*Macaca nemestrina nemestrina*) in West Sumatra. *Primates,* **32,** 15–31.

Paterson, H. E. H. (1985). The recognition concept of species. In *Species and speciation,* ed. E. S. Vrba, pp. 21–9. Transvaal Museum Monograph, Number 4. Pretoria: Transvaal Museum.

Petit, O. & Thierry, B. (1992). Affiliative function of the silent bared-teeth display in moor macaques (*Macaca maurus*): further evidence for the particular status of Sulawesi macaques. *International Journal of Primatology,* **13,** 97–105.

Phillips-Conroy, J. E. & Jolly, C. M. (1986). Changes in the structure of the baboon hybrid zone in the Awash National Park, Ethiopia. *American Journal of Physical Anthropology,* **71,** 337–50.

Phillips-Conroy, J. E., Jolly, C. J. & Brett, F. L. (1991). Characteristics of hamadryas-like male baboons living in anubis baboon troops in the Awash hybrid zone, Ethiopia. *American Journal of Physical Anthropology,* **86,** 353–68.

Raubenheimer, D. & Crowe, T. M. (1987). The recognition species concept: is it really an alternative? *South African Journal of Science,* **83,** 530–4.

Roth, V. L. (1990). Insular dwarf elephants: a case study in body size estimation and ecological inference. In *Body size and mammalian paleobiology: estimation and biological implications*, ed. J. Damuth & B. J. MacFadden, pp. 151–79. Cambridge: Cambridge University Press.

Samuels, A. & Altmann, J. (1986). Immigration of a *Papio anubis* male into a group of *Papio cynocephalus* baboons and evidence for an *anubis–cynocephalus* hybrid zone in Amboseli, Kenya. *International Journal of Primatology*, **7**, 131–8.

Shotake, T. (1981). Population genetical study of natural hybridization between *Papio anubis* and *P. hamadryas*. *Primates*, **22**, 285–308.

Sody, H. J. V. (1949). Notes on some primates, carnivora and the babirusa from the Indo-Malayan and Indo-Australian regions (with descriptions of 10 new species and subspecies). *Treubia*, **20**, 121–90.

Sondaar, P. Y. (1977). Insularity and its effects on mammal evolution. In *Major patterns in vertebrate evolution*, ed. M. Hecht, P. Goody & B. Hecht, pp. 671–707. New York: Plenum Press.

Southwick, C. H. & Southwick, K. L. (1983). Polyspecific groups of macaques on the Kowloon Peninsula, New Territories, Hong Kong. *American Journal of Primatology*, **5**, 17–24.

Struhsaker, T. T., Butynski, T. M. & Lwanga, J. S. (1988). Hybridization between redtail (*Cercopithecus ascanius schmidti*) and blue (*C. mitis stuhlmanni*) monkeys in the Kibale Forest, Uganda. In *A primate radiation: evolutionary biology of the African guenons*, ed. A. Gautier-Hion, F. Bourlière, J. Gautier & J. Kingdon, pp. 477–97. Cambridge: Cambridge University Press.

Sugawara, K. (1988). Ethological study of the social behavior of Hybrid baboons between *Papio anubis* and *P. hamadryas* in free-ranging groups. *Primates*, **29**, 429–48.

Supriatna, J. (1991). Hybridization between *Macaca maurus* and *Macaca tonkeana*: a test of species status using behavioral and morphogenetic analyses. PhD thesis, University of New Mexico.

Suryobroto, B. & Iwamoto, M. (1985). Dermatoglyphics of Sulawesi macaques: a preliminary report. *Kyoto University Overseas Research Report of Studies on Asian Non-human Primates*, **4**, 87–103.

Templeton, A. R. (1989). The meaning of species and speciation: a genetic perspective. In *Speciation and its consequences*, ed. D. Otte & J. A. Endler, pp. 3–27. Sunderland, MA: Sinauer Press.

Thorington, R. W., Jr & Groves, C. P. (1970). An annotated classification of the Cercopithecoidea. In *Old World monkeys: evolution, systematics, and behavior*, ed. J. R. Napier & P. H. Napier, pp. 629–47. New York: Academic Press.

Watanabe, K. & Brotoisworo, E. (1989). Present situation of Sulawesi macaques. *Kyoto University Overseas Research Report of Studies on Asian Non-Human Primates*, **7**, 43–61.

Watanabe, T., Hamada, Y., Suryobroto, B. & Iwamoto, M. (1987). Somatometrical data of Sulawesi macaques and Sumatran pig-tails collected in 1984 and 1986. *Kyoto University Overseas Research Report of Studies on Asian Non-Human Primates*, **6**, 49–56.

Whitten, A. J., Mustafa, M. & Henderson, G. S. (1987). *The ecology of Sulawesi*. Jogjakarta, Indonesia: Gadjah Mada University Press.

Part II

Population biology, ecology and conservation

4

A comparison of ecological strategies of pig-tailed macaques, mandrills and drills

J. O. CALDECOTT, A. T. C. FEISTNER
AND E. L. GADSBY

Introduction

Pig-tailed macaques (*Macaca nemestrina*) live in South-East Asia, on the Greater Sunda Islands of Sumatra and Borneo, and on the Asian mainland in the Malay peninsula, Thailand, and Burma. At the opposite end of the Old World, in West Central Africa, from Nigeria to the Congo Republic, dwell the mandrill (*Mandrillus sphinx*) and the drill (*M. leucophaeus*). These three species are all highly mobile and often travel and forage on the ground within moist tropical forests; they are large-bodied, short-tailed and sexually dimorphic.

As information on mandrills and drills gradually accumulated it became clear that these species closely resemble each other in all easily detectable features of their morphology, ecology, and behaviour; a fact that is consistent with their parapatric distribution (Grubb, 1973). Additionally, Feistner (1989) pointed out that mandrills, although different from other baboons, could be regarded as socio-ecological analogues of pig-tailed macaques. Similarly, Gadsby's (1990) research on drills has also suggested major similarities between drills and pig-tails.

The present chapter is an inter-generic comparison amongst the three species. In making such a comparison there are numerous difficulties in terms of the available data, their comparability and interpretation. Additionally, the size of the geographical range is quite different for each species, as are the variety of forest habitats that are occupied and the degree to which they are seasonal. Given these fundamental uncertainties, we restrict ourselves to describing major common features and contrasts between on the one hand pig-tails and South-East Asian forests, and on the other mandrills, drills and West Central African forests. This chapter describes some of the similarities and differences between the three species, and discusses the ecological forces that may have created those features.

Previous studies

The three species are more often glimpsed as they vanish into the undergrowth, or are seen as cadavers over cooking fires, than they are observed behaving naturally in the forest. As a result, they have a low profile in the primatological literature from the 1960s and 1970s, when the ground rules of primate behaviour and ecology were being worked out using more readily observable models. Nevertheless, some outline observations were published, usually in the form of comparisons with better known sympatric species: pig-tails (Bernstein, 1967; Rodman, 1979; Crockett & Wilson, 1980; MacKinnon & MacKinnon, 1980), mandrills (Sabater Pi, 1972; Jouventin, 1975) and drills (Struhsaker, 1969; Gartlan, 1970; Gartlan & Struhsaker, 1972).

In the 1980s longer-term and more intensive studies of all three species have been initiated or undertaken: pig-tails (Caldecott, 1986 *a,b*; Robertson, 1986; Oi, 1987, 1990 and chapter 16, this volume), mandrills (Hoshino *et al.*, 1984; Hoshino, 1985; Lahm, 1986; Kudo, 1987; Harrison, 1988; Feistner, 1989) and drills (Gadsby, 1990; Oates *et al.*, 1990). However, sustained observation and habituation of these species has not yet been achieved because visibility at ground level in tropical forest is often less than 20 m and their habits are semi-terrestrial. Thus, researchers have typically recorded an average of fewer than five useful animal contacts per month in natural settings. For these reasons, detailed information on social behaviour is only available from less complex environments such as a 70 ha patch of natural forest within an oil-palm plantation in which pig-tails were studied (Bernstein, 1967; Caldecott 1986*a,b*), or a 5.3 ha enclosure within natural gallery forest in which a large semi-free-ranging group of mandrills was observed (Feistner 1989; Feistner, Cooper & Evans, 1992). These sources, nonetheless, provide important data that form the basis for interpreting other fragmentary information from more representative situations.

Comparisons between species
Group size and composition

There are difficulties in defining 'group size' in these three species since known individuals may be encountered in small parties or in very large aggregations at different times. This may represent temporary fissioning and fusing of several kinds of social units thus making it difficult to establish group size in the wild. Nevertheless, we are confident that the 'basic social unit' of pig-tails and mandrills, and probably drills, can be described as a group comprising an average of 20–40 individuals. We believe, therefore, that the smaller and larger aggregations encountered in the forest represent 'sub-groups' and 'super-groups' derived in each case from one or more such groups. Working with this terminology, then, we find that the ratio of adult males to adult females within groups is highly skewed towards females in all three species, indeed more so than in other papionins. In pig-tails, the intra-group male to female ratio is about 1:8 (Caldecott, 1986a), in mandrills (and probably drills) around 1:9 (Gartlan, 1970; Jouventin, 1975; Feistner, 1989). As implied by the skewed sex ratio, most adult males live as peripherals or solitaries (though not, apparently, in all-male bands). The proximate mechanism for this 'shedding' of males from groups in pig-tails involves displays and antagonism by the group-living adult males towards adolescent and other adult males. Relations among males seem to be more relaxed in mandrills and drills.

Sexual and other social behaviour

Pig-tail, mandrill and drill juveniles associate with one another. Large juveniles also associate, as do adolescent males. Thus, juveniles play, travel, forage and sit near one another but avoid adolescent and adult males (agonistic interactions in this context are rarer in mandrills than in pig-tails). The relationship between group-living adult males and females varies significantly depending on the female's reproductive state and particularly on the status of her perineal swelling (another feature common to all three species). Thus, group-living adult males and swollen (periovulatory) females are strongly associated with one another in all three species.

Group-living adult male pig-tails form and maintain consortships with swollen females (Caldecott, 1986a), and rather similar mate-guarding behaviour by alpha males is seen in mandrills (Feistner, 1989). The main difference between the two species lies in the peripheralisation of the pair during consortship in pig-tails but not in mandrills. Otherwise, consortship behaviour of both species is very similar, with maintenance of proximity (though no herding), synchronisation of movement, and regular grooming and mounting.

These relations contrast with those between adult males and unswollen females, which are characterised by an absence of grooming, mounting, or sustained proximity. Female sexual attractiveness and proceptivity is thus strongly linked to the late follicular phase of the menstrual cycle (Caldecott, 1986a,b). Sexual behaviour of wild female pig-tails is oestrus-like in its sudden onset, intensity, and rapid cessation, strongly resembling that in mandrills (Feistner, 1989, 1991).

Pig-tails are multiple-mount, rather than single-mount ejaculators, and six species of macaques can be divided into two groups on this basis. Such a distinction correlates with females being selective or promiscuous copulators, with intra-group adult sex ratio, and also with several related features such as the extent of 'paternalistic' behaviour (Caldecott, 1986b; Table 4.1). Caldecott (1986b) concluded that multiple-mount ejaculation is appropriate to prolonged consortships, since it rewards the female with continued attention of that male as a sexual partner while in principle denying other males access to her. This is typical of species with high adult male:adult female group ratios. In contrast, single-mount ejaculation is related to brief consortships and female promiscuity, and associated with low male to female intra-group adult ratios (attributed ultimately to male confusion over the paternity of individual infants). In both situations, the ultimate 'choice' is seen as a female one, with male ejaculatory behaviour being adapted to female fidelity to the individual male during her late follicular phase.

The habitat-quality/social behaviour model

The direction of reproductive decisions in macaques, in an evolutionary sense, has been tentatively attributed to habitat quality by Caldecott (1984, 1986b). According to his model, low male:female ratios may be a sign that inter-sexual feeding competition is not sufficiently important to the reproductive interests of females for them to encourage a reduced presence of adult males within groups. This outcome would imply that feeding competition outweighs factors such as the advantage of 'paternalistic' male behaviour and added protection from predators by males.

The generalisation of this model to other papionins, such as the mandrill, is problematic, since this species is a single-mount ejaculator despite its skewed adult sex ratio (Feistner, 1989, 1991). However, Feistner (1989) has argued that seasonality of food supply may be the determining factor, since reproduction in mandrills and sympatric guenons (*Cercopithecus* spp.) is seasonal (Feistner, 1990). Variability in rainfall influences food supply, mainly fruit and arthropods, which in turn affects reproductive behaviour in both taxa (Butynski, 1988). Thus, if periovulatory female mandrills are available only for a short period in the year, male reproductive effort must be concentrated during this time. Evidence for synchronised ovulation in female mandrills (Feistner, 1989) implies that multiple periovulatory and proceptive females may require insemination during a short period. Even where females are selective in terms of sexual partner, multiple consecutive consortships should have selected for male mandrills to deliver relatively

Table 4.1. *Attributes related to mating system in macaques and mandrills*

Attribute	Group 1 macaques	Mandrills	Group 2 macaques
Female mate choice	Selectivity	Selectivity	Promiscuity
Consortship duration	Hours to days	Days	Minutes
Copulation pattern	Multiple-mount	Single-mount	Single-mount
Inter-male relations	Antagonistic	Antagonistic?	Relaxed
Mean adult sex ratio	1 male : 1.7 to 8.0	1 male : 8.0	1 male : 0.9 to 1.2
Paternalism	Generally weak	Generally weak	Generally strong
Male emigration	Common	Common	Rare
Inbreeding	Relatively low	Relatively low ?	Relatively high
Species example	*M. nemestrina*	*M. sphinx*	*M. sylvanus*

Macaque data from Caldecott, 1986*b*. Mandrill data from Feistner, 1989.

large volumes of semen in the course of single-mount ejaculations. Feistner's (1989) refinement of the original model adds to it a new and complex variable, that of environmental seasonality. Since Caldecott (1986*b*) did not consider seasonal influences, a case can be made for re-examining the evidence from macaque societies taking into account adult male:female ratios, ejaculatory patterns, female selectivity, and seasonality of food supply at the species, sub-species, and population levels.

Ecological implications of social behaviour

Available data suggest that pig-tails form one-male harem units that are defended but not imposed (Caldecott, 1986*a*). This is consistent with what is known about mandrills and drills. An explanation for this can be made in ecological terms, since these three species all show features of an ecological strategy that may reflect adaptation to food-poor or marginal habitats. The basic premise for this is that dispersal of surplus resident adult and sub-adult males from a group results in a reduction in feeding competition. This would ultimately be influenced by female reproductive interests and choices, and may be followed if two conditions are met: (1) if the benefits of increased food access outweigh increased predation risk; and (2) if males are unable to sequester or coerce females because the environment is physically complex enough to allow females to evade males when they wish to do so.

Adult male pig-tails guard their groups, at least when in the open (crossing roads, raiding crops, etc.), and adult male mandrills and drills are very vigilant and play an important role in group cohesion. Increasing the number of males within a group would not necessarily improve group security, however, and particularly not during normal foraging travel, when the group is typically widely dispersed in conditions of poor visibility. Meanwhile, the second condition mentioned above is fulfilled for all significantly dimorphic forest-dwelling monkeys, since smaller females are able to access parts of the environment where large males are unable to follow (thin branches, etc.; Caldecott, 1986a; Feistner, 1989).

South-East Asian and West Central African forests compared

The remainder of this chapter focuses on whether it is reasonable to describe the habitats occupied by pig-tails, mandrills, and drills as 'food-poor or marginal' and, if so, what common features of the three species' foraging strategies can be attributed to this.

Sundaland

South-East Asia comprises part of the Asian mainland (Indochina, Burma, Thailand, and peninsular Malaysia or Malaya) and the largest archipelago in the world (the Malay archipelago), which stretches for over 5000 km from Malaya to Australia. About half-way along the archipelago, however, it ceases to be truly 'Asian' in a biogeographical sense and becomes first 'Wallacean', and then 'Australasian', in terms of its plant and animal communities.

The western part of the archipelago is distinctive and comprises Malaya, Borneo, Sumatra, and Java, which all lie on a common continental shelf called 'Sundaland'. They are separated at present by seas of less than 200 m depth, although at times during the Pleistocene they were joined by dry land during periods of high latitude glaciation. The forming and breaking of land bridges had a great influence on dispersal patterns, colonisation, and speciation in this area (Wallace, 1869; Whitten *et al.*, 1984), Malaya, Borneo, and Sumatra, especially, have a number of ecological features in common. They are grouped in West Malesia by plant biogeographers (van Steenis, 1950) and all share a typical 'Sundaic' form of lowland rainforest, which is characteristically dominated by members of one tree family, the Dipterocarpaceae (Whitmore, 1984).

This abundance of dipterocarps is fundamental to the ecology of terrestrial Sundaland, and ecological studies in this region are often concerned with the impact of dipterocarp dominance on animal communities (Caldecott, 1988, 1991; MacKinnon *et al.*, in press). West Malesia, Sundaland, and dipterocarp dominance all end at the outer edges of Borneo and Sumatra, and at the Kra Isthmus in south Thailand on the Malay peninsula, and sharp ecological and taxonomic changes are seen as these boundaries are crossed (Whitmore, 1984; Whitten, Muslimin & Henderson, 1987).

One of these changes is in the papionin community, since there is a Sundaic pig-tail sub-species (*M. n. nemestrina*) in the Malay peninsula, Borneo, and Sumatra that is replaced at the Kra Isthmus by an Indochinese sub-species (*M. n. leonina*). Beyond the coastal edges of Sundaland and West Malesia, meanwhile, pig-tails do not occur on Sulawesi (east of Borneo) or the Mentawai Islands (west of Sumatra), but they are replaced by other similar species (at least four macaque species on Sulawesi and *Macaca pagensis* on Mentawai). Pig-tails are extinct on Java and there, and on the Lesser Sunda Islands further east, only the long-tailed macaque (*M. fascicularis*) occurs.

The ecological significance of dipterocarps to the large mammal community needs to be explained. The trees produce abundant oil-rich, wind-dispersed seeds, which are often quite large. They are not toxic or unpalatable and many species are used as a source of dietary fat by humans. These seeds are known to be eaten by pig-tails (Robertson, 1986). The problem lies, however, in their pattern of availability, since the Dipterocarpaceae are almost all mast-fruiting species, which rely on satiation of seed predators during fruiting incidents to ensure seed survival and regeneration (Janzen, 1974). Additionally, many other families, including many with animal-dispersed fruits, also show masting (Medway, 1972; Leighton & Leighton, 1983). Fruiting across species in any one (large) area is synchronous and because dipterocarp populations mast irregularly, but seldom more often than at 2–4 year intervals, the supply of their seeds cannot, in general, influence the biomass of pig-tails or other vertebrate frugivores.

This constraint is thought to be one reason why the frugivore community in Sundaic dipterocarp forests is, in general, relatively species-poor and of low biomass. There are, however, some species that have adapted to the use of dipterocarp seeds as food sources, and one example is the bearded pig, *Sus barbatus*, whose populations are known to 'erupt' lemming-like on the rare occasions that dipterocarps mast in consecutive years (Caldecott, 1988, 1991, Table 4.2).

The biomass of less exceptional Sundaic frugivores is strongly correlated with the proportion of total basal area of a forest that is contributed by a relatively small number of non-dipterocarp tree genera (Caldecott, 1980, 1986*a*, 1991; Bennett & Caldecott, 1989). This effect is two-fold: (1) an abundance of trees belonging to genera known to produce sugary, digestible fruits or non-toxic seeds is 'good' for frugivores; (2) an abundance of dipterocarp trees is associated with mast-fruiting and an irregular food supply, so must be considered 'bad' for frugivores.

Forests must, therefore, be distinguished in terms of both the density and the nature of fruit sources (controlling average food supply), and also the prevailing phenological patterns (controlling timing of food presentation). Sundaic dipterocarp forests have both a low density of 'good' fruit trees and a very sporadic supply of edible fruits, and would be expected to have very low average 'harvestable productivity' (see below) for papionins and other frugivores.

By contrast, non-dipterocarp Sundaic forests tend to have many more fruit trees and much more continuous fruiting activity. These include the unusual *Koompassia*-Burseraceae forests of western Malaya (where Bernstein

Table 4.2. *Ecogeography of macaques and pigs in and around Sundaland*

Macaques	Pigs
Natural state within Sundaland	
Macaca n. nemestrina adapted to less productive dipterocarp forests and non-'edge' habitats: a large-bodied, highly mobile, low-biomass species. *Macaca fascicularis* adapted to more productive 'edge' habitats: a small-bodied, sedentary, high-biomass species	*Sus b. barbatus* adapted to less productive dipterocarp forests and non-'edge' habitats: a large-bodied, highly mobile, low-biomass species. *Sus Scrofa* adapted to more productive 'edge' habitats: a small-bodied, sedentary, high-biomass species
Natural state at periphery of Sundaland	
Macaca n. leonina and *M. arctoides* replace *M. n. nemestrina* at Kra isthmus in south Thailand. *Macaca fascicularis* entirely replaces *M. nemestrina* in the Philippines	Intermediate *S. b. ahoenobarbus* replaces *S. b. barbatus* in Palawan; small-bodied *S. b. philippensis* and *S. b. cebifrons* replace *S. b. barbatus* elsewhere in the Philippines
Disturbed state within Sundaland	
Macaca fascicularis replaces *M. nemestrina* in disturbed habitats in Malaya, Sumatra and Borneo	*Sus scrofa* replaces *S. barbatus* in disturbed habitats in Malaya and Sumatra

From Caldecott (1991).

(1967) and Caldecott (1986*a,b*) worked on pig-tails), and a wide range of riverine, coastal and island forest formations and forest fringes.

West Central Africa

Forest habitats in West Central Africa differ from those in Sundaland in a number of important ways. They are continental, so biogeographical patterns have to be deciphered with reference to riverine barriers, advancing and retreating forests and savannas, altitude, and human influences, rather than in terms of rising and falling sea levels.

Prior to human interference, a continuous forest block stretched across equatorial Africa (broken only by the Dahomey Gap, which separates the main forest block from the Upper Guinea forest block to the west) with savanna to the east and north, and sea to the south and west. Despite this continuity, various regional communities are recognised in the Lower Guinea block: southern Nigeria, Cameroon, West Equatorial Africa (south of the Sanaga river), and the Congo basin (Oates, 1986). The distinct

Cameroon biota is centred on Mount Cameroon between the Cross and Sanaga rivers and includes the island of Bioko. The drill occurs exclusively within this zone, as do two other cercopithecines, the red-eared guenon (*Cercopithecus erythrotis*) and Preuss's guenon (*C. preussi*) (Oates, 1988). The mandrill is found only to the south of the Sanaga river.

Although dipterocarps do exist in West Central Africa, they are a trivial feature of the flora in terms of species diversity and abundance, and no single family dominates West Central African forest formations in a way comparable to the dipterocarps in Sundaland (Richards, 1952). Mast-fruiting also is not a prominent feature of West Central African forests, though marked seasonality in rainfall is associated with seasonal fruit production. Few climates in West Central Africa are so continuously wet as those of Sundaland. If fruiting records from Gabon are compared with those from Malaysia and Sumatra, in Africa there is high intra-annual but low inter-annual variability in fruit production, whereas in South-East Asia there is high variability both within and between years (Terborgh & van Schaik, 1987).

Apparent differences in phytochemical features, making West Central African vegetation relatively less toxic and more digestible than Sundaic vegetation, should also be mentioned. These factors combine to increase the relative hospitality of an average West Central African forest to frugivores in comparison to an average Sundaic one (see below).

Harvestable productivity

Only a subset of the primary productivity of any habitat is 'harvestable' by the animals living within it, and it is this that determines the potential biomass of animals supported in each habitat. Harvestable productivity varies, to some extent independently of the gross primary productivity, as a function of secondary compounds, fibre content, etc., but it is difficult to quantify the difference in the absence of detailed studies of the flora and its biochemical composition.

Some work has been done that explains in these terms the very much higher biomasses of folivorous primates in several moist equatorial African forests compared with several Asian forests (Davies, Bennett & Waterman, 1988; Waterman *et al.*, 1988). In these studies, selected African forests were shown to have lower average foliar concentrations of digestion inhibitors than did selected Asian ones, implying that similar gross primary productivity across sites was less important than the proportion of that productivity that was harvestable by animals.

This probably represents another factor influencing populations of Sundaic frugivores because, if dipterocarp forests have relatively high levels of toxins and digestion inhibitors, this may influence both the availability of alternative foods during periods of reduced fruiting, and the palatability or toxicity of animal, especially invertebrate, prey. Although the quantitative significance of this factor is unknown, we suspect it to be important.

Digestion, diet and competition

All the papionins share with the other cercopithecines a diet of fruits, seeds, tubers and animal matter. However, among the baboons (*Papio*, *Theropithecus*, and *Mandrillus*), mandrills and drills have teeth and jaws that are particularly well adapted to a frugivorous diet. They have broad, high-crowned incisors, used to remove rinds and husks, with comparatively small cheek teeth that have low, rounded cusps and are used to crush the soft inner parts of forest fruits and tubers (Jolly, 1970). The large temporal musculature, meanwhile, is set in a way appropriate for incisal nibbling, and both mandrills and drills can be described as 'strippers and pithers', since many food items (e.g. grass stems, roots, tubers, stems, twigs, bark, and many fruits) are prepared in this way (A.T.C. Feistner, personal observations and E L. Gadsby, personal observation).

The simple stomach of these species precludes foregut fermentation to denature toxins or to counteract the inhibition of digestion by the high tannin and fibre content of food. This is an underlying constraint on all similar mammals, and it implies that high biomass of rainforest primate frugivore-faunivores should be a reliable indicator of high harvestable productivity for papionins. It also seems reasonable to suppose that the presence of many sympatric frugivore-faunivore species represents the subdivision of a common resource amongst competing specialists whose foraging strategies must therefore complement one another. This is relevant when comparing the very different circumstances of competition affecting Sundaic pig-tails and West Central African mandrills and drills.

Pig-tails living in dipterocarp forests are doing so at very low biomass and essentially in the absence of sympatric papionin competitors. Sundaic non-dipterocarp forests tend to be densely occupied by long-tailed macaques (*M. fascicularis*). North of the Kra Isthmus in the Malay peninsula, where there is a sharp decline in the abundance and diversity of dipterocarps, the pig-tail is joined by several other macaques living sympatrically.

The situation of Sundaic pig-tails can be contrasted with those of mandrills and drills that live with a large number of sympatric frugivores.

Table 4.3. *Cercopithecines sympatric with pig-tails, mandrills and drills*

Species	Location	Sympatric cercopithecines
Pig-tail	Borneo	None as long-term residents in dipterocarp forest
	Thailand	Long-tailed macaque (*Macaca fascicularis*), stump-tailed macaque[a] (*M. arctoides*), rhesus macaque[a] (*M. mulatta*), Assamese macaque (*M. assamensis*)
Mandrill	Gabon	Grey-cheeked mangabey (*Lophocebus albigena*), white-collared mangabey[a] (*Cercocebus torquatus*), greater spot-nosed guenon (*Cercopithecus nictitans*), moustached guenon (*C. cephus*), crowned guenon (*C. pogonias*), de Brazza monkey[a] (*C. neglectus*), sun-tailed guenon[a] (*C. solatus*), talapoin monkey[a] (*Miopithecus talapoin*)
	Rio Muni and Cameroon	*L. albigena, C. torquatus, C. nictitans, C. cephus, C. pogonias, C. neglectus, M. talapoin*
Drill	Bioko	*C. nictitans, C. pogonias*, red-eared guenon (*C. erythrotis*), Preuss's guenon[a] (*C. preussi*)
	Nigeria and Cameroon	*L. albigena, C. torquatus, C. nictitans, C. pogonias, C. erythrotis, C. preussi*, mona monkey (*C. mona*)

[a]Semi-terrestrial species.
Data from: Caldecott (1986a), Feistner (1989), Gadsby (1990), Butynski & Koster (1990).

For example, 82% of all primary consumers at M'Passa in Gabon are frugivorous (Emmons, Gautier-Hion & Dubost, 1983) and primates, particularly cercopithecines (*Cercopithecus* and *Miopithecus*) make up a large proportion. In Gabon and Rio Muni, mandrills are sympatric with eight cercopithecine monkeys as well as two apes, and in Cameroon with at least seven monkeys and two apes (Table 4.3). In particular, several other primates are semi-terrestrial and may thus be in greater competition than exclusively arboreal species.

The occurrence of gorillas (*Gorilla gorilla*) and chimpanzees (*Pan troglodytes*) within the distribution of mandrills and drills should not be overlooked, especially as western lowland gorillas have been found to be essentially frugivorous (Williamson, 1988; Williamson *et al.*, 1990). Papionins and apes also feed on the same terrestrial herbaceous vegetation (A. T. C.

Feistner, personal observations). In Gabon, there was 44% overlap in diet between gorillas (*G. g. gorilla*) and mandrills (Lahm, 1986; Williamson, 1988), while in Cameroon a 46% dietary overlap was calculated between mandrills and mangabeys (*Cercocebus torquatus torquatus*; Hoshino, 1985; Mitani, 1989; Feistner, 1989).

Constraints on harvestable productivity

The foregoing discussion has led us to the following conclusions:

1. Although moist tropical forests are highly productive, they vary in the extent to which this productivity can be harvested by simian primates.
2. Average harvestable productivity of a forest for any frugivore-faunivore depends on the overall fruit availability and hence the density of trees that produce usable fruit. Levels of toxicity and indigestibility may affect both fruit and non-fruit food sources.
3. Actual harvestable productivity is the proportion of production that can be used by cercopithecines to support average biomass, and this will not include surplus fruit during masting events.
4. Tropical moist forests, which have a high density of trees that produce edible fruit, and/or a continuous overall supply of edible fruit, and/or an abundant supply of alternative food sources, will support high biomasses of frugivore-faunivores, such as cercopithecines.
5. For any single forest papionin, actual harvestable productivity will also be affected by populations of other frugivores, particularly other cercopithecines.
6. Sundaic dipterocarp forests can be described as absolutely marginal habitats for papionins, mainly because fruit supply is sparse and irregular. They are so marginal, in fact, that they support only one species, the pig-tailed macaque, living at extremely low biomass.
7. West Central African moist forests, in contrast, are inferred to be very hospitable habitats for cercopithecines and have been colonised by, or have supported the evolution of, numerous species. The presence of all these species, however, living in communities of high total biomass, means that actual harvestable productivity for any one species may be quite low due to competition for food.
8. In West Central African forests, therefore, it would not be surprising to find a very generalist species such as a papionin living as if it were in a marginal habitat. This case seems to be represented by the mandrill and drill.

In effect, then, it seems that pig-tails, mandrills, and drills are all living in marginal, food-poor environments, whether this is due to the absence of fruiting trees, the presence of toxins, the abundance of competitors, or the fact that fruit is available at intervals too widely spaced to be useful.

Ecological strategies and foraging tactics

The overall average harvestable productivity of a habitat is relevant to, but does not explain, the ecological strategies and behavioural tactics used by monkeys to survive in that habitat. To go further, we need to consider some other factors affecting the way in which harvestable food items are made available to forest-dwelling monkeys.

The structure of the forest is one such important factor, since in tropical moist forest most of the productivity that might be expected to be harvestable by papionins occurs in the canopy. Frugivorous monkeys are thus, inevitably, primarily harvesting arboreal food resources. The most important 'units' of food supply are, therefore, fruiting trees. These can be accessed by travelling through the canopy or by climbing up from the ground.

Travelling rapidly through the canopy on the tops of branches is energetically expensive and to do it well demands a number of adaptations, such as high inter-membral index (for jumping), long tail (for balance), and small body size (for route choice). These adaptations simultaneously constrain efficient travel on the ground, which is associated with low inter-membral index, short tail, and large body size. The main alternatives are thus either arboreal or terrestrial travel between arboreal food sources.

Even where an adaptation has been made to terrestriality, arboreal competence cannot be compromised too far, since the canopy must still be accessed for food. Most forest-dwelling monkeys are strongly arboreal, although all can travel short distances on the ground. This implies that the 'best place to be' is in the trees, where most of the food is located. Moreover, since long-distance arboreal travel is energetically costly, an alternative strategy is to occupy a small area of dense food sources, become intimately familiar with where and when the sources will ripen and the best routes to reach them, and then to harvest all the best ones before competitors do.

This is the 'classic' ecological strategy of the forest-dwelling cercopithecine, although there is variation because these animals are intelligent and behaviourally flexible, and the forest is inconstant. At times of widespread fruiting, for example, long-tailed macaques can be encountered in Bornean dipterocarp forest far from their normal location in riverine non-dipterocarp forests (Caldecott, 1986*a*), which illustrates the need for long-term field studies if ecological strategies are to be fully understood.

If the forest canopy tends to be occupied by resident frugivorous monkeys that are very efficient at locating ripe, edible fruits, any other monkey with similar dietary requirements in such a forest has only two choices in an evolutionary sense: to join the arboreal community as a specialist consumer, or to adapt (or stay adapted) to terrestrial travel. The former option might involve radical reduction in body size (such as in *Miopithecus talapoin*) or adaption to a particular subset of arboreal habitats (such as riverine or 'edge' forests as in *Cercopithecus mona, C. neglectus, Cercocebus torquatus* and *Macaca fascicularis*) or even both (as in *M. talapoin*).

The other option, that of long-distance terrestrial travel between arboreal food sources, is characteristic of the pig-tail, mandrill, and the drill. In the case of the Sundaic forest macaque, this is interpreted as being due to a lack of harvestable resources in the canopy to support arboreality by any cercopithecine. In the two forest baboons it is interpreted to have happened because there is not enough harvestable productivity left by competitors in the canopy to support them as arboreal specialists.

Pig-tails, mandrills, and drills are, therefore, believed to survive in the forests by means of rapid, energy-efficient terrestrial travel between arboreal food sources. However, since canopy food sources or opportunities to harvest them are likely to be relatively infrequent, the role of terrestrial food sources is likely to be correspondingly increased. Pig-tails, mandrills, and drills are all known to spend considerable time foraging terrestrially on invertebrates, ground vegetation, fungi and fallen fruit. These kinds of foods are distributed differently from the large 'food supply units' represented by fruiting trees. The forest contains many items that can be eaten by papionins, but most of them are small, scattered, and cryptic and demand continual exploratory foraging to locate them, rather than being clumped and obvious, and therefore, amenable to prolonged exploitation. This is inferred to be the main reason why pig-tails, mandrills, and drills typically forage in a dispersed manner, spread out over several hundred metres, and spend much of their time picking through the leaf litter, rotten wood, etc.

All three species supplement access to major arboreal food sources by continuous terrestrial foraging travel. Although all will stay to exploit a particularly large or rich food resource, aggregating in and around fruiting trees and remaining nearby for days, they do not show the feasting and resting pattern typical of arboreal species such as gibbons (*Hylobates* spp.) and long-tailed macaques (Raemaekers, 1979; MacKinnon & MacKinnon,

1980). Much more usually, a pig-tail or mandrill group will locate a food source, exploit it hastily, and them move on.

Since a group of pig-tails, mandrills or drills is usually moving quite fast and is widely dispersed in a habitat of poor visibility, the role of vocal behaviour becomes very important. Individuals of these species are all extremely vocal, with characteristic contact calls that they emit regularly during the day (Gartlan, 1970; Caldecott, 1986a; Kudo 1987; Feistner, 1989), although this can be influenced by hunting pressure. In pig-tails, variants of these calls are produced on encountering large food sources, and may act to recruit nearby individuals. This would make sense if the source was too large for a single individual to exploit without getting left behind by the group, especially if nearby animals were related. Since pig-tails, mandrills, and drills all form sub-groups during foraging travel, the supposition is that these may be composed of matrilineally-related individuals (Caldecott, 1986a; A.T.C. Feistner, unpublished observations).

Food-finding tactics that involve continuous foraging travel would be expected to translate directly into long day-range lengths and large home ranges, which are all characteristics of pig-tails, mandrills, and drills. Therefore, although in a fruit-rich *Koompassia*-Burseraceae forest a pig-tail group can survive in 50–60 hectares, in a dipterocarp forest it will need an area one or two orders of magnitude more extensive (Caldecott, 1985, 1986a). In a competitor-rich West Central African forest, a mandrill group may travel up to 8 km each day and use a home range of 40–50 km² (Jouventin, 1975).

In all forests then, groups of these three species seem routinely to travel far more, and cover far more ground overall, than other groups of cercopithecines that may be present in the same habitat. All three can be said to be using a terrestrial, mobile, continuous-foraging strategy, with opportunistic exploitation of major arboreal food sources coupled with a pattern of sub-grouping and rich vocal behaviour.

Sub-grouping and super-grouping in these monkeys is presumably controlled by food supply. In this respect there appears to be convergence between pig-tails and mandrills, although nothing is yet known of the extent of temporal stability in membership of mandrill groups. That aggregation of groups is influenced by food supply is supported by the indication that the large congregations of mandrills, numbering 100–300+ individuals, are more frequently observed during periods of relatively low fruit availability (Jouventin, 1975; Hoshino *et al.*, 1984; Feistner, 1989). It seems possible that mandrills make a dietary switch at such times, and increase the amount of terrestrial herbaceous food sources (such as

Marantaceae) in the diet. The ubiquitous nature of these resources, in contrast to highly localised arboreal resources, allows large groups of mandrills to forage together. Although super-grouping involving similar numbers of animals also occurs in drills and pig-tails, this was reported to be associated with high local fruit availability in drills (Gadsby, 1990), and no correlation in either direction was evident in pig-tails (Caldecott, 1986a).

Adult male pig-tails neither travel nor feed in parties, except when paired with consorts, and thus represent independent foraging units, whether group-living, peripheral, or solitary. Jouventin (1975) suggested that male mandrills feed largely terrestrially, and that only a single dominant male mandrill is found in a group of mandrills because of feeding competition between adult males foraging on the forest floor. If this were the case, a male mandrill might also be considered to represent an independent foraging unit. Small groupings of 2–6 monkeys have not been reported in mandrills but such 'parties' appear to be a normal foraging unit in pig-tails. Such small foraging units have been reported among drills on Bioko (Butynski & Koster, 1990).

Conclusions

Behavioural ecology

Pig-tails, mandrills, and drills show many similarities in their behavioural ecology, including:

1. Extreme bias towards females in intra-group adult sex ratio.
2. Sub-grouping and super-grouping tendencies.
3. Adaptation to terrestrial travel.
4. Very long day range lengths.
5. Very large home range areas.
6. Prolonged daily spells of foraging travel.
7. High dispersal of individuals during foraging travel.
8. Rich vocal behaviour with a strong emphasis on contact calling.

These similarities are attributed to a common diet, determined by similar digestive systems, interacting with low levels of harvestable productivity in the habitats where they live. The latter is attributed either to the prevailing patterns of primary and secondary productivity in the forest or to the density of competing arboreal species with similar dietary requirements. These have similar net results, and for all three species selection pressures are interpreted to favour the shedding of 'surplus' adult males from groups, and the adoption of foraging strategies appropriate to marginal or food-poor habitats. The latter assertion implies the need for a much more detailed and quantitative understanding of harvestable productivity in primate habitats, taking into account species composition, phenological patterns, and phytochemical constraints as modifiers of gross production of potential foods.

Implications for conservation

This discussion would not be complete without briefly mentioning the conservation circumstances of the three species, which in part derive from their common ecological strategies. Pig-tails are relatively widely distributed and occur in several legally protected forest areas. Mandrills have a smaller distribution and occur in few protected areas. Because of their mobile foraging tactics, however, both are vulnerable to intense human predation around the edges of those protected areas since they are prone to raid crops for food, and mandrills are a preferred bushmeat. Due to their low overall density, this means that their populations can easily be 'drained' out of even very large areas of conserved habitat. This problem is seldom recognised, for example in terms of hunting and trade regulations, since both species are often considered as 'common' (because widespread) and annoying agricultural 'pests'.

Nevertheless, the conservation circumstances of pig-tails and mandrills are relatively favourable compared with those of the drill, which is one of Africa's most endangered primates (Oates, 1986; Lee, Thornback & Bennett, 1988). This is because of severe hunting pressure and habitat fragmentation throughout most of its limited geographical range (Gadsby, 1990; Gadsby, Feistner & Jenkins, 1994). Although the drill occurs within two large National Parks (Cross River in Nigeria and Korup in Cameroon), these have only recently been constituted and hunting of drills has not yet been effectively suppressed within them. While enforcement efforts remain inadequate, there is the real risk that drills may become extinct in one or more sectors of these parks. This possibility is being addressed through complementary *ex-situ* measures in Nigeria, involving the salvage of captive animals from villages to establish a captive breeding population, which is showing excellent early success. This will constitute an important resource for research, for coordination with international *ex-situ* zoo breeding programmes through the exchange of individuals and information, for local conservation education, and would allow the possibility of eventual reintroduction to the wild in areas where extinction has occurred (Gadsby *et al.*, 1994).

Acknowledgements

The authors thank Dr Bob Cooper, Dr Siân Evans, Dr Marty Fujita, and Mr Peter Jenkins Jr for significant input to the ideas expressed in this chapter.

Julian Caldecott's work on pig-tails was supported by the United States National Cancer Institute and the Department of Anatomy of the University of Cambridge, and his later residence in Nigeria was supported by the World Wide Fund for Nature (WWF), the Commission of the European Communities and the United Kingdom's Overseas Development Administration.

Anna Feistner's work on mandrills was supported by the Leverhulme Trust. Most of the work took place at the Centre International de Recherches Médicales de Franceville (CIRMF), Gabon, which is 70% supported by the Republic of Gabon and 30% by Elf Gabon. CIRMF also funded the field station at the Lopé Reserve where wild primates were studied.

Liza Gadsby's work on drills was carried out with Peter Jenkins Jr. Work in Nigeria was supported by Wildlife Conservation International, WWF-US and WWF-UK. Work in Cameroon is Fauna and Flora Preservation Society (FFPS) project no. 90/1/1. The Drill Rehabilitation and Breeding Centre in Nigeria, where captive observations are ongoing, is funded by Pandrillus with assistance from the FFPS.

References

Bennett, E. L. & Caldecott, J. O. (1989). Rainforest primates of peninsular Malaysia. In *Ecosystems of the world*, vol. 14b, Tropical rainforest ecosystems, ed. H. Leith & M. J. A. Werger, pp. 355–63. Amsterdam: Elsevier.

Bernstein, I. S. (1967). A field study of the pig-tail monkey. *Primates*, **8**, 217–28.

Butynski, T. M. (1988). Guenon birth seasons and correlates with rainfall and food. In *A primate radiation: evolutionary biology of the African guenons*, ed. A. Gautier-Hion, F. Bourlière, J.-P. Gautier & J. Kingdon, pp. 284–322. Cambridge: Cambridge University Press.

Butynski, T. M. & Koster, S. H. (1990). *The status and conservation of forests and primates on Bioko Island (Fernando Poo), Equatorial Guinea.* Chicago: World Wildlife Fund-US and Chicago Zoological Society.

Caldecott, J. O. (1980). Habitat quality and populations of two sympatric gibbons on a mountain in Malaya. *Folia Primatologica*, **33**, 291–309.

Caldecott, J. O. (1984). Coming of age in *Macaca. New Scientist*, **1369**, 10–12.

Caldecott, J. O. (1985). Feeding and foraging in the pig-tailed macaque: some comparisons with sympatric species. In *Current perspectives in primate social dynamics*, ed. D. M. Taub & F. A. King, pp. 152-8. New York: Van Nostrand Reinhold.

Caldecott, J. O. (1986a). *An ecological and behavioural study of the pig-tailed macaque.* Contributions to Primatology, vol. 21. Basel: S. Karger.

Caldecott, J. O. (1986b). Mating patterns, societies and the ecogeography of macaques. *Animal Behaviour*, **34**, 208-20.

Caldecott, J. O. (1988). *Hunting and wildlife management in Sarawak.* Gland, Switzerland: World Conservation Union (IUCN).

Caldecott, J. O. (1991). Eruptions and migrations of bearded pig populations. *Bongo*, Berlin **18**, 233–43.

Crockett, C. M. & Wilson, W. L. (1980). The ecological separation of *Macaca nemestrina* and *M. Fascicularis* in Sumatra. In *The macaques: studies in ecology, behaviour and evolution*, ed. D. G. Lindburg, pp. 148–81. New York: Van Nostrand Reinhold.

Davies, A. G., Bennett, E. L. & Waterman, P. G. (1988). Food selection by two South-East Asian Colobine monkeys (*Presbytis rubicunda* and *Presbytis melalophos*) in relation to plant chemistry. *Biological Journal of the Linnaean Society*, **34**, 33–56.

Emmons, L. H., Gautier-Hion, A. & Dubost, G. (1983). Community structure of the frugivorous-folivorous forest mammals of Gabon. *Journal of Zoology*, **199** 209–22.

Feistner, A. T. C. (1989). Behaviour of a social group of mandrills (*Mandrillus sphinx*). PhD thesis, University of Stirling.

Feistner, A. T. C. (1990). Reproductive parameters in a semi-free-ranging group of mandrills. In *Baboons: behaviour and ecology, use and care*, ed. M. T. de Mello, A. Whitten & R. W. Byrne, pp. 77–8. Brasilia: University of Brasilia Press.

Feistner, A. T. C. (1991). Aspects of reproduction in female mandrills, *Mandrillus sphinx*. *International Zoo Yearbook*, **31**, 170–8.

Feistner, A. T. C., Cooper R. W. & Evans, S. (1992). The establishment and reproduction of a semi-free-ranging group of mandrills. *Zoo Biology*, **11**, 385–95.

Gadsby, E. L. (1990). The status and distribution of the drill (*Mandrillus leucophaeus*) in Nigeria. Unpublished report to WCI, WWF-US, WWF-UK, and the Nigerian Government.

Gadsby, E. L., Feistner, A. T. C. & Jenkins, P. D., Jr (1994). Coordinating conservation for the drill (*Mandrillus leucophaeus*), endangered in forest and zoo. In *Creative conservation: interactive management of wild and captive animals*, ed. P. J. S. Olney, G. M. Mace & A. T. C. Feistner, pp. 439–54. London: Chapman & Hall.

Gartlan, J. S. (1970). Preliminary notes on the ecology and behaviour of the drill. In *Old World monkeys: evolution, systematics and behavior*, ed. J. R. Napier & P. H. Napier, pp. 445–80. New York: Academic Press.

Gartlan, J. S. & Struhsaker, T. T. (1972). Polyspecific associations and niche separation of rain forest anthropoids in Cameroon, West Africa. *Journal of the Zoological Society of London*, **168**, 221–66.

Grubb, P. (1973). Distribution, divergence and speciation of the drill and mandrill. *Folia Primatologica*, **20**, 161–77.

Harrison, M. J. S. (1988). The mandrill in Gabon's rain forest – ecology, distribution and status. *Oryx*, **22**, 218–28.

Hoshino, J. (1985). Feeding ecology of mandrills (*Mandrillus sphinx*) in Campo Animal Reserve, Cameroon. *Primates*, **26**, 248–73.

Hoshino, J., Mori, A., Kudo, H. & Kawai, M. (1984). Preliminary report on the grouping of mandrills (*Mandrillus sphinx*) in Cameroon. *Primates*, **25**, 295–307.

Janzen, D. H. (1974). Tropical blackwater rivers, animals, and mast-fruiting by the Dipterocarpaceae. *Biotropica*, **6**, 69–103.

Jolly, C. (1970). The large African monkeys as an adaptive array. In *Old World monkeys: evolution, systematics and behavior*, ed. J. R. Napier & P. H. Napier, pp. 139–74. New York: Academic Press.

Jouventin, P. (1975). Observations sur la socio-écologie du mandrill. *La Terre et la Vie*, **29**, 493–532.

Kudo, H. (1987). The study of vocal communication of wild mandrills in Cameroon in relation to their social structure. *Primates*, **28**, 289–308.

Lahm, S. (1986). Diet and habitat preference of *Mandrillus sphinx* in Gabon: implications of foraging strategy. *American Journal of Primatology*, **11**, 19–26.

Lee, P. C., Thornback, J. & Bennett, E. L. (1988). *Threatened primates of Africa: the IUCN red data book*. Gland, Switzerland: World Conservation Union (IUCN).

Leighton, M. & Leighton, D. R. (1983). Vertebrate responses to fruiting seasonality within a Bornean rain forest. In *Tropical rain forest: ecology and management*, ed. S. L. Sutton, T. C. Whitmore & A. C. Chadwick, pp. 181–96. Oxford: Blackwell Scientific Publications.

MacKinnon, J. R. & MacKinnon, K. S. (1980). Niche differentiation in a primate community. In *Malayan forest primates: ten years' study in tropical rain forest*, ed. D. J. Chivers, pp. 167–90. New York: Plenum Press.

MacKinnon, K., Hatta, G., Halim, H. & Mangalik, A. (in press). *The ecology of Kalimantan*. Jogjakarta: Gadjah Mada University Press.

Medway, Lord (1972). Phenology of a tropical rain forest in Malaya. *Biological Journal of the Linnaean Society*, **4**, 117–46.

Mitani, M. (1989). *Cercocebus torquatus*: adaptive feeding and ranging behaviours related to seasonal fluctuations of food resources in the tropical rain forest of south-western Cameroon. *Primates*, **30**, 307–23.

Oates, J. F. (1986). *IUCN/SSC primate specialist group action plan for African primate conservation: 1986–90*. Gland and Cambridge: IUCN/WWF.

Oates, J. F. (1988). The distribution of *Cercopithecus* monkeys in West African forests. In *A primate radiation: evolutionary biology of the African guenons*, ed. A. Gautier-Hion, F. Bourlière, J.-P. Gautier & J. Kingdon, pp. 79–103. Cambridge: Cambridge University Press.

Oates, J. F., White, D., Gadsby, E. L. & Bisong, P. (1990). *Conservation of gorillas and other species. Appendix 1 of Cross River National Park (Okwangwo division): plan for developing the park and its support zone*. Godalming: World Wide Fund for Nature.

Oi, T. (1987). Sexual behaviour of the wild pig-tailed macaques in west Sumatra. *Kyoto University Overseas Research Report of Studies on Asian Non-Human Primates*, **6**, 67–80.

Oi, T. (1990). Population organization of wild pig-tailed macaques (*Macaca nemestrina nemestrina*) in west Sumatra. *Primates*, **31**, 15–31.

Raemaekers, J. J. (1979). Ecology of sympatric gibbons. *Folia Primatologica*, **31**, 227–45.

Richards, P. W. (1952). *The tropical rain forest: an ecological study*. Cambridge: Cambridge University Press.

Robertson, J. M. Y. (1986). On the evolution of pig-tailed macaque societies. PhD thesis, University of Cambridge.

Rodman, P. S. (1979). Skeletal differentiation of *Macaca fascicularis* and *Macaca nemestrina* in relation to arboreal and terrestrial quadrupedalism. *American Journal of Physical Anthropology*, **51**, 51–62.

Sabater Pi, J. (1972). Contributions to the ecology of *Mandrillus sphinx* Linnaeus

1758 of Rio Muni (Republic of Equatoria Guinea). *Folia Primatologica*, **17**, 304–19.

Struhsaker, T. T. (1969). Correlates of ecology and social organization among African cercopithecines. *Folia Primatologica*, **11**, 80–118.

Terborgh, J. & van Schaik, C. P. (1987). Convergence vs. nonconvergence in primate communities. In *Organization of communities, past and present*, ed. J. H. R. Gee & P. S. Giller, pp. 205–26. Oxford: Blackwell Scientific Publications.

van Steenis, C. G. G. J. (1950). The delimitation of Malaysia and its main plant geographical divisions. *Flora Malesiana*, **1**, 70–5.

Wallace, A. R. (1869). *The Malay archipelago*. London: Macmillan.

Waterman, P. G., Ross, J. A. M., Bennett, E. L. & Davies, A. G. (1988). A comparison of the floristics and leaf chemistry of the tree flora in two Malaysian rain forests and the influence of leaf chemistry on populations of colobine monkeys in the Old World. *Biological Journal of the Linnaean Society*, **34**, 1–32.

Whitmore, T. C. (1984). *Tropical rain forests of the Far East*. Oxford: Clarendon Press.

Whitten, A. J., Damanik, S. J., Anwar, J. & Hisyam, N. (1984). *The ecology of Sumatra*. Jogjakarta: Gadjah Mada University Press.

Whitten, A. J., Muslimin Mustafa & Henderson, G. S. (1987). *The ecology of Sulawesi*. Jogjakarta: Gadjah Mada University Press.

Williamson, E. A. (1988). *Behavioural ecology of western lowland gorillas in Gabon*. PhD thesis, University of Stirling.

Williamson, E. A., Tutin, C. E. G., Rogers, M. E. & Fernandez, M. (1990). Composition of the diet of lowland gorillas at Lopé in Gabon. *American Journal of Primatology*, **21**, 265–77.

5

Population ecology of rhesus macaques in tropical and temperate habitats in China

C. H. SOUTHWICK, ZHANG YONGZU, JIANG HAISHENG, LIU ZHENHE AND QU WENYUAN

Introduction

Rhesus macaques (*Macaca mulatta*) have the most extensive geographical range of any non-human primate, extending from Pakistan and Afghanistan in western Asia to China and Vietnam in the east. In the sub-continent, rhesus range from 16° N latitude in southern India (Fooden, 1981) to 35° 35′ N latitude in Pakistan (Roberts, 1977, cited in Wolfheim, 1983), and approximately 35° N latitude in Afghanistan (Puget, 1971). The ecology of rhesus groups seems to differ considerably in these latitudinal extremes, but there have been no comparable studies in either tropical or temperate habitats in the Indian sub-continent by the same team of investigators.

Most field studies of rhesus in India and Nepal have involved sub-tropical environments (Southwick, Beg & Siddiqi, 1965; Lindburg, 1971; Makwana, 1979; Seth & Seth, 1983; Southwick & Siddiqi 1988), while those in Pakistan and Afghanistan have been in temperate habitats (Pearl *et al.*, 1987). As a result, the extensive literature on rhesus ecology contains few direct comparisons of rhesus in tropical and temperate habitats. This chapter attempts to provide such comparisons from field studies on two rhesus populations in China occupying latitudinal extremes.

In China, natural rhesus populations extend from south-eastern Hainan Island, latitude 18° 10′ N, to the Henan-Shanxi border, north-west of Zhengzhou and north of the Yellow River, latitude 35° 10′ N (Fig. 5.1). As recently as 1987, a natural population of rhesus lived in the vicinity of Xinglung, approximately 160 km north-east of Beijing at a latitude of 40° 24′ N, but the last known individuals of this population were killed in 1987 by a local hunter (Zhang *et al.*, 1989).

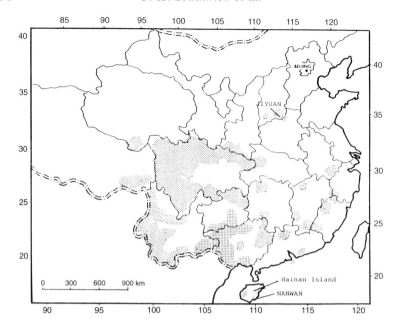

Fig. 5.1. Distribution of the rhesus monkey (*Macaca mulatta*) in China (from Tan, 1985).

Study areas

Nanwan

The tropical habitat, Nanwan Nature Reserve, is located on the south-east coast of Hainan Island in the South China Sea at latitude 18° 23′ N and longitude 110°. The reserve occupies a peninsula, which is an island from the standpoint of monkey habitat (Fig. 5.2). The area of the peninsula (or island) is approximately 10 km with a ridge of small mountains attaining an elevation of 255 m across the length of the peninsula. The climate is tropical monsoon, with annual rainfall averaging 1575 mm and a rainy season from May to October. Monthly mean temperatures range from 22.2 °C in January to 28.1 °C in July.

The peninsula was once the site of a pineapple plantation and mixed farming, but has now reverted to secondary vegetation with only small agricultural fields. A botanical survey identified 388 species of vascular plants, which included 336 species of angiosperms belonging to 98 families. The monkeys on Nanwan have been observed to eat 120 species of these plants (Jiang *et al.*, 1991).

The primary vegetation types on Nanwan are mixed shrubs and herbs,

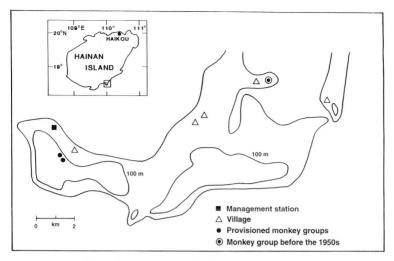

Fig. 5.2. Nanwan peninsula, Hainan Island.

and low secondary evergreen monsoon forest. The mixed shrub and herb community is dominated by *Carmona microphylla, Dodonea viscosa, Cymbopogon tortilis,* and *Heteropogon contortus.* The monsoon forest, with a long dry season, consists of low, shrubby trees, mostly less than 10 m tall. Dominant species are *Vatica astrotricha, Lithocarpus corneus, Cleistanthus saichikii* and *Coelodepas hainanensis.* These low trees provide dense tangles of cover for the monkeys and are generally impenetrable for human beings. A few trails have been cut through the forest for human access.

There are three villages, with about 4500 people, on the peninsula near the reserve; they have a total of about 140 ha of farm fields. The Government has constructed a tourist centre at the edge of the reserve, where monkeys are given rice twice daily, and tourists may also feed the monkeys. Of 20 rhesus groups on the peninsula, only two are habituated to people and come to the feeding arena.

Jiyuan

The temperate site, Jiyuan Nature Reserve, is on the Henan-Shanxi border, approximately 180 km north-west of Zhengzhou. The topography consists of rugged mountains to an elevation of 1960 m, dissected by steep rocky canyons with precipitous walls and vertical cliffs. The mountains represent the southern end of the Taihang mountains, which extend 700 km to the north-east.

The Jiyuan area has marked seasonal contrast – hot, rainy monsoon summers, and severely cold, dry winters. The July mean temperature is 26°C, but daily temperatures may rise above 40°C. The January mean temperature is − 1°C, with − 20°C not uncommon. Mean annual rainfall is 641 mm, with approximately 70% of this recorded from June to August; several snowfalls occur in winter, and canyon waterfalls, often higher than 10 m, freeze completely. The nearest village is several kilometres from the reserve, but individual farms occur in and around the reserve.

The primary forests of the Jiyuan area were cut and burned completely in the 1950s. Natural secondary forests dominated by oaks (*Quercus variabilis*, *Q. baronii*, and *Q. aliena*) now cover most of the reserve, interspersed with conifers (*Pinus armandii* and others). Other woody species include *Acer mono* (maple), *Carpinus turezaninowii* (hornbeam), *Diospyros kaki* (persimmon), *Morus cathyana* (mulberry), and *Gleditsia heterophylla* (honey locust). Understorey plants include *Vitis amurensis* (grape), *Rubus* spp. (blackberry), *Craetaegus* spp. (hawthorn), and *Lespedeza bicolor* (bush clover).

In summary, Nanwan is tropical in climate and vegetation, reminiscent of islands in the Caribbean, whereas Jiyuan is temperate, more similar to the mountains and canyons of Arizona or Colorado.

Taxonomy

Both populations are clearly *Macaca mulatta*, but the Nanwan rhesus have been given sub-specific status by some authors. Allen (1938) and Ellerman & Morrison-Scott (1951) named the Hainan rhesus as *Macaca mulatta brachyurus* or *M. m. brevicauda*. The rhesus of Hainan are noticeably small, perhaps 10% to 15% smaller in each age-class than typical rhesus of mainland Asia, and the tail is approximately 20% shorter. However, Hill (1974) considered these differences to be within the range of *Macaca mulatta mulatta*.

At Jiyuan the monkeys are heavy-bodied with thick fur, resembling Japanese macaques in body configuration. They appear to have the normal tail length of typical rhesus, and have not been given sub-specific status.

Methods

The monkeys of Nanwan, protected since 1965, have been observed for 25 years, and studied intensively since 1981 by Jiang Haisheng. The Jiyuan monkeys, protected since 1982, have been studied since 1981 by Qu *et al.* (1993). Monkeys at both sites are provisioned with rice, with two groups coming regularly for provisioned food at both locations. The provisioned groups have been the focus of detailed data collection on group compositions, social behaviour, birth phenology and birth rates. Provisioning has been provided in the morning around 8.00 a.m. in an open area of the forest edge at both Nanwan and Jiyuan. Observational conditions on provisioned groups have been good for group counts, sex and age classifications, and individual recognition.

In addition to the study of provisioned groups, forest surveys have been conducted at both sites to locate unprovisioned groups. At Nanwan, these have involved trails cut through the dense scrub forests, or traversing some ridgetops, shorelines or edges where monkeys emerge from the forest. At Jiyuan, forest surveys have required walking along canyon bottoms and ridgetops; rugged topography has been the constraint at Jiyuan rather than dense scrub forest. Group counts were possible when groups crossed streams in canyon bottoms, or when they scaled steep cliffs with open visibility. When disturbed, monkeys often climbed canyon walls, providing good census opportunities.

Individuals were age-classified as adults, sub-adults, juveniles, or infants, and sexes were identified in adults and sub-adults.

Results

The Nanwan (tropical) and Jiyuan (temperate) populations are comparable in the following ways: (1) both populations are free ranging, (2) both occupy forest habitats; (3) both have two provisioned groups, although the majority of both populations consists of wild, unprovisioned groups; and (4) both populations are refugia-type populations, isolated by several hundred km from other rhesus populations.

The two populations are substantially different in the following ways:

1. Nanwan rhesus occupy a tropical evergreen scrub forest, whereas Jiyuan rhesus occupy a temperate deciduous forest.
2. Nanwan rhesus have moderate topography in a coastal environment, with maximum elevation only 255 m above sea level, whereas the Jiyuan rhesus occupy rugged mountains and canyons with elevations extending from 240 m to 1960 m.
3. Virtually every aspect of their climate and biological communities are different: Nanwan gives the immediate impression of suitable rhesus habitat, whereas Jiyuan presents the impression of a harsh and unsuitable rhesus habitat, or an environment at the adaptational extremes for rhesus.

4. The provisioned Nanwan rhesus groups are more exposed daily to tourists and are more commensal with humans than the Jiyuan provisioned groups, who have no exposure to tourists and only limited contact with humans.

Population sizes and densities

In 1987, the Nanwan populations consisted of approximately 1200 monkeys living on the restricted peninsula of 10 km^2; hence there was a population density of 120 rhesus/km^2. The Jiyuan population had an estimated population of slightly more than 2000 monkeys living in an unrestricted area of approximately 280 km^2, hence a population density of 7.2 rhesus/km^2. The Jiyuan population, though larger in total size, had a much broader area and lower overall population density than the restricted Nanwan population.

Table 5.1 summarizes some of the main climatic and demographic differences between the Nanwan and Jiyuan populations, and these differences identified in 1987 are highlighted in the text below.

Average group sizes

The Nanwan population consisted of 20 groups, with an average group size of 59 individuals per group (range of group size 20–110). The two provisioned groups at Nanwan were close to the average, 52 and 67 individuals per group in 1987.

The Jiyuan population consisted of approximately 25 groups, with an average group size of 82, and a range of 35–125. The two provisioned groups at Jiyuan were 65 and 104 individuals per group in 1987.

Home ranges

The Nanwan groups were moderately crowded, with an average home range of 0.37 km^2. These home ranges were concentrated on the southern and western portions of the peninsula, away from the mainland of Hainan Island and also away from the villages, with the exception of the two provisioned groups, which came near the western-most village (Fig. 5.2). Different group home ranges at Nanwan varied in size from around 0.10 km^2 to 0.72 km^2.

At Jiyuan home ranges were enormous; groups moved over mountains and ridges into complex canyon systems and were difficult to follow.

Table 5.1. *Ecological comparisons of Nanwan and Jiyuan Nature Reserves, and rhesus monkey populations*

Trait	Nanwan	Jiyuan
Habitat	Tropical evergreen scrub forest	Temperate deciduous forest
Latitude	18° 23′ N	35° 10′ N
Mean January temp. (°C)	22.2	−1
Mean July temp. (°C)	28.1	26
Annual rainfall (mm)	1575	641
Topography	Hills to 255 m	Mountains to 1962 m
Estimated Rhesus population size	*ca* 1200	*ca* 2000
Estimated population density/km	120	7.2
Average group sizes	59	82
Average birth rates (%)	77.8	50.7
Average home ranges (km²)	0.37	16

Exceptional work by local field workers was able to locate and identify groups, and establish estimates of home range sizes. These varied from 11 to 22 km², the largest ever recorded for rhesus groups. The average home range of six groups studied at Jiyuan was 16 km².

Natality

Annual birth rates at Nanwan averaged 77.8% from 1978 to 1984 and 50% at Jiyuan. The Nanwan population showed an unusual pattern of birth rates alternating yearly between high and low, with 1979, 1981 and 1983 showing high rates of 83–100% in the provisioned groups, whereas the alternate years of 1980, 1982, and 1984 showed low rates of 58–78% (Fig. 5.3).

In Jiyuan, birth rate data are available on only the two provisioned groups in 1987. These showed an annual birth rate for one group of 31.5% and for the other group of 64.0%, resulting in a combined birth rate of 50.7%. The group with the annual birth rate of 64% had been provisioned for three winters, whereas the group with the low birth rate of 31.5% had been provisioned for only one winter.

Population growth rates

The rhesus populations at both Nanwan and Jiyuan have been increasing, but actual rates of increases are known only for Nanwan. From 1965 to

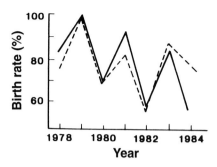

Fig. 5.3. Annual birth rates of the Nanwan rhesus population, 1978–84.

1984, the Nanwan population increased from 115 animals to 930, an annual rate of increase of 12.7%. From 1984 to 1987, the rate of growth slowed to 8.9% per year.

Food habits

Detailed studies of foods and feeding behaviour of the Nanwan and Jiyuan monkeys have not been made, but some differences are obvious. The Nanwan rhesus have a much greater diversity of plants in their environment (336 species of angiosperms have been identified at Nanwan), and green leafy vegetation with many fruiting species are available throughout the year. At Jiyuan, the plant community is less diverse (probably less than half the number of angiosperm species compared to Nanwan), and the deciduous trees are leafless from approximately November to May. Fruits are much less abundant at Jiyuan. In the winter and early spring, the rhesus at Jiyuan forage on bark, buds, twig ends, and roots; the only green vegetation available in winter at Jiyuan is from coniferous trees.

Discussion

The results show both the adaptability of rhesus macaques to a wide range of environmental conditions, and the ecological nature of some of these adaptations. Rhesus in cold temperate environments tend to have larger group sizes, much larger home ranges, but lower reproductive rates, and probably lower rates of population growth, although the latter point is not fully established. The larger group sizes could provide some protection against cold temperatures in the sense that they enable more individuals to huddle together during resting, sleeping and feeding. Large group sizes may also be an adaptation to sparse resources, providing more efficient foraging in habitats where food is hard to find.

Large home ranges are a common response to patchy resources, in habitats where widespread foraging has benefits in finding food supplies that are scattered and limited (Altmann, 1974).

Lower birth rates indicate that fewer females have young at one-year intervals, and more females have young at two-year intervals. The common theme in all of these patterns may be energy conservation in relation to winter survival.

Several studies in India, Pakistan, and Afghanistan, although not undertaken by the same individuals, as in our direct comparative studies, show similar comparative patterns. Rhesus monkeys in the Dunga Gali forests of northern Pakistan also live in rugged mountainous terrain with cold, snowy winters, and hot summers (Pearl *et al.*, 1987). Like the Jiyuan rhesus, those of Dunga Gali are heavy bodied and well furred. They live in groups of between 20 and 100 monkeys or more; home ranges are large, one main study group had a home range of 8 km^2 (Richard, 1985); reproductive rates are low, with an annual birth rate of 0.38 and an average inter-birth period of over 2 years (Richard, 1985; Pearl *et al.*, 1987). In both Dunga Gali and Jiyuan, winter foods consisted of bark, buds and roots. Virtually all of the ecological characteristics of rhesus in northern Pakistan and northern China make sense as adaptations to harsh winter conditions.

In Afghanistan, detailed ecological studies of rhesus macaques have not been done, but preliminary field work by Puget (1971) showed rhesus in very large groups, averaging over 120 monkeys, with large home ranges and winter food habits involving bark and roots.

In India and Nepal, where most rhesus field studies have been done (e.g. Southwick, Beg & Siddiqi, 1965; Mukherjee, 1969; Lindburg, 1971; Roonwal & Mohnot, 1977; Makwana, 1979; Seth & Seth, 1983; Teas, 1983; Pirta, 1984; Malik, 1986; Southwick & Siddiqi, 1988), rhesus group sizes tend to average 15 to 50 monkeys, although larger groups do occur. Groups usually have home ranges of less than 1 km^2; they consume green vegetation year round and do not forage on bark or roots; they typically have annual birth rates around 0.75 to 0.85. Most females give birth to one young every year. Rhesus in India show annual rates of population growth ranging from 6% to 20% if protected and living in a suitable environment.

Latitudinal comparisons of primate ecology have also been made of the Japanese macaques by Azuma (1985). A population of Japanese monkeys in a cold environment at Shimokita (41° 30′ N latitude), have low population densities and very large home ranges compared with a southern population at Yakushima (30° 20′ N). The northern monkeys also had a lower birth rate (0.28) than the southern monkeys (0.44). There was not, however, any significant difference in average group sizes, and no consistent differences in the rates of population growth.

Conclusions

Rhesus populations in China extend from tropical monsoon forests to temperate deciduous forests with cold, snowy winters. Rhesus in the tropics have higher population densities, smaller home ranges, smaller groups sizes, higher birth rates, and higher rates of population growth.

In the temperate populations, lower population densities, larger home ranges, larger group sizes, and lower birth rates are all characteristics that seem to be adaptive for winter survival and energy conservation.

Acknowledgements

We are indebted to officials of the Nanwan and Jiyuan Nature Reserves for permits to work in these areas, and to Government Officers of the Guandong and Henan Provincial Governments. We also thank David Manry for field assistance at Jiyuan in 1987. In 1987 and 1988, the participation of Charles Southwick and Yongzu Zhang in this work was supported by a grant from the National Geographic Society to the Academia Sinica, Beijing, China.

References

Allen, G. M. (1938). *Mammals of China and Mongolia.* Part I. New York: American Museum of Natural History.

Altmann, S. A. (1974). Baboons, space, time and energy. *American Zoologist*, **14**, 221–48.

Azuma, S. (1985). Ecological biogeography of Japanese monkeys (*Macaca muscatta* Blyth) in the warm and cold-temperate forest. In *Recent mammalogy of China and Japan*, ed. Kawamichi, pp. 1–5. Kanrin, Inuyama, Japan: Primate Research Institute.

Ellerman, J. R. & Morrison-Scott, T. C. S. (1951). *Checklist of Palaearctic and Indian mammals*, 1758–1946. London: British Museum of Natural History.

Fooden, J. (1981). Taxonomy and evolution of the *sinica* group of macaques. 2. Species and subspecies accounts of the Indian Bonnet macaque, *Macaca radiata. Fieldiana Zoology*, **9**, 1–52.

Hill, W. C. O. (1974). *Primates: Comparative anatomy and taxonomy*; Cercocebus, Macaca, Cynopithecus, vol. VII. New York: John Wiley & Sons.

Jiang Haisheng, Liu Zhenhe, Zhang Yongzu & Southwick, C. (1991). Population ecology of rhesus monkeys (*Macaca mulatta*) at Nanwan Nature Reserve,

Hainan, China. *American Journal of Primatology*, **25**, 207–17.

Lindburg, D. G. (1971). The rhesus monkey in North India: an ecological and behavioral study. In *Primate behavior: developments in field and laboratory research* II, ed. L. A. Rosenblum pp. 1–106. New York: Academic Press.

Makwana, S. C. (1979). Field ecology and behaviour of the rhesus macaque *Macaca mulatta*. I. Group composition, home range, roosting sites and foraging routes in the Asarori forest. *Primates*, **19**, 483–92.

Malik, I. (1986). Time budgets and activity patterns in free-ranging rhesus monkeys. In *Primate ecology and Conservation*, ed. J. G. Else & P. C. Lee, pp. 105–23. Cambridge: Cambridge University Press.

Mukherjee, R. P. (1969). A field study on the behaviour of two roadside groups of rhesus macaque (*Macaca mulatta*, Zimmerman) in northern Uttar Pradesh. *Journal of the Bombay Natural History Society*, **66**, 47–56.

Pearl, M., Melnick, D., Goldstein, S. & Richard, A. (1987). All-weather monkeys. *Animal Kingdom*, **90**, 32–41.

Pirta, R. S. (1984). Cooperative behaviour in rhesus monkeys (*Macaca mulatta*) living in arban and forest areas. In *Current primate researchers*, ed. M. L. Roonwal, S. M. Mohnot & N. S. Rathmore, pp. 271–83. Jodhpur: University of Jodhpur.

Puget, A. (1971). Observations sur le macaque rhesus (*Macaca mulatta*; Zimmerman, 1780), en Afghanistan. *Mammalia*, **35**, 199–203.

Qu, Wenyuan, Zhang, Yongzu, Manry, D. & Southwick, C. H. (1993). Rhesus monkeys (*Macaca mulatta*) in the Taihang mountains, Jiyuan County, Henan, China. *International Journal of Primatology*, **14**, 607–21.

Richard, A. (1985). *Primates in nature*. New York: W. H. Freeman.

Roonwal, M. L. & Mohnot, S. M. (1977). *Primates of South Asia: ecology, socio-biology and behaviour*. Cambridge, MA: Harvard University Press.

Seth, P. K. & Seth, S. (1983). Population dynamics of free-ranging rhesus monkeys in different ecological conditions in India. *American Journal Of Primatology*, **5**, 61–7.

Southwick, C. H., Beg, M. A. & Siddiqi, M. R. (1965). Rhesus monkeys in north India. In *Primate, behavior: field studies of monkeys and apes*, vol. I, ed. DeVore, pp. 111–59. New York: Holt, Rinehart & Winston.

Southwick, C. H. & Siddiqi, M. F. (1988). Partial recovery and a new population estimate of rhesus monkey populations in India. *American Journal Of Primatology*, **16**, 187–97.

Tan Bangjie (1985). The status of primates in China. *Primate Conservation*, **5**, 63–81.

Teas, J. (1983). Ecological considerations important in the interpretation of census data on free-ranging monkeys in Nepal. In *Perspectives in primate biology*, ed. P. K. Seth, pp. 211–5. New Delhi: Today and Tomorrow's Printers and Publishers.

Wolfheim, J. (1983). *Primates of the world*. Seattle: University of Washington Press.

Zhang Yongzu, Quan, G., Lin, Y. & Southwick, C. (1989). Extinction of rhesus monkeys (*Macaca mulatta*) in Xinglung, North China. *International Journal of Primatology*, **10**, 375–81.

6

Demography and ecology of Barbary macaques (*Macaca sylvanus*) in two different habitats

N. MÉNARD AND D. VALLET

Introduction

The distribution of Barbary macaques is limited to Morocco and Algeria and includes populations of various sizes more or less isolated from each other by fragmentation of suitable habitats (Taub, 1977; Fa *et al.*, 1984). The colonisation of different habitat types probably illustrates how very adaptable the species is. However, we can compare the degree of success of the Barbary macaques in their different habitats. At present, monkey density is the only factor available for estimating the state of most Barbary macaque populations. Density varies from 1 to 70 monkeys per km², the lowest values being for thermophilous scrub habitats and the highest values for cedar forests (Deag, 1974; Taub, 1977; Fa, 1984). Large variations in density are also observed between areas of the same habitat type, for example from 8 to 70 monkeys/km² for different populations living in cedar forest. In most cases, the ecological factors that determine these differences are poorly understood without a long-term study. Abundance of resources is generally recognised as one of the principal factors limiting monkey density (Iwamoto, 1978; Southwick, Siddiqi & Oppenheimer, 1983; Altmann, Hausfater & Altmann, 1985). Yet, the carrying capacity of an environment is, in general, difficult to estimate for omnivorous non-human primates. One must consider quantity, quality and availability of resources, and then evaluate their use by the animals over the whole annual cycle (Bourlière, 1979). This type of study is rare but studies have been done in tropical forest (Hladik, 1977; Oates, 1977; Waser, 1977; Gautier-Hion, Gautier & Quris, 1981), in savanna (Norton, Rhine & Wynn, 1987) and in temperate habitats (Ménard, 1985; Ménard & Vallet, 1986, 1988). It is widely supposed that the high eclecticism of the diet of the Barbary macaques is a strategy that allows them to exploit different habitats (Deag, 1974; Drucker, 1984; Fa, 1984).

As shown by Ménard, Vallet & Gautier-Hion (1985), demographic parameters differ according to the habitat colonised by the species. This chapter describes the ecology and demography of Barbary macaque groups living in two different forest habitats and tries to identify environmental factors which might explain the observed differences.

Methods

Two parks were studied: one at Tigounatine in the cedar-oak forest of the Djurdjura National Park (4° 8′ E, 36° 27′ N) and the other in the deciduous oak forest at Akfadou (4° 33′ E, 36° 27′ N). The study of demography and group dynamics was carried out between February 1983 and July 1990 in the course of observations organised so as to cover the birth season and part of the mating season. In each group, individuals were identified by coloured ear tags and/or morphological characteristics and their presence recorded during each day of observation. Group composition was established using age/sex classes previously defined on the basis of morphology and degree of sexual maturity (Ménard *et al.*, 1985, see also Tables 6.1 and 6.2). Group size and composition are given for July of each year (i.e. the end of the birth season) and for February 1983.

Dates of birth, sex and affiliations of infants were recorded; birth rate for 1982 was estimated from group composition as determined in February 1983. Data on birth rate and infant mortality rate were uncertain because it was not possible to confirm for 1982 and 1989 whether or not four females at Tigounatine and two at Akfadou had given birth then lost their infants prior to the start of observations. Unweaned infants and juveniles socially dependent on their mothers were presumed not to have survived if they had disappeared. Individuals who left the study groups were only considered to be 'emigrants' if subsequently relocated, otherwise they were recorded as disappearances.

The study of diet and available resources was carried out from February 1983 to July 1987 using methods detailed elsewhere (Ménard, 1985; Ménard & Vallet, 1986, 1988). However, the majority of data quantified was from 1983 and 1984; the collection of these data was structured to be representative of an annual cycle. Observations on animals were conducted six days per month on average. 'Scan samples' (Altmann, 1974) were recorded every 15 min on five animals, and the type of activity and the foods consumed noted.

Study areas were mapped using a grid system of 50 m or 100 m quadrats depending on visibility in the habitat. The study of available resources was carried out each month. Acorn production was estimated along 10 transects with a total length of 500 m. Transects were also used to estimate shrub cover. For the herbaceous layer, vegetation sampling was carried out in 1 m quadrats placed at the intersections of a 200 m grid covering the group's home ranges. Plant species were recorded along with the degree of ground cover and the height of plants. Leaf volume was calculated for each herbaceous species and used to estimate the relative abundance of their available resources. The phenological state of all plants was also recorded.

Indices of specific diversity (D) of resources and of diet were calculated for each vegetation layer according to Simpson's formula (Levins, 1968): $D = (\Sigma P_i^2)^{-1}$, where P is the frequency of different species in each vegetation layer, or the frequency of different species eaten over the year. D can vary from 1 to N, where $D = 1$ when a single species is present in the vegetation layer or in the diet, and $D = N$ when N species are of equal importance.

The degree of selectivity exercised by the monkeys was estimated by dividing the rate of consumption for each species by its rate of relative abundance. The species are thus classified: (1) preferred, when their rate of consumption is more than 1.5 times the relative rate of abundance; (2) avoided, when the rate of consumption is less than 0.5 times the relative rate of abundance; (3) neutral, when the rate of consumption is more or less equal to the relative rate of abundance. Certain species that were rare both in the habitat and the diet were not classified.

The spatial use of the habitat was quantified from 1983 to 1990. In the course of observations, the centre of the group was positioned with reference to the grid system every 30 min, 6125 data points at Tigounatine and 4275 at Akfadou.

Results

Demography

Size and structure of groups

Tables 6.1 and 6.2 show size and structure of the groups studied from February 1983 to July 1990. Group size varied from 13 to 88 individuals. The initial Tigounatine group increased from 38 to 88 animals and then split into three new groups (SM, LO and UL) of 50, 24 and 13 animals, respectively that had together reached a total of 106 individuals by 1990. At the same time, the Akfadou group increased from 33 to 53 animals. The mean annual growth rate of the groups from July 1983 to July 1990 was

18.6% at Tigounatine and 5.1% at Akfadou. In each case, more than 75% of this increase was accounted for by the natality/mortality balance, while the immigration/emigration balance had less effect (Ménard & Vallet, 1993*a,b*).

At both sites, groups had a multimale structure. The mean overall sex ratio (males:females) and the mean adult sex ratios were relatively balanced, varying from 1:0.7 to 1:1.2 (range 1:0.6 to 1:1.4) and from 1:0.9 to 1:1.2 (range 1:0.8 to 1:1.9), respectively. Group UL was an exception and temporarily had a single-male group structure with one adult male for five adult females in 1989 after the splitting of the initial Tigounatine group. The proportion of immatures did not differ between the groups, varying from 0.41 to 0.59 at Tigounatine and 0.42 to 0.58 at Akfadou according to the year.

Natality

Seasonality

Ménard & Vallet (1993*b*) have shown that almost all births occurred between 14 April and 8 July in Algeria, with a significant earlier median birth date at Akfadou than at Tigounatine (3 May and 19 May, respectively). At both sites, primiparous females and multiparous females without any previous infant (whether they did not give birth the year before or whether they had lost their previous infant) gave birth significantly earlier than multiparous females accompanied by their previous infant (Fig. 6.1) (ANOVA, $F = 4.07$; $p < 0.05$). This difference did not vary according to the year (ANOVA, $F = 2.00$; $p > 0.05$).

Birth rate and female fertility

The mean birth rate over nine years varied from 0.56–0.58 ($N = 97–101$ infants) at Tigounatine to 0.63–0.65 ($N = 66–68$ infants) at Akfadou with considerable inter-annual variations, ranging from 0.14 to 0.80 and from 0.33 to 1.00, respectively. There was no significant difference between the sites (for a more detailed analysis see Ménard & Vallet, 1993*b*).

Primiparous females averaged 5.5 years old at Tigounatine and 5.3 years old at Akfadou (ranging from 4 to 8 years). Females of 3 years of age never gave birth and 4 year olds only rarely (Table 6.3). Females showed their maximum reproductive rate between 8 and 15 years and, seemingly, declined thereafter.

The mean inter-birth interval was not significantly correlated with the

Table 6.1. *Demographic parameters of the Tigounatine groups from February 1983 to july 1990*

A. *Initial group*

Classes	Age (years)	02-83	07-83	07-84	07-85	07-86	07-87	07-88
Adults								
♂♂	>5	7	9	9	10	16	20	26
♀♀	>5	7	9	9	14	17	19	23
	4	2	1	5	4	3	4	0
Sub-adults								
♂♂	4	2	0	2	4	2	1	1
	3	0	2	3	2	3	1	5
♀♀	3	1	5	4	3	4	0	6
Juveniles								
♂♂	2	2	3	2	3	1	5	4
	1	3	2	3	1	5	4	2
♀♀	2	5	4	3	4	0	6	4
	1	4	3	5	0	7	5	5
Infants								
♂♂	0–1	2	3	1	6	5	3	5
♀♀	0–1	3	5	1	8	6	5	7
Group size		38	46	47	59	69	73	88
Overall sex ratio ♂:♀		1:1.4	1:1.4	1:1.4	1:1.3	1:1.2	1:1.1	1:1
Adult sex ratio ♂:♀		1:1.3	1:1.1	1:1.6	1:1.8	1:1.3	1:1.2	1:0.9
Proportion of immatures		0.58	0.59	0.51	0.53	0.48	0.41	0.44

Date (month-year)

B. *New groups*

Classes	Age (years)	SM		LO		UL	
		07-89	07-90	07-89	07-90	07-89	07-90
Adults							
♂♂	> 5	14	15	7	9	1	6
♀♀	> 5	12	14	7	7	3	5
	4	2	2	0	1	2	1
Sub-adults							
♂♂	4	3	2	1	2	0	0
	3	2	2	2	0	0	0
♀♀	3	2	1	1	0	1	3
Juveniles							
♂♂	2	2	0	0	2	0	0
	1	0	4	2	1	0	1
♀♀	2	1	2	0	0	3	0
	1	2	3	0	3	0	2
Infants							
♂♂	0–1	5	3	1	4	1	3
♀♀	0–1	5	7	3	1	2	0
Group size		50	55	24	30	13	21
Overall sex ratio ♂:♀		1:0.9	1:1.1	1:0.8	1:0.7	1:5.5	1:1.1
Adult sex ratio ♂:♀		1:1	1:1.1	1:1	1:0.9	1:5	1:1
Proportion of immatures		0.44	0.44	0.42	0.43	0.54	0.43

Table 6.2. *Demographic parameters of the Akfadou group from February 1983 to July 1990*

Classes	Age (years)	Date (month-year)								
		02-83	07-83	07-84	07-85	07-86	07-87	07-88	07-89	07-90
Adults										
♂♂	>5	8	9	10	7	7	9	12	11	13
♀♀	>5	6	6	8	10	11	12	9	11	15
	4	0	2	2	1	2	0	4	4	0
Sub-adults										
♂♂	>4	1	1	3	0	3	1	2	2	2
	3	1	3	3	3	1	2	2	2	5
♀♀	3	2	2	1	2	0	4	4	0	1
Juveniles										
♂♂	2	3	3	3	1	2	2	2	5	3
	1	3	3	1	2	2	2	5	3	4
♀♀	2	2	1	2	0	4	4	0	1	2
	1	1	2	0	4	4	1	1	2	0
xx	2	0	1							
	1	1								
Infants										
♂♂	0-1	3	3	2	2	3	8	3	4	4
♀♀	0-1	2	3	6	4	2	2	2	0	4
Group size		33	39	41	36	41	47	46	45	53
Overall sex ratio ♂:♀		1:0.6–1:0.7	1:0.7–1:0.8	1:0.9	1:1.4	1:1.3	1:1	1:0.8	1:0.7	1:0.7
Adult sex ratio ♂:♀		1:0.8	1:0.9	1:1	1:1.6	1:1.9	1:1.3	1:1.1	1:1.4	1:1.2
Proportion of immatures		0.58	0.56	0.51	0.50	0.51	0.55	0.46	0.42	0.47

xx, represents sex unknown.

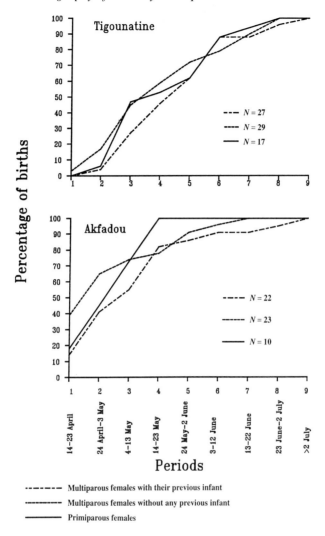

Fig. 6.1. Distribution of births over time depending on the reproductive status of the mothers. The birth season is divided into nine 10-day periods: Period 1: 14–23 April; period 2: 24 April–3 May; period 3: 4–13 May; period 4: 14–23 May; period 5: 24 May–2 June; period 6: 3–12 June; period 7: 13–22 June; period 8: 23 June–2 July; period 9: >2 July.

sites (1.4 years, $N = 85$, at Tigounatine and 1.3 years, $N = 52$, at Akfadou), the reproductive status of the mother or the sex of the previous surviving infant (ANOVA, $p > 0.3$, NS). A one-year interval was found in the majority of cases (57% and 79%, respectively, at Tigounatine and at

Table 6.3. *Age-specific birth rates of wild Barbary macaques in Algeria*

Age (years)	Reproductive females × years	Birth rate
3	43	0.00
4	37	0.03
5	32	0.56
6	23	0.57
7	18	0.67
8	14	0.71
9	8	0.75
10–15	56	0.71
>15	17	0.65

Reproductive females × years is the cumulative number of females observed over the years.

Akfadou) and females rarely failed to give birth for more than two years (Table 6.4).

At both sites, the interval between births was shorter after the death of an infant (1.2 years) than when the previous infant survived (1.5 and 1.4 years at Tigounatine and at Akfadou, respectively) but the difference was not significant (ANOVA, $F = 3.39$, $p = 0.06$).

Sex ratio at birth

Females tended to give birth to more infant females at Tigounatine than at Akfadou between 1982 and 1990 (55%, $N = 96$ and 46%, $N = 63$, respectively, Table 6.5), but the difference was not significant (Wilcoxon, $p > 0.2$). Higher inter-annual variations occurred at Akfadou (0–75%) than at Tigounatine (44–63%).

Mortality

Infant mortality rate was 0.23–0.27 ($N = 78$–82) at Tigounatine and 0.38–0.40 ($N = 58$–60) at Akfadou, with great inter-annual variations (0–0.67 and 0–0.83, respectively). The differences were not significant (for a more detailed analysis see Ménard & Vallet, 1993*b*).

Mortality rate was 0.26 for infant males and 0.24 for infant females at Tigounatine and 0.30 and 0.48, respectively, at Akfadou (Table 6.6), but did not differ significantly according to the sex of the infant (Wilcoxon, $p > 0.05$). Mortality of the infant females tended to be higher at Akfadou. Mortality did not differ significantly between infants of primiparous mothers and those of multiparous mothers at either site (Wilcoxon, $p > 0.05$).

At both sites, the majority of infant deaths occurred between June and

Table 6.4. *Percentage of different lengths of inter-birth intervals and mean length of inter-birth interval (years)*

	Tigounatine						Akfadou					
	Interval (years)						Interval (years)					
	1	2	3	4	Mean	N	1	2	3	4	Mean	N
Overall intervals	57	42	1	0	1.4	85	79	17	0	4	1.3	52
After first birth	38	62	0	0	1.6	21	78	22	0	0	1.2	9
After subsequent births	63	35	2	0	1.4	64	79	16	0	5	1.3	43
With previous infant surviving	47	52	1	0	1.5	64	71	26	0	3	1.4	34
Male	31	65	4	0	1.7	26	62	33	0	5	1.4	21
Female	58	42	0	0	1.4	38	85	15	0	0	1.2	13
After infant loss	86	14	0	0	1.2	21	94	0	0	6	1.2	18

Table 6.5. *Percentage of females at birth from 1982 to 1990.*
Four infants with unknown sex were excluded. Number of
infants is given in parentheses

Year	Tigounatine		Akfadou	
1982	60	(5)	40	(5)
1983	63	(8)	50	(6)
1984	50	(2)	75	(8)
1985	57	(14)	67	(6)
1986	55	(11)	63	(8)
1987	56	(9)	25	(12)
1988	58	(12)	33	(6)
1989	59	(17)	0	(4)
1990	44	(18)	50	(8)
Overall	55	(96)	46	(63)

Table 6.6. *Infant mortality rate. Number of infants is given*
in parenthesis

	Tigounatine		Akfadou	
All infants	0.23–0.27	(78–82)	0.38–0.40	(58–60)
Males	0.26	(31)	0.30	(27)
Females	0.24	(42)	0.48	(23)
Born from primiparous	0.24	(17)	0.50	(10)
Born from multiparous	0.25	(56)	0.40	(43)

November (78% at Tigounatine and 77% at Akfadou), whereas infant mortality was relatively low during the winter months. Springtime mortality was higher at Akfadou than at Tigounatine (39% vs 6%) with the greatest disparity in June, when 24% vs 0% of annual deaths occurred (Ménard & Vallet, 1993*b*).

Transfers and disappearances

Migration rates

At Tigounatine 55 transfers or disappearances were noted, including 29 immigrants, 14 emigrants, 3 transfers between the new groups SM, LO and UL, and 9 disappearances. At Akfadou, 20 transfers or disappearances were noted, including 7 immigrants, 3 emigrants and 10 disappearances (Fig. 6.2).

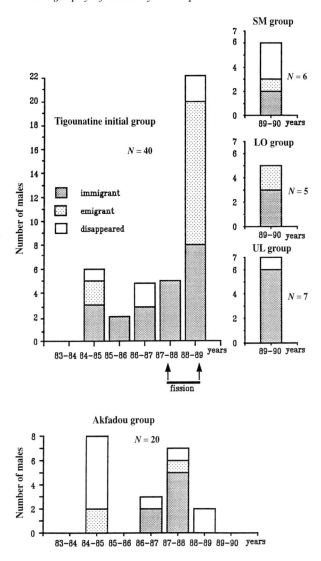

Fig. 6.2. Distribution of transfers of adult and sub-adult males in the two habitats.

Ménard & Vallet (1993*b*) found that the rate of emigration for males over 3 years old neither differed between groups nor showed a relationship with the increase in group size. Nevertheless, the overall proportion of immigrants was significantly higher in the Tigounatine group than in the Akfadou group. The proportion of immigrants to the Akfadou group did not vary according to group size, whereas the Tigounatine group received the greatest proportion of immigrants as it grew in size in the period leading up to fission. Indeed, 45% ($N = 29$) of all immigration occurred during the fission process, between 1987 and 1989. As a result of transfers, only three of the nine resident adult males observed in 1983 at Tigounatine still belonged to one or other of the three new groups in 1990, while at Akfadou two of the nine resident males still belonged to the group in 1990.

Age of migrants

At Tigounatine 78% ($N = 46$; excluding animals which had disappeared) of migrants were adult males (> 5 years old), 54% being more than 8 years old, whereas 22% were sub-adult males (3–5 years old) (Table 6.7). Nevertheless, the rate of immigration did not differ significantly with the age of animals (Kolmogorov–Smirnov $D_{max} = 0.20$, $N = 17$, $p > 0.05$). At Akfadou all migrants ($N = 10$) were adult males, most of them over 8 years old. Even if ten other animals that had disappeared were considered as emigrants, 75% of the migrants were adult males. Given uncertainties regarding the status of animals that had disappeared, the relationship between age and the rate of emigration could not be tested.

At Tigounatine five of the 16 males born between 1980 and 1985 emigrated before adulthood, three disappeared, while eight others (5–9 years old) still belonged to one or other of the three groups in July 1990. At Akfadou seven out of the 14 males born between 1980 and 1985 disappeared before adulthood, whereas seven others (from 5–10 years old) belonged to the group in 1990. Therefore, at both sites, 50% of males reached adulthood in their natal group and some were still in the same group when 9 years old.

Pattern of transfers

At Tigounatine, at least 11 of the 23 males that left the group joined neighbouring groups, one in one group and ten in a second group; three others were only temporary emigrants. At Akfadou, only two out of 13 males that disappeared could be located, each in a different neighbouring group; one was a temporary emigrant.

At both sites, immigrants stayed in the studied groups for periods ranging from less than one month to more than five years; twenty-six (72%) stayed at least a year. Out of a total of 40 observed migrants of different identity, five were seen to transfer twice and all of the emigrants that left their original group for more than a month (17) were never again seen with that group. Therefore, males that transfer several times during their life probably visit several different groups.

Table 6.7. *Age-specific male migrations. Rate of emigration: number of emigrants/group's number of males × years*

Age of migrants (years)	No. of migrants	No. of immigrants	No. of emigrants	No. of males × years	Emigration rate
Tigounatine					
3–4	2	1	1	20	0.05
4–5	8	6	2	14	0.14
5–8	11	8	3	46	0.07
>8	25	14	11	66	0.17
Total	46	29	17	146	0.12
Akfadou					
3–4	0	0	0	16	0.0
4–5	0	0	0	12	0.0
5–8	3	3	0	24	0.0
>8	7	4	3	41	0.07
Total	10	7	3	93	0.03

No. of males × years is the cumulative number of males observed over the years.

Periods of transfers

Almost all transfers occurred outside the birth season, except in the cases of two migrants that temporarily visited the studied groups during June in Akfadou. At least 40% and 62.5% of transfers (at Tigounatine and at Akfadou, respectively) took place between October and February, during the mating season.

Habitat characteristics and use by monkeys

Availability of resources

Ménard & Vallet (1988) have described several differences between the two habitats studied (see also Table 6.8).

1. The Akfadou forest, at a lower altitude than that at Tigounatine, has a lower rainfall, a drier summer and less winter snow.
2. The ranges of the two groups are both forested but the forest of Tigounatine has 74% tree cover dominated by two evergreen species (*Cedrus atlantica* and *Quercus ilex*), while Akfadou has 93% tree cover composed mostly of two deciduous species (*Quercus faginea* and *Quercus afares*).

Table 6.8. *Ecological characteristics of the two sites*

Habitat type	Tigounatine–Djurdjura (cedar–oak forest)	Akfadou (deciduous oak forest)
Altitude (m)	1200–1900	800–1300
Annual temperatures, 1914–38		
Mean minima (°C)	+4.5	+9.0
Mean maxima (°C)	+16.3	+21.1
Mean annual precipitations (mm)		
1914–38	1410	1010
1968–76	1000	600
Snowy period	December–March	January–February
Composition of the habitat		
Forest	74%:	93%:
	Cedrus atlantica, 50%	*Quercus afares*, 54%
	Quercus ilex, 24%	*Quercus faginea*, 36%
Scrub areas	2%	5%
Grassland areas	24%	2%
Specific diversity		
Trees	1.80	1.08
	(D = 0–5)	(D = 0–4)
Shrubs	3.10	1.61
	(D = 0–26)	(D = 0–8)
Ground vegetation	9.68	9.26
	(D = 0–240)	(D = 0–152)
Monthly percentage herb cover	0–49	0–33
<20%	August–October	April–March
	February–March	

Data on temperatures and precipitations are provided by Seltzer (1946) and the Division d'études de recherche hydraulique of Algeria.

3. Lists of 271 and 164 plant species have been recorded for Tigounatine and Akfadou, respectively. There were 65 species of herbaceous plants common to the two habitats, representing 27% ($N = 240$) of species at Tigounatine and 43% ($N = 152$) at Akfadou. Moreover, species diversity is greater at Tigounatine than at Akfadou for all vegetation types. Consequently, the Tigounatine forest represents a habitat buffered against a drastic fall in the production of any single food species, a wider choice of other species being available to make up the temporary deficit.

4. Marked seasonal variations are found in both habitats, but the production of plant resources is more evenly spread over the course of the year at Tigounatine due to phenological asynchrony among the various species (particularly shrubs).

Spring re-growth at Akfadou begins about three weeks earlier than at Tigounatine. Spring production is abundant but concentrated in two months (April–May) when oak and *Cytisus triflorus* flower and proliferation of caterpillars takes place. At Tigounatine, production extends until July owing to new leaf growth and flowering of shrubs (April–June), oak flowering and the proliferation of caterpillars (June). Herbaceous plant cover is greatest in May in both habitats but is higher at Tigounatine than at Akfadou (49% vs 33%). It is greatly reduced (< 20%), or negligible by the end of spring, but the dry summer begins earlier and lasts longer at Akfadou as the autumn rains are much less than at Tigounatine. Moreover, when the lower strata of vegetation (herbaceous cover and shrubs) are covered with snow the two evergreen tree species at Tigounatine provide an important food source absent from the deciduous oak forest of Akfadou.

For both habitats, most inter-annual variation in resources was related to the occasional springtime proliferation of caterpillars that feed on oak leaves, resulting in little or no acorn production in the autumn of the same year, when generally less than 32% of oaks fruit and then yield only 3–4 acorns per tree. Invasions of caterpillars occurred in 1983, 1986 and 1987 in Tigounatine, and in 1983, 1984 and 1985 in Akfadou.

At first sight, the Akfadou oak forest with its denser structure looks as if it could provide more food for a population of monkeys than the more open cedar–oak forest of Tigounatine. Yet, the results show that, overall, the deciduous oak forest is a habitat with poorer and less reliable nutritional resources than the mixed cedar–oak forest, and where the dry summer, particularly the month of June, is always a critical period even in years of good acorn production.

Diet

The Barbary macaques that we studied modified their diet to suit different habitats (Ménard, 1985; Ménard & Vallet, 1986). At Akfadou 63% of food taken came from trees and shrubs, compared with 41% at Tigounatine (Ménard & Vallet, 1988). This probably reflected the difference in extent of tree and shrub cover between the two habitats. Similarly, the proportion of herbaceous plants in their diet was higher at Tigounatine, where there was more herbaceous cover, than at Akfadou (59% vs 37%). The macaques of both study groups were essentially seed and leaf eaters (see Table 6.9); seeds and leaves together represented an annual average of 75% and 59% of time spent feeding at Tigounatine and Akfadou, respectively. At both sites, acorns and leaves of herbaceous plants made up the largest part of seed and leaf consumption. However, at Tigounatine the proportion of tree leaves

Table 6.9. *Diet and food selectivity*

Habitat type	Tigounatine–Djurdjura (cedar–oak forest)	Akfadou (deciduous oak forest)
Dietary composition (%)		
Seeds	26.7	32.1
Acorns	14.2	26.2
Herbs	11.3	5.9
Leaves	48.1	27.3
Trees	12.1	0.1
Shrubs	0.9	8.7
Herbs	35.1	18.5
Animal prey	5.6	10.5
Caterpillars	5.5	9.5
Lichens	1.9	14.2
Roots	7.7	6.9
Others	5.9	10
Specific diversity (D)		
Trees	2.06	2.02
	($D = 0$–4)	($D = 0$–3)
Shrubs	3.40	2.12
	($D = 0$–19)	($D = 0$–6)
Ground vegetation	7.78	2.33
	($D = 0$–101)	($D = 0$–63)
Selectivity (%)		
Preferred species		
Proportion (N)[a]	9 (24)	6 (10)
Abundance on home range	20.3	3.1
Avoided species		
Proportion (N)	8 (21)	11 (17)
Abundance on home range	46.6	33.0
Neutral species		
Proportion (N)	2 (5)	1 (2)
Abundance on home range	25.8	59.8
Non-classified		
Proportion (N)	82 (221)	82 (132)
Abundance on home range	7.3	4.1

D, is diversity of species according to Simpson's formula (Levins, 1968).
[a]Number of species in parentheses.

was more important because of the winter consumption of cedar leaves, with lichens being the comparable winter food at Akfadou. Animal prey consisted mainly of caterpillars. In parallel with annual variations of available foods, the amount of caterpillars eaten by the macaques varied according to the year and when the former were absent the monkeys supplemented their diet with flowers of oak and *Cytisus triflorus* at Akfadou and herbaceous leaves at Tigounatine.

Seasonal variations in diet are marked in both habitats with two folivorous phases, one in spring and one in winter, and one granivorous phase in summer and autumn. According to the month, the monkeys use 8 to 27 plant species of which only 2 to 3 were important staples that made up 51–93% of the diet at Akfadou and 43–85% of the diet at Tigounatine.

In connection with the higher species diversity of food resources at Tigounatine than at Akfadou we noted that species diversity in all three categories of food plants at Tigounatine was higher than at Akfadou (Table 6.9).

Selectivity

Ménard & Vallet (1988) have shown that out of 271 species of plants listed at Tigounatine and 161 at Akfadou 47% of the former and 48% of the latter are found in the monkeys' diet over the course of a year. Overall the Tigounatine monkeys avoided a smaller proportion of available species than those at Akfadou (8% vs 11%) whereas a larger proportion of species were selected (9% vs 6%). Species avoided at Tigounatine represent a larger proportion of resources than do those avoided at Akfadou (47% vs 33%), whereas the species selected at Tigounatine represent a resource seven times as abundant as those selected at Akfadou (20% vs 3%). It, therefore, appears that the degree of selectivity exercised on plant foods by the macaques is much greater at Tigounatine.

Home range use

As Fig. 6.3 indicates, the curve representing the area of home range flattens out at the end of 1984 for both sites after 1600 and 1800 observations, evenly distributed over a year, at Tigounatine and Akfadou, respectively. In 1984, 74.2% (2.79 km^2) and 77.1% (3.27 km^2) of the 1990 home range size was then known at Tigounatine and Akfadou, respectively.

The initial Tigounatine group, which had varied from 46 to 73 individuals between 1983 and 1988 used an area of 3.76 km^2 (Fig. 6.3). The three groups that emerged from the fission of the initial group, a total of 106 animals, continued to use the same area but as three overlapping home

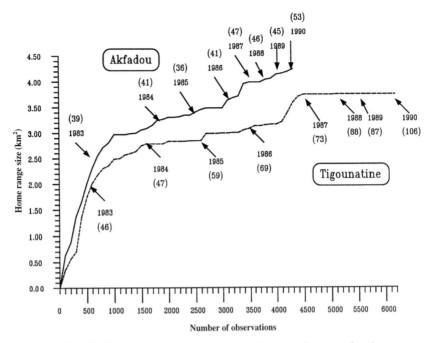

Fig. 6.3. Cumulative curves of the home range size over the years for the groups studied at Tigounatine and at Akfadou. Corresponding number of troop members is indicated in parentheses. For 1989 and 1990, home range corresponds to the total area used by the three new groups at Tigounatine.

ranges. The degree of overlap of the three ranges was still large in 1989–90 with between 48% and 80% of any range being shared. However, the three home ranges still lay within the limits of the home range of the initial group.

The Akfadou group, with 39 to 45 animals between 1983 and 1990, used an area of 4.24 km^2 as their home range. This range expanded progressively as they explored new areas.

Even though the total number of animals for the Tigounatine groups had more than doubled between July 1984 and July 1990, the area taken up by their home ranges had expanded relatively little (46%). An abrupt expansion was noted in 1987 when the acorn production was negligible and the monkeys moved into new areas of grassland to look for food. At Akfadou, however, the home range expanded by 40% between July 1984 and July 1990, even though the size of the group stayed relatively stable. Thus, the density of monkeys at Tigounatine was twice as high as that at Akfadou (28.2 individuals/km^2 vs 12.5 individuals/km^2). At both sites this is probably an underestimate because of the degree of overlap with neighbouring groups, but as this differs little between the two sites the comparison is still valid.

Discussion

Resource availability and macaque use

The two groups of Barbary macaques studied live in habitats with totally different ecological characteristics; there are large seasonal variations in food availability, linked with the temperate climate, as well as large inter-annual variations. In response to habitat type, the Barbary macaque adjusts its diet according to the nutritional resources available.

In the cedar–oak forest of Tigounatine, there is less tree cover and there are more extensive areas of open grassland than in the deciduous oak forest of Akfadou. However, Tigounatine shows a higher number and diversity of plant species than Akfadou. At the same time, the Barbary macaques at Tigounatine have a higher diet diversity and eat more leaves and seeds of herbaceous plants and fewer acorns than those living in the oak forest. In both habitats, monkeys are able to shift from one month to the next between a carnivorous diet based on caterpillars, to a folivorous diet composed of herbaceous leaves, or to a granivorous diet based on acorns. Moreover, there are almost no caterpillars in the diet in some years, and the contribution of acorns can be spread over 4–10 months depending on their production.

Although the Barbary macaque is considered to be an eclectic feeder, the animals are selective in the two habitats studied, as only 48% (Tigounatine) and 47% (Akfadou) of available plant species are actually included in the diet. The degree of food selectivity by the animals, higher overall at Tigounatine than at Akfadou, seems to be a response to the conditions of a less limiting habitat. Moreover, the resources preferred by the monkeys at Tigounatine are three times more abundant in their home range than at Akfadou. In addition to these differences in overall food availability, resource availability is more evenly distributed throughout the year at Tigounatine than at Akfadou, owing mainly to the shorter and less intense summer at the former site.

Demography

In the two habitats studied, the groups showed the multimale–multifemale structure typical of the species, with a relatively balanced sex ratio, the exception being the temporary formation of a one-male group. The proportion of immature animals, which varies from 0.41 to 0.54 according to the group and the year, is comparable to that found by Mehlman (1989) in Morocco (0.47).

Inter-habitat variation of seasonal resource distribution affects the birth season (which is earlier by about 15 days at Akfadou than at Tigounatine) and the critical period for infant survival. The latter, which is during the dry season at both sites, when more than 77% of infant deaths occurred, is a month shorter at Tigounatine than at Akfadou. In addition, infant mortality is particularly high in June at Akfadou, the beginning of the dry season when resources are severely reduced. However, infant mortality is low during the winter.

The demographic parameters are subject to large inter-annual variation and inter-group differences were not statistically significant. Nevertheless, the proportion of females at birth tended to be higher and infant mortality lower (especially for females) at Tigounatine than at Akfadou (18.6% vs 5.1%) and is essentially the result of differences in the natality : mortality ratio. The high rate of increase of the Tigounatine group is similar to that of provisioned macaque groups (13 to 21%, Koford, 1965; Drickamer, 1974; Sugiyama & Ohsawa, 1982; Malik, Seth & Southwick, 1984; Paul & Kuester, 1988), whereas at Akfadou it is comparable with that of wild macaque groups (0–4.9%, Dittus, 1977; Teas et al., 1981; Sugiyama & Ohsawa, 1982; Southwick & Siddiqi, 1988). This last point, as well as the aforementioned ecological differences, suggest that the cedar–oak forest is a more favourable habitat for the development of monkey groups than the oak forest of Akfadou. The fact that the Tigounatine group uses a smaller home range, and has a population density more than double (28 vs 13 individuals/km^2) that at Akfadou, supports this hypothesis. Moreover, the carrying capacity of the habitat does not seem to have been reached at Tigounatine, since the number of monkeys continues to increase without a noticeable change in the size of the area used.

Nevertheless, it is difficult to show that the ecological factors particular to each habitat are responsible for the observed demographic differences because of the large inter-annual variations in resource availability. As highlighted by Dittus (1977) and Altmann, Hausfater & Altmann (1988), infant mortality is the parameter most sensitive to environmental change. Ménard & Vallet (1993b) have indicated that in both these habitats, mortality was positively correlated with the frequency of caterpillar plagues and subsequent reduction of the acorn crop. The conclusion that can be drawn from this is that the more frequent the caterpillar invasions of a habitat, the worse the habitat becomes for the Barbary macaque population. These results illustrate the importance of long-term studies, especially in the case of species inhabiting temperate habitats with strong seasonality and strong inter-annual variations in resource production.

The comparison of demographic parameters between wild and captive groups clearly shows the effect of food availability on female fertility (age of first breeding, birth rate, inter-birth interval) and on infant mortality. The birth rate is lower for the studied groups (0.56–0.65) than for captive groups (0.75; Paul & Kuester, 1988) and is also lower whatever the age of the females; age at first breeding is later (5.2 and 5.3 years, this study, vs 4.5–4.9 in captivity; Fa, 1984; Paul & Kuester, 1988). The fertility pattern as a function of age differs little between the wild (this study) and captivity (Paul & Kuester, 1988), showing highest fertility between 8 and 15 years. However, the decline in fertility appears less obvious in the studied groups than in captive groups, probably because of lower female life expectancy in the wild.

The most frequent inter-birth interval in the study groups is one year (57% and 79% of cases at Tigounatine and Akfadou, respectively). Intervals of two years are, however, more frequent than in captivity (17–42%, this study, vs 12%, Burton & Sawchuk, 1982; Paul & Thommen, 1984). As Paul & Thommen (1984) observed, the reproductive status of the mother or the sex of the infant did not appear to affect the interval between births. Nevertheless, the loss of an infant tended to reduce the ensuing interval. As Altmann, Altmann & Hausfater (1978) emphasised, the effect of such an occurrence should be less marked in a seasonally breeding species, whereas the interval is cut shorter in the case of a species that breeds nearly all year round, such as *Papio cynocephalus* (Altmann *et al.*, 1977). However, the loss of an infant has the effect of bringing forward the time of birth within the season, as was noted by Paul & Thommen (1984).

Dispersal

The different rates of increase of the two groups are expressed in different patterns of dispersion. The large increase of the Tigounatine group led to its division into three new groups. This division facilitates the introduction of males from outside the old group and explains the higher immigration rate at Tigounatine than at Akfadou. Contrary to the suggestion made by Paul & Kuester (1988), rate of emigration is not influenced by group size.

In both study groups, the male Barbary macaques do not migrate before they are 3.5 years old and, unlike other macaque species, they migrate late.

Fifty per cent reach sexual maturity in their natal group and some are still breeding there at the age of 9; most of the observed migrants are over 5 years old. However, the emigration rate does not vary significantly according to age. In the study groups, the migration rate of sub-adult males (0.05 at age 3–4 years) is lower than in captivity (0.136; Paul & Kuester, 1988), whereas the migration rate of adults over 8 is higher (0.17, this study, vs <0.07; Paul & Kuester, 1988). Contrary to the hypothesis of Paul & Kuester (1985, 1988) according to which 'if Barbary macaques avoid mating with close relatives, migration rate in the wild should be higher than in Salem [captive colony]', the average annual migration rate observed (0.12–0.16, Ménard & Vallet, 1993b) differs little from that of captive groups (0.07). In fact, as Melnick & Kidd (1983) suggested, because of a longer inter-birth interval and a shorter female life expectancy in the wild, the length of lineages is relatively shorter and the number larger for captive groups of comparable size. Therefore, group composition, more than group size, has a determining role to play in the male migration rate, insofar as this avoids matings with close kin. Nearly all migrations took place during the mating season and males joined neighbouring groups, as has already been observed in other species (Pusey & Packer, 1987).

Acknowledgements

The authors thank the Ministère de l'Hydraulique, de l'Environnement et des Forêts and Mr Gassi (Director of the Djurdjura National Park, Algeria) for permission to carry out this field work. We greatly appreciate the hospitality and logistic help of Mr Messaoud and his staff. We also thank D. Dérian for collaboration in collecting home range data during spring 1990. Our special gratitude to Dr Annie Gautier-Hion (CNRS) and Professor F. Bourlière for their persistent support at all stages of this work. This research was supported in part by the Centre National de la Recherche Scientifique (France), the Directions pour la Recherche Scientifique (Algeria), the Institut National d'Enseignement Supérieur de Biologie (Tizi Ouzou, Algeria) and National Geographic Society grant no. 3766–88.

References

Altmann, J. (1974). Observational study of behavior: sampling methods. *Behaviour*, **49**, 227–66.

Altmann, J., Altmann, S. A., Hausfater, G. & McCuskey, S. A. (1977). Life history of yellow baboons: physical development, reproductive parameters, and infant mortality. *Primates*, **18**, 315–30.

Altmann, J., Altmann, S. A. & Hausfater, G. (1978). Primate infant's effect on mother's future reproduction. *Science*, **201**, 1028–9.

Altmann, J., Hausfater, G. & Altmann, S. A. (1985). Demography of Amboseli baboons, 1963–1983. *American Journal of Primatology*, **8**, 113–25.

Altmann, J., Hausfater, G. & Altmann, S. A. (1988). Determinants of reproductive success in savannah baboons, *Papio cynocephalus*. In *Reproductive success*, ed. T. H. Clutton-Brock, pp. 403–18. Chicago: University of Chicago Press.

Bourlière, F. (1979). Significant parameters of environmental quality for nonhuman primates. In *Primate ecology and human origins: ecological influences on social organization*, ed. I. S. Bernstein & E. O. Smith, pp. 23–46. New York: Garland STPM Press.

Burton, F. D. & Sawchuk, L. A. (1982). Birth intervals in *M. sylvanus* of Gibraltar. *Primates*, **23**, 140–4.

Deag, J. M. (1974). A study of the social behaviour and ecology of the wild Barbary macaque *Macaca sylvanus* L. PhD thesis, University of Bristol.

Dittus, W. P. J. (1977). The social regulation of population density and age–sex distribution in the toque macaque. *Behaviour*, **63**, 281–322.

Drickamer, L. C. (1974). A ten-year study of reproductive data for free-ranging *Macaca mulatta*. *Folia Primatologica*, **21**, 61–80.

Drucker, G. R. (1984). The feeding ecology of the Barbary macaque and cedar forest conservation in the Moroccan Moyen Atlas. In *The Barbary macaque: a case study in conservation*, ed. J. E. Fa, pp. 135–64. New York: Plenum Press.

Fa, J. E. (1984). Habitat distribution and habitat preference in Barbary macaques (*Macaca sylvanus*). *International Journal of Primatology*, **5**, 273–85.

Fa, J. E., Taub, D. M., Ménard, N. & Stewart, P. J. (1984). The distribution and current status of the Barbary macaque in North Africa. In *The Barbary macaque – a case study in conservation*, ed. J. E. Fa, pp. 79–111. New York: Plenum Press.

Gautier-Hion, A. Gautier, J. P. & Quris, R. (1981). Forest structure and fruit availability as complementary factors influencing habitat use by a troop of monkeys (*Cercopithecus cephus*). *Revue d'Écologie (Terre Vie)*, **35**, 511–36.

Hladik, C. M. (1977). A comparative study of the feeding strategies of two sympatric species of leaf monkeys: *Presbytis senex* and *Presbytis entellus*. In *Primate ecology: studies of feeding and ranging behaviour in lemurs, monkeys and apes*, ed. T. H. Clutton-Brock, pp. 324–54. London: Academic Press.

Iwamoto, T. (1978). Food availability as a limitation factor on population density of the Japanese monkey and Gelada baboon. In *Recent advances in primatology*, ed. D. J. Chivers & J. Hebert, pp. 287–303. London: Academic Press.

Koford, C. B. (1965). Population dynamics of rhesus monkeys on Cayo Santiago. In *Primate behavior: field studies of monkeys and apes*, ed. I. DeVore, pp. 160–74. New York: Holt, Rinehart & Winston.

Levins, R. (1968). *Evolution in changing environments*. Princeton: Princeton University Press.

Malik, I., Seth, P. K. & Southwick, C. H. (1984). Population growth of free-ranging rhesus monkeys at Tughlaqabad. *American Journal of Primatology*, **7**, 311–21.

Ménard, N. (1985). Le régime alimentaire de *Macaca sylvanus* dans différents habitats d'Algérie: I Régime en chênaie décidue. *Revue d'Écologie (Terre Vie)*, **40**, 451–66.

Ménard, N. & Vallet, D. (1986). Le régime alimentaire de *Macaca sylvanus* dans différents habitats d'Algérie: II Régime en forêt sempervirente et sur les sommets rocheux. *Revue d'Écologie (Terre Vie)*, **41**, 173–92.

Ménard, N. & Vallet, D. (1988). Disponibilités et utilisation des ressources par le magot (*Macaca sylvanus*) dans différents milieux en Algérie. *Revue d'Écologie (Terre Vie)*, **43**, 201–50.

Ménard, N. & Vallet, D. (1993a). Dynamics of fission in a wild Barbary macaques group (*Macaca sylvanus*) in Algeria. *International Journal of Primatology*, **14**, 479–500.

Ménard, N. & Vallet, D. (1993b). Population dynamics of *Macaca sylvanus* in Algeria: an 8-year study. *American Journal of Primatology*, **30**, 101–8.

Ménard, N., Vallet, D. & Gautier-Hion, A. (1985). Démographie et reproduction de *Macaca sylvanus* dans différents habitats en Algérie. *Folia Primatologica*, **44**, 65–81.

Mehlman, P.T. (1989). Comparative density, demography, and ranging behavior of Barbary macaques (*Macaca sylvanus*) in marginal and prime conifer habitats. *International Journal of Primatology*, **10**, 269–92.

Melnick D.J. & Kidd, K.K. (1983). The genetic consequences of social group fission in a wild population of rhesus monkeys (*Macaca mulatta*). *Behavioral Ecology and Sociobiology*, **12**, 229–36.

Norton, G.W., Rhine, R.J. & Wynn, R.D. (1987). Baboon diet: a five year study of stability and variability in the plant feeding and habitat of the yellow baboons (*Papio cynocephalus*) of Mikumi National Park, Tanzania. *Folia Primatologica*, **48**, 78–120.

Oates, J.F. (1977). The guereza and its food. In *Primate ecology, studies of feeding and ranging behaviour in lemurs, monkeys and apes*, ed. T.H. Clutton-Brock, pp. 275–321. London: Academic Press.

Paul, A. & Kuester, J. (1985). Intergroup transfer and incest avoidance in semifree-ranging Barbary macaques (*Macaca sylvanus*) at Salem (FRG). *American Journal of Primatology*, **8**, 317–22.

Paul, A. & Kuester, J. (1988). Life-history patterns of Barbary macaques (*Macaca sylvanus*) at Affenberg Salem. In *Ecology and behavior of food-enhanced primate groups*, ed. J.E. Fa & C.H. Southwick, pp. 199–228. New York: Alan R. Liss.

Paul, A. & Thommen, D. (1984). Timing of birth, female reproductive success and infant sex ratio in semifree-ranging Barbary macaques (*Macaca sylvanus*). *Folia Primatologica*, **42**, 2–16.

Pusey, A.E. & Packer, C. (1987). Dispersal and phylopatry. In *Primate societies*, ed. B.B. Smuts, D.L. Cheney, R.M. Seyfarth, R.W. Wrangham & T. Struhsaker, pp. 250–66. Chicago: University of Chicago Press.

Seltzer, P. (1946). *Le climat d'Algérie*. Alger: Carbonel.

Southwick, C.H. & Siddiqi, M.F. (1988). Partial recovery and a new population estimate of rhesus monkey population in India. *American Journal of Primatology*, **16**, 187–97.

Southwick, C., Siddiqi, M.F. & Oppenheimer, J.R. (1983). Twenty-year changes in rhesus monkey populations in agricultural areas of northern India. *Ecology*, **64**, 434–9.

Sugiyama, Y. & Ohsawa, H. (1982). Population dynamics of Japanese monkeys with special reference to the effect of artificial feeding. *Folia Primatologica*, **39**, 238–63.

Taub, D.M. (1977). Geographic distribution and habitat diversity of the Barbary macaque *M. sylvanus* L. *Folia Primatologica*, **27**, 108–33.

Teas, J., Richie, T. L., Taylor, H. G., Siddiqi, M. F. & Southwick, C. H. (1981). Natural regulation of rhesus monkey populations in Kathmandu, Nepal. *Folia Primatologica*, **35**, 117–23.

Waser, P. (1977). Feeding, ranging and group size in the mangabey *Cercocebus albigena*. In *Primate ecology: studies of feeding and ranging behaviour in lemurs, monkeys and apes*, ed. T. H. Clutton-Brock, pp. 183–222. London: Academic Press.

7

Macaca nigrescens: grouping patterns and group composition in a Sulawesi macaque

A. KOHLHAAS AND C. H. SOUTHWICK

Introduction

Primates exhibit a wide variety of social and grouping patterns (Crook & Gartlan, 1966; Eisenberg, Muckenhirn & Rudran, 1972). The genus *Macaca* is generally characterized as occurring in large stable multimale–multifemale groups. In part this characterization is due to the many studies on commensal species, especially *M. mulatta*, *M. fascicularis*, and *M. fuscata*, which often receive food supplements from humans (Fa & Southwick, 1988). Relatively fewer studies have been done on the forest-dwelling species and especially those wholly dependent on natural food sources. Some variation in macaque grouping patterns has been found in recent studies of *M. fuscata* (Fukuda, 1989), *M. nemestrina* (Caldecott, 1986; Robertson, 1986; Oi, 1990), and *M. sylvanus* (Ménard *et al.*, 1990). The purpose of this chapter is to describe grouping patterns of *Macaca nigrescens*, a little-known species of forest macaque endemic to the island of Sulawesi.

Macaca nigrescens is one of seven little-studied macaque species endemic to the island of Sulawesi (formerly Celebes) in Indonesia. Virtually this entire species occurs within Dumoga–Bone National Park in the northern peninsula of Sulawesi (Fig. 7.1). Much previous research on the Sulawesi macaques has focused on the taxonomy of the group with continuing debate as to whether the Sulawesi macaques constitute one (Thorington & Groves, 1970), four (Groves, 1980), or seven species (Fooden, 1969, 1976, 1980; Albrecht, 1978; Takenaka *et al.*, 1987*a*,*b*). Here, we follow the taxonomy described by Fooden (1969) in recognizing *M. nigrescens* as a separate species. The alternative would be to consider it to be conspecific and a sub-species of *M. nigra*, thus *M. nigra nigrescens* (Groves, 1980). Although there is evidence of gene flow between the Sulawesi macaques (Groves, 1980; Ciani *et al.*, 1989), most recent studies accept the separation

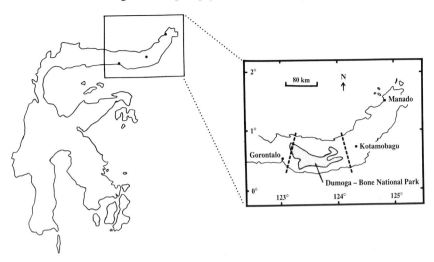

Fig. 7.1. Dumoga–Bone National Park and the geographical range of *Macaca nigrescens*. Dashed lines indicate approximate boundaries of *M. nigrescens* range.

of seven species of Sulawesi macaques (Hamada *et al.*, 1988; Watanabe, Lapasere & Tantu, 1991) and the recognition of *M. nigrescens* as distinct from *M. nigra*.

Very little has yet been published on *M. nigrescens*, in the wild. Bismark (1982*a*,*b*) published a 15-day study of *M. nigrescens* grouping patterns and ecology. His groups ranged in size from 9 to 28 individuals with a mean of 16.5; the sex ratio averaged 1 : 2.2 (male : female). Groups were seen to split into sub-groups of 1 to 17 individuals for foraging. Watanabe & Brotoisworo (1982) observed *M. nigrescens* for five days and recorded multimale–multifemale groups of 20–40 individuals. Sugardjito *et al.*, (1989) reported average group sizes of 6.2 to 22 individuals in different areas of *M. nigrescens'* range. Most commonly, small groups of 6–8 individuals were seen; the adult male : female ratio was 1 : 2.1. Only 14.6% of adult females had infants and only 24.7% of the population was immature (Sugardjito *et al.*, 1989).

From an ecological point of view, *M. nigrescens* and the other Sulawesi macaques are of special interest because they are forest-dwelling, non-commensal macaques that evolved in isolation from the other anthropoid primates. *Macaca nigrescens* is especially interesting because it is one of the more derived forms of Sulawesi macaque, and one of the least studied. The long geographical isolation of Sulawesi from mainland Asia and the rest of the Indonesian/Malaysian archipelago has resulted in a high degree of endemism, including all of its macaque species and a large proportion of its other fauna and flora. This has also resulted in less floral and faunal diversity as compared to other South-East Asian forests on mainland and continental shelf Asia (Borneo, Sumatra, and Java). Therefore, in contrast to most of the other South-East Asian primates, the Sulawesi macaques (1) have no primate competitors, except as they occur parapatrically in some narrow hybrid zones, (2) have relatively few predators, and (3) experience less floral diversity than primates in the other South-East Asian forests (Whitmore & Sidiyasa, 1986), especially in the number of dipterocarp species.

In this chapter, we describe the grouping patterns observed in a natural population of *M. nigrescens* and discuss the possible implications of these patterns, especially regarding macaque ecology. The discussion also includes consideration of the problems of censusing primates, especially those in large groups, in tropical rainforest.

Methods

Study site

This study was conducted in Dumoga–Bone National Park, North Sulawesi, Indonesia (Fig. 7.1). The specific site was located at 0° 30′ N and 124° E, 100–200 m above mean sea level (AMSL), and near the village of Toraut. We established 22 km of trails intersecting every 100–200 m in a 140 ha area bordered by the Toraut and Tumpah rivers on three sides. The natural vegetation is lowland rainforest: larger tree species included *Dracontomelon dao, Ficus* spp., *Celtis philippensis, Sandoricum koetjapi*, and *Bischoffia javanica*. Other abundant tree species included *Pometia pinnata, Nephelium mutabile, Myristica* spp., *Polyalthia* spp., and *Aglaia* spp. Habitat types ranged from relatively undisturbed rainforest to patches of grasses and other secondary growth (*Piper aduncum, Mallotus* spp., *Macaranga* spp.). Although the majority of the site was 'relatively undisturbed' rainforest, with some trees 30–40 m in height, it must be noted that this forest was characterized by many small gaps that might be natural or remnants of earlier disturbance. We do not know the extent of earlier human disturbances. Small-scale disturbance by humans has occurred in the form of palm leaf, fruit, and rattan collection and, more rarely, palm and small tree harvesting. A few grass and secondary growth areas are the remnants of recent farms or gardens.

Data collection and analysis

Data were collected from April 1989 to June 1990 on a partially habituated population of *M. nigrescens*. This 14-month study period followed six months of field work in 1987 and 1988 on site selection, trail construction, and preliminary data collection. 'Partially habituated' as used here indicates that the monkeys usually tolerated our presence as long as we were at sufficient distance (typically 20–30 m) and degree of visibility. However, their tolerance levels varied from approaching us within 5–10 m to intolerance, in which they would flee at the first glimpse or sound of our presence. Observation times were always the maximum allowed by the monkeys, daylight and weather, and consistent with collecting behavioral data. These times ranged from a few seconds to 12 hours. We avoided creating undue stress to the monkeys and would terminate observations on monkeys overtly disturbed by our presence.

In this study, 'group' was used to refer to any aggregation of monkeys and even to single monkeys. Two basic methods were used to obtain group size estimates. Firstly, when groups traveled they often progressed through a bottle-neck in the vegetation and a reasonably accurate count could be obtained at that time. This is a *progression count*. Secondly, when a group was observed but was not seen to move via a bottle-neck, group size was estimated by calculating the conservative sum of the monkeys that had been seen individually throughout the observation time. This type of count is henceforth referred to as an *observation count*. Progression counts are believed to be the most accurate; observation counts are the best alternative in the absence of a progression count. Observation counts should increase in accuracy with observation duration. However, their accuracy will be limited by the observer's ability to distinguish individuals. As the monkeys in this study were not identified individually, this was of special concern. Keeping track of their movements made up for some of this limitation. Each group size estimate was coded according to the method used (0 = observation count, 1 = progression count) and sighting duration (1 = ≤ 15 min, 2 = 15 min to 1 h, 3 = 1–3 h, 4 = > 3 h).

During any sighting, monkeys were identified as specifically as possible to an age–sex category. The categories and their physical descriptions are as follows:

Adult male – large size, robust, crown hair usually long, male genitalia.
Adult female – 50–75% size of adult male, usually has some visible pink skin by ischial callosities and perineally.
   (a) Estrous – skin swollen and red around ischial callosities and perineally.
   (b) Non-estrous – skin dull-colored and not swollen around ischial callosities, may or may not have a small swelling perineally.

Sub-adult male – between adult female and adult male size, not as robust as adult male, male genitalia.

Sub-adult female – same size as adult female, lacking obvious nipples, no pink skin by ischial callosities or perineally.

Juvenile – from infant to adult female size, moves independently.

Infant – small size, carried during travel, spends majority of time on belly of adult female,

(a) Dark-faced – older infant with a dark face.

(b) White-faced – very young infant with a white or gray face.

Results

The following results are based on 372 separate group sightings that occurred from April 1989 to June 1990. These sightings were of individuals in three basic social groups that used the study area. One social group was consistently in the area; two others moved into and out of the study area. Observed groups ranged in sized from 1 to 62 individuals and averaged 13.7 monkeys. Most of these observed groups were of 20 or fewer individuals ($N = 280$, 75%) and nearly half were of 10 or fewer individuals ($N = 168$, 45%) (Fig. 7.2). On 54 occasions (15% of sightings) single monkeys were seen. Even if single individuals are not considered as 'groups', 71% of the groups still consisted of 20 or fewer individuals and 36% consisted of 10 or fewer individuals. These data indicate that smaller groups were more common. However, two factors must still be considered. Firstly, it is often difficult to obtain complete group counts in tropical forests. This factor will be considered later in the results. Secondly, if a group splits and reforms during the day, then observers have two or more chances of locating sub-groups after a split compared with only one chance of locating a single, unified group or 'super-group.' This should be partially offset by greater ease of locating a larger group once one is in suitable proximity, due to a larger distribution and larger number of individuals to see and hear.

Group size and seasonality

Monthly group size averages ranged from 9 to 17.7 (Fig. 7.3). Even though a wide range of group sizes was seen each month, there was a significant difference in group sizes between months ($N = 15$, Kruskal–Wallis $= 24.6$, $P < 0.05$). These differences, however, did not follow any clear pattern.

Environmental conditions may affect group size either directly through interference in travels or by affecting energetic costs in extreme weather, or

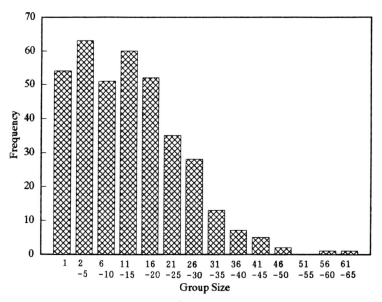

Fig. 7.2. Frequency distribution of *Macaca nigrescens* observed group sizes.

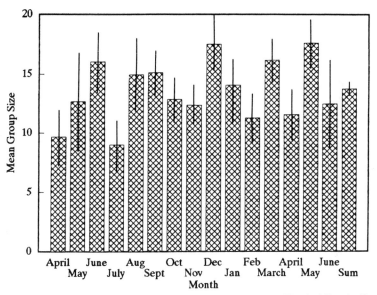

Fig. 7.3. Monthly mean group sizes of *Macaca nigrescens*. Vertical line indicates standard error; Sum is the average of all group sizes.

indirectly by affecting vegetation and fruiting cycles. The weather at Toraut was usually mild with little variation in daily maximum and minimum temperatures. Monthly rainfall varied from 93 mm to 281.5 mm and averaged 199 mm. The majority of the rain (78.1%) fell during the day. However, regression analysis showed no significant correlation or relation between monthly rainfall amounts and group size averages.

Group composition

During this study, the population averaged 59.6% adults and sub-adults, 32.5% juveniles and infants, and 7.9% were unclassified (Fig. 7.4). Therefore, there were approximately twice as many mature as immature individuals. The adult sex ratio was 1 : 1.70 males to females. The infant to adult female ratio was 0.21.

Group composition was similar in groups of different sizes except for very small groups (Fig. 7.5). Positively identified single individuals were always adults and were usually adult males (59.3% of all identifications, 84.2% of positive identifications). Groups of 2–5 monkeys also had a high proportion of adult–sub-adults (65.6%) and fewer juveniles–infants (12.3%). Small groups also had higher percentages of unclassified individuals (22.1% in groups of 2–5, 16.7% of singles). This may be indicative of greater difficulty in seeing monkeys in small groups due to their greater wariness when in small groups. The compositional differences noted for adult–sub-adult, juvenile–infant, and unclassified were significant for groups of 1 and 2–5 when compared to the population average (chi-square = 51.7 and 39.0, respectively, $P < 0.001$). All other group sizes were not significantly different in composition from the population average.

No compositional changes were expected seasonally because the data from all group sizes were pooled for each month and thus reflected population (not group) composition. However, significant differences versus the overall average of adult–sub-adults, juvenile–infants, and unclassified were seen for the months of September (chi-square = 20.1, $P < 0.001$), October (chi-square = 8.6, $P < 0.05$), February (chi-square = 7.0, $P < 0.05$), and June 1990 (chi-square = 11.3, $P < 0.01$). These differences are most likely due to the changes in the number of unclassified individuals during those months and not reflective of true compositional changes. A true difference may have occurred in June 1990 when more juvenile–infants were seen, but this month also had a small sample size.

Some seasonal variation in births is indicated by higher proportions of

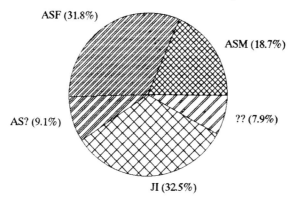

Fig. 7.4. Population composition of *Macaca nigrescens*. A, adult; S, sub-adult; J, juvenile; I, infant; M, male; F, female; ?, unclassified (age or sex).

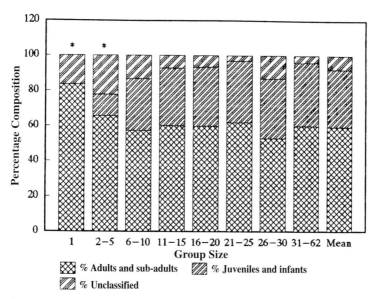

Fig. 7.5. Percentage composition of different-sized groups of *Macaca nigrescens*. *, significant difference from mean, chi-square; $P < 0.001$.

infants and white-faced infants from February to June (Fig. 7.6). Infants are born with white faces, which gradually darken in the succeeding few months. Therefore, white-faced infants are the best indicators of birth time. These were seen sporadically from April until 6 June 1989, then not again until 30 September 1989. The percentage of infants and white-faced infants remained low through January 1990 and then increased from February to

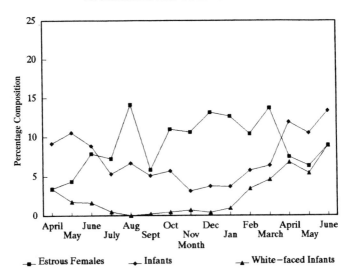

Fig. 7.6. Monthly percentage representation of estrous females, infants and white-faced infants in the Toraut *Macaca nigrescens* population.

June 1990. These data must still be regarded cautiously as the actual number of white-faced infants seen in any group was always low, usually one or two, rarely three or four, and only once (24 April 1990) were five seen in various stages of white through gray. However, Fig. 7.6 also shows that the percentage of estrous females varies inversely with the percentage of infants; this is consistent with a birthing peak from February to June in the year of our study.

Effect of count method and sighting duration

One concern in this study was how to obtain accurate group counts in a tropical forest. Often large groups were spread over 100 m or more and, therefore, only a portion were visible at one time. Also, since the monkeys were not individually recognized, two individuals of the same age and sex might be counted as one. Figure 7.7 shows our mean group size estimates obtained for each count method at four duration intervals. The average group size was 23.8 for progression counts and 7.5 for observation counts. Group size estimates were also higher when sighting duration was longer (Fig. 7.7). There was a significant difference in group sizes between the different sighting durations (Kruskal–Wallis = 218.4, $P < 0.001$). It should be noted that progression counts occurred more commonly in longer duration sightings. For the 142 progression counts, 57.7% were of more

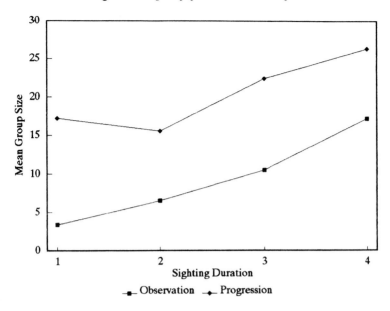

Fig. 7.7. Mean group size estimates obtained by progression and observation counts at four sighting durations. $1 = \leq 15$ min, $2 = 15$ min to 1 h, $3 = 1$ to 3 h, $4 = > 3$ h.

than 3 h and 86.6% were of more than 1 h. For the 230 observation counts, 40.4% were less than 15 min and 65.2% were of less than 1 h.

Discussion

Recent population studies of several primate species have shown their grouping patterns to be much more flexible than was indicated in some of the first review papers on grouping patterns. While some species or populations within species may retain a rigid composition, it is now apparent that many alter their group size to cope with ecological or social conditions. Beauchamp & Cabana (1990) showed that group size varies in many primates. In the Cercopithecidae, variability increased with large mean group sizes. The results here indicate that *M. nigrescens* is among those whose group size fluctuates.

This was the first study to monitor an *M. nigrescens* population over one year. In the previous, shorter studies, a variety of group sizes was also seen but of a smaller range. Bismark (1982a) found group sizes of 9 to 28, averaging 16.5 monkeys, and he recorded foraging sub-groups of 1 to 17, averaging 4.5 monkeys. During a primate survey of North Sulawesi,

Sugardjito *et al.* (1989) reported mean group sizes of 6.2 to 22 individuals in different parts of *M. nigrescens*' range. The average group size at Toraut was 7.3 individuals. We recorded groups of 1 to 62 monkeys, averaging 13.7 individuals. The longer-term nature of this study and 'partial habituation' of the monkeys may explain why we found some groups of much larger size compared to both previous studies. We did not distinguish sub-groups from groups, which explains our lower limit and lower average compared to those of Bismark (1982*a*). Our higher average compared with the 1987–8 surveys of Sugardjito *et al.* (1989) may be due to several reasons: (1) increased habituation in the present study resulting in longer and better observations, (2) a possibly depressed population in 1987–8 due to droughts in 1982 and 1987 and then consequent revival in 1989–90, and (3) ecological conditions in 1989–90 may have favored larger group sizes compared with those of 1987–8.

Several recent studies have shown similar group size variability to occur in other *Macaca* spp. Caldecott (1986) had average group sizes of 30 (1 group, 3 counts) and 45–55 (4 groups, 20 counts, range = 15–96) for *M. nemestrina* at two locations in Malaysia. These groups were seen to split and rejoin and forage in separated parties. Very large group sizes were the result of temporary fusion of two or more groups. Robertson (1986) studied *M. nemestrina* in Sumatra, Indonesia. Like Caldecott (1986), he saw a complex multi-leveled society with splitting and regrouping, although his categories were somewhat different. In contrast to Caldecott (1986) and Robertson (1986), Oi (1990) had two troops of *M. nemestrina* that did not form sub-groups, whereas another troop frequently split. Ménard *et al.* (1990) studied *M. sylvanus* in Algeria where party size ranged from 7 to 58 individuals, most frequently 15–20 individuals. Fukuda (1989) recorded one troop of Japanese macaques (*M. fuscata*) that habitually sub-divided in contrast to neighboring troops that did not. Seasonal changes in sub-group size were also noted.

It appears that some populations or groups of several macaque species may routinely sub-divide. This is most likely due to ecological factors, especially food supply, but may involve other factors. Why one troop of *M. fuscata* (Fukuda, 1989) or *M. nemestrina* (Oi, 1990) habitually sub-divided, whereas neighboring troops did not, may be partially explained by social factors within the troops. In the case of *M. nigrescens*, it appears that we have large troops that sub-divide for foraging and other daily activities. Whether this occurs in all troops is not yet known. An alternative explanation would be that we have small groups that coalesce for feeding or progressions. Further examination of our behavioral and ecological data should reveal whether this is a viable alternative.

Population composition in *M. nigrescens* was consistent with expected ratios; that is, 32.5% juveniles and infants, and an adult–sub-adult sex ratio of 1 : 1.70 males to females. The actual ratio is probably higher in the direction of females as many of the unclassified animals were believed to be females. Group composition was similar to population composition except for very small groups (2–5) and solitary individuals. For these, the proportions of adult–sub-adults and especially males increased, the proportion of juvenile–infants decreased, and the proportion of unclassified increased. These proportions are probably due to the need for adult–sub-adult males to emigrate, the increased protection afforded juvenile–infants in larger groups, and increased wariness of observers by smaller groups, respectively.

The increased sightings and proportions of infants, especially white-faced infants, and decreased proportions of estrous females from February to June indicate a seasonal birth trend at that time. The low actual numbers of white-faced infants at any one time preclude a conclusive statement of seasonality but it is certainly indicated. This seasonal birthing pattern likely coincides with seasonal ecological patterns in food supply. Alternatively, it could be due to synchronous cycling following a past disruptive event, such as the 1987 drought mentioned by Sugardjito *et al.* (1989), which might have temporarily halted reproduction.

The structure of tropical forests makes observations difficult. This, coupled with large group size and incomplete habituation, almost ensures that all group members will not be seen. Progression counts and increased sighting durations apparently increased our group counts. However, this does not invalidate observation counts or short duration sightings. Since we always attempted to maximize sighting times, this result is likely due to an increased ability to observe large groups. That is, large groups were more amenable to observation. They definitely showed less alarm behavior and less flight distance than small groups. Further more, longer sighting durations were more likely to contain a progression count.

Conclusions

1. *Macaca nigrescens* associates in groups of various sizes, from single individuals to groups of more than 60 monkeys.
2. Most commonly, *M. nigrescens* is seen in groups of 20 or less.
3. Groups of five or less consist of a higher proportion of adult–sub-adults, especially adult–sub-adult males, and a lower proportion of juvenile–infants compared to the general population.
4. Single individuals were always adult–sub-adults and were usually males.

5. Increased sightings of infants, especially white-faced infants, and decreased sightings of estrous females indicate a slight increase in births from February to June.
6. Progression counts and longer sighting durations yielded higher group counts. Most likely, these results are partially due to behavioral factors and not just to methodology. That is, larger groups seem more amenable to observation, which has resulted in longer observation times and consequently increased probability of seeing a progression.

Acknowledgements

We thank the Indonesian Institute of Sciences (LIPI), the Department of Forestry (PHPA), and Universitas Indonesia for their support of this project. Special mention goes to T. Hainald, R. Palete, S. Somadikarta and M. Susanto. Special thanks also go to the staff of Dumoga–Bone National Park, especially Ubus W. Maskar (Director), C. Hamid, H. Hamid, Junaid, M. W. Lela and S. Winenang. We also wish to thank our associates in the Sulawesi Primate Project (SPP), especially S. Baker, J. Erwin, J. Sugardjito, J. Supriatna and J. Tanasale. We appreciate the financial support provided by the National Geographic Society to Dr J. Erwin, the U.S. Public Health Service grant number RR04391 to the University of Colorado, and the World Wildlife Fund for Nature.

References

Albrecht, G. H. (1978). *The craniofacial morphology of the Sulawesi macaques: multivariate approaches to biological problems.* Contributions to Primatology, vol. 13, ed. F. S. Szalay. Basel: S. Karger.

Beauchamp, G. & Cabana, G. (1990). Group size variability in primates. *Primates*, **31**, 171–82.

Bismark, M. (1982a). Ekologi dan tingkahlaku *Macaca nigrescens* di Suaka Margasatwa Dumoga, Sulawesi Utara. *Laporan*, no. 392. Balai Penelitian Hutan.

Bismark, M. (1982b). Pengaruh perladangan dan pengambilan kayu gergajian terhadap Yakis (*Macaca nigrescens*) di Suaka Margasatwa Dumoga, Sulawesi Utara. *Duta Rimba*, 346–8.

Caldecott, J. O. (1986). *An ecological and behavioural study of the pig-tailed macaque.* Contributions to Primatology, vol. 21, ed. F. S. Szalay. Basel: S. Karger.

Ciani, A. C., Stanyon, R., Scheffrahn, W. & Sampurno, B. (1989). Evidence of gene flow between Sulawesi macaques. *American Journal of Primatology*, **17**, 257–70.

Crook, J. H. & Gartlan, J. S. (1966). Evolution of primate societies. *Nature*, **210**, 1200–3.

Eisenberg, J. F., Muckenhirn, N. A. & Rudran, R. (1972). The relation between ecology and social structure in primates. *Science*, **176**, 863–74.

Fa, J. E. & Southwick, C. H. (eds.) (1988). *Ecology and behavior of the food-enhanced primate groups.* New York: Alan R. Liss.

Fooden, J. (1969). *Taxonomy and evolution of the monkeys of Celebes (Primates: Cercopithecidae).* Bibliotheca Primatologica, vol. 10. Basel: S. Karger.

Fooden, J. (1976). Provisional classification and key to living species of macaques (Primates: *Macaca*). *Folia Primatologica*, **25**, 225–36.

Fooden, J. (1980). Classification and distribution of living macaques (*Macaca* Lacepede, 1799). In *The macaques: studies in ecology, behavior and evolution*, ed. D. G. Lindburg, pp. 1–9. New York: Van Nostrand Reinhold.

Fukuda, F. (1989). Habitual fission–fusion and social organization of the Hakone Troop T of Japanese macaques in Kanagawa Prefecture, Japan. *International Journal of Primatology*, **10**, 419–39.

Groves, C. P. (1980). Speciation in *Macaca*: the view from Sulawesi. In *The macaques: studies in ecology, behavior and evolution*, ed. D. G. Lindburg, pp. 84–124. New York: Van Nostrand Reinhold.

Hamada, Y., Watanabe, T., Takenaka, O., Suryobroto, B. & Kawamoto, Y. (1988). Morphological studies on the Sulawesi macaques. I. Phyletic analysis of body color. *Primates*, **29**, 65–80.

Ménard, N., Hecham, R., Vallet, D., Chikhi, H. & Gautier-Hion, A. (1990). Grouping patterns of a mountain population of *Macaca sylvanus* in Algeria – a fission-fusion system? *Folia Primatologica*, **55**, 166–75.

Oi, T. (1990). Population organization of wild pig-tailed macaques (*Macaca nemestrina nemestrina*) in West Sumatra. *Primates*, **31**, 15–31.

Robertson, J. M. Y. (1986). On the evolution of pig-tailed macaque societies. PhD thesis, University of Cambridge.

Sugardjito, J., Southwick, C. H., Supriatna, J., Kohlhaas, A., Baker, S., Erwin, J., Froehlich, J. & Lerche, N. (1989). Population survey of macaques in northern Sulawesi. *American Journal of Primatology*, **18**, 285–301.

Takenaka, O., Hotta, M., Takenaka, A., Kawamoto, Y., Suryobroto, B. & Brotoisworo, E. (1987a). Origin and evolution of the Sulawesi macaques. 1. Electrophoretic analysis of hemoglobins. *Primates*, **28**, 87–98.

Takenaka, O., Hotta, M., Kawamoto, Y., Suryobroto, B. & Brotoisworo, E. (1987b). Origin and evolution of the Sulawesi macaques. 2. Complete amino acid sequences of seven B chains of three molecular types. *Primates*, **28**, 99–109.

Thorington, R. W., Jr & Groves, C. P. (1970). An annotated classification of the Cercopithecoidea. In *Old world monkeys: evolution, systematics and behavior*, ed. J. R. Napier & P. H. Napier, pp. 629–47. New York: Academic Press.

Watanabe, K. & Brotoisworo, E. (1982). Field observation of Sulawesi macaques. *Kyoto University Overseas Research Report of Studies on Asian Non-human Primates*, **2**, 3–9.

Watanabe, K., Lapasere, H. & Tantu, R. (1991). External characteristics and associated developmental changes in two species of Sulawesi macaques, *Macaca tonkeana* and *M. hecki*, with special reference to hybrids and the borderland between the species. *Primates*, **32**, 61–76.

Whitmore, T. C. & Sidiyasa, K. (1986). Composition and structure of a lowland rainforest at Toraut, Northern Sulawesi. *Kew Bulletin*, **41**, 747–56.

8

Socio-ecological dynamics of Japanese macaque troop ranging

T. MARUHASHI AND H. TAKASAKI

Introduction

Japanese macaques live in 'troops', in which the female members are matrilineally related. The troop is a social unit, and when a troop becomes large it may split into two troops, each of which retains the matrilineal structure. Because resources are limited, competition arises between troops. Therefore, the troop is also an ecological unit of survival. In the Japanese macaque multimale–female troop structure, males transfer between troops whereas females remain. Their socio-ecological dynamics may be interpreted as a complex of reproductive strategies of males and competition over land and food resources between matrilines.

Longitudinal inter-troop and inter-matriline socio-ecological dynamics can be studied only where troops are continuously distributed in a well-preserved habitat. The long-term studies of the Yaku macaque (*Macaca fuscata yakui*, an island sub-species of the Japanese macaque) conducted in Yakushima (Maruhashi 1980, 1981, 1982, 1991; Furuichi, 1983; Yamagiwa, 1985; Maruhashi, Yamagiwa & Furuichi, 1986; Mitani, 1986; Oi, 1988; Sprague, 1989; Okayasu, 1991) have provided rare data that are usable in this line of analyses.

In the study population, where small troops are continuously distributed, social changes, such as fission, takeover, and extinction, are not rare. Such changes often depend on the troop size. When the troop size is small, takeover is likely to take place. When large, fission is more likely to take place. Troop extinction may even occur if the home range shrinks. Troop size (N) and home range area (R) in Japanese macaques have a correlation expressed by the equation $QR = \alpha N$, where Q is a parameter of habitat quality and α is a constant (Takasaki, 1981a,b, 1984). This chapter aims to provide a rough socio-ecological scenario for the changes in ranging of the study troop by using this model as an integrating tool.

Terminology

Food patch

A food patch in this study signifies a tree or vine in which a monkey feeds regardless of the plant part eaten.

Troop male and non-troop male

At the study site a troop usually consists of < 30 monkeys. The average sex ratio (the number of adult males per adult female) is 0.9 during the non-mating season, remarkably higher than in other habitats (Takasaki & Masui, 1984). Most males stay in troops even during the non-mating season. In other words, there are few solitary males. Males who have been staying in a particular troop are defined as 'troop males', and newly approaching males who have not been seen during the preceding non-mating season as 'non-troop males'.

Takeover

Non-troop males appear around a troop mostly during the mating season, and immigration may take place. Although usually used for one-male groups in other primate species, here 'takeover' is applied to cases in which a non-troop male immigrates as the new alpha male of the troop.

Fission

Troop 'fission' occurs when a lower-ranking matriline forms a new troop that begins to range independently. A non-troop male who is dominant to the alpha male of the original troop often joins as the alpha male of the new troop.

Brief history of study groups

Long-term study of a Yaku macaque population on the western slope of Mt Kuniwari in Yakushima (130° 25′ E, 30° 20′ N) has continued since 1975 (for earlier studies, see Maruhashi, 1980, 1981, 1982; Maruhashi *et al.*, 1986; see also Sprague, Suzuki & Tsukahara and Hill & Okayasu, Chapters 20 & 21, this volume).

Figure 8.1 illustrates the history of Ko troop, which ultimately split into four troops. Ko troop, consisting of 47 monkeys in 1976, fissioned twice, in the 1977–8 and 1978–9 mating seasons, to form three troops: A, H, and M. In 1979, A troop had 27 monkeys, H troop 15, and M troop 22 (Maruhashi, 1982). In the 1986 mating season, A troop fissioned into two further troops, T and P (Tsukahara, 1990).

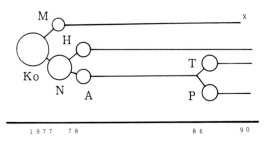

Fig. 8.1. Diagrammatic fission history of the study troops. (The circle size roughly represents the relative troop size. M troop became extinct (x) in 1989.)

The alpha male of the branched M troop was a non-troop male dominant to the alpha male of Ko/N troop (Yamagiwa, 1985). The alpha male of A troop was also dominant to the alpha male of N/H troop. When A troop fissioned, the alpha male of P troop was also dominant to the alpha male of A/T troop. Therefore, as a rule, the non-troop male who became the alpha male of a branched troop was dominant to the alpha male of the original troop. The dominance between two neighboring troops was determined by the dominance between the alpha males (Maruhashi, 1982), as has been reported for provisioned Japanese macaque troops (Kawanaka, 1973).

The dominant M troop succeeded to the central part of the original Ko troop's home range. However, N troop used the peripheries and home range overlaps with neighboring troops. This unstable ranging of N troop probably triggered the successive fission to form H and A troops. The home ranges of the resulting three troops were mostly confined within the home range of the former Ko troop. Owing to the frequent interactions with neighboring troops, H troop's home range fluctuated. After a takeover the ex-alpha male of H troop transferred as the alpha male of M troop. This reversed the dominance relationship between H and M troops, so that H troop then invaded the home range of M troop. A troop, which had retained the same alpha male since its formation, also gradually expanded onto M troop's home range.

M troop once attained a maximum size of 28 monkeys. However, it gradually declined in both troop size and home range area owing to the invasion from the neighboring H and A troops. In the summer of 1989, there were only three monkeys in M troop, one adult male, one adult female, and an offspring. Although the troop size was small, they still moved together exchanging contact calls with one another. M troop almost lost its exclusive home range, and its members foraged while avoiding encounters with neighboring troops. In the 1989 mating season, one female and one adult male disappeared, and the last remaining mother and her 4-year-old female offspring joined the neighboring H troop. Thus M troop became extinct (Takahata, 1994).

Troop fission occurs as the consequence of competition between matrilines over land and food resources. Dominance between matrilines determines the grouping within new troops. Although no analysis is feasible for the two fission cases originating from Ko troop with regard to dominance between matrilines, on the fission of A troop the lowest-ranking matriline split off with the new alpha male to form the branched P troop. Similar fission cases were reported from another troop in Yakushima (Oi, 1988) and from provisioned Japanese macaque troops (Koyama, 1970), although in the provisioned troops the remnant members of the original troops later upset the dominance relationships. This issue requires further research as the Yaku macaque population in the wild does not hold the youngest ascendency rule reported from provisioned troops (Kawamura, 1958), in which the dominance among sisters is in the reversed order of their birth. This may differentiate the mechanism of dominance formation between matrilines in the Yaku population (see Hill & Okayasu, Chapter 21, this volume). Nevertheless, dominance between troops is still dependent on the dominance between the alpha males.

Troop size–home range area model

The equation $QR = \alpha N$ is a model that correlates troop size (N) and home range area (R) and incorporates a parameter of habitat quality (Q) (Takasaki, 1981a,b, 1984). This model states that the home range area is proportional to the troop size, and the per capita range area ($R/N = \alpha/Q$) is inversely proportional to the habitat quality. This model, though first devised as an empirical descriptive model (Furuichi, Takasaki & Sprague, 1982), is also interpreted in terms of energy. The left-hand side may be regarded as a measure of the energy harvest of the home range area, and the right-hand side as the consumption by the monkeys (Takasaki, 1984).

Because the habitat quality is correlated with geographical parameters such as the annual mean temperature and solar radiation received, the per capita range area is expected to show a north–south cline. In data from wild Japanese macaques, however, the per capita range area roughly classifies Japanese macaque troops into two groups, those living in deciduous broadleaf forest habitats (including those living in secondary deciduous vegetation in the south) and those living in evergreen broadleaf forest habitats (Takasaki, 1981a,b). In other words, assuming a fixed value for Q

among each group (Q_D for the 'deciduous' group and Q_E for the 'evergreen' group), the equation $QR = \alpha N$ correlates the variable troop size and home range data area of the Japanese macaque across its major habitat vegetation zones (Fig. 8.2). Incidentally, the per capita range area is about ten times larger in the deciduous group (i.e. $Q_E = 10Q_D$, although the primary production of the forest differs by only 2–3 fold; Kira and Shidei, 1967).

For a fixed value of Q, the equation may be rewritten as $R = kN$, where k is a constant ($= \alpha/Q$). In this form the home range area is directly proportional to the troop size. A statistical examination (Takasaki, 1984) has shown that this proportional relationship between troop size and home range area is valid for the Yakushima population, which lives in a relatively well-preserved habitat.

The equation $QR = \alpha N$ implies that the troop becomes extinct ($N = 0$) when the home range shrinks to nil ($R = 0$). By assuming fluctuating deviations from the equation, dynamics in troop ranging associated with troop fission and takeover can also be integrated into the model. For the fluctuation of the point (N, R) much depends on the dominance relationships between troops. Boundaries of home ranges of the study troops were independent of natural features such as landscape, and distribution or discontinuity of particular vegetation types, but were socially determined by the presence of neighboring troops. Although troops overlap some of their home ranges, they also retain exclusive areas as their own. Expansion or shrinkage of home range is much affected by the dominance relationships between troops. Dominant troops expand their home ranges, while less dominant troops contract.

Encounters between two troops often result in fierce aggression, in particular during the period after troop fission when their dominance relationships are still unstable. In this period the home range overlap is exclusive. As the exclusive areas are determined by the relationships between the troops, the overlap decreases in area. In this critical period, the territorial prosperity of each troop depends on its alpha male, who determines the dominance between troops.

Feeding behaviour and troop ranging

The Yaku macaque, living in broadleaf evergreen forest in the warm temperate zone, feeds on various fruits all year round. Its habitat, where even some sub-tropical plant species grow, has a high fruit production and supports the highest population density in the entire distribution range of the Japanese macaque.

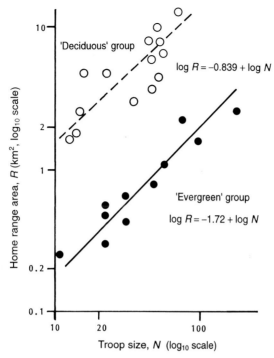

Fig. 8.2. Troop size (N) and home range area (R) data from the Japanese macaque's entire distribution range. (Two groups distinguished by the vegetation type, broadleaf deciduous or broadleaf evergreen, show a good fit to the regression lines expected from the equation $QR = \alpha N$, from Takasaki, 1981*b*.)

There are five notable features in the feeding behaviour of the Yaku macaque (Maruhashi, 1986): (1) The duration of stay in a food patch is almost constant. (2) Different food items are chosen one after another. (3) Feeding in a food patch is basically practiced singly, although members of the same matriline occasionally co-feed. (4) When a troop travels, members of the same matriline often move together. (5) Monkeys frequently exchange the contact 'coo' call while foraging.

The home range contains a few patches that are repeatedly used, and some that are used only infrequently. Each member of a troop seems to know the species of each tree and its location in the home range. This mental map is traditional knowledge, transmitted from mothers to their daughters.

In the broadleaf evergreen forest in Yakushima, trees of the same species are usually found nearby. This allows monkeys of a troop, who travel together, to visit food patches one after another by differentiating the food species. They usually disperse in different food patches when feeding, and avoid inter-individual encounter in a feeding tree, which may result in aggression. This avoidance of encounter during feeding is performed by the frequent exchange of 'coo' calls.

The per capita food patch number, calculated as per capita range area, which allows the most efficient dispersed intra-troop feeding in a particular habitat is represented by the constant k in the proportional relationship between troop size (N) and home range area ($R = kN$). In other words, when the home range area falls short of the area estimated by the equation, for example, inter-individual strain over food will be intensified. The deviation from the value predicted by the equation gives a measure of the intensity of intra-troop competition over food. When strained, any system may transform itself to reduce the strain. A strained troop, depending on its size, intra-troop inter-matriline relationships, and inter-troop relationships, may either expand or contract its home range or it may fission.

Social dynamics and troop ranging

When the troop size (N) and home range area (R) data for A, H, and M troops are plotted over the five-year period 1979 to 1984, the proportional relationship still approximately holds (Fig. 8.3). When the home range area increased/decreased, the troop size also increased/decreased. In other words, a troop's survival and prosperity depended on whether or not the troop retained and expanded its size and home range area.

Now let us reconstruct, from the viewpoint of feeding ecology, the process of M troop's home range shrinkage followed by the decrease in troop size, which eventually led to its extinction. The process had two stages: initially the troop reduced its home range without any decrease in troop size (Fig. 8.1, 1979–82). Subsequently, the decreases in troop size followed (1982–4). In the first stage, the number of usable food patches also decreased following the decrease in the home range area. As a result, the frequency of feeding on repeatedly used food patches increased, and in some patches overgrazing deformed or even killed the trees.

When the home range area became small, the distance of travel between feeding bouts became shorter, although the data have not yet been fully analysed. So the number of usable food patches per unit home range decreased, and inter-individual encounters while feeding increased. Subordinate monkeys were often suppressed by their dominants.

Shrinkage of the home range was also followed by decreased birth rate and increased infant mortality. The females became unable to take in

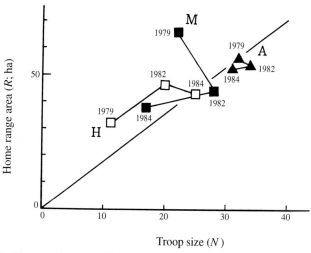

Fig. 8.3. Change of size and home range area in three study troops. (*N* and *R* for 1979, 1982, and 1984 are plotted for A, H, and M troops.)

enough energy for reproduction. As the inter-individual strain was intensified, the lowest-ranking matriline began to forage independently away from the other members of the troop. The members of the middle-ranking matriline continued to forage in the usual fashion of intra-troop dispersed feeding, therefore they were most affected by the reduced home range area, and when the home range shrank further, monkeys of the middle-ranking matriline disappeared.

The above is the probable scenario for the extinction case of M troop. After the reversal of the dominance relationship between M and H troops, M troop's exclusive range area decreased. Even after the middle-ranking matriline disappeared, the home range continued to decrease in area. In 1989, with only three monkeys remaining, no exclusive area was left for M troop. In the 1989 mating season, the last two members merged with the neighboring H troop; thus M troop became extinct.

Reproductive strategies and troop ranging

Fission is likely to occur in a large troop, while takeover is likely to occur in a small troop. In both cases, non-troop males are involved. Also both are linked to the ranging behavior of matrilines which make up the troop. These are socio-ecological phenomena related to the reproductive strategies of males and females.

The Japanese macaque has a distinct mating season. In the Yaku sub-species, it usually starts in September and ends in February. At the beginning of the mating season few non-troop males appear, and females copulate with troop males. The estrus of each female is not synchronized, and each female is in a different estrous phase. Troop males compete over females who are in the ovulatory phase. Higher-ranking males probably have more access to them and more chances to impregnate them. In Japanese macaques fertilization mostly occurs at the first estrus in a mating season (Takahata, 1980). Therefore, the offspring born in the following birth season are mostly of the troop males. In the middle term of the mating season, many non-troop males appear, and females mate with both troop and non-troop males. In this period, however, although the females continue with estrous cycling (Hill & Okayasu, Chapter 21, this volume), the probability of impregnation is low. It is puzzling why the non-troop males come, and why the females continue to cycle even after conception.

In Yakushima, where small troops were distributed continuously, there were few solitary males. In the mating season, however, non-troop males outnumbered troop males: where were they from? Non-alpha high-ranking troop males have two options to improve their reproductive success: (1) to stay in the troop and upset the rank; or (2) to leave the troop and join another troop as a high-ranking male, or as the new alpha male through takeover or troop fission. For the maximum reproductive success, they seek the opportunity to become alpha males. Therefore males frequently monitor neighboring troops, assessing the probability of takeover or fission. This is the underlying condition for large numbers of non-troop males in the mating season (Sprague, 1989).

Each non-troop male singly approached a troop and the troop males usually defended cooperatively. The cooperative defense against the non-troop male by the troop males depended on the number of troop males, which was often correlated with the troop size. The probability of success of approach or copulation by non-troop males depended on the cooperative defense by the troop males (Yamagiwa, 1985). Many non-troop males approached a small troop which was likely to be easily taken over, and they might have copulated with the females. By contrast, few non-troop males approached a medium-sized troop which was unlikely to be taken over or split, and extra-troop mating was rare. In a large troop, however, the lowest-ranking matriline might have foraged independently from the rest of the troop, which is a phenomenon that precedes fission (Oi, 1988), and again many non-troop males may approach.

In Yakushima, when a large troop consisting of > 30 monkeys underwent a fission, for example, first a troop consisting of 10 monkeys of the lowest-ranking matriline split off, and then often another fission took place in the remaining larger troop. From the male's point of view, becoming the alpha male of a small, branched new troop is not much different from taking over a small troop. In either case the male becomes the alpha male, the position providing him with the best opportunity for reproductive success in the following mating season(s).

From the female's point of view, it is the home range that provides the nutritional resources for reproduction. For maximizing their reproductive success, the females need to attract alpha male candidates who would protect their home range better. Therefore, their post-conception estrus may be functionally interpreted.

After fission, the branched troop will be more dominant to the troop retaining the original troop's males, as inter-troop dominance depends on the dominance relationship between the non-troop male who joined the new troop as its alpha male and the original alpha male. The newly formed dominant troop takes over the exclusive area of its original troop. It may often occupy a home range larger than expected from the troop size immediately after fission. Subsequently, the new dominant troop increases its troop size. The troop composed of the remaining former higher-ranking matrilines becomes a troop subordinate to the new troop. It is restricted to areas overlapping with the home ranges of neighboring troops. The resulting social strains make it split again successively. Therefore, when a larger troop undergoes fission, successive troop fission into smaller troops may follow in a few years.

Both takeover and fission provide a male with the opportunity to become the alpha male. For females, both are directly linked to the competition over land and resources between troops or matrilines. Takeover provides the females in the troop with a new alpha male who secures their home range better than the former alpha male. Fission provides a low-ranking matriline with an opportunity to upset the dominance relationships with the other matrilines.

Summary and conclusions

The socio-ecological dynamics of the Yaku macaque troop reviewed above demonstrates that perpetuating matrilines are competing over the home range. The correlation between troop size (N) and home range area (R), incorporating a parameter of habitat quality (Q), $QR = \alpha N$, gives a good fit to the observed values of N and R (Takasaki, 1981b; Maruhashi, 1982, 1986). The left-hand side of this equation can be interpreted as the energy harvest from range area, and the right-hand side as the consumption by the monkeys. Although for the monkeys' survival the inequality, $QR > \alpha N$, suffices, we find that the equation holds instead of the inequality (Takasaki, 1984). The monkeys do not have the surplus area allowed in $QR > \alpha N$. Their ranging behavior is a socio-ecological outcome of the tight balance between intra-troop and inter-troop expansion/contraction pressures. The intra-troop pressure is due to individually spaced foraging and density of food patches, while the inter-troop pressure is determined by dominance between troops.

A small troop, for its proliferation, needs a strong alpha male to expand its home range area; it may go through takeover to obtain a better alpha male. A low-ranking matriline in a new troop may split off and obtain its own alpha male to form a new troop. The new alpha male is usually dominant to the alpha male of the original troop, and the new branched troop becomes dominant to the troop retaining the rest of the original troop. These two types of troop transformation have the same socio-ecological background. A matriline requires a large, stable home range for proliferation. The prolonged estrus and post-conception estrus of Japanese macaque females may function to attract many non-troop males who can become alpha male candidates. Adult males may strive for the alpha position, which probably provides them with high reproductive success. The requirements of both sexes are met in takeover of a small troop or fission of a large troop.

The incidence of such troop transformations depends on the probability of appearance of non-troop males who are dominant to the alpha male of the troop. This probability depends on the number of troops and number of non-troop males per unit area in a particular habitat. A great variation may exist in the patterns of socio-ecological dynamics between different regional populations. In Yakushima, there are more than 100 troops, and many of their home ranges partly overlap. In such a habitat the frequency of troop transformation is probably much greater than in some sites where only a few isolated troops live. Indeed, in Yakushima the average troop size is < 30, considerably smaller than in other habitats of the Japanese macaque (Takasaki & Masui, 1984).

Transformation of a troop will affect its neighboring troops as well. Transformation may be propagated in the form of waves of expansion/contraction of the home range. After the fission of Ko troop (Maruhashi, 1982), the neighboring two troops also fissioned in the next few years (Oi, 1988). Repeated troop transformations will give a spatial distribution of troops within a population (Fig. 8.4). Some genealogically distant troops and patches of sibling troops live in mosaics. An actual spatial distribution

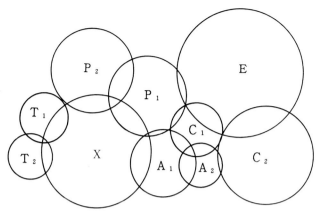

Fig. 8.4. A schematic model of genealogical and distributional structure of a regional population of the Japanese macaque. (Each circle indicates a troop A, C, E, T or P, and the circle size roughly represents the relative troop size. The same troop name initials indicates that the troops originated from the same large troop by fission.)

of troops observed in Yakushima (Maruhashi, 1982) gave a similar pattern.

Fission, takeover, or extinction of a troop occurs with the probability expected from the conditions of a particular troop. Although some chance factors may determine the density of a troop, more importantly the transformation is led by socio-ecological factors. Oversized or undersized troops deviating from the ecological expectation of the equation $QR = \alpha N$ tend to transform in such a way as to reduce the deviation. Moreover, the underlying social motif of the transformation is the competition over land and food resources between matrilines in combination with the dominance relationships between males.

Acknowledgements

We thank all those who shared the field-work in Yakushima, without whose contributions in various ways the long-term data on which this chapter is based would not have been accumulated. T. Hidaka and many of the local inhabitants helped our field-work in Yakushima. T. M.'s earlier studies were supported in part by the Cooperative Fund of the Kyoto University Primate Research Institute, Monbusho (Japan Ministry of Education, Science and Culture) Grant-in-Aid for Encouragement of Young Scientists A (no. 63740367), and Ishida Foundation Research Grant (no. 59-283). For preparation of this chapter H. T. was supported in part by the Monbusho Grant-in-Aid for Encouragement of Young Scientists A (no.

02954069). Earlier drafts benefited from comments by Y. Takahata, D. A. Hill, N. Okayasu, and D. S. Sprague. To these people and institutions we are very grateful.

References

Furuichi, T. (1983). Interindividual distance and influence of dominance on feeding in a natural Japanese macaque troop. *Primates*, **24**, 445–55.

Furuichi, T., Takasaki, H. & Sprague, D. S. (1982). Winter range utilization of a Japanese macaque troop in a snowy habitat. *Folia Primatologica*, **37**, 77–94.

Kawamura, S. (1958). [Matriarchal social order in the Minoo-B troop: a study of the rank system of the Japanese monkeys.] (In Japanese.) *Primates*, **2**, 149–56. (English translation in: Imanishi, K. & Altmann, S. A. (eds.) (1965). *Japanese monkeys*. Atlanta: Emory University Press.)

Kawanaka, K. (1973). Intertroop relationships among Japanese monkeys. *Primates*, **14**, 113–59.

Kira, T. & Shidei, T. (1967). Primary production and turnover of organic matter in different forest ecosystems of western Pacific. *Japanese Journal of Ecology*, **17**, 70–80.

Koyama, N. (1970). Changes in dominance rank and division of a wild Japanese monkey troop in Arashiyama. *Primates*, **11**, 335–90.

Maruhashi, T. (1980). Feeding behavior and diet of the Japanese monkeys (*Macaca fuscata yakui*) on Yakushima Island, Japan. *Primates*, **21**, 241–60.

Maruhashi, T. (1981). Activity patterns of Japanese monkeys (*Macaca fuscata yakui*) on Yakushima Island, Japan. *Primates*, **22**, 1–14.

Maruhashi, T. (1982). An ecological study of troop fissions of Japanese monkeys (*Macaca fuscata yakui*) on Yakushima Island, Japan. *Primates*, **23**, 317–37.

Maruhashi, T. (1986). [Feeding ecology of the Yaku macaque.] (In Japanese.) In [*Wild Japanese monkeys of Yakushima*], ed. T. Maruhashi, J. Yamigiwa & T. Furuichi, pp. 13–59. Tokyo: Tokai University Press.

Maruhashi, T. (1991). [Social dynamics of the Japanese monkey troops: socioecology of the Yaku macaque.] (In Japanese.) In [*Social histography of monkeys*], ed. T. Nishida, K. Izawa & T. Kano, pp. 109–27. Tokyo: Heibonsha.

Maruhashi, T., Yamagiwa, J. & Furuichi T. (1986). [*Wild Japanese monkeys of Yakushima.*] (In Japanese.) Tokyo: Tokai University Press.

Mitani, M. (1986). Voiceprint identification and its application to sociological studies of wild Japanese monkeys (*Macaca fuscata yakui*). *Primates*, **27**, 397–412.

Oi, T. (1988). Sociological study on the troop fission of wild Japanese monkeys (*Macaca fuscata yakui*) on Yakushima Island. *Primates*, **29**, 1–19.

Okayasu, N. (1991). [Social role of the alpha female in a wild Japanese macaque of Yakushima.] (In Japanese.) In [*Social histography of monkeys*], ed. T. Nishida, K. Izawa & T. Kano, pp. 455–90. Tokyo: Heibonsha.

Sprague, D. S. (1989). Male intertroop movement during mating season among the Japanese macaques of Yakushima Island, Japan. PhD thesis, Yale University.

Takahata, Y. (1980). The reproductive biology of a free-ranging troop of Japanese monkeys. *Primates*, **21**, 303–29.

Takahata, Y., Suzuki, S., Okayasu, N. & Hill, D. (1994). Troop extinction and fusion in wild Japanese monkeys of Yakushima Island, Japan. *American*

Journal of Primatology, **33**, 317–22.

Takasaki, H. (1981*a*). On the deciduous–evergreen zonal gap in the per capita range area of Japanese macaque troop from north to south: a preliminary note. *Physiology and Ecology, Japan*, **18**, 1–5.

Takasaki, H. (1981*b*). Troop size, habitat quality and home range area in Japanese macaques. *Behavioral Ecology and Sociobiology*, **19**, 277–81.

Takasaki, H. (1984). A model for relating troop size and home range area in a primate species. *Primates*, **25**, 22–7.

Takasaki, H. & Masui, K. (1984). Troop composition data of wild Japanese macaques reviewed by multivariate method. *Primates*, **25**, 308–13.

Tsukahara, T. (1990). Initiation and solicitation in male–female grooming in a wild Japanese macaque troop on Yakushima Island. *Primates*, **31**, 147–56.

Yamagiwa, J. (1985). Socio-sexual factors of troop fission of the Japanese monkeys (*Macaca fuscata yakui*) on Yakushima Island, Japan. *Primates*, **26**, 105–20.

9

Riverine refuging by wild Sumatran long-tailed macaques (*Macaca fascicularis*)

C. P. VAN SCHAIK, A. VAN AMERONGEN AND
M. A. VAN NOORDWIJK

Introduction

The long-tailed macaque (*Macaca fascicularis*) is a widely distributed species throughout the Malesian region (see Rodman, 1991). It inhabits undisturbed dry land forests but is often reported to sleep in river-edge trees (Kurland, 1973; Aldrich-Blake, 1980; Fittinghoff & Lindburg, 1980; Wheatley, 1980). This peculiar habit, dubbed riverine refuging, has also been observed in a number of other small or medium-sized forest primates (talapoin *Miopithecus talapoin*) (Gautier-Hion, 1973); De Brazza's monkey *Cercopithecus neglectus* (Gautier-Hion & Gautier, 1978), and vervets *C. aethiops* (Lancaster, 1972)), although it is less strict in the latter two species. Since it must inevitably lead to some increase in travel costs between food sources and refuge, it raises the question of what benefits these monkeys derive from roosting along river edges.

Fittinghoff & Lindburg (1980) have hypothesised that riverine refuging in Bornean long-tails served to defend the home range against groups using the opposite river bank. This explanation is unlikely to be a general one because at our study site (Ketambe): (1) groups on opposite banks of the river often virtually ignored each other (in contrast to the exchanges of displays and mutual withdrawal when two groups met in the forest) and (2) groups also slept along the river in those parts where they occupied both banks. Hence, we have to consider alternative explanations for the refuging by long-tailed macaques. In this chapter, we evaluate the possibility that the monkeys are trading off the avoidance of predation and minimisation of energy loss.

At least in principle, animals can avoid predation during their inactive periods in two contrasting ways. They can either hide, and so avoid being detected by a searching predator, or withdraw to places where it is difficult for a predator to approach them undetected. Many small-bodied primates, both nocturnal and diurnal, follow the first strategy. They sleep in tree hollows or in dense vegetation tangles (see e.g. Dawson, 1979; Bearder, 1987). Among the larger diurnal primates only the orang-utan (*Pongo pygmaeus*) follows this strategy. The nesting habit of all the largest age–sex classes is best interpreted as serving the function of avoiding detection by a searching predator (Sugardjito, 1983). Most other large primates, however, live in sizeable groups, which precludes the use of hiding places. Instead, they roost in places where they maintain vigilance throughout the night so as to see or feel the approach of a predator. This means roosting in exposed places (in high tree crowns or on cliffs), a behaviour clearly associated with predation avoidance (Kummer, 1971; van Schaik & Mitrasetia, 1990). Therefore, choosing trees on the river's edge to sleep in may serve to minimise the risk of predation, probably because it allows animals to escape from approaching predators through the water (Gautier-Hion, Gautier & Quris, 1981).

Such contrasting strategies also have important energetic consequences. These are expected to be strongest for diurnal species because they are inactive during the cool night when they cannot rely on heat generated by their activity to compensate for heat loss. Indeed, nests used by small primates may not only serve as a hiding place against predators but also help to conserve energy, since several of these species go into nightly torpor (e.g. Dawson, 1979). Sleeping on exposed cliffs or in tree crowns is thus likely to cause considerably greater heat loss. This should be particularly disadvantageous for small-bodied primates because the lower critical temperature decreases with increasing body size (Schmidt-Nielsen, 1975). This means that if long-tails can limit energy loss during the night because temperatures near the river are warmer, this may more than compensate for energetic losses due to increased travel.

Where safe roosting places are scarce their distribution may impose strong constraints on the range use of the animals, as in cliff-sleeping hamadryas baboons (*Papio hamadryas*; Sigg & Stolba, 1981). Theoretical considerations (Hamilton & Watt, 1970; Anderson, 1978; Orians & Pearson, 1979; Schoener, 1979) lead us to expect a linear decrease in intensity of use and an increase in food selectivity as the distance from the refuge increases. There are, however, two essential differences between the classical refuging system and the present one: (1) in this case the refuge is not a point in space but a linear series of points, and (2) rather than visit only one or a few food sources during one foraging trip, the monkeys visit many sources during the day before returning to the refuge. Hence, it is not clear to what extent these predictions apply to riverine refuging. In particular, it seems necessary also to consider the confounding effects of diurnal variation in activities and food choice, since they may affect the patterns observed in relation to distance from the refuge.

In this study, data were collected on Sumatran long-tailed macaques in order to evaluate these various ideas. First, observations were made on roost trees and the monkeys' behaviour in them to assess the importance of predation avoidance. Second, some aspects of the microclimate of these trees were measured and compared with emergent trees in the inland forest in which pig-tailed macaques (*Macaca nemestrina*) and Thomas leaf monkeys (*Presbytis thomasi*) roosted. Finally, aspects of range use and activity budgets relevant to the refuging habit were examined. Other determinants of range use by these monkeys will be discussed in a subsequent publication.

Methods

Habitat

The study area is located along the Alas river in the Gunung Leuser National Park, Sumatra, Indonesia. It is covered by undisturbed tropical rainforest, as described by Rijksen (1978) and van Schaik & Mirmanto (1985). It consists of six riverine terraces of varying age and productivity and the slopes connecting them, as well as the lower parts of the adjacent mountain range. In the latter, three types of topographical unit are distinguishable: ridges, slopes, and steep-sided gullies.

Details on methods used to estimate the availability of food (fruit, insects and young leaves) for each of the ten topographical units are given by van Schaik (1985). A grid system of quadrats of 50 m × 50 m each was drawn on a 1 : 4000 map of the study area. Each quadrat was assigned to one of the ten topographical units. In order to characterise the productivity of a quadrat the mean value for the whole topographical unit was used. This is justified because the within-unit variance in leaf litter-fall was far less than the between-unit variance, and the variation within a topographical unit was not related to the forest's basal area, and thus its phase in the natural regeneration cycle (van Schaik & Mirmanto, 1985). As to fruit, there was great consistency in the differences between topographical units in samples taken in different parts of the study area.

Ranging and activities

For this chapter, data on four different groups (K, H, A, and I; see map in van Schaik & van Noordwijk, 1985) were used. Only those periods in which

Table 9.1. *The number of observation hours of the four groups reported in this study. The periods when data on ranging behaviour were collected are underlined*

Group	Period	Hours
Ketambe	Dec. 1976–Aug. 1977	670
	Feb. 1980–Apr. 1981	<u>508</u>
	Dec. 1981–Jan. 1983	<u>543</u>
House	Nov. 1976–Aug. 1977	238
	Jan. 1980–May 1981	<u>1407</u>
	Sep. 1981–Feb. 1983	<u>1062</u>
Antara	Mar. 1980–Dec. 1980	<u>399</u>
	Nov. 1981–Dec. 1982	<u>233</u>
Insulinde	Apr. 1982–Jan. 1983	<u>516</u>

the group's activities were recorded alongside its ranging have been employed (Table 9.1). The group's activities were recorded after each five minutes, using the description rest, travel, feed (on clumped food, usually fruit, sometimes leaves, rarely insects), and forage (on dispersed food, usually insects, young leaves and other miscellaneous vegetable material, occasionally fruit). Definitions follow van Schaik *et al.* (1983). To assess food choice, the foraging behaviour of an animal was frequently followed where the nature of all new foraging attempts was recorded until it moved out of view (cf. Terborgh, 1983). All groups were studied using dawn-to-dusk tracking; samples of groups A, H, and K were collected in blocks of five or six consecutive days. Data on all groups covered all seasons so no correction for seasonal variation in range use was needed.

In principle, only the main group, i.e. the largest part of the social group that travelled as one unit, was followed. Each half hour the group's position was marked and all movements noted on a 1 : 4000 map of the study area.

Roost trees

Continuous records of temperature and relative humidity were obtained simultaneously in roost trees and emergent inland trees by using two calibrated Lambrecht's thermohygrographs (type 252ua) placed on exposed branches in standard weather boxes.

For 68 roost trees known to be used by one or more groups, 32 different characters of potential importance were recorded. In addition, the same variables were recorded for 15 randomly selected river-side trees (> 15 cm diameter and < 10 m from the river).

Statistics

In order to detect relationships among ranging parameters, correlations between quadrat features (e.g. vegetation type) and some aspects of the animals' use of them are usually calculated. The statistical evaluation of these relationships is problematic because the size and hence the number of quadrats is largely arbitrary. As a conservative approximation one quarter of the number of quadrats was used as the number of independent observations, because this is equivalent to using only those quadrats without immediate neighbours. However, in addition, consistency of outcome in the different groups was examined.

Results

General characteristics of roost trees

Not all trees along the river's edge were used for roosting. Each group used only a few (two to six) trees frequently. Roost trees tended to have open crowns. A mere 3% of the branches in which the monkeys roosted were immediately contiguous with other trees, and 12% of them could be reached by jumping from a neighbouring tree. Therefore, it is difficult for a predator to reach the average roost branch without being detected. Of the roost branches, 72% extended above the river. It was noted in a number of instances that roost trees were no longer used when the river under them became dry, temporarily in the dry season, or permanently when the river had changed its course. There were, however, no significant differences in these characteristics between roost trees and random trees. Although this may imply that all trees along the river have the features desired by the monkeys, it is more likely that other factors, in particular their position in the home range and the distance to the nearest other roost tree, force the monkeys to use less suitable ones. Therefore, a more precise comparison was made, namely between frequently used roost trees ($n = 12$) and trees used only occasionally ($n = 15$), including only trees in the centres of the home ranges to exclude the influence of spatial position. In comparison with rarely used trees, frequently used ones tended to have (1) a more open crown structure, (2) a higher proportion of roost branches from which the main trunk was visible and (3) fewer tangles of lianas in their crown. Also (4) they were surrounded less frequently by higher neighbouring trees and (5) they never had roost branches that could be reached by jumping from adjacent trees. Although none of these separate differences were significant, their cumulative effect is consistent with the idea that a group's roost tree gives more safety by providing good visibility for the monkeys and by being difficult to enter from a side or above by a predator. In addition, the monkeys could jump into the river if necessary.

Behaviour in roost trees

Does the behaviour of the monkeys in the roost trees confirm the suggestion that they remain vigilant at night in order to detect possible predators? The monkeys generally roosted in dense huddles of up to four animals, usually a mother with one or more of her immature offspring (the youngest of which sits closest to her), although adult and sub-adult males frequently sat alone. Animals of unequal size sat together ventro-ventrally, while those of equal size used to sit together dorso-ventrally. The great majority (87%, $n = 92$) of the animal samples after dark (only one sample per evening, counting only the largest individual of huddles) sat facing the main trunk, while a mere 2% sat with their back to the main trunk. The roosting monkeys tended to sit on the thinner outer parts of the branches. To illustrate this, only 12% (of 33 records) of the huddles sat on the rising parts of branches.

At least some of the animals remained alert at night, although they probably have a light sleep (cf. Bert *et al.*, 1967), as witnessed by occasional bouts of intensive alarm calling and agonistic interactions during all hours of the night, and copulation vocalisations during the early hours of the evening.

Microclimate in roost trees

Within the flat parts of the forest no spatial differences were detected in the minimum temperature at undergrowth level (1 m height). Vertical differences, however, were significant (Table 9.2). The minimum temperature in the treetops was lower. It appears that the temperature difference increases with the height of the (emerging) tree crown, but it is also clear that at a height of about 25 m the difference in the inland tree is about 0.6 °C larger than along the river. Within riverine roost trees there was no discernible vertical gradient.

The roost branches could be anywhere between 5 and 40 m high. The primate species that roost in inland forest often roost at heights of 40 m or more. If we take into account that monkeys roosting along the river in effect roost within the canopy, whereas those roosting inland roost above it, the minimum temperature in the riverine roost tree would be 0.5 to 1.0 °C higher than in the inland roost tree.

Table 9.2. *The difference between average minimum temperatures*
(AvT < min >) in the high canopy and the undergrowth for three different
trees in the forest and along the river (as indicated by distance of river)

Distance to river (m)	Height of measurement (m)	AvT < min > undergrowth (°C)	Average difference undergrowth– canopy (°C)	n
0	22.5	20.77	0.22**	38
125	25.5	20.68	0.80***	31
140	32.5	21.03	1.11***	34

**$p < 0.01$; ***$p < 0.001$; n indicates number of measurements.

Effects of distance from river on range use

The effect of the refuging habit on the monkeys' range use was dramatic.
There was a continuous decrease in the intensity of use away from the river
in all groups (Fig. 9.1). In fact, the monkeys spent more than half of their
time at less than 100 m from the river. Obviously, the correlation between
each activity and distance from the river was also negative (Table 9.3).

As one moves away from the river and enters higher terraces, productivity,
and hence the amount of food available, decreases. Does this confounding
effect of productivity explain the steady decrease in use further inland? This
proposal is supported in part, since if we control for the effect of fruit
production on time spent feeding and of leaf litter production on foraging
time, the correlations with distance from the river are no longer significant,
although they all remain negative (Table 9.4). Note, however, that our
procedure for testing for statistical significance is quite conservative.

The theoretical models for refuging suggest that this pronounced
zonation may affect the distribution of activities and the duration of visits
to food sources. Time budgets relative to the distance from the river show
that the strongest effect on activity was found in the zone closest to the river,
where the animals spent relatively more of their time resting and less time
foraging and travelling (Fig. 9.2). It was also expected that the monkeys
would pay many short visits to food sources or foraging areas near the river,
whereas visits further inland would be less frequent but longer. There was
no significant positive correlation between distance from the river and
feeding time per visit, although a trend existed for foraging time per visit
(Table 9.5). Groups varied in size over the years, and as group size
increased, they tended to spend more time further inland (Table 9.6).

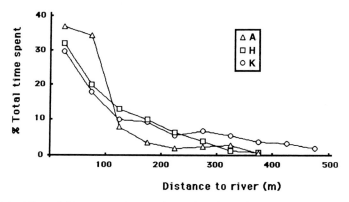

Fig. 9.1. Effect of distance to river on intensity of use (indicated by percentage of time spent), for three groups at our study site (A = Antara; H = House; K = Ketambe).

Table 9.3. *Effect of distance from river on total time spent (in each activity) and on frequency of entering quadrats: Pearson correlation coefficients with distance to river (degrees of freedom: House 33; Ketambe 35; Antara 22)*

Group	Time	Feed	Forage	Rest	Travel	Enter
House	−0.493**	−0.432**	−0.430**	−0.428**	−0.500**	−0.522**
Ketambe	−0.426**	−0.340*	−0.453**	−0.357*	−0.444**	−0.484**
Antara	−0.345	−0.300	−0.280	−0.343	−0.322	−0.365
Combination test	26.42***	19.63**	22.69***	20.95**	26.23***	31.30***

$*p < 0.05$; $**p < 0.01$; $***p < 0.001$.

Table 9.4. *Effect of distance from river on time spent feeding and foraging when partialised for production of fruit and leaf litter: Pearson correlation coefficients with distance to river (degrees of freedom: House 32; Ketambe 34; Antara 21)*

Group	Feed (fruit)	Forage (leaf litter)
House	−0.162	−0.340*
Ketambe	−0.132	−0.229
Antara	−0.141	−0.067
Combination test	4.97	10.07

$*p < 0.05$.

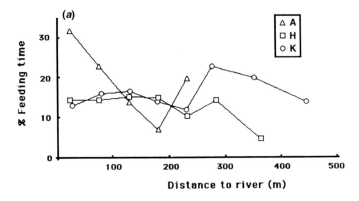

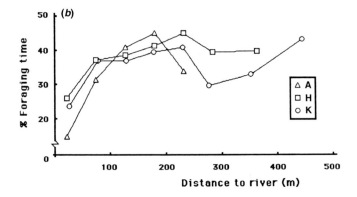

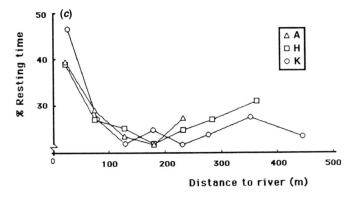

Fig. 9.2. Effect of distance to river on the time budget of group activities:
(a) percentage feeding time; (b) percentage foraging time; (c) percentage resting
time; (d) percentage travelling time (A = Antara; H = House; K = Ketambe).

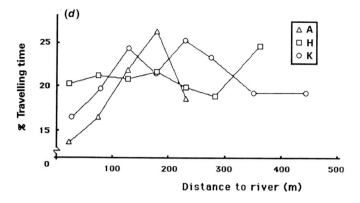

Fig. 9.2 (*cont.*)

Table 9.5. *Effect of distance from river on feeding and foraging per visit of quadrat: Pearson correlation coefficients with distance from river, also when partialised for production of fruit and leaf litter (see Tables 9.3 and 9.4 for degrees of freedom)*

Group	Feed per visit	Feed per visit (fruit)	Forage per visit	Forage per visit (leaf)
House	−0.228	+0.069	+0.190	+0.166
Ketambe	−0.057	+0.109	+0.023	+0.044
Antara	−0.345	−0.188	+0.201	+0.236
Combination test	8.40		4.97	5.12

Table 9.6. *Mean distance from the river during the morning hours (until 12.00) over three sample periods during which both groups (House and Ketambe) increased in size (based on entering frequency of the quadrats)*

	Period 1			Period 2			Period 3		
	size	dist	s	size	dist	s	size	dist	s
House	29	94.0	319	33	128.4	1607	40	133.5	2583
Ketambe	24	154.7	772	31	199.7	645	37	230.1	1412

Size = group size; dist = mean distance (m); s = sum of entering frequencies over all quadrats. Period 1 was from 1976 to 1977; period 2 from 1980 to 1981; period 3 was in 1982 (see also Table 9.1).

Diurnal variation in range use, activities and food choice

The results on range use show that the refuging habit has a strong impact on range use, but not as expected from the refuging models. The pronounced zonation of the range may be an artifact of diurnal patterns in the long-tails' behaviour rather than a direct effect of the position of the refuge. Hence, how time of day affects the interaction between distance from the river and maintenance activities requires investigation.

There was a clear diurnal pattern in the four main activities in all groups (Fig. 9.3). After the departure from the roost tree, resting remained low but tended to rise after 3–4 p.m. Travel time peaked in the early morning and showed a gradual decline during the rest of the day. Time spent feeding on clumped food (mainly fruit) tended to show a high peak in the early morning and a somewhat lower peak in the late afternoon. Foraging for dispersed food rose to a high level around 8 a.m., then remained high until around 3 p.m., after which a gradual decrease set in. Time of day also affected food choice during foraging. During the second half of the afternoon, the monkeys increasingly selected vegetation matter (Fig. 9.4).

Although the monkeys usually set out in a direction at right angles to the river, they had reached the maximum distance away from the river at 10 or 11 a.m. (Fig. 9.5), and spent many of the afternoon hours near the river. However, this does not mean that they covered most of the day's journey before noon, because half the total day journey length was reached around noon (see also travel time in Fig. 9.3).

In conclusion, the monkeys spent much of the afternoons, when their resting increased and their foraging decreased and they became less selective, near the river. The question arises as to why the monkeys return to the riverine strip so early.

It might be argued that they do so in order to reduce their heat load by swimming or sitting in the cool air above the river. Indeed, it was often observed that the monkeys rested on the shady branches above the water, where it is much cooler during the warm afternoon hours. Young monkeys occasionally swam during the afternoon, perhaps indicating that they had built up a heat load during their travel through the forest. However, there were also rivulets in the forest where they could (and sometimes did) swim. Moreover, there was no correlation between the maximum temperature in a given day and the time the group returned within 25 m from the river in H ($r = 0.09, n = 108$) and K ($r = 0.17, n = 44$). Group A returned even later on hotter days ($r = 0.42, n = 24, p < 0.05$).

In general, there were very few significant correlations between weather

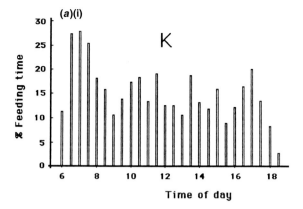

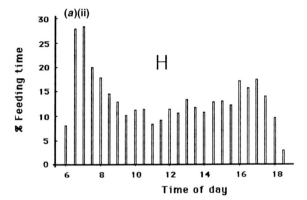

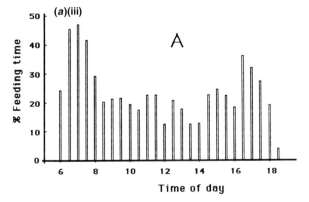

Fig. 9.3. Diurnal patterns of group activities: (*a*) percentage feeding time;
(*b*) percentage foraging time; (*c*) percentage resting time; (*d*) percentage travelling
time (A = Antara; H = House; K = Ketambe).

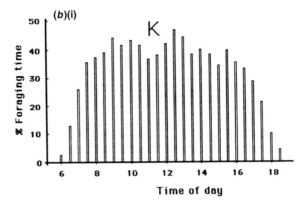

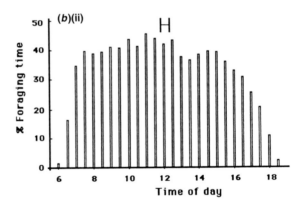

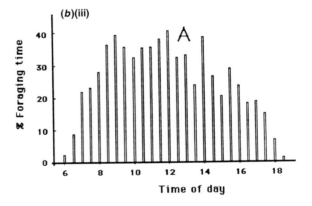

Fig. 9.3 (*cont.*)

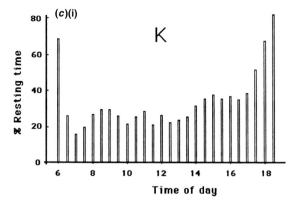

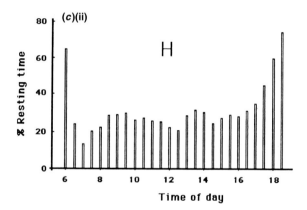

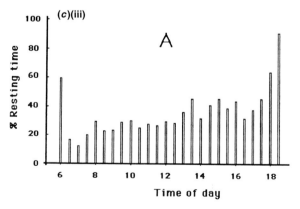

Fig. 9.3 (*cont.*)

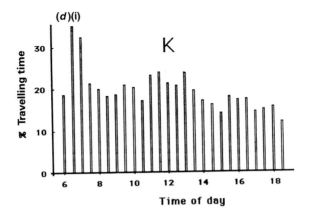

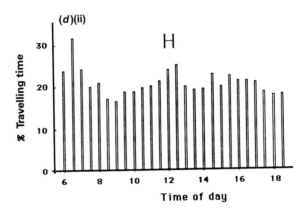

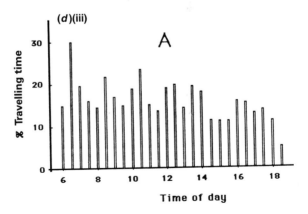

Fig. 9.3 (*cont.*)

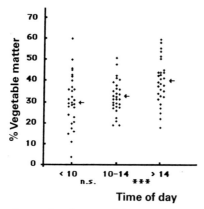

Fig. 9.4. The median (arrows) and range of the monthly means in the percentage of vegetable matter taken during foraging in relation to time of day. Included are data from group H ($n = 17$ months), K ($n = 7$), and A ($n = 5$). Differences between time blocks were evaluated using the sign test (n.s., not significant; ***, $p < 0.001$).

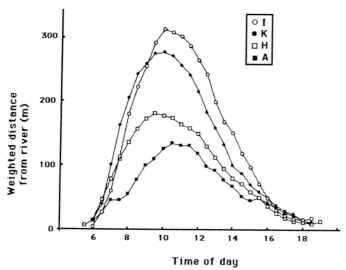

Fig. 9.5. Mean distance of river, weighted for time spent at each visit to a quadrat, in relation to time of day for all groups with activity data (see Table 9.1).

variables and group activities or ranging parameters, and those that were found could be confounded by seasonal trends. In order to control for this possibility, ranging variables of two groups (K and H) followed simultaneously on 17 mornings (to 12:00) within a two-month period were compared. If weather influenced the behaviour of the monkeys, then the distance covered, or the patterning of bouts of rest and movement should be positively correlated between the two groups. The total distance covered, however, was not positively correlated between the two groups ($r = -0.253$, $n = 17$), and neither was the proportion of the journey covered in each of the three two-hour blocks ($r = -0.093$ for percentage covered up to 8:00; $r = 0.069$ for percentage between 08:00 and 10:00. and $r = -0.356$ for percentage covered between 10:00 and 12:00).

Discussion

The advantage of riverine refuging

Roost trees along the river tended to be difficult to enter. They had open crowns and branches extending out over the river. The monkeys usually sat on the thinner branches far from the crown, facing the main trunk. This form of roosting along the river is consistent with the predation avoidance hypothesis. However, these results do not explain why only riverside trees are used to roost, because inland emergents might serve this function equally well. Our suggestion is that roosting along the river has the additional advantage of reducing energy loss during the night. Because minimum temperatures in the study area are around 20 °C, so considerably below the critical temperature of about 28 °C (Schmidt-Nielsen, 1975), a monkey resting in a tree crown will lose heat. The following calculations attempt to express the reduction of heat loss due to roosting along the river in terms of travel costs.

A monkey in a roost tree will lose heat mainly through convection, evaporation and radiation. Heat loss due to convection will be ignored here, because wind speeds are generally low during the night and not likely to be different between riverine and inland roost trees. Differences in evaporative heat loss due to the saturation of the air they breathe are to be expected when the air varies in humidity among roost trees. Often differences in relative humidity between riverine and inland tree crowns are negligible, but during the driest periods the difference becomes larger (maximum difference from at least 95% relative humidity above the river to 75% in inland trees). Occasionally there might, therefore, be an extra heat loss caused by roosting in an inland tree because exhaled air is fully saturated with water and the animal supplies the heat for the forced evaporation. At a temperature of around 20 °C a 3 kg monkey will have a metabolic rate of around $1.8 \times$ basal rate, and thus needs some 2.75 litre of oxygen per hour, or around 55 litres of air per hour (see Schmidt-Nielsen, 1975). The air exhaled by a monkey will contain 2.24 g of water per hour. The air it inhales will, at 20 °C contain 0.96 g of water, at 100% relative humidity (RH). A 20% difference in RH of the air will make a difference of 0.19 g of water, or about 110/cal per hour. Over the course of the night, with a constant 20% difference in RH, a monkey roosting along the river saves 1/kcal (1 kcal = 4.184 kJ).

Radiative heat loss is probably the most common source of energetic cost. We make the following assumptions, all based on Moen (1973). A 3 kg monkey has a radiative surface area of about 29 dm^2, which it can reduce by perhaps 40% through huddling together with others, leaving 17.4 dm^2. The night lasts 12 hours, and the sky is usually overcast, so we can assume that its radiant temperature equals that of the air. The animal's radiant temperature equals 9.5 + (0.75 × air temperature). Now the monkey's radiative heat loss (HL) over the night can be calculated as:

$$HL = 0.58 \times 17.4 \times (T_m - T_a), \text{ (Kleiber, 1975)}$$

where T_m is the temperature of the monkey and T_a is the air temperature. Therefore, the monkey's radiative heat loss increases by 2.5 kcal for each 1 deg.C decrease in air temperature. With a mean temperature difference of between 0.5 and 1 deg.C, this amounts to around 2 kcal for a whole night.

Hence, by roosting along the river the average 3 kg monkey may gain up to about 3 kcal. This heat loss is equivalent to walking for up to 600 m, and so a monkey should be prepared to walk to the river to roost in a strip up to 300 m from the river. If this result is approximately correct, it does explain why long-tail groups inhabiting riverine areas roost along the river, but not why they are restricted to the riverine zone (one home range deep).

Two additional factors that would make riverine refuging advantageous remain to be discussed. First, it is possible that jumping into the water is a more effective way to escape arboreal predators than leaping into a neighbouring tree. However, where crocodiles infest the waters, as in some coastal swamps, this is not necessarily true. Second, the thermal advantages of riverine refuging may be minimal in lowland swamp forest, at least during the flooding stage. One possibility worth exploring is that mosquito densities are actually lower along the river due to the air movements caused by the running water.

Riverine refuging and distribution

In comparison with inland forest, riverine forests tend to have denser canopy and ground cover, greater liana density, more numerous but generally smaller fruit sources, and greater leaf production and hence insect availability (van Schaik & Mirmanto, 1985, for Ketambe). These differences conspire to make it harder for the long-tails to maintain a positive energy balance in the inland forest: lower fruit production, more costly locomotion and cooler nights. Moreover, the effective productivity of the inland areas may be further reduced by the presence of pig-tailed macaques, whose diet probably overlaps extensively with that of the long-tails (cf. Rodman, 1991).

Can the energy balance approach also account for distribution patterns of the species in Sumatra? Long-tailed macaques have the lowest upper altitudinal limit of any diurnal primate in Sumatra (van Strien, 1985; C. P. van Schaik, unpublished data.) being very rare over 900 m above sea level; productivity and night-time temperatures both decline with elevation. Deviations from the general trend are clearly related to productivity. In northern Sumatra, the species is absent in narrow, unproductive valleys as low as 250 m (in the upper Bengkung, for instance), but occurs in riverine forests of the Karo highlands on fertile volcanic soils as high as 1400 m. From this we can derive the prediction that populations in productive lowland rainforests may show a continuous distribution of home ranges without riverine refuging, just like they do in mangroves and swamp forests. The Simeulue Island long-tailed macaques (*Macaca fascicularis fusca*) occur in all habitats and, therefore, do not show consistent riverine refuging (Sugardjito *et al.*, 1989). In further testing this prediction, one should remember that the remnant groups in forest fragments often sleep at the forest's edge.

The long-tail's abundance in forest of high productivity may also account for its frequent occurrence in secondary forests and landscapes heavily modified by humans. Structural and productivity patterns of secondary or heavily exploited forests resemble those of riverine forests in many respects. Therefore, it is not surprising that long-tails often attain high densities in secondary forests (Crockett & Wilson, 1980) and cling tenaciously to their riverine habitats after humans have cleared most of it. It is probably the preference for fertile riverine forests and the capacity to live in secondary forests on these soils that underlies the correlation between human and long-tail densities, rather than a direct dependence on human activities for a substantial portion of their diet, as suggested by the 'weed macaque' hypothesis of Richard, Goldstein & Dewar (1989).

Refuging and range use

The Ketambe monkeys' range use was strongly influenced by their refuging habit. They spent much of their time near the river, where they also tended to have a different time budget. Here we discuss to what extent theoretical models of refuging apply to the situation we studied. Hamilton & Watt

(1970) were the first to predict that food sources further away from the refuge should be larger or of higher quality to balance the high travel costs incurred in reaching them. Hence, the monkeys should spend more time feeding per visit in distant parts of the range. In previous studies of primate refuging systems this prediction was confirmed (Wheatley, 1980; Sigg & Stolba, 1981). In Wheatley's (1980) study of Bornean long-tailed macaques, the result was directly attributable to a correlation between tree crown size and distance from the river. We could not confirm this prediction in our study, perhaps because the long-tails forage continuously as they move between fruit trees rather than in discrete bouts.

This study has indicated that the riverine strip was exploited quite heavily, since total time spent feeding and foraging decreased with distance from the river. Nevertheless, the time budget is different in the riverine strip. The proportion of time spent resting was higher than elsewhere, while time spent foraging was lower. This pattern cannot be equated with the differentiation of a trampling zone and arena found in other refuging systems (Hamilton & Watt, 1970) because of the heavy feeding and foraging in the riverine strip. It seems best explained by the monkeys' tendency to spend much of their late afternoons along the river, and by the reduced foraging and increased resting during that time of day. This diurnal patterning is also found in other monkey species that use forest trees for roosting (e.g. Terborgh, 1983). However, the question still remains 'why should the monkeys spend so much time near the river'? The amount of time spent near the river decreases as the range becomes less productive (Fig. 9.5); group I went farthest inland and inhabited a poor mountain slope, group A, which inhabited a highly productive low terrace, stayed near the river and returned early, whilst groups H and K were intermediate in both respects. Group A was also small, and smaller groups in which competition for food was less severe (van Schaik et al., 1983) spent more time near the river as indicated by within-group comparisons (Table 9.6).

All these patterns are compatible with the idea that the monkeys prefer to stay close to the river as much as possible, unless hunger leads them to exploit the areas away from it more heavily. The most likely explanation for this pattern is that they prefer to be in familiar areas, in which they have the best information about the presence and absence of predators (van Schaik, 1985).

Acknowledgements

We thank the Indonesian Institute of Sciences (LIPI) for permission, and Universitas Nasional (Jakarta) for sponsoring this research. The Indonesian

Forest Protection and Nature Conservation Service (PHPA) gave us permission to work in Ketambe. Idrusman Ariga, Johan Mouton and especially Rob de Boer and Isolde den Tonkelaar helped with data collection. Financial support was provided by the Netherlands Foundation for the Advancement of Tropical Research (WOTRO) and Dobberke Foundation.

References

Aldrich-Blake, F. P. G. (1980). Long-tailed macaques. In *Malayan forest primates*, ed. D. J. Chivers, pp. 147–65. New York: Plenum Press.

Anderson, M. (1978). Optimal foraging area: size and allocation of search effort. *Theoretical Population Biology*, **13**, 397–409.

Bearder, S. K. (1987). Lorises, bushbabies, and tarsiers: diverse societies in solitary foragers. In *Primate societies*, ed. B. B. Smuts, D. L. Cheney, R. M. Seyfarth, R. W. Wrangham & T. Struhsaker, pp. 11–24. Chicago: Chicago University Press.

Bert, J., Ayats, H., Martino, A. & Collomb, H. (1967). Le sommeil nocturne chez le babouin, *Papio papio*. *Folia Primatologica*. **6**, 28–43.

Crockett, C. M. & Wilson, W. L. (1980). The ecological separation of *Macaca nemestrina* and *M. fascicularis* in Sumatra. In *The macaques: studies in ecology, behavior and evolution*, ed. D. G. Lindburg, pp. 148–81. New York: van Nostrand Reinhold.

Dawson, G. A. (1979). The use of time and space by the Panamanian tamarin, *Sanguinus oedipus*. *Folia Primatologica*, **31**, 253–84.

Fittinghoff, N. A. Jr & Lindburg, D. G. (1980). Riverine refuging in East Bornean *Macaca fascicularis*. In *The macaques: studies in ecology, behavior and evolution*, ed. D. G. Lindburg, pp. 182–214. New York: van Nostrand Reinhold.

Gautier-Hion, A. (1973). Social and ecological features of talapoin monkeys – comparisons with sympatric cercopithecines. In *Comparative ecology and behavior of primates*, ed. R. P. Michael & J. H. Crook, pp. 147–50. New York: Academic Press.

Gautier-Hion, A. & Gautier, J.-P. (1978). Le singe de Brazza: une stratégie originale. *Zeitschrift für Tierpsychologie*, **46**, 84–104.

Gautier-Hion, A., Gautier, J. P. & Quris, R. (1981). Forest structure and fruit availability as complementary factors influencing habitat use by a troop of monkeys (*Cercopithecus cephus*). *Revue d'écologie (Terre et Vie)*, **35**, 511–36.

Hamilton, W. J. & Watt, K. E. F. (1970). Refuging. *Annual Review of Ecology Systematics*, **1**, 263–86.

Kleiber, M. (1975). *The fire of life*, 2nd edn. New York: Krieger Publishing Company.

Kurland, J. A. (1973). A natural history of kra macaques (*Macaca fascicularis* Raffles 1821) at the Kutai Reserve, Kalimantan Timur, Indonesia. *Primates*, **14**, 245–62.

Kummer, H. (1971). *Primate societies*. Oxford: Aldine-Atherton.

Lancaster, J. B. (1972). Play-mothering: the relations between juvenile females and young infants among free-ranging vervet monkeys. In *Primate socialization*, ed. F. E. Poirier, pp. 83–104. New York: Random House.

Moen, A. N. (1973). *Wildlife ecology, an analytical approach*. San Francisco, CA: W. H. Freeman.

Orians, G. H. & Pearson, N. E. (1979). On the theory of central place foraging. In *Analysis of ecological systems*, ed. D. J. Horn, R. D. Mitchell & G. R. Stairs, pp. 155–77. Columbus: Ohio State University Press.

Richard, A. F., Goldstein, S. J. & Dewar, R. E. (1989). Weed macaques: the evolutionary implications of macaque feeding ecology. *International Journal of Primatology*, **10**, 569–94.

Rijksen, H. D. (1978). *A field study on Sumatran Orang Utans (Pongo pygmaeus abelii* Lesson 1827). Wageningen: Veenman & Zonen.

Rodman, P. S. (1991). Structural differentiation of microhabitats of sympatric *Macaca fascicularis* and *M. nemestrina* in East Kalimantan, Indonesia. *International Journal of Primatology*, **12**, 357–75.

Schmidt-Nielsen, D. (1975). *Animal physiology*. Cambridge: Cambridge University Press.

Schoener, T. W. (1979). Generality of the size-distance relation in models of optimal foraging. *American Naturalist*, **114**, 902–14.

Sigg, H. & Stolba, A. (1981). Home range and daily march in a Hamadryas baboon troop. *Folia Primatologica*, **36**, 40–75.

Sokal, R. R. & Rohlf, F. J. (1981). *Biometry*. San Francisco, CA: W. H. Freeman.

Sugardjito, J. (1983). Selecting nest sites by Sumatran orang-utans *(Pongo pygmaeus abelii)* in the Gunung Leuser National Park, Indonesia. *Primates*, **24**, 467–74.

Sugardjito, J., van Schaik, C. P., van Noordwijk, M. A. & Mitrasetia, T. (1989). Population status of the Simeulue monkey *(Macaca fascicularis fusca)*. *American Journal of Primatology*, **17**, 197–207.

Terborgh, J. (1983). *Five new world primates*. Princeton: Princeton University Press.

van Schaik, C. P. (1985). The socio-ecology of Sumatran long-tailed macaques *(Macaca fascicularis)*. I. Costs and benefits of sociality. PhD thesis, University of Utrecht.

van Schaik, C. P. & Mirmanto, E. (1985). Spatial variation in the structure and litterfall of a Sumatran rain forest. *Biotropica*, **17**, 196–205.

van Schaik, C. P. & Mitrasetia, T. (1990). Changes in the behaviour of wild long-tailed macaques *(Macaca fascicularis)* after encounters with a model python. *Folia Primatologica*, **55**, 104–8.

van Schaik, C. P. & van Noordwijk, M. A. (1985). Interannual variability in fruit abundance and the reproductive seasonality in Sumatran long-tailed macaques *(Macaca fascicularis)*. *Journal of Zoology, London*, **13**, 533–49.

van Schaik, C. P., van Noordwijk, M. A., de Boer, R. J. & den Tonkelaar, I. (1983). The effect of group size on time budgets and social behaviour in wild long-tailed macaques *(Macaca fascicularis)*. *Behavioral Ecology and Sociobiology*, **13**, 173–81.

van Strien, N. J. (1985). The Sumatran rhinoceros – *Dicerorhinus sumatrensis* (Fisher, 1814) – in the Gunung Leuser National Park, Sumatra, Indonesia: its distribution, ecology and conservation. PhD thesis, Agricultural University, Wageningen.

Wheatley, B. P. (1980). Feeding and ranging of East Bornean *Macaca fascicularis*. In *The macaques: studies in ecology, behavior and evolution*, ed. D. G. Lindburg, pp. 215–46. New York: van Nostrand Reinhold.

10

A comparison of wild and food-enhanced long-tailed macaques (*Macaca fascicularis*)

B. P. WHEATLEY, D. K. HARYA PUTRA
AND M. K. GONDER

Introduction

The common long-tailed or crab-eating macaque (*Macaca fascicularis*) of South-East Asia is widespread from mainland Asia to the Philippine and Indonesian archipelago (Fig. 10.1). Hill (1974) considers it one of the most widely distributed catarrhine monkeys. Extreme variations can be expected within the species because of the differing strength and extent of diverse selective forces at work on its populations. Tail length and body size presumably fluctuate according to Bergmann's and Allen's Rules. A comparison of relative tail length in *M. fascicularis* populations on the Malay peninsula from the map in Fooden (1971) shows that northern populations above 8° latitude have significantly shorter tails than populations south of this latitude (Median Test; $p < 0.001$). There is no question that physical differences between the various populations of *M. fascicularis* account for the unnecessary proliferation of species names. References to this species as *M. cynomolous* or *M. irus* still appear in the literature today despite the corrections by Blanford (1887), Miller (1942) and Fooden (1964). Not only has the species more sub-species than any other macaque (Napier & Napier, 1967) but also one of the commonest sub-species of *M. fascicularis* has had 15 different names assigned to four different genera (Chasen, 1940).

Have the same mistakes been made with behaviour as taxonomists of old made with morphology? Have we held steadfastly to our assumed 'wild' behaviour unfettered by human contact as we have to our equally unfettered and assumed 'wild' habitat? If the identification of a species' morphological similarities is so complex, then just how difficult is the determination of behavioural similarities when they have not been studied. It is important, therefore, to examine a species in a variety of habitats ranging from the wild to food-enhanced areas and to the laboratory.

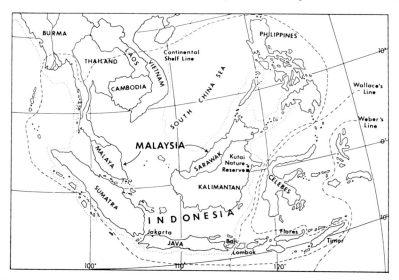

Fig. 10.1. Approximate distribution of the crab-eating macaque (*Macaca fascicularis*).

The attraction of studying a species in the wild is well known and often stated. Generally, there is a desire to observe and understand how behaviours evolved over millions of years 'undisturbed' by humans. Contrastingly, studying a species in food-enhanced environments 'disturbed' by humans is not so readily appreciated. Who can say that *M. fascicularis*, with its relatively recent evolutionary history, has had no influence from humans to the present day? The swidden agriculture of South-East Asian peoples over thousands of years has opened up new habitat areas for the species just as naturally disturbed riverine habitat has kept them going. Observations on wild populations are hampered by problems of individual recognition and behaviour. If we were to rely on information gleaned from wild populations, our knowledge of a species' repertoire would be woefully inadequate. Captive studies provide many advantages, from the manipulation of variables in a controlled setting to detailed observations extending into physiological and psychological areas. All avenues of investigation are valid and necessary to understand a species adequately.

The origin and speciation of *M. fascicularis* is an interesting question. The selective effects of malaria on the ecological isolation and speciation of this and other macaque species have been hypothesised (Wheatley, 1980a). Briefly, malaria may delimit the geographical ranges of some macaques, especially the allopatric species of *M. mulatta* and *M. fascicularis*. Furthermore, a developed genetic immunity to this disease in the latter species may have facilitated its divergence from an ancestral macaque and its dispersal into South-East Asia. These, and other hypotheses, are currently under study by several researchers, especially Takenada & Takenada (1990).

The species is sexually dimorphic; Washburn (1942) reported an average weight of 4.8 kg for 15 males and 3.0 kg for 11 females collected in Sabah. Adult females of *M. fascicularis fascicularis* are, therefore, 62.5% of the body weight of adult males. Schultz (1956) reported that adult females were 63.9% of the weight of adult males. This ratio may be age dependent, since it steadily declines from 88% at birth to 50% body weight of males at age 7 (Spiegel, 1956). *Macaca fascicularis* is, therefore, more sexually dimorphic than *M. mulatta*, with females at 69% of male body weight (Schultz, 1956).

This chapter compares *M. fascicularis* in Kalimantan with those of Bali. The sub-species of these populations are known as *M. f. fascicularis* and *M. fascicularis submordax*, respectively. The latter designation for the Balinese sub-species follows Sody (1949), who split Hill's (1974) sub-species *M. irus mordax* for both Javanese and Balinese populations by retaining *mordax* for the Javanese and giving the new sub-species name of *submordax* to the Balinese forms (Fig. 10.2). This differentiation is supported by recent electrophoretic examinations of blood proteins (Kawamoto & Ischak, 1981; Kawamoto, Nozawa & Ischak, 1981; Kawamoto, Ischak & Supriatna, 1984; Kawamoto & Suryobroto, 1985).

Study sites

Kalimantan

Research in East Kalimantan extended from October 1974 to June 1976. The study area was at the Hilmi Oesman Memorial Research Station in the Kutai Nature Reserve (Fig. 10.1). It is located at approximately latitude 0° 32′ N and longitude 117° 28′ E, about 17 km west of the Makassar Strait.

Bali

Research in Bali (Fig. 10.1) was carried out from June to August 1986. The study area was in the Monkey Forest at the village of Ubud, about 20 km north of Denpasar, the provincial capital. The 1990 research season was from July to September.

Habitats

Kalimantan

The Kalimantan site was chosen to study a *M. fascicularis* population as undisturbed by humans as possible. An in-depth analysis of the habitat

Fig. 10.2. A Balinese *M. fascicularis* mother. Photo by B. Wheatley.

most frequented by the study group was made. The Kutai Nature Reserve is tropical lowland mixed dipterocarp forest or tropical lowland evergreen rainforest. The study troop of monkeys spent 75% of its time in 27 ha. One of these hectares was randomly chosen and intensively analysed. The selected hectare was 50 m from the Sengata River and it was subject to periodic flooding. All trees and lianas with a diameter greater than or equal to 4.5 cm at chest height were identified by a local assistant and numbered with a tag. Sixty-three per cent of the 1179 trees and lianas in the sample hectare belonged to the smallest diameter class interval, between 4.5 and 12.6 cm. Ten species representing 63% of *all* trees and lianas and 85% of the total basal area in the hectare are given in Table 10.1. Figure 10.3 gives a profile diagram of a randomly selected strip 10 m wide and 100 m long within the sample hectare. The profile diagram gives tree height, width, and

Table 10.1. *The ten most dominant trees in the sample hectare as determined by basal areas. The numbers refer to those given in Fig. 10.3*

Number	Scientific name	Relative dominance (%)	Frequency
18	*Macaranga pruinosa*	29	197
13	*Callicarpa farinosa*	16	140
6	*Pterospermum* sp.	11	26
7	*Ficus* sp.	8	1
19, 20, 21, 22	*Phoebe* sp.	6	72
14	*Hyonocarpus* (?) sp.	5	36
17	*Leea* sp.	5	28
28	*Vitex* sp.	3	156
33	*Eusideroxylon zwageri*	1	45
3	*Notaphoebe opaca*	1	29

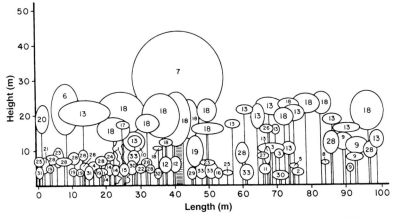

Fig. 10.3. Profile diagram of a sample strip of forest 10 m wide and 100 m long. See Table 10.1 for key to tree species numbers.

crown depth in cross-section. The tree numbers in the diagram are identified in Table 10.1. The Shannon–Wiener diversity measure for canopy trees (between 16 and 30 m tall) is 1.3. For understorey trees the measure is 3.0 and for the entire hectare 3.4. An analysis of vertical structure using crown depth shows three strata with little overlap between tree types.

This hectare's features are typical of secondary forests: high density of short, small trees, domination of pioneer trees and indications of human hand-logging on ironwood (*Eusideroxylon zwageri*). The main cause of secondary forest is swidden agriculture, but natural disturbances can also

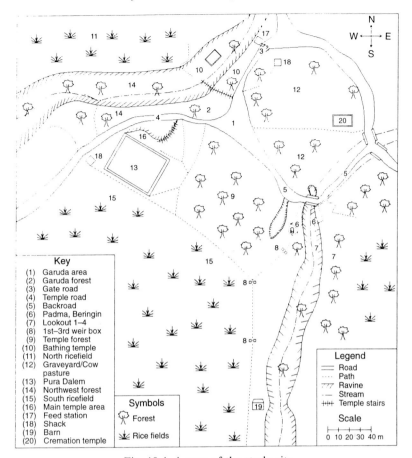

Fig. 10.4. A map of the study site.

be important. Poor drainage of B horizon clay soils, shallow root systems and weak tree stems, and frequent flooding of riverine areas contribute to the felling of trees during storms (see Wheatley, 1978).

Bali

The Monkey Forest site at Ubud, Bali was selected for detailed behavioural observations on habituated *M. fascicularis*. The frequent visitation of tourists to Ubud and the location of the Monkey Forest in a sacred area have guaranteed habituation and protection of these monkeys. The Monkey Forest is just south of Ubud and is surrounded by rice fields (Fig. 10.4). It is about four hectares in area and contains three temples and a graveyard. The largest temple is the Pura Dalem or Death Temple, the next largest is the Cremation temple and the smallest is the Bathing temple. According to informants these temples and the forest have been undisturbed for at least four centuries. A detailed floristic analysis has not yet been undertaken. Climatological data from Bali were obtained at the weather station in the nearby village of Celuk.

Ranging and diet

Kalimantan

The study troop entered 125 different hectares. The home range was thus 1.25 km^2. The average daily range, however, was 19 ha. Most of the troop's time each day (62% of time between 6.00 and 1.00 p.m.) was spent in streamside secondary forest. This could be because the fruit in their diet was more frequently available and abundant in secondary forests. Surveys of fallen fruit on 1800 m of trails, twice a month over a period of a year and a half, showed that fruit was significantly more frequent and continually available in streamside areas than in non-streamside areas (Mann–Whitney–Wilcoxon Tests). Over half the trees found in these areas contained edible fruit. The most utilised food source was *Callicarpa farinosa*, which occurs in very high densities and always has fruits available. The fruits are small and individual monkeys spent less than 5 min per feeding bout on them per tree but they have one of the highest crude protein percentage (11.35%) of any fruit tested. Individual trees of this species fruit several times per year. The study troop appeared to monitor the larger fruit trees in the primary rainforest. When these trees such as *Koordersiodendron pinnatum*, *Dracontomelon mangiferum*, *Cratoxyglon* sp. or *Ficus* sp. fruited, the troop would spend 20 minutes or more per feeding bout per tree. The animals tended to forage singly and widely dispersed in the trees of the secondary forest but as a troop and less dispersed in the larger trees of the primary rainforest. A number of significant correlations on ranging were noted. As the distance from a stream increased so did the feeding duration per tree whereas as the average time spent in a hectare fell so did feeding frequency.

Another ranging pattern was the dispersal from their refuge or sleeping tree each morning and return to it each afternoon. This pattern has been called central refuging (Hamilton & Watt, 1970; Fittinghoff & Lindburg, 1980). Animals returned to their primary refuge tree 70% of the time. This tree, *Intsia palembanica*, was the largest, loftiest and most visible of any tree in the troop's home range that hung out over the Sengata River. Six other refuge trees used by the troop also hung out over the water. Riverine refuging is probably related to avoidance of predators, the distribution and abundance of food, efficient foraging strategy, predictability of congregating in one location, detection of trespassing conspecifics and the maintenance of relatively exclusive home ranges. The importance of sleeping on the ends of small branches of the refuge tree is that an approaching predator can be detected and avoided by jumping into the water below and swimming off to safety. Hoogerwerf (1970) found hairs from *M. fascicularis*, *Presbytis aygula* and *P. cristatus* in all three of the panther fecal specimens that he obtained in Java. Clouded leopards were seen in the study area. Ridley (1906) also noted that *M. fascicularis* never slept in small bushy trees for fear of being surprised by snakes at night. The troop mobbed a sun bear on three occasions and the researcher on one occasion when he climbed into a tree. During these mobbings the animals continuously alarm-called for up to 30 min and approached within four to five metres. The bear always moved off. The refuge tree was located at the base of a ridge bisecting the animals' home range and was also equidistant from the two small streams along which the animals foraged. The troop generally appeared to forage upstream or downstream on a daily basis in the secondary forests and then return to the refuge tree in the afternoon after monitoring the area for fruit. After harvesting most of the available fruit in one area, the troop would alternate to the other unexploited secondary forest the next day. For more information on ranging and diet, refer to Wheatley (1980*b*).

Diet was estimated by using Struhsaker's (1974) method, and Table 10.2 lists the proportion of different items. Most notable is the predominance of fruit. The animals usually preferred ripe fruit. For example, only the red fruits of *C. farinosa* were eaten, but only immature *Eusideroxylon zwageri* fruits were eaten. Small amounts of kaolin clay were also eaten.

An activity profile was obtained at the end of the study by following an adult male for three days. Moving occupied 45% of his time, followed by resting (42%) and feeding (13%) (see Table 10.2).

Bali

Troop ranges were small given the small size of the forest (Fig. 10.4). The home range of the three troops averaged seven hectares. Ranging patterns of these troops were significantly different. The probability that these ranges were the same is effectively zero ($p < 0.0001$) with chi-square $=$ 417; d.f. $= 30$; $n = 399$. These results showed the basic integrity of the troops; that is, they are not sub-groups of a single group. The core areas of each troop are shown in Fig. 10.5, with one troop ranging between the other

Table 10.2. *Descriptive summary of* **M.** fascicularis *at the two study sites in the Kutai Nature Reserve and in Bali*

	Kutai 1974–76	Bali 1986	Bali 1990
Location	Latitude 0°32′N Longitude 117°28′E	Latitude 8°30′S Longitude 115°15′E	
Habitat	Secondary forest Tropical lowland evergreen rainforest	Forest refugia surrounded by rice fields	
Elevation (m above sea level)	20–100		230
Temperature (mean/month; °C)	25.5 ($n = 6$)		26.4 ($n = 6$)
Humidity (mean/month; %)	75 ($n = 6$)		87.5 ($n = 11$)
Rainfall (mean/month; mm)	198 ($n = 42$)		181 ($n = 17$)
Home range (mean)	1.25 km²		7 ha
Density (km²)	25		1111
Daily range (mean; m)	1900		450
Troop size	42, 29, 27, 22 ($n = 4$)	See Table 10.4	
Mean troop size	30 ($n = 1$)	23 ($n = 3$)	31 ($n = 3$)
Troop composition	Multimale	See Table 10.4	
Socionomic sex ratio	1:3.3	1:6.75	
Breeding sex ratio	1:2	1:2.5	
Activity (%) adult male: Travel, rest, feed	45, 42, 13		25, 65, 10
Arboreality (%)	97		25–50
Diet (%)			See Table 10.3
Fruit	87	32[a] 18[b]	
Insect	4	29 12	
Flower	3	5 2	
Grass	2	23 11	
Leaf	2	11 5	
Other	2	0 1	23 (peanuts) 19 (sweet potatoes)

[a]Excluding food from human sources; [b]including food from human sources.

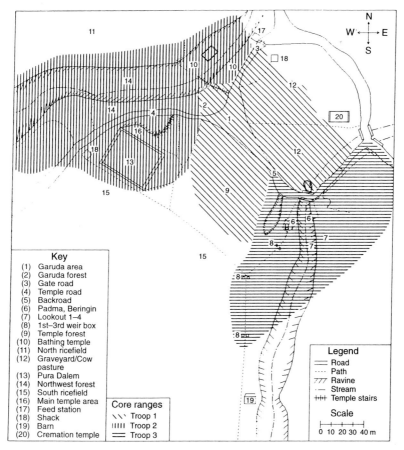

Fig. 10.5. Core ranges (troops 1–3).

Key
(1) Garuda area
(2) Garuda forest
(3) Gate road
(4) Temple road
(5) Backroad
(6) Padma, Beringin
(7) Lookout 1–4
(8) 1st–3rd weir box
(9) Temple forest
(10) Bathing temple
(11) North ricefield
(12) Graveyard/Cow pasture
(13) Pura Dalem
(14) Northwest forest
(15) South ricefield
(16) Main temple area
(17) Feed station
(18) Shack
(19) Barn
(20) Cremation temple

Core ranges
\\\ Troop 1
||||| Troop 2
≡ Troop 3

Legend
═══ Road
···· Path
/// Ravine
·— Stream
+++ Temple stairs

Scale
0 10 20 30 40 m

two troops. The core range was determined by scan samples taken at half-hour intervals on the three troops. Areas were defined according to the keys given in Figs. 10.4 and 10.5. A troop had to frequent an area for greater than 50% of the time for it to be included as part of its core range. The other two troops would together spend less than 50% of their time in that area. The daily range of all three troops was 450 m as determined from scan samples and a grid system overlay on a map. Troop 1 averaged 650 m over seven days, troop 2 averaged about 400 m over ten days and troop 3 averaged 350 m over nine days.

Troop 1's core area dominated the roadside areas. The most important area for tourist feeding was the Garuda area where three roads intersect. In contrast, the core area of troop 3 was basically restricted to a gorge that was an area of refuge when troop 1 displaced and chased them. Troop 3 spent most of its day seemingly 'trying' to enter the tourist areas when troop 1 was elsewhere. The scan samples showed that troop 1 was on or near the road at twice the frequency of troops 2 and 3.

The average troop diameter for all three troops was 44 m. The average for troop 3 was 32 m, whereas the average for troops 1 and 2 was 47 and 53 m, respectively. The difference was statistically significant with $t = 52.5$; $p < 0.001$, d.f. $= 387$.

The scan samples on arboreality show that, on average, over 50% of troop 3 is arboreal whereas the average for troop 1 is 25%. The mean number of arboreal individuals in troop 1 is 5.6 whereas the mean for troops 2 and 3 is 9.2 and 9.0 individuals, respectively. A t-test of the number of arboreal individuals is statistically significant with $t = 18.9, p < 0.0001$, d.f. $= 364$.

The diet for the main study troop, troop 1, in 1986 was again estimated using Struhsaker's (1974) method. Results are presented in Table 10.2; numbers to the left under the horizontal role line exclude foods from human sources, whereas those to the right include them. Fifty-eight per cent of this troop's food came from human sources. The peanuts were provided by tourists and local vendors. The sweet potatoes, *Ipomoea reptan*, were provisioned by local guards. Dietary proportions are also given in Table 10.2, with all human sources of food excluded from the table. Fruits are predominant in the diet, but insects and grass also form a significant proportion of the diet. The most unusual dietary behaviour was temple-licking. *Ad libitum* samples of animals who temple-licked showed that females licked significantly longer than males (Mann–Whitney–Wilcoxon statistic, corrected for ties, $z = -1.66, m = 7, n = 26, n = 33, p = 0.0485$). The mean duration of temple-licking for females was 8.1 min, whereas the mean duration for males was 3.5 min. This behaviour is probably related to sodium deficiencies, especially as an earlier study by Takenaka (1986) showed marginal deficiencies in the blood plasma of monkeys in a nearby area.

A more accurate estimation of diet was made in 1990 using scan samples on all three troops. Results are presented in Table 10.3 with data classified by food type, food part, food identification and by age–sex class. Unknown foods are not included. 'Tourist' food consisted primarily of peanuts and bananas sold in the forest by local vendors. Provisioned foods, such as sweet potatoes and papaya leaves, were only rarely given and are not tabulated. 'Offering' refers to rice and other token food items left each day on shrines and other holy places in the forest. The genus of grass eaten over 90% of the time is either *Cynodon* sp. or *Zoisia* sp.

Table 10.3. *Diet for all three troops in 1990*

A. *Food types*

	Tourist	Tree	Insect	Grass	Offering	Vine	Shrub	Herb
Adult male	55	22	17	12	8	3	1	0
Adult female	158	112	74	42	18	15	12	4
Mother	94	43	8	17	1	0	2	1
Juvenile	214	134	83	39	18	22	10	4
Infant	41	9	2	3	2	0	0	0
Total	562	320	184	113	47	40	25	9

B. *Food parts*

	Fruit	Seed	Leaf	Stem	Flower	Root and bark
Adult male	39	33	19	7	2	1
Adult female	164	97	67	24	15	5
Mother	60	50	36	7	5	1
Juvenile	184	140	87	25	23	2
Infant	26	17	6	0	1	0
Total	473	337	215	63	46	9

C. *Food identification*

	Peanut	Banana	Grass	Ficus	Coconut	Tamarind
Adult male	15	31	12	3	4	2
Adult female	83	78	42	26	33	11
Mother	50	44	17	4	9	0
Juvenile	115	99	39	22	27	10
Infant	17	23	3	2	3	0
Total	280	275	113	57	76	23

The observation of cultural behaviour and tool use for *M. fascicularis* at Ubud was first reported by Wheatley (1988). The behaviours consist of the washing and peeling of sweet potatoes and cassava roots and the rubbing of objects with detached leaves. In the summer of 1986, provisioned sweet potatoes formed the bulk of troop 1's diet. All of the 1361 sweet potatoes eaten were peeled with the incisor and canine teeth first. Sweet potatoes were also rubbed and/or washed when water was available. The washing, peeling and eating of cassava roots was also observed; the fibrous core of cassava roots was not eaten. The rubbing of objects with leaves ripped from trees was observed 17 times. The rubbed objects were generally hard round ones or worm-like. The latter were also rolled in the hands prior to being eaten. In some cases, leaf-rubbing appeared to be a cleaning behaviour, in others, it seemed to immobilise worm-like objects or it appeared to be meaningless or playful. All of these rubbing, washing and peeling behaviours appeared to be socially learned. Younger animals commonly watched and copied the behaviours of older animals.

An activity profile was determined from the 400 scan samples on the three troops. Percentages for adult males were compared with those of the wild troop in Kalimantan. The percentages for rest, travel and feed are 65, 25 and 10, respectively (Table 10.2).

Social organisation

Troop size, composition and immigration

Kalimantan

The largest troop numbers counted for four troops were 42 for the main study troop and 29, 27 and 22 for three other nearby ones. The study troop averaged 30 individuals and all troops were multimale. Table 10.4 gives average age–sex composition, socionomic and breeding sex ratio for the study troop.

Male replacements in this species were first reported by Wheatley (1982) for this study troop. The term replacement refers to a change in leadership in which a strange adult male successfully takes over a resident male's harem position. Two replacements of the alpha rank by strange adult males were seen. Both were very aggressive. The first occurred when one of the four immigrating males chased the resident alpha male out of the refuge tree. The other three resident adult males and ten other animals left the study area. Fourteen months later there was another adult male replacement. One of the two immigrating adult males succeeded in replacing the resident alpha male, albeit without ousting him from the troop. This aggressive replacement took two months and occurred just after the troop's mating peak. During this replacement redirected aggression on mothers with young infants was seen but there was no infanticide. A mating peak occurred just about one month prior to one of the adult male replacements. Adult females give a unique, staccato call during copulation that appears to advertise sexual receptivity. These calls may help to control the quality of the female's mate by advertising her sexual receptivity and intensifying male–male competition (Wheatley, 1980*b*).

In contrast to adult male immigration, sub-adult immigration was less aggressive and took place outside the mating peak.

Bali

Table 10.4 gives the size and composition of the three troops in Bali. There are, in addition, about 15 other individuals, seven of which are adults forming a loose all-male group. The total population size in 1990 was 111 individuals. In 1986, it was 69, and 1978, it was 31 according to Koyama, Asnan & Natsir (1981). Since the population presently ranges through a total of about 10 hectares, population density is $1111/km^2$.

Neither immigration nor replacement has been seen. In 1986, there were no extra-troop adult males. An adolescent male, in troop 1, whose permanent canines erupted in 1986, had risen to alpha male by 1990 after the former alpha male was killed in a fall.

Inter-troop interactions

Kalimantan

Only two inter-troop interactions were observed on the same side of the river. They were very aggressive and the resident troop chased the trespassers out of the area. Only nine cases of home range overlap were recorded. The amount of overlap was about 20% of the study troop's home range.

Inter-troop interactions across the river were interesting. Another troop slept across the river from the study troop and they tended to avoid each other. For 57 days of data, the probability that they showed up across the river from each other was 0.105. The troops gave each other branch-bouncing displays which consisted of bouncing on a limb that was too large to be shaken, while uttering loud 'Ho!' vocalisations. Displays occurring as a response to extra-troop challenges, such as inter-troop contact, observer presence, noise or alien males, accounted for 44% of all known contexts.

Table 10.4. *Age–sex composition of the three troops at Ubud*

Troops	n	Adult		Adolescent		Juvenile		Old infant		Young infant	
		M	F	M	F	M	F	M	F	M	F
A. 1986											
1	26	2	10	5	1	4	0	2	2		
2	27	1	10	2	2	3	2	3	2	1	1[a]
3	16	1	7	0	2	1	0	2	2		1
B. 1990											
1	28	2	12	0	2	2	5	3	1	0	1
2	22	2	12	4				0	1	0	4[a]
3	25–28	2	7	3	3	3–4	6–8				

The three troops are ranked by their order of supplanting one another. M, male; F, female.

[a] Sex is unknown.

The largest single context was inter-troop displays. The alpha male did 90% of the branch bounces and interposed himself between the source of the disturbance and the rest of the troop. When the alpha male of the study troop exhibited this behaviour the adult males usually sat within a metre of each other. This was significant because adult males rarely associated with each other. During the mating peak in April, the copulatory rate was greater when both troops were opposite each other across the river than when either troop was by itself. When both troops were present, the copulatory rate was 2.0/h ($n = 24$ h) for the study troop and 1.39 for the other troop ($n = 23$ h). The rate was 1.56 ($n = 36$ h) for the former and 1.31 ($n = 13$ h) for the latter when only one troop was present at the sleeping tree.

Bali

During the summer of 1986, there were 22 cases of aggressive inter-troop encounters. This was defined as contact or non-contact aggression between members of different troops whereby aggressors supplant the losers. The encounter rate was 1/15.5 hours of observation or about one every three days. Most encounters occurred where human food sources were located, for example, on the roads where tourists provided food. There were only two cases in 1986 where troop 2 fought back against troop 1 instead of its usual retreat. The first case, on 14 July, was a 19 min encounter of solid screeching, scream threats and calling by all individuals in both troops, except for one high-ranking female of troop 1 who appeared merely to sit and watch. Each troop formed a line of adult animals shoulder to shoulder and faced the other troop. Troop 2 advanced through the forest 50 m on the ground in this phalanx formation while troop 1 retreated. Generally, the adult females charged and slapped other adult females whereas the adult males lunged at each other. All infants and even weaned young juveniles were on the bellies of their mothers during the entire encounter. This included two mothers from troop 2 with 19 day old infants who lunged and slapped at the adults of troop 1 as much as the other adults. Half-way through the encounter, the alpha male of troop 2 retreated 10 m and gave a 'wahoo' call. Troop 1 then quickly advanced but two adult females charged the alpha male of troop 1 who retreated 3 m. The beta male charged these females but then retreated 3 m. The two females held off both males until the alpha male and all other members of troop ran up to support them. The troop 1 males withdrew while the females of both troops continued to lunge and slap for another minute until troop 1 climbed into a nearby tree to groom. Three individuals of troop 1 were wounded; the beta male and a juvenile male had cut tails. A mid-ranking adult female also had a cut tail and a cut chest.

The second encounter between the troops occurred 15 days later on 29 July. It was very similar to the previous one and lasted 16 min. Again, the beta male of troop 1 was the most noticeable and aggressive in charging, lunging, slapping and biting. Troop 2 was at the provisioning place and all adults lined up in a phalanx. The beta male of troop 1 charged in and the alpha male of troop 2 threatened two of his own oestrous females. He then charged at the beta male. Again, the mothers with 35-day-old infants were as persistent if not more so, in their attacks as any other animal. They threatened and slapped the beta male. Again, the alpha male of troop 2 retreated, gave the 'wahoo' call and chased the alpha male of troop 1. A mother with her infant also threatened the alpha male of troop 1. Both alpha males copulated during this encounter. The total number of occurrences of contact aggression was at least 26, which is a rate of 97.5/h. There were two individuals who were upended and each one was mobbed by four to five animals, none of which was an adult male. Both animals were extensively bitten. One juvenile male from troop 1 could barely walk and had copious amounts of blood on his back, side and tail. It was surprising that he survived. Two other adult females of troop 1 also had wounds but only on their tails.

There were two deaths in troop 1, both infants. One was an unweaned, old infant female who died on 22 July, and another was born on 20 or 21 July and had disappeared by 26 July. The mother of the old infant was seen on the morning of 21 July with blood on the right side of her mouth, on her stomach and on her upper right thigh. She also had a hurt foot. Three other adult females also had fresh wounds. An adolescent male had a hurt right wrist and an old infant male had a wound on his lower right leg. The infant female died the next morning. A tourist was seriously bitten on the forearm when he approached the dying infant.

The newborn was first seen early in the morning of 21 July. Efforts to observe this infant and the mother were met by 11 threats from adult females and the alpha male and contact aggression. The newborn had disappeared by 26 July when the mother was seen to have fresh blood from five punctures on her stomach and breasts. There were also cuts on her left elbow and left ear and three patches of missing hair on her back and side. She appeared not to have lactated. The seriousness and extent of the wounds suggested another inter-troop encounter between the dates of the two previously described. The only other possible cause of such wounds may have been dogs. These are known to interact with monkeys but they have never posed a serious threat to any monkey. Monkeys, in fact, will attack dogs. Given the vulnerability of infants it is noteworthy that their mothers are actively involved in the fighting. One explanation for these events is that troop 2 with seven infants had a greater nutritional need than troop 1 with only four infants (Table 10.4). No extra-troop or immigrating males were ever seen in the Monkey Forest during the summer of 1986.

Aggressive inter-troop encounters often took place in the summer of 1990. The dominance order to troops remained the same (see Table 10.4), but there were no serious injuries. Adult females were less involved than adult and sub-adult males. This is probably a reflection of the four-fold increase of such males from 1986. Adult females still threatened off adult males from other troops.

Intra-troop interactions

Kalimantan

Half-hour focal samples showed that adult females formed the stable core of the study troop. Adult females groomed significantly more frequently than did adult males (Mann–Whitney–Wilcoxon Test, $z = -2.53, m = 5, n = 13, p = 0.006$). They constituted one third of the troop yet did 61% of all grooming bouts observed and 55% of the total grooming time. This adult female cohesiveness is also seen in their sociability indices. For example, the probability of adult males being within 2 m of each other when adult females are present is significantly less (chi-square $= 9.88, p < 0.01$). Adult females were within 2 m of each other significantly more frequently than all other troop members (Mann–Whitney–Wilcoxon Test, $z = 1.97, m = 13, n = 17, p = 0.03$). Adult females approached each other at a rate of 1.87/h whereas they approached adult males at a rate of 0.8/h. New mothers and infants were especially popular objects of visitation.

Adult males were very uncohesive. They threatened significantly more frequently than did adult females (Mann–Whitney–Wilcoxon Test, $z = 2.56, m = 5, n = 13, p = 0.005$). Correcting the data with the socionomic sex ratio, a male to male visitation rate was 23% of the adult male to adult female visitation within one metre. Out of a total of 21 recorded threats, 20 were done by adult males.

The four adult males and the two sub-adult males formed a strict linear dominance hierarchy. The eldest adult male ranked number four and the sub-adults ranked below him. The alpha male performed over twice as many copulations as each of the other males. After he was replaced, he did not copulate and the new alpha male did 75% of all copulations.

Twenty-two newborns were seen in three different troops at the study site. Births were recorded in every month except November and December. The alpha male also did 90% of the branch bounces during both inter- or extra-troop challenges and in intra-troop contexts.

Bali

The focal animal data have not been analysed yet, however, it is clear that females form the core of the troop. Females are arranged in a clear-cut agonistic dominance hierarchy that has remained remarkably stable from 1986 to 1990. Adult females can wield considerable power through the use of scream threats and appeal aggression. Female coalitions are formed very quickly. On several occasions, females have been seen to threaten and chase alpha males of their own troop as well as alpha males of other troops. The intra-troop occasions usually involve infants. The alpha male of troop 1, for example, lip-smacked to the second-ranked adult female when she supported the sixth-ranked female. After the second-ranked female had stopped the alpha male's aggression, she went over to him, slapped him in the face and groomed him. The other female plus two other high-ranking adult females also approached and groomed him.

The most dramatic finding of 1990 was the observation of kidnappings (Fig. 10.6). There were four infant deaths. One infant died from a wound on his back and the other three probably of starvation. Two of the latter three newborns were kidnapped from their mothers by non-lactating females. The first observed death was a newborn male from troop 2 on 3 August. He was seen with a gaping wound on his back on 29 July. The second newborn was last seen on the afternoon of 9 August. The newborn female from troop 1 was first seen nursing on the morning of 4 August and then again on 6 August. The mother ranked number 13 out of 15 in the female hierarchy. An adult named Tubby, ranked number 3, was seen with the newborn on the morning of 7 August and she kept it until she disappeared. Tubby was an older non-lactating female. Continuous follows of Tubby on the morning and afternoon of the kidnapping showed that other individuals pulled or made a grab for the newborn at a rate of approximately 10/h over a five-hour period. The next morning this rate had dropped to 5/h over a $3\frac{1}{2}$ h period. A coalition of females and the beta male constantly threatened individuals, including the mother, away from the baby. The mother managed to approach Tubby and groom her baby at least once, but she was unable to retrieve it. The mother came into oestrus and copulated with peripheral males from 8 to 22 September. The third newborn died on the afternoon of 20 August. This troop 3 newborn was first seen on the afternoon of 17

Fig. 10.6. A successful kidnapping in troop 2 in progress. The mother later retrieved her infant. Photo by B. Wheatley.

August. The mother ranked number 7 out of 11 in the troop's female hierarchy. The infant stayed on its non-lactating mother the entire time and was never seen to nurse. No wounds were visible on the dead infant and its sex was unknown. The fourth infant was last seen on the afternoon of 5 September. The newborn male from troop 3 was first sighted soon after his birth on 2 September. That morning two females, a young juvenile, and a sub-adult female ranked number 3, named Cleft-chin, grabbed the infant but the mother, ranked number 6, held on. Cleft-chin had kidnapped the infant by the afternoon of 2 September and was surrounded by the females ranked numbers 2 and 7. The mother attempted unsuccessfully to grab her baby and ran down the road to feed on food offered by tourists. The next morning the top three ranked females all had the baby and threatened away the mother, although she did manage to groom her baby. Cleft-chin had the baby almost the entire time over the next two days and the mother was chased and bitten by old juvenile and sub-adult males, and several others, on two separate occasions. The mother was not able to approach and retrieve her baby. It was never seen to nurse and probably died of starvation.

Kidnapping may be a significant source of infant mortality. All but one of the infants born during the study period died and two of the four newborns died of starvation as a result of kidnapping. Some kidnappings were only temporary and did not result in death to the infant. One female in troop 2 was an habitual kidnapper. She acquired infants 23 times, occasionally using force to do so. This non-lactating female ranked in the middle of the female hierarchy. Only once did the kidnapper take an infant from a female who outranked her. None of these kidnappings lasted longer than three hours. Observations on these non-lethal kidnappings show that a lethal kidnapping requires a coalition of kin or others to assist the kidnapper by chasing or threatening the mother away.

The reproductive data for 1986 showed that all of the adults in the study troop copulated. Two old juvenile males also copulated. The data on these and two adult males in other troops showed that 82% of the total number of mounts to ejaculation ($n = 169$) was of the single mount to ejaculation type and that 18% was of the multiple mount to ejaculation type. It is interesting that of the latter type, the alpha male did 34.5% and the old juvenile male did 10.3%. The average time of copulation to ejaculation was longer for the third ranked male at 8.1 s as compared with the top two ranked males who both had a mean of 7.6 s. For all copulations the alpha male did 50.9%, the beta male did 31.1%, the old juvenile male did 18.0% and another younger male did 0.01%. Out of a total of 126 copulations to ejaculation, 97% had a copulatory call, whereas only four cases had no call. More details are given by Wheatley (1991).

Conclusion

Fieldwork in these two habitats has led to two basic conclusions. First, that the macaque's habitat is naturally or humanly disturbed or both. Second, that the results in the two habitats can be seen as complementary. Research in Kalimantan emphasised the ecological adaptations, whereas that in Bali emphasised their social behaviour and organisation. In a sense the observation conditions of wild macaques in the rain forest precluded detailed behavioural research while the habituated temple monkeys at Ubud facilitated such detailed behavioural study. These studies have provided us with a more detailed and complete picture of *Macaca fascicularis* than a study in only one habitat could provide.

Almost every study of this species of macaque has emphasised its preference for riverine, secondary forest (Southwick and Cadigan, 1972; Wilson and Wilson, 1975, 1977; Rijksen, 1978; Aldrich-Blake, 1980; Crocket and Wilson, 1980; Johns and Skorupa, 1987). It seems quite probable that the species' rapid and recent colonisation of South-East Asia was facilitated by the changes in Sundaland throughout the Pleistocene.

These changes were climatic ones that repeatedly submerged and exposed wide expanses of the present-day continental shelf (Fig. 10.1). Submerged river systems are still visible on the floor of the South China Sea. The arrival of humans and their slash and burn agriculture gradually led to an increasing role in the production of riverine secondary forests. Swidden agriculture is still the most common cause of secondary forests today (Whitmore, 1975). A sub-species, such as *fascicularis*, with the ability to depend on habitats markedly disturbed by humans, has been termed a weed-species by Richard, Goldstein & Dewar, (1989). They also tentatively identified *M. mulatta*, *M. radiata* and *M. sinica* as weed-species. 'This dependence on human resources', they say, is not 'a fall from grace', but simply an integral part of their ecological strategy. With this viewpoint in mind, we now turn to our investigation of *M. fascicularis* in the Monkey Forest. The data from these temple monkeys can be used to formulate more detailed hypotheses as well as to guide research on 'wild' monkeys. The burden of proof falls on those researchers who disbelieve that the same selective mechanisms operate on both 'wild' and food-enhanced populations.

The centuries of association between humans and Balinese macaques at the Monkey Forest and the greatly increased amount of time that they spend resting compared with the Kutai macaques is, presumably, related to their use of tools and cultural behaviour (Asquith, 1989). The peeling of the cortex of cassava and sweet potato helps to avoid prussic acid poisoning.

The small home and daily ranges of the Balinese troops allow for a detailed study of inter-troop competition. Troop 1 is the dominant troop controlling roadside areas where tourists are most likely to provision food. Over 80% of the scan samples, for example, indicate that troop 1 is on or near the road in contrast to fewer than 40% for troop 3. A comparison between troops 1 and 2 also shows that troop 1 is on or near a road 67% of the time compared with 33% for troop 2. There are also major dietary differences between the troops. Troop 1, for example, feeds more frequently than troop 3. Troop 1 feeds on over twice the number of tourist foods, especially peanuts and bananas, than does troop 3. Troop 3 feeds on twice the number of insects than troop 1, almost as if making up for the lost protein of the peanuts that troop 1 obtains. Angst (1975) also found similar inter-troop competition over tourist food in the Balinese temple monkeys at Sangeh and Pulaki.

The infant deaths, so far, appear to be due to wounds, probably from inter-troop encounters and from starvation due to intra-troop kidnappings by females. The very high population density of over $1100/km^2$ is probably another contributing cause of such aggression. It is tempting to speculate on the adaptive function of the kidnapping and deaths of two infants of low-ranking mothers in two different troops by high-ranking adult females. In other words, infanticide by females can have some adaptive consequences, such as the maintenance of the highest ranking matrifocal group over the lower ranking groups. The elucidation of the proximate and ultimate causes and especially the untangling of the process of selection *of* from selection *for* in regards to infant mortality by kidnapping, however, is complicated (Sober, 1984; Bernstein, 1987). Further work on inter- and intra-troop resource competition in Balinese macaques is continuing.

Acknowledgements

We are very grateful to the people and country of Indonesia for the opportunity to conduct research in Bali. We thank the Indonesian Institute of Sciences and the Research and Development Centre for Biology for sponsoring our research. We also thank Udayana University and the University of Indonesia. The following individuals deserve our thanks: Moertini Atmowidjojo, T. Hainald, Dr J. Sugardjito, Dr Soetikno Wirjoatmodjo, Dr I. Gde Suyatna, Widjaya, Scetoto, Chuck Darsono and family, P. Houghton, Dr O. Smith and J. Kujawski. We are very grateful for the support of the Fulbright Program and the Center for Field Research. We are also thankful for the grant support of Earthwatch and its many tireless workers and volunteers who made this research successful.

References

Aldrich-Blake, F. P. G. (1980). Long-tailed macaques. In *Malayan forest primates: ten years' study in tropical rain forest*, ed. D. J. Chivers, pp. 147–65. New York: Plenum Press.

Angst, W. (1975). Basic data and concepts on the social organization of *Macaca fascicularis*. In *Primate behavior: developments in field and laboratory research*, vol. 4, ed. L. A. Rosenblum, pp. 325–88. New York: Academic Press.

Asquith, P. (1989). Provisioning and the study of free-ranging primates: history, effects, and prospects. *Yearbook of Physical Anthropology*, **32**, 129–58.

Bernstein, I. S. (1987). The evolution of nonhuman primate social behavior. *Genetica*, **73**, 99–116.

Blanford, W. T. (1887). Critical notes on the nomenclature of Indian mammals. *Proceedings of the Zoological Society of London*, pp. 620–38.

Chasen, N. C. (1940). A handlist of Malaysian mammals. *Bulletin of the Raffles Museum, Straits Settlements*, **15**, 1–209.

Crockett, C. M. & Wilson, W. L. (1980). The ecological separation of *Macaca nemestrina* and *M. fascicularis* in Sumatra. In *The macaques: studies in ecology, behavior and evolution*, ed. D. G. Lindburg, pp. 148–81. New York: Van Nostrand Reinhold.

Fittinghoff, N.A., Jr & Lindburg, D.G. (1980) Riverine refuging in East Bornean *Macaca fascicularis*. In *The macaques: studies in ecology, behavior and evolution*, ed. D.G. Lindburg, pp. 182–214. New York: Van Nostrand Reinhold.

Fooden, J. (1964). Rhesus and crab-eating macaques: intergradation in Thailand. *Science*, **143**, 363–5.

Fooden, J. (1971). Report on primates collected in western Thailand, January–April, 1967. *Fieldiana Zoology*, **59**, 1–62.

Hamilton, W.J. III & Watt, K.E.F. (1970). Refuging. *Annual Review of Ecology and Systematics*, **1**, 263–86.

Hill, W.C.O. (1974). *Primates. Comparative anatomy and taxonomy*. VII. *Cynopithecinae*. New York: John Wiley & Sons.

Hoogerwerf, A. (1970). *Udjung Kulon*. Leiden: E.J. Brill.

Johns, A.D. & Skorupa, J.P. (1987). Responses of rain-forest primates to habitat disturbance: a review. *International Journal of Primatology*, **8**, 157–91.

Kawamoto, Y. & Ischak, T.B. (1981). Genetic differentiation of the Indonesia crab-eating macaque (*Macaca fascicularis*). I. Preliminary report on blood protein polymorphism. *Primates*, **22**, 237–52.

Kawamoto, Y., Ischak, T.B. & Supriatna, J. (1984). Genetic variations within and between troops of the crab-eating macaque (*Macaca fascicularis*) on Sumatra, Java, Bali, Lombok and Sumbawa, Indonesia. *Primates*, **25**, 131–59.

Kawamoto, Y., Nozawa, K. & Ischak, T.B. (1981). Genetic variability and differentiation of local populations in the Indonesia crab-eating macaque (*Macaca fascicularis*). *Kyoto University Overseas Report of Studies on Indonesian Macaque*, **1**, 15–39.

Kawamoto, Y. & Suryobroto, B. (1985). Gene constitution of crab-eating macaques (*Macaca fascicularis*) on Timor. *Kyoto University Overseas Research Report of Studies on Asian Non-human Primates*, **4**, 35–40.

Koyama, N., Asnan, A. & Natsir, N. (1981). Socio-ecological study of the crab-eating monkeys in Indonesia. *Kyoto University Overseas Research Report of Studies on Indonesian Macaque*, **1**, 1–10.

Miller, G.S., Jr (1942). Zoological results of the George Vanderbilt Sumatran Expedition, 1936–1939. Part V. Mammals collected by Frederick A. Ulmer, Jr on Sumatra and Nias. *Proceedings of the Academy of Natural Sciences, Philadelphia*, **94**, 107–67.

Napier, J.R. & Napier, P.H. (1967). *A handbook of living primates*. New York: Academic Press.

Richard, A.F., Goldstein, S.J. & Dewar, R.E. (1989). Weed macaques: the evolutionary implications of macaque feeding ecology. *International Journal of Primatology*, **10**, 569–94.

Ridley, H.N. (1906). The menagerie at the botanic gardens. *Royal Asian Society of Great Britain & Ireland, Straits Branch Singapore Journal*, **46**, 133–94.

Rijksen, H.D. (1978). A fieldstudy on Sumatran orang utans (*Pongo Pygmaeus Abelii*, Lesson 1827). *Mededelingen Lanbouwhogeschool Wageningen, Nederland*, 78–2, H. Veenman & B.V. Zonen.

Schultz, A.H. (1956). Post-embryonic age changes. In *Primatologica*, vol. I, ed. H. Hofer, A.H. Schultz & D. Starck, pp. 887–964. Basal: S. Karger.

Sober, E. (1984). *The nature of selection*. Cambridge, MA: Bradford Books, MIT Press.

Sody, H. (1949). Notes on some primates, carnivora, and the babirusa from the Indo–Malayan and Indo–Australian regions. *Treubia*, **20**, 121–90.

Southwick, C. H. & Cadigan, F. C. Jr (1972). Population studies of Malaysian primates. *Primates*, **13**, 1–18.

Spiegel, A. (1956). Über das Körperwachstum der Javamakaken. *Zoologischer Anzeiger*, **156**, 1–8.

Struhsaker, T. T. (1969). Correlates of ecology and social organization among African cercopithecines. *Folia Primatologica*, **11**, 80–118.

Struhsaker, T. T. (1974). *The red colobus monkey.* Chicago: University of Chicago Press.

Takenada, A. & Takenada, O. (1990). Multiplication of α-globin genes in higher non-human primates. Abstract, XIIIth Congress of the International Primatological Society, 18–24 July 1990, Nagoya & Kyoto, Japan, p. 63.

Takenaka, O. (1986). Blood characteristics of the crab-eating monkeys (*Macaca fascicularis*) in Bali Island, Indonesia: implications of water deficiency in west Bali. *Journal of Medical Primatologica*, **15**, 97–104.

Washburn, S. L. (1942). The skeletal proportions of langurs and macaques. *Human Biology*, **14**, 444–72.

Wheatley, B. P. (1978). Riverine secondary forest in the Kutai Nature Reserve, East Kalimantan, Indonesia. *Malayan Nature Journal*, **32**, 19–29.

Wheatley, B. P. (1980a). Malaria as a possible selective factor in the speciation of macaques. *Journal of Mammalogy*, **61**, 307–11.

Wheatley, B. P. (1980b). Feeding and ranging of East Bornean *Macaca fascicularis*. In *The macaques: studies in ecology, behavior and evolution*, ed. D. G. Lindburg, pp. 215–46. New York: Van Nostrand Reinhold.

Wheatley, B. P. (1982). Adult male replacement in *Macaca fascicularis* of East Kalimantan, Indonesia. *International Journal of Primatology*, **3**, 203–19.

Wheatley, B. P. (1988). Cultural behavior and extractive foraging in *Macaca fascicularis*. *Current Anthropology*, **29**, 516–19.

Wheatley, B. P. (1991). The role of females in inter-troop encounters and infanticide among Balinese *Macaca fascicularis*. In *Primatology today*, ed. A. Ehara, T. Kimura, O. Takenaka & M. Iwamoto, pp. 169–72. Amsterdam: Elsevier.

Whitmore, T. C. (1975). *Tropical rainforests of the Far East.* Oxford: Clarendon Press.

Wilson, C. C. & Wilson, W. L. (1975). The influence of selective logging on primates and some other animals in East Kalimantan. *Folia Primatologica*, **23**, 245–74.

Wilson, C. C. & Wilson, W. L. (1977). Behavioral and morphological variation among primate populations in Sumatra. *Yearbook of Physical Anthropology*, **20**, 207–33.

11

Inter-regional and inter-seasonal variations of food quality in Japanese macaques: constraints of digestive volume and feeding time

N. NAKAGAWA, T. IWAMOTO, N. YOKOTA AND
A. G. SOUMAH

Introduction

The Japanese macaque (*Macaca fuscata*) is distributed in Japan from the Shimokita peninsula (41° 31′ N) in the north to Yakushima Island (30° 30′ N) in the south. Its range covers a diversity of habitats which vary from sub-tropical forest in the Yakushima lowlands through the warm- and cool-temperate forests to the sub-arctic (sub-alpine) forest in the mountainous areas of central and northern Honshu. While most of the non-human primate species have all or a substantial part of their geographical range within the tropics, the Japanese macaque and the rhesus macaque (Southwick *et al.*, 1991) are unique. This broad distribution inevitably causes inter-regional variation in diet and substantial seasonal variation. To date, the food habits of the Japanese macaque during each season have been examined in various regions only as basic ecological data. Comparisons reveal inter-regional variation in diet (Suzuki, 1965; Uehara, 1975, 1977; Maruhashi, 1980). Generally speaking, however, these studies have done no more than record food species, edible portions, and relative frequency of items in the diet (e.g. Shimokita (Izawa & Nishida, 1963; Furuichi, *et al.*, 1982); Kinkazan (Izawa & Nishida, 1963; Izawa, 1983; Satoh, 1988); Shiga (Suzuki, 1965; Wada & Ichiki, 1980); Hakusan (Izawa, 1982); Arashiyama (Huffman, 1984); Takasakiyama (Itani, 1956)). Even when quantity of feeding has been determined by systematic sampling, time spent feeding on each food item (Yakushima (Maruhashi, 1980, 1986); Kinkazan (Nakagawa, 1989*a*)) has been the measure used. Although time spent feeding on each item may reflect the item's contribution to the total quantity of food eaten, it does not necessarily do so (Richard, 1985), as shown by Hladik (1977). A more reliable measure of feeding is by quantifying weight intake or calorie intake of each item. Iwamoto (1974, 1982) reported weight and nutrient intake of each item and total nutritional intake in Japanese macaques.

Several bioenergetic studies on Japanese macaques have been conducted since Iwamoto's (1982) study (Shimokita (Watanuki, 1984; Watanuki & Nakayama, 1986); Kinkazan (Nakagawa, 1989*b*); Takasakiyama (Soumah & Yokota, 1991)). Such studies include nutritional analyses of food items as well as data on food quantity eaten. In the present study, quantity of plant food consumed by monkeys during autumn and in winter is compared for animals in two different habitat types (cool-temperate deciduous forest zone and warm-temperate evergreen forest).

Study area

The four Japanese macaque localities chosen for study were Kinkazan Island and Takasakiyama in warm-temperate evergreen forest, and Koshima Islet and Shimokita peninsula within cool-temperate deciduous forests (Fig. 11.1).

Climatic and vegetational features, as well as population features of the monkeys in the four habitats, are shown in Table 11.1.

Methods

In the first three habitats, focal animal sampling was employed to record feeding bout duration, or amount of food eaten during a part of a feeding bout or whole feeding bout (see Kinkazan (Nakagawa, 1989*b*); Takasakiyama (Soumah & Yokota, 1991); Koshima (Iwamoto, 1982)). In Shimokita, the scanning method was employed to estimate activity budget and feeding rate of subject individuals (see Watanuki & Nakayama, 1986).

The main plant food of non-pregnant, non-lactating females was used for analysis in this study. A main food item is defined as a food where proportion of time spent feeding to total feeding time is more than 0.5%. Number of food items used for analysis and proportion of time spent feeding to total feeding on all of these items was as follows: nine items out of 13 main food items, 93.93% (in autumn at Kinkazan); 15 out of 21 items, 88.65% (in winter at Kinkazan); 19 items, 57.75% (in autumn at Koshima); 21 items, 61.10% (in winter at Koshima); 24 out of 29 items, 85.18% of time spent feeding on natural food items (in winter at Takasakiyama). In the case of Shimokita, only the nine most preferred items for which time spent feeding comprised 75.2% of total feeding time were used.

Food samples were clipped in the field and weighed fresh. They were brought to the laboratory and their dry weights were measured. These were later analyzed for crude protein, lipids, crude fibre, ash, and soluble

Fig. 11.1. Locations of the four study sites.

carbohydrates (for methods, see Iwamoto, 1982). Calorific content of each food item was calculated by multiplying each gram of carbohydrate (crude fibre + soluble carbohydrate), protein, and lipid by 4.15, 5.65, and 9.40 kcal, respectively (Maynard *et al.*, 1979).

Plant food composition, rate of intake, water content, and nutritional content of each of the food items are shown in Appendix Tables 11.A1–11.A6.

Results

Animals need not only adequate energy (or calories) but also amino acids, fatty acids, vitamins, minerals, and trace elements from their diet. However, they should avoid overingestion of secondary compounds from their food sources. The present study focuses on calories and protein.

Daily energy requirement (DER in kcal/individual/per day) is calculated as $2 \times$ BMR (basal metabolic rate: $70 \times W^{3/4}$, where W is body weight in kg; see Iwamoto, 1988). Daily protein requirement (DPR) is 2.54 g dw/kg body weight per day (where dw is dry weight) for food with digestibility of 83.4% (Robbins & Gavan, 1966; Milton, 1979). If a non-pregnant,

Table 11.1. Some features of the four habitats selected for this study

	Location	Annual mean temperature (°C)	Annual rainfall (mm)	Deepest amount of snow (cm)	Vegetational zone	Dominant species	Name of troops in this region	Study period	Detailed description
Shimokita peninsula	41° 30′ N 141° 00′ E	9.6	1407	90	Cool-temperate	*Fagus crenata* *Quercus mongolia*	A_1^a, A_2, O (south-west)	Jan. 1982– Mar. 1982	Izawa & Nishida (1963)
						Thujopsis dolabrata	A, I, M*, Z (north-west)	Dec. 1982– Mar. 1983 Dec. 1983– Apr. 1984 Dec. 1984– Mar. 1985 Nov. 1985– Dec. 1985	Izawa (1982)
Kinkazan Island	38° 16′ N 141° 35′ E	11	1500	10	Cool-temperate	*Fagus crenata* *Abies firma*	A*, B_1, B_2, C, D	Nov. 1987	Izawa (1983) Nakagawa (1989a)
Takasakiyama	33° 15′ N 131° 32′ E	15	1625	10	Warm-temperate	*Quercus glauca* *Quercus serrata*	A^a, B*,^{a}, C^a	Feb. 1988– Mar. 1988	
Koshima	31° 17′ N 131° 22′ E	17	2500	0	Warm-temperate	*Machirus thunbergii* *Daphniphyllum*	Main*,^{b} Maki^b	Oct. 1977– Feb. 1978	Iwamoto (1974)

*Indicates troop for this study; [a]provisioned troop; [b]provisioned troop, but unprovisioned during the conduct of this study.

non-lactating female of 8.5 kg body weight is adopted as a reference individual according to Iwamoto (1988), the daily energy requirement and daily protein requirements are estimated at about $70\,000/Y$ kcal and $1800/Y$ g (where Y is digestibility; %), respectively.

Nutritional content

Because animals can not ingest food infinitely to satisfy the above nutritional requirements, the amount of food that a monkey can process in a day is restricted by gut volume and/or food passage rate. Although no studies on gut volume and food passage rate of Japanese macaques have been conducted, the amount of food that a monkey can process in a day has been estimated at about 300 g dw on the basis of measurements of actual daily food intake (Mori, 1979a; Iwamoto, 1982). As actual daily food intake of the Japanese macaques in November at Kinkazan was estimated to average about 310 g dw (Nakagawa, 1989b), this result supports the above estimation. Japanese macaques must ingest $70\,000/Y$ kcal energy and $1800/Y$ g protein under the constraint of 300 g dw digestive volume. To meet the above calorie and protein requirements, therefore, the necessary condition is that calorific contents (C_1 kcal/g) and protein contents (P_1 g/g) of each of their food items meet the following two expressions, respectively.

$$C_1 > 70\,000/300\,Y \tag{11.1a}$$
$$P_1 > 1800/300\,Y \tag{11.1b}$$

As food digestibility is known to correlate with its fibre content, Iwamoto (1988) transformed Mitchell's (1964) equation, and calculated the digestibility of each food item from the following transformed equation:

$$Y = -1.43X + 89.86 \tag{11.2}$$

where Y = digestibility and X = crude fibre (%).

When substituting equation (11.2) in expressions (11.1a) and (11.1b), respectively, the necessary condition is that caloric content (C_1) and protein content (P_1) of each food item meets the following two equations, respectively.

$$X \leq -163.17/C_1 + 62.84 \tag{11.3a}$$
$$X \leq -4.20/P_1 + 62.84 \tag{11.3b}$$

In other words, when the monkeys eat 300 g dw of an item that meets these equations, they can meet both calorie and protein requirements. Equation (11.3b) coincides with the fact that the protein : fibre ratio is

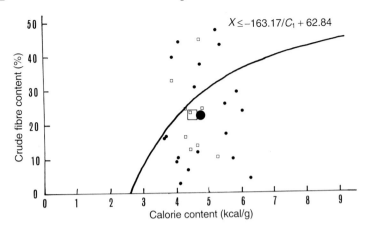

Fig. 11.2. Relationship between calorie content and fibre content of food items of Japanese macaques inhabiting Kinkazan (closed circles) and Koshima (open squares). Larger symbols indicate the averages.

important as a predictor of food selection (Milton, 1979; McKey *et al.*, 1981). The relationship between calorie content and fibre content of food items of Japanese macaques inhabiting Kinkazan and Koshima in autumn is shown in Fig. 11.2.

Calories: total intake

Figure 11.3 describes the percentages of food items which satisfy equation (11.3a) (Fig. 11.3(*a*)) and calorie content of food items (Fig. 11.3(*b*)) in each season at each site. In autumn, 77.8% of food items (7 out of 9 items) at Kinkazan and 68.4% (13 out of 19 items) at Koshima met the equation. The average calorie content and crude fibre content in each region (4.54 kcal/g, 22.59% in Kinkazan; 4.82 kcal/g, 22.51% in Koshima) sufficiently met the equation. Inter-regional differences in these contents were not significant (Mann–Whitney U-test, $p > 0.05$).

In winter, the percentage of food items that met the equation decreased to 60.0% (9 out of 15 items) at Kinkazan and 52.4% (11 out of 21 items) at Koshima. The values of Shimokita (55.6%; 5 out of 9 items) and of Takasakiyama (54.2%; 13 out of 24 items) were also similar to these values. The average calorie content and crude fibre content in each region (4.21 kcal/g, 25.12% in Shimokita; 4.05 kcal/g, 23.14% in Kinkazan; 4.10 kcal/g, 25.46% in Takasakiyama; 4.18 kcal/g, 20.98% in Koshima) barely met the equation or were slightly insufficient, whereas inter-regional differences in these contents were not significant in all combinations of the

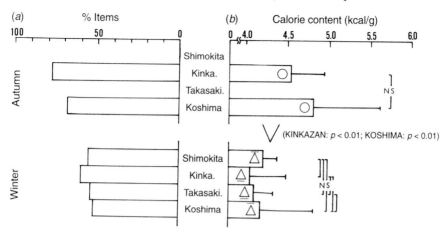

Fig. 11.3. Percentages of food items which met the equation (11.3a) (*a*) and calorie content of food items (*b*). Symbols in the columns indicate the following. Circles: averages sufficiently met the equation. Triangles below bar: averages barely met the equation. Triangles above bar: averages were slightly insufficient for meeting the equation. *P* indicates where inter-seasonal comparisons were significant. NS indicates where inter-regional comparisons were not significant.

regions (Mann–Whitney U-test, $p > 0.05$). Crude fibre content was not significantly different (Mann–Whitney U-test, $p > 0.05$).

Protein: total intake

Figure 11.4 describes the percentages of food items that met equation (11.3b) (Fig. 11.4(*a*)) and protein content of food items (Fig. 11.4(*b*)) in each season at each site.

In autumn, 44.4% of food items (4 out of 9 items) at Kinkazan and 31.6% (6 out of 19 items) at Koshima met the equation; these values were lower than equivalent values of calorie content in autumn. Whereas the averages of protein content and crude fibre content at Kinkazan (11.30% and 22.59%, respectively) barely met the equation, those in Koshima (8.80% and 22.51%, respectively) did not meet it at all.

Moreover, in winter, 46.7% of food items (7 out of 15 items) at Kinkazan, 23.8% (5 out of 21 items) at Koshima, 33.3% (3 out of 9 items) at Shimokita, and 37.5% (9 out of 24 items) at Takasakiyama met the equation, and these values were also lower than equivalent values for calorie content in winter. Whereas the averages of protein content and crude fibre content in Shimokita (10.17% and 22.12%), in Kinkazan (9.91% and 23.14%), and in Takasakiyama (11.83% and 25.46%) barely met the equation or were slightly insufficient, those in Koshima (8.19% and

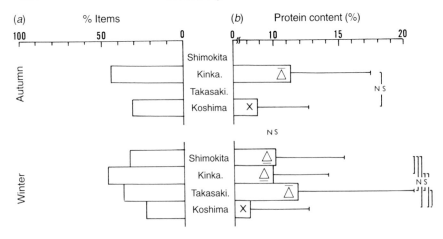

Fig. 11.4. Percentages of food items which met the equation (11.3b) (*a*) and protein content of food items (*b*). Cross: averages did not meet the equation at all. See also legend for Fig. 11.3. NS indicates where inter-regional and inter-seasonal comparisons were not significant.

21.00%) did not meet it at all. Neither inter-regional or inter-seasonal differences in protein content were significant in all combinations (Mann–Whitney U-test, $p > 0.05$).

Speed of nutritional intake

Another constraint to be considered when the monkeys ingest $70\,000/Y$ kcal of energy and $1800/Y$ g of protein is the time involved. It is very difficult to determine the upper limit of the proportion of time spent feeding during the active period. The proportion of time spent feeding in Japanese macaques is normally 20–30% of total active time (28.6% in Koshima, Kuroki, 1975; 27.8% in Takagoyama, Yotsumoto, 1976; 23.5% in Yakushima, Maruhashi, 1980; 26.7% in Shiga, Wada & Tokida, 1981). In Kinkazan, however, the monkeys increased the proportion of time spent feeding from 34.4% at the end of November 1984 to 66.3% at the beginning of February 1985 to protect against the deterioration of the food environment during winter (Nakagawa, 1989*a*). Moreover, it reached 73.0% at the end of November 1987, and 70.0% at the end of February 1988 (Nakagawa, 1989*b*). These values are considered to be extremely high for the time spent feeding in Japanese macaques. Compared with other non-human primate species, only the gelada showed similar values (Iwamoto, 1979). When these are taken into account, the upper limit of the proportion of time spent feeding in Japanese macaques is estimated to be 70% in this study. In addition, their active period is considered to be 10.5 h from the average of daytime

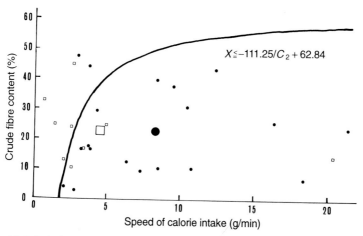

Fig. 11.5. Relationship between speed of calorie intake and fibre content of food items of Japanese macaques inhabiting Kinkazan (closed circles) and Koshima (open squares). Larger symbols indicate the averages.

duration in the middle of the study period in both habitats, although the daytime duration changes with the region and the season. Therefore, the upper limit of time spent feeding in a day is estimated at 440 min. Japanese macaques must ingest $70\,000/Y$ kcal energy and $1800/Y$ g protein under the constraint of 440 min feeding time. To meet the above calorie and protein requirements, therefore, the necessary condition is that the rate of calorie intake (C_2 kcal/min) and the rate of protein intake (P_2 g/min) of their food items meet the following two equations, respectively, according to the equivalent calculations for nutritional content.

$$X \leq -111.25/C_2 + 62.84 \tag{11.4a}$$
$$X \leq -2.86/P_2 + 62.84 \tag{11.4b}$$

In other words, when the monkeys spend 440 min feeding on items that meet these equations, they can meet calorie and protein requirements, respectively. The relationship between rate of calorie intake and fibre content of feed items of Japanese macaques inhabiting Kinkazan and Koshima in autumn is shown in Fig. 11.5.

Rate of intake

Calories

Figure 11.6 describes the percentages of food items that met the equation (11.4a) (Fig. 11.6(*a*)) and rate of calorie intake of food items (Fig. 11.6(*b*)) in each season at each site.

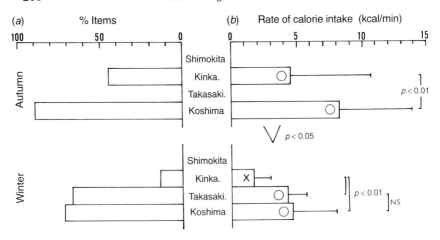

Fig. 11.6. Percentages of food items which met the equation (11.4a) (a) and rate of calorie intake of food items (b). See also legends for Figs. 11.3 and 11.4.

In autumn, 44.4% of food items (4 out of 9 items) at Kinkazan and 89.5% (17 out of 19 items) at Koshima met the equation. The average rate of calorie intake and fibre content in each region (4.59 kcal/min, 22.59% in Kinkazan; 8.30 kcal/min, 22.51% in Koshima, respectively) were sufficient to meet the equation. The rate of calorie intake in Koshima was significantly higher than that in Kinkazan (Mann–Whitney U-test, $p < 0.01$).

In winter, however, the proportion of food items that met the equation decreased to 13.3% (2 out of 15 items) at Kinkazan and to 71.43% (15 out of 21 items) at Koshima; Takasakiyama (66.7%, 16 out of 24 items) was similar to Koshima. Whereas the average rate of calorie intake and fibre content at Koshima (4.67 kcal/min, 20.98%) and at Takasakiyama (4.28 kcal/min, 25.46%) were sufficient to meet the equation, those at Kinkazan (1.72 kcal/min, 23.14%) did not satisfy it. Both in autumn and in winter, the average rate of calorie intake at Koshima as well as at Takasakiyama was significantly higher than that at Kinkazan (Mann–Whitney U-test, $p < 0.01$). In addition, both at Koshima and at Kinkazan, it was significantly higher in autumn than in winter (Mann–Whitney U-test, $p < 0.05$).

Protein

Figure 11.7 describes the percentages of food items which met the equation (11.4b) (Fig. 11.7(a)) and rates of protein intake of food items (Fig. 11.7(b)) in each season at each site.

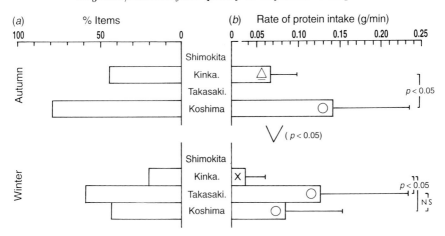

Fig. 11.7. Percentages of food items which met the equation (11.4b) (*a*) and rate of protein intake of food items (*b*). See also legends for Figs. 11.3 and 11.4.

In autumn, 44.4% of food items (4 out of 9 items) at Kinkazan and 78.95% (15 out of 19 items) at Koshima met the equation, and these values were equal to or lower than equivalent values for calorie intake in autumn. The average rate of protein intake and fibre content at Koshima (0.1431 g/min, 22.51%) was sufficient to meet the equation, although those at Kinkazan (0.0672 g/min, 22.59%) were slightly insufficient.

In winter, however, the proportions of food items that met the equation decreased to 20.0% (3 out of 15) at Kinkazan and to 42.9% (9 out of 21 items) at Koshima. Those at Takasakiyama (58.3%, 14 out of 24 items) were similar to those at Koshima. Whereas the average rate of protein intake and fibre content in Koshima (0.1276 g/min, 25.46%) was sufficient to meet the equation, those in Kinkazan (0.0369 g/min, 23.14%) were far below the values that met the equation. Both in autumn and in winter, the average rate of protein intake both in Koshima and Takasakiyama was significantly higher than in Kinkazan (Mann–Whitney U-test, $p < 0.05$). Both in Koshima and in Kinkazan, intake in autumn was significantly higher than that in winter (Mann–Whitney U-test, $p < 0.05$).

Discussion

Bioenergetic studies of Japanese macaques demonstrated that they met calorie and protein requirements in autumn both at Koshima and Kinkazan, but that they barely do so in winter at Koshima and that their intake was much lower than the requirement in winter at Kinkazan (Iwamoto, 1982; Nakagawa, 1989*b*). It was more difficult to meet the protein requirement than the calorie requirement (Iwamoto, 1982; Nakagawa, 1989*b*). The results of food quality naturally reflected these findings.

Briefly, the proportion of food items that met the equation was generally lower for protein than for calories in terms of both content and rate intake. This signifies a greater difficulty in meeting the protein requirements.

In autumn, regardless of the constraints of 300 g dw digestive volume and 440 min feeding time, the monkeys could meet the requirement because, for both calorie and protein, the nutritional content was high and there was a high rate of intake of food items. As compared with Kinkazan, however, the protein content of plant food was lower in Koshima, and the proportion of plant food that met the equation was lower. This lower proportion was a result of the monkeys spending more time foraging for protein-rich insects (about 65% dw) in order to meet their protein requirement (see Fig. 11.4a); the time remaining made it possible to forage on insects intensively. In Koshima, protein-rich plant food tends to be exhausted soon (Iwamoto, 1982) because of the abnormally high population density resulting from long-term artificial provisioning or foraging on insects.

In winter, both at Koshima and at Kinkazan, it became more difficult to meet the nutritional requirement because of lower nutritional contents and a lower speed of intake in terms of both calories and protein. However, the degree of food deficiency in winter showed great differences between Koshima and Kinkazan. Whereas the food deficiency was slight in Koshima, it was severe in Kinkazan, where the monkeys ingested only 55% of their calorie requirement and 38% of their protein requirement. This difference is due to the following reasons: slight food deficiency in Koshima resulted from low nutritional content; severe food deficiency in Kinkazan resulted not only from the low nutritional content but also from a low speed of intake. Considering the similar high speed of intake in Takasakiyama to that in Koshima, it is suggested that, because there is plenty of food with high speed of intake even in winter in warm-temperate forest, the monkeys have more time to spare, and the constraint of feeding time is left out of consideration. Food deficiency in warm-temperate forest in winter is brought about by the constraint of digestive volume. However, the food deficiency in winter in cool-temperate forest is brought about only partly by the constraint of feeding time. When we consider food categories, the main food in winter is evergreen-broad leaves in warm-temperate forest (19.2% of total feeding time at Koshima, and 29.9% of total time spent feeding on natural food at Takasakiyama), but in cool-temperate forest it consisted of bark, winter buds and twigs (34.8% of total feeding time at Kinkazan, and at least 75.2% at Shimokita). The following differences in quality between these two food categories reflect the above results: evergreen leaves allow a high rate of calorie intake (5.58 kcal/min on average, based on 14 main food items at Koshima and at Takasakiyama), but have low calorie (4.19 kcal/g) and high fibre contents (24.39%); buds, bark, and twigs only allow a low rate of nutritional intake (1.79 kcal/min; 0.0383 g/min in protein on average, based on eight main food items in Kinkazan). The determining factors of food deficiency in winter were different in the two areas.

Yakushima, which is the southernmost limit of distribution of Japanese macaques, is covered with broad-leaf evergreen forest including sub-tropical forest elements, and has high fruit production throughout the year. Although the monkeys utilised a substantial amount of leaves in March–April when fruit production decreased, the proportion of time spent feeding on fruits compared with total feeding time still amounted to 20% (Maruhashi, 1986). The centre of distribution of most macaque species is generally in the tropics, where there may be much more fruit production throughout the year than in Yakushima. Although leaf-eating may increase somewhat during the off-crop season in other *Macaca* spp., bark-eating and winter-bud-eating rarely occur (e.g. *Macaca nemestrina*, see Caldecott, 1986; *M. fascicularis*, see Wheatley, 1982; *M. mulatta*, see Lindburg, 1977). Then, large fruit-eating and leaf-eating generally show a high rate of intake, so that these monkeys would rarely suffer from food deficiency; if it occurs at all it would probably result from low nutritional content. However, because *M. mulatta* (at high altitudes and high latitude), *M. thibetana* and *M. sylvanus* depend on bark, winter buds, and grass in winter (e.g. *M. thibetana*, see Zhao, Deng & Xu, 1991; *M. mulatta*, see Southwick *et al.*, 1991; *M. sylvanus*, see Drucker, 1984), they may suffer from food deficiency resulting from a low rate of nutritional intake.

Acknowledgements

We express our thanks to Dr Y. Watanuki and Ms Y. Nakayama for their generous permission to use their valuable unpublished data. We are grateful to Mr D. Yoshiba for kindly revising the English of this paper.

This study was financed by the Scientific Research Fund of the Ministry of Education, Japan and by the Cooperative Research Fund of the Primate Research Institute, Kyoto University.

Appendix: Food items in autumn and winter at Kinkazan and Koshima, and in winter at Takasakiyama and Shimokita

For each of the food species consumed at the different study sites, the plant food composition, speed of intake, water content, and nutritional content are tabulated.

Table 11.A1. *Plant food composition, speed of intake, water content and nutritional content of each of the food items in autumn at Kinkazan*

Species name	Japanese name	Part eaten[a]	DUR[b] (%)	dw[c] (%)	Speed of calorie intake (kcal/ min)	Speed of protein intake (g dw/ min)	Nutritional content[d] Water content (%)	Crude protein (%)	Ether extract (%)	Crude fibre (%)	Soluble carbo- hydrate (%)	Ash (%)	Calories (kcal/g)
Rubus microphyllus	Nigaichigo	Lf*	0.64	0.46	2.17	0.0715	75.24	14.62	8.01	12.66	56.19	8.52	4.44
Lonicera japonica	Suikazura	Fr	0.14	0.12	2.78	0.0689	70.34	11.25	8.95	8.97	64.84	5.99	4.54
Berberis thunbergii	Megi	Lf*	4.35	5.23	3.47	0.1090	71.33	13.44	5.20	16.37	56.56	8.32	4.28
		Fr	0.37	0.50	4.13	0.1030	48.61	11.26	7.27	12.07	64.88	4.52	4.51
Rosa multiflora	Noibara	Fr*	2.85	4.93	5.00	0.0967	51.99	8.28	4.12	24.49	58.04	5.07	4.28
		Lf	0.16										
Viburnum dilatatum	Gamazumi	Fr*	3.03	2.69	2.67	0.0356	67.12	5.94	7.52	23.45	58.67	4.42	4.45
Torreya nucifera	Kaya	Se*	3.88	2.90	2.67	0.0875	70.87	17.33	20.38	10.13	47.45	4.71	5.28
Clematis apiifolia	Botanzuru	Se*	9.90	4.36	1.43	0.0564	19.85	18.98	10.50	24.62	41.87	4.03	4.82
Oplismenus uadulatifolius	Chijimizasa	*	2.52	0.70	0.73	0.0118	—	6.29	0.52	32.77	51.17	9.25	3.89
Malus tschonoskii	Ourajironoki	Fr*	3.72	23.79	0.0464	65.80	1.06	11.72	14.18	70.61	2.43	2.43	4.68
Carpinus tschonoskii	Inushide	Se*	63.04	53.14	2.70	0.0898	20.71	15.78	10.37	44.63	24.65	4.57	4.74
Perilla frutescens	Remonegoma	Se	1.86										

Quercus serrata	Konara	Se	1.62		
Viscum album	Yadorigi	Fr	0.53		
Schisandra repanda	Matsubusa	Fr	0.64	0.78	84.80
Fagus crenata	Buna	Se	0.14		
Boehmeria spicata	Koakaso	Se	0.04		
Trifolium repens	Shirotsumekusa	Lf	0.09		
Ganoderma lucidium	Mannendake		0.23		
Coriolus versicolor	Kawaradake		0.05		
Seaweeds			0.18		

*Indicates the main food items used for analysis in this study. [a] Lf: leaf; Fr: fruit; Se: seed; YLf: young leaf; Br: bark; Bu: bud; Tw: twig; St: stem; Sh: shoot; Rz: rhizome; Fl: flower; [b] DUR: proportion of time spent feeding on each food item to the total time spent feeding on all food items, including animal material. In the case of Takasakiyama, time spent feeding on artificial food was excluded from the total time spent feeding. [c] dw: proportion of dry weight intake of each food item to the total dry weight intake. When dw% was calculated, food items of which the unit weight could not be estimated were excluded. [d] shown on a dry weight basis.

Table 11.A2. *Plant food composition, speed of intake, water content and nutritional contents of each of the food items in winter at Kinkazan*

Species name	Japanese name	Part eaten[a]	DUR[b] (%)	dw[c] (%)	Speed of calorie intake (kcal/min)	Speed of protein intake (g dw/min)	Water content (%)	Crude protein (%)	Ether extract (%)	Crude fibre (%)	Soluble carbohydrate (%)	Ash (%)	Calories (kcal/g)
Oplismenus uadulatifolius	Chijimizasa		43.72	42.93	1.20	0.0194	41.97	6.29	0.52	32.77	51.17	9.25	3.89
Zanthoxylum piperitum	Sansho	Br*	4.20	4.73	1.49	0.0393	—	11.10	2.32	23.26	57.73	5.59	4.21
		Bu*	9.31	4.37	0.64	0.0247	44.29	16.77	2.75	19.90	55.21	5.37	4.32
		Tw*	2.70	2.72	1.34	0.0294	—	9.27	1.38	38.49	47.51	3.35	4.22
Zoysia japonica	Shiba	*	7.00	8.74	1.47	0.0581	67.93	14.79	2.44	10.34	53.90	18.53	3.73
Carpinus tschonoskii	Inushide	Se*	6.57	4.80	0.96	0.0122	16.67	5.33	1.88	54.32	34.91	3.56	4.18
Castanea crenata	Kuri	Bu*	3.74	1.73	0.64	0.0112	43.80	7.73	6.66	23.66	56.57	5.38	4.39
		Se	0.02										
		St	3.39										
Celastrus orbiculatus	Tsuruumemodoki												
Celtis sinensis	Enoki	Br*	2.38	9.58	3.56	0.0658	21.06	7.38	3.27	32.80	45.95	10.60	3.99
Schizophragma hydrangeoides	Iwagarami	Bu*	2.44	3.82	2.03	0.0396	62.41	8.05	3.16	16.87	64.22	7.70	4.12
Grasses													
Ixeris debilis	Ojishibari	*	2.27	0.83	0.49	0.0211	83.97	18.37	4.50	11.19	55.85	10.09	4.24
Izeris sitonifera	Iwanigana												
Trifolium repens	Shirotsumekusa												
Chamaele decumbens	Sentouso												
Callicarpa japonica	Murasakishikibu	Br	2.14										
Lonicera japonica	Suikazura	Lf*	1.58	0.24	0.21	0.0065	69.55	13.44	5.20	16.37	56.67	8.32	4.28
Rhus ambigua	Tsutaurushi	Br&											
		Bu	1.54										

Symplocos chinensis	Sawafutagi	Br	1.37										
Cornus kousa	Yamaboushi	Bu*	0.95	4.00	3.18	0.0701	58.02	9.45	6.02	16.20	60.79	7.54	4.29
Trachelospermum asiaticum	Teikakazura	Lf*	0.75	2.23	2.62	0.0408	58.21	6.21	3.80	25.66	53.27	11.06	3.98
Euonymus fortunei	Tsurumasaki	Lf	0.68	1.57									
Fraxinus lanuginosa	Aodamo	Br	0.56										
Cornus macrophylla	Kumanomizuki	Bu*	0.54	0.47	1.16	0.0262	50.14	9.63	5.77	16.12	60.27	8.21	4.26
Sagina maxima	Hamatsumekusa	*	0.50	2.91	4.79	0.0886	53.09	4.83	1.76	9.17	43.05	41.19	2.61
Coriolus versicolor	Kawaradake		0.22										
Torreya nucifera	Kaya	YLf	0.04										
Ilex macropoda	Aohada	Br	0.03										
Rubus microphyllus	Nigaichigo	Bu	0.03										
Unknown			0.15										

For footnotes see Table 11.A1.

Table 11.A3. Plant food composition, speed of intake, water content and nutritional content of each of the food items in autumn at Koshima

Species name	Japanese name	Part eaten[a]	DUR[b] (%)	dw[c] (%)	Speed of calorie intake (kcal/min)	Speed of protein intake (g dw/min)	Water content (%)	Crude protein (%)	Ether extract (%)	Crude fibre (%)	Soluble carbohydrate (%)	Ash (%)	Calories (kcal/g)
Daphniphyllum teijsmanni	Himeyuzuriha	Lf	0.35	0.25	13.46	0.2475	66.91	7.76	3.75	26.83	55.84	5.82	4.22
Machilus thunbergii	Tabunoki	Lf	0.15	0.35	9.76	0.1936	53.20	8.53	4.63	30.61	50.81	5.42	4.30
Neolitsea sericea	Shirodamo	Fr*	0.75	2.80	21.21	0.3728	51.83	10.65	37.82	23.80	22.03	5.70	6.06
Ardisia sieboldi	Mokutachibana	Fr*	5.15	10.20	8.59	0.0804	68.76	3.81	1.07	10.36	80.00	4.76	4.07
Pittosporum tobira	Tobera	Fr*	1.05	2.25	10.49	0.2233	48.48	9.75	7.96	30.80	48.36	3.13	4.58
Euonymus japonicus	Masaki	Fr*	5.05	1.45	2.20	0.0498	54.90	14.24	40.12	4.27	37.10	4.27	6.29
Quercus glauca	Arakashi	Fr*	4.15	3.90	2.91	0.0249	66.28	3.50	0.47	2.85	90.21	2.97	4.10
Ligustrum japonicum	Nezumimochi	Fr*	6.50	15.45	6.48	0.1026	70.83	7.38	11.20	12.04	64.73	4.45	4.66
Michelia compressa	Ogatamanoki	Fr*	6.90	18.70	12.43	0.2007	70.54	8.69	23.66	43.20	21.02	3.43	5.38
Podocarpus macrophylla	Inumaki	Fr*	1.35	7.35	18.44	0.2616	44.24	6.17	3.88	6.81	80.73	2.41	4.35
Elaeocarpus japonicus	Kobanmochi	Fr*	0.70	1.80	7.36	0.1290	65.79	3.21	1.63	9.37	79.53	6.26	4.02
Dendropanax trifidus	Kakuremino	Fr*	0.85	2.30	9.54	0.1262	63.10	6.28	14.07	37.50	36.42	5.73	4.75
Cinnamomum camphora	Kusunoki	Fr*	5.90	5.10	4.40	0.1181	32.37	15.74	30.28	29.45	21.97	2.56	5.87
Eurya emarginata	Hamahisakaki	Fr*	1.80	1.90	3.82	0.0408	68.82	4.39	3.34	43.96	41.59	6.72	4.11
Ficus erecta	Inubiwa	Lf*	0.50	0.60	3.90	0.1326	82.54	12.51	3.64	16.03	47.37	20.45	3.68
		Tw*	0.85	1.25	8.48	0.1063	60.10	4.90	1.70	39.80	43.80	9.80	3.91
Callicarpa japonica	Murasakishikibu	Lf	0.40	0.35	4.08	0.1208	81.49	12.58	3.79	22.02	54.01	7.00	4.25
Rhus succedanea	Hazenoki	Fr*	0.70	2.60	16.43	0.2931	58.85	9.87	25.11	25.80	37.17	2.05	5.53
Zanthoxylum ailanthoides	Karasusansho	Fr*	7.50	5.50	3.05	0.0592	12.58	10.20	22.50	47.64	14.05	5.61	5.25

			1.90	3.40	10.85	0.1914	64.15	10.18	31.19	10.03	44.59	4.01	5.77
Akebia quinata	Akebi	Fr*											
Ampelopsis brevipedunculata	Nobudo	Lf*	0.60	0.30	3.38	0.1144	84.05	12.43	2.98	16.49	48.23	19.87	3.67
Kadsura japonica	Sanekazura	Fr*	5.55	3.85	3.82	0.0922	78.27	13.36	27.39	17.45	35.50	6.30	5.53

For footnotes see Table A11.1.

Table 11.A4. *Plant food composition, speed of intake, water content and nutritional content of each of the food items in autumn at Koshima*

Species name	Japanese name	Part eaten[a]	DUR[b] (%)	dw[c] (%)	Speed of calorie intake (kcal/min)	Speed of protein intake (g dw/min)	Water content (%)	Nutritional contents[a]					
								Crude protein (%)	Ether extract (%)	Crude fibre (%)	Soluble carbohydrate (%)	Ash (%)	Calories (kcal/g)
Daphniphyllum teijsmanni	Himeyuzuriha	Lf*	4.67	18.50	13.46	0.2475	66.91	7.76	3.75	26.83	55.84	5.82	4.22
Machilus thunbergii	Tabunoki	Lf*	7.00	15.93	9.76	0.1936	53.20	8.53	4.63	30.61	50.81	5.42	4.30
Ilex integra	Mochinoki	Lf*	2.93	2.83	5.49	0.0701	68.70	5.48	8.30	31.91	45.19	9.12	4.29
Symplocos lucida	Kuroki	Lf*	3.73	8.93	7.73	0.1218	64.30	5.94	3.69	13.93	60.35	16.09	3.77
Ficus wightiana	Akou	Lf*	0.50	0.60	2.96	0.1010	59.82	13.83	2.16	21.96	51.91	10.14	4.05
Ardisia sieboldii	Mokutachibana	Fr*	0.13	0.67	8.59	0.0804	68.67	3.81	1.07	10.36	80.00	4.76	4.07
		Se*	5.67	3.80	2.89	0.0271	68.67	3.81	1.07	10.36	80.00	4.76	4.07
Pittosporum tobira	Tobera	Fr	0.10	0.50	10.49	0.2233	48.48	9.75	7.96	30.80	48.36	3.13	4.58
		Se*	4.80	2.33	2.15	0.0453	48.48	9.75	7.96	30.80	48.36	3.13	4.58
Euonymus japonicus	Masaki	Fr*	1.77	0.83	2.20	0.0498	54.90	14.24	40.12	4.27	37.10	4.27	6.29
Quercus glauca	Arakashi	Fr*	10.93	9.17	2.91	0.0249	66.28	3.50	0.47	2.85	90.21	2.97	4.10
		Se*	0.60	0.20	2.58	0.0146	66.28	2.39	2.78	2.41	90.00	2.42	4.23
Ligustrum japonicum	Nezumimochi	Fr*	2.57	3.70	6.48	0.1026	70.83	7.38	11.20	12.04	64.73	4.45	4.66
Podocarpus macrophylla	Inumaki	Se	0.03	0.03	3.61	0.0512	44.24	6.17	3.88	6.81	80.73	2.41	4.35
Elaeagnus pungens	Nawashirogumi	Fi*	0.77	0.27	1.16	0.0460	69.81	17.69	3.50	17.00	58.52	3.29	4.46
Camellia japonica	Yabutsubaki	Fl*	4.57	12.87	9.27	0.1191	84.43	5.39	2.24	14.58	74.11	3.68	4.20
Ficus erecta	Inubiwa	Lf	0.40	0.60	3.90	0.1326	82.54	12.51	3.64	16.03	47.37	20.45	3.68
		Sh*	0.93	0.40	1.52	0.0191	60.10	4.90	1.70	39.80	43.80	9.80	3.91
		Tw*	1.83	4.77	8.48	0.1063	60.10	4.90	1.70	39.80	43.80	9.80	3.91

Callicarpa japonica	Murasakishikibu	Lf	0.03	0.07	4.08	0.1208	81.49	12.58	3.79	22.02	54.01	7.00	4.25
Rhus succedanea	Hazenoki	Lf	0.17	0.17	3.09	0.0690	79.14	9.32	4.10	16.60	61.98	8.00	4.17
Trachelospermum asiaticum	Teikakazura	Lf	0.33	0.20	2.50	0.0428	60.10	6.90	2.50	22.30	60.00	8.30	4.04
Paper kadzura	Fuutoukazura	Lf	0.40	0.27	2.60	0.0690	76.70	10.00	3.00	14.40	56.10	16.50	3.77
		St*	0.83	0.80	1.39	0.0369	85.32	10.53	1.28	36.25	42.37	9.57	3.98
Unidentified sp. (1)		Lf	0.10	0.07	1.86	0.0490	76.70	10.00	3.00	14.40	56.10	16.50	3.77
Unidentified sp. (2)		St	0.20	0.20	2.32	0.0764	65.00	13.40	2.30	36.25	38.48	9.57	4.07
Miscanthus sinensis	Susuki	Lf*	0.87	1.10	5.53	0.0990	71.64	7.02	2.41	27.50	51.89	11.18	3.92
		Sh*	4.03	1.63	1.51	0.0352	68.30	9.50	5.20	27.00	46.90	11.40	4.09
		Rz	0.07	0.03	1.55	0.0362	70.10	9.27	2.00	32.00	46.73	10.00	3.98
Cyperus brevifolius	Himekugu	Lf	0.13	0.13	2.82	0.0505	71.64	7.02	2.41	27.50	51.89	11.18	3.92
Carex sp.	Suge	Lf	0.23	0.33	4.08	0.0730	71.64	7.02	2.41	27.50	51.89	11.18	3.92
Polygonum chinese	Tsurusoba	Lf	0.10	0.13	3.09	0.0995	54.04	12.92	1.95	13.81	60.70	10.62	4.01
		Fl*	0.90	0.43	1.47	0.0501	86.50	14.30	2.87	22.55	52.80	7.48	4.20
Bohemeria holosericea	Oniyabumao	St	0.03	0.07	3.76	0.0682	84.80	7.10	1.00	28.10	56.30	7.50	3.91
Lingularia tussilaginea	Tsuwabuki	Rz*	0.50	0.57	4.28	0.0275	77.80	2.50	1.90	12.60	73.50	9.50	3.89
Balanophora tobiracola	Kiretsuchitorimochi	Fl, St	0.43	0.97	6.50	0.1497	74.21	10.05	3.13	21.28	62.94	2.60	4.36
Auricularia auricula-judae	Kikurage		0.07	0.13	13.08	0.3000	74.21	10.00	3.13	12.00	72.27	2.60	4.36
Gloipeltis furcata	Fukurofunori	*	0.70	1.10	4.83	0.2319	80.00	12.67	0.25	15.54	30.17	41.37	2.64

For footnotes see Table 11.A.1.

Table 11.A5. *Plant food composition, speed of intake, water content and nutritional content of each of the food items in winter at Takasakiyama*

Species name	Japanese name	Part eaten[a]	DUR[b] (%)	dw[c] (%)	Speed of calorie intake (kcal/min)	Speed of protein intake (g dw/min)	Water content (%)	Crude protein (%)	Ether extract (%)	Crude fibre (%)	Soluble carbohydrate (%)	Ash (%)	Calories (kcal/g)
Aucuba japonika	Aoki	Lf*	2.89	3.72	5.54	0.1775	62.29	14.16	6.83	36.42	35.31	7.27	4.42
Quercus glauca	Arakashi	Lf*	5.65	6.26	4.49	0.1119	49.86	10.41	3.44	31.65	47.14	7.36	4.18
		Tw*	6.66	5.25	3.09	0.0345	43.24	4.52	0.52	43.06	47.29	4.61	4.05
Camellia sinensis	Cha	Lf	0.15	0.06	1.70	0.0733	60.83	18.45	2.43	17.04	55.57	6.51	4.28
Eurya japonica	Hisakaki	Lf*	1.52	1.23	3.20	0.0680	64.67	8.68	3.69	13.91	64.38	9.34	4.09
Cinnamomum camphora	Kusunoki	Lf-Bu*	11.55	9.90	3.49	0.1020	70.89	12.26	1.27	22.02	59.51	4.94	4.20
		Lf*	9.25	7.18	3.20	0.0242	47.82	3.21	5.81	17.34	67.18	6.46	4.24
		Tw*	15.05	10.14	2.77	0.0254	43.60	3.89	2.61	44.72	46.06	2.72	4.23
Celtis sinensis	Enoki	Br*	7.96	9.17	3.98	0.0585	35.87	5.23	1.04	44.47	33.82	15.44	3.56
Ligustrum japonicum	Nezumimochi	Lf*	1.65	2.80	7.03	0.1329	60.26	8.09	4.31	16.32	66.10	5.18	4.28
Rosa multiflora	Noibara	Lf-Bu*	0.71	0.66	3.97	0.2531	74.41	28.11	3.00	9.64	51.65	7.60	4.41
Oenanthe javanica	Seri	Lf-St*	0.50	0.47	3.67	0.2791	90.73	30.52	3.27	9.20	38.54	18.47	4.01
Neolitsea sericea	Shirodamo	Lf-Bu	0.03	0.00	0.53	0.0173	76.07	14.30	2.93	19.37	59.32	4.08	4.35
Machilus thunbergii	Tabunoki	Lf	0.41	0.53	5.30	0.1014	45.51	8.10	3.54	26.62	56.36	5.38	4.23
		Tw*	1.06	1.00	3.82	0.0299	61.18	3.28	2.14	38.57	53.19	2.82	4.19

Species	Common name	Part											
Camellia japonica	Yabutsubaki	Lf*	1.09	1.38	5.02	0.0990	61.73	8.11	2.62	20.96	61.09	7.22	4.11
		Fl*	0.75	1.25	6.77	0.0869	84.43	5.39	2.24	14.58	74.11	3.68	4.20
Cinnamomum japonicum	Yabunikkei	Lf*	1.37	1.40	4.29	0.0759	41.01	7.56	3.76	41.00	44.34	3.34	4.32
Galium spurium	Yaemuguru	Lf-St*	1.30	1.18	3.58	0.2309	89.71	26.20	4.41	11.64	40.60	17.15	4.06
Artemisia princeps	Yomogi	Lf-St*	0.82	0.85	4.18	0.2851	86.41	28.37	3.16	14.74	39.80	13.92	4.16
Trifolium repens	Shirotsumekusa	Lf-St	0.01	0.04	11.60	0.5972	85.26	22.24	4.95	14.72	47.90	10.19	4.32
Veronica persica	Oinufugure	Lf-St*	0.50	0.86	6.74	0.2622	88.64	15.60	4.67	14.71	50.05	14.97	4.01
Stellaria media	Hakobe	Lf-St*	0.66	1.14	6.52	0.4016	87.85	23.70	4.27	11.00	39.76	21.27	3.85
Akebia quinata	Akebi	Tw	0.08	0.04									
Cirsium sp.	Azami	Lf	0.07	0.21									
Styrax japonica	Egonoki	Bu	0.03	0.01									
Setaria sp.	Enokorogusa	Lf	0.81	1.02									
Petasites japonicus	Fuki	Lf	0.05	0.09									
Rubus buergeri	Fuyuichigo	Tw	0.23	0.13									
Clematis apiifolia	Botanzuru	Lf	0.21	0.26									
Carex sp.	Suge	Lf*	3.51	3.91	4.24	0.0759	71.64	7.02	2.41	27.50	51.89	11.18	3.92
Acer palmatum	Irohamomiji	Bu	1.33	0.58									
Ophiopogon japonicus	Jonohige	Lf	0.31	0.16									
Vicia angustifolia	Karasunoendo	Lf	1.11	2.59									
Symplocos tanakae	Kuroki	Lf	0.01	0.02									
Rubus hirsutus	Kusaichigo	Tw	0.62	0.37									
Deutzia scabra	Marubautsugi	Bu	0.30	0.07									
Euonymus sieboldianus	Mayumi	Tw	0.35	0.18									
Stauntania hexaphylla	Mube	Tw*	2.98	3.65	4.85	0.0369	53.46	3.11	0.63	57.97	34.79	3.51	4.08
Aphananthe aspera	Mukunoki	Br*	1.80	1.24	2.31	0.0418	54.15	6.26	1.81	30.84	39.92	21.18	3.46
		Tw	0.45	0.49									
Ilex chinensis	Nanaminoki	Tw	0.75	0.34									
Sonchus oleraceus	Nogeshi	Lf	0.13	0.44									
Pittosporum tobira	Tobera	Lf	0.47	1.06									
Lonicera japonica	Suikazura	Lf*	0.83	0.96	4.78	0.1502	71.33	13.44	5.20	16.37	56.67	8.32	4.28

Table 11.A5. (cont.)

Species name	Japanese name	Part eaten[a]	DUR[b] (%)	dw[c] (%)	Speed of calorie intake (kcal/ min)	Speed of protein intake (g dw/ min)	Nutritional content[a]						
							Water content (%)	Crude protein (%)	Ether extract (%)	Crude fibre (%)	Soluble carbo-hydrate (%)	Ash (%)	Calories (kcal/g)
Trachelospermum asiaticum	Teikakazura	Lf*	5.12	1.46	1.12	0.0191	60.10	6.90	2.50	22.30	60.00	8.30	4.04
Deutzia crenata	Utsugi	Bu	0.29	0.09									
Unidentified sp.		Lf	0.32	0.55									
Unknown			7.54										

For footnotes see Table 11.A1.

Table 11.A6. *Plant food composition, speed of intake, water content and nutritional content of each of the food items in winter at Shimokita. Data from Y. Nakayama and Y. Watanuki (unpublished)*

Species name	Japanese name	Part eaten[a]	DUR[b] (%)	dw[c] (%)	Speed of calorie intake (kcal/min)	Speed of protein intake (g dw/min)	Water content (%)	Crude protein (%)	Ether extract (%)	Crude fibre (%)	Soluble carbohydrate (%)	Ash (%)	Calories (kcal/g)
Morus bombycis	Yamaguwa	Br*	15.90				44.69	7.63	3.91	27.28	52.59	8.59	4.11
		Bu*	9.50				46.67	12.96	4.61	25.31	51.21	5.91	4.34
Actinidia arguta	Sarunashi	Bu*	14.70				65.60	9.74	3.37	22.09	59.39	5.41	4.25
Schizophragma hydrangeoides	Iwagarami	Br*	15.90				60.99	6.51	3.46	17.20	65.42	7.41	4.12
Fraxinus lanuginosa	Keaodamo	Bu*	5.50				41.36	6.38	7.30	26.64	56.22	3.46	4.49
Euonymus sp.	Tsuribana	Br*	3.30				50.07	7.35	3.08	13.78	68.50	7.29	4.12
		Bu*	4.30				60.64	16.08	3.84	22.15	50.69	7.24	4.29
Hydorangia petiolaris	Tsuruajisai	Bd*	3.30				69.80	20.17	2.42	19.87	48.87	8.67	4.22
Acanthopanax sciadoptylloides	Koshiabura	Br*	2.80				44.38	4.69	3.94	24.78	55.39	11.20	3.92
Others			24.80										

For footnotes see Table 11.A1.

References

Caldecott, J. O. (1986). *An ecological and behavioural study of the pig-tailed macaque.* Contributions to Primatology, vol. 21. Basel: S. Karger.

Drucker, G. R. (1984). The feeding ecology of the Barbary macaque and cedar forest conservation in the Moroccan Moyen Atlas. In *The Barbary macaque – a case study in conservation,* ed. J. E. Fa, pp. 135–64. New York: Plenum Press.

Furuichi, T., Takasaki, H. & Sprague, D. S. (1982). Winter range utilization of a Japanese macaque troop in a snowy habitat. *Folia Primatologica,* **37,** 77–94.

Hladik, C. M. (1977). A comparative study of the feeding strategies of two sympatric species of leaf monkey: *Presbytis senex* and *Presbytis entellus.* In *Primate ecology,* ed. T. H. Clutton-Brock, pp. 481–501. London: Academic Press.

Huffman, M. A. (1984). [Plant food and feeding behavior of Japanese monkeys in Arashiyama: annual change of plant food.] (In Japanese.) *Arashiyama Natural History Institute Report,* **3,** 55–65.

Itani, J. (1956). [The food habits of the Japanese monkeys in Takasakiyama.] (In Japanese.) *Primates Research Group,* **3,** 1–14.

Iwamoto, T. (1974). A bioeconomic study on a provisioned troop of Japanese monkeys (*Macaca fuscata fuscata*) at Koshima Islet, Miyazaki. *Primates,* **15,** 241–62.

Iwamoto, T. (1979). Feeding ecology. In *Ecological and sociological studies of Gelada baboons,* ed. M. Kawai, pp. 279–310. Basel: S. Karger.

Iwamoto, T. (1982). Food and nutritional condition of free ranging Japanese monkeys on Koshima Islet during the winter. *Primates,* **23,** 153–70.

Iwamoto, T. (1988). Food and energetics of provisioned wild Japanese macaques (*Macaca fuscata*). In *Ecology and behavior of food-enhanced primate groups,* ed. J. E. Fa & C. H. Southwick, pp. 79–94. New York: Alan R. Liss.

Izawa, K. (1982). [*Ecology of the Japanese monkeys.*] (In Japanese.) Tokyo: Dobutsu-sha.

Izawa, K. (1983). The ecological study of wild Japanese monkeys living in Kinkazan Island, Miyagi Prefecture: a preliminary report. (In Japanese with English summary.) *Bulletin of the Miyagi University of Education,* **18,** 24–46.

Izawa, K. & Nishida, T. (1963). Monkeys living in the northern limits of their distribution. *Primates,* **4,** 67–88.

Kuroki, K. (1975). A quantitative study of the daily activity patterns of wild Japanese monkeys, *Macaca fuscata fuscata,* at Koshima Islet. Master's thesis, Kyushu University.

Lindburg, D. G. (1977). Feeding behaviour and diet of rhesus monkeys (*Macaca mulatta*) in a Siwalik forest in north India. In *Primate ecology,* ed. T. H. Clutton-Brock, pp. 223–49. London: Academic Press.

Maruhashi, T. (1980). Feeding behavior and diet of the Japanese monkey (*Macaca fuscata yakui*) on Yakushima Island, Japan. *Primates,* **21,** 141–60.

Maruhashi, T. (1986). [Feeding ecology of Japanese monkeys in Yakushima Island.] (In Japanese.) In [*The wild Japanese monkeys on Yakushima Island*], ed. T. Maruhashi, J. Yamagiwa & R. Furuichi, pp. 13–59. Tokyo: Tokai Daigaku Shuppankai.

Maynard, L. A., Loosli, J. K., Hintz, H. F. & Warner, R. G. (1979). *Animal nutrition,* 7th edn. New York: McGraw-Hill Book Co.

McKey, D. B., Gartlan, J. S., Waterman, P. G. & Choo, G. M. (1981). Food selection by black colobus monkeys (*Colobus satanas*) in relation to plant chemistry. *Biological Journal of the Linnaean Society*, **16**, 115–46.

Milton, K. (1979). Factors influencing leaf choice by howler monkeys: a test of food selection by generalist herbivores. *American Naturalist*, **114**, 362–78.

Mitchell, H. H. (1964). *Comparative nutrition of man and domestic animals.* New York: Academic Press.

Mori, A. (1979*a*). An experiment on the relation between the feeding speed and the caloric intake through leaf eating in Japanese monkeys. *Primates*, **20**, 185–95.

Mori, A. (1979*b*). Analysis of population changes by measurement of body weight in Koshima troop of Japanese monkeys. *Primates*, **20**, 371–97.

Nakagawa, N. (1989*a*). Feeding strategies of Japanese monkeys against the deterioration of habitat quality. *Primates*, **30**, 1–16.

Nakagawa, N. (1989*b*). Bioenergetics of Japanese monkeys (*Macaca fuscata*) on Kinkazan Island during winter. *Primates*, **30**, 441–60.

Richard, A. F. (1985). *Primates in nature.* New York: W. H. Freeman.

Robbins, R. C. & Gavan, J. A. (1966). Utilization of energy and protein of a commercial diet by rhesus monkeys (*Macaca mulatta*). *Laboratory Animal Care*, **16**, 286–91.

Satoh, S. (1988). [The monkeys of Kinkazan A-troop.] (In Japanese.) *The Japanese monkeys in Miyagi Prefecture*. **3**, 6–29.

Soumah, A. G. & Yokota, N. (1991). Female rank and feeding strategies in a free-ranging provisioned troop of Japanese macaques. *Folia Primatologica*, **57**, 191–200.

Southwick, C., Zhang, Y., Jiang, H. & Qu, W. (1991). Comparative ecology of Rhesus population at latitudinal extremes in China. In *Primatology today*, ed. A. Ehara, T. Kimura, O. Takenaka & M. Iwamoto, pp. 25–8. Amsterdam: Elsevier.

Suzuki, A. (1965). An ecological study of wild Japanese monkeys in snowy areas, focused on their food habits. *Primates*, **6**, 31–72.

Uehara, S. (1975). The importance of the temperate forest elements among woody food plants utilized by Japanese monkeys and its possible historical meaning for the establishment of the monkey's range. A preliminary report. In *Contemporary primatology*, ed. S. Kondo, M. Kawai & A. Ehara, pp. 392–400. Basel: S. Karger.

Uehara, S. (1977). [A biogeographic study of adaptation of Japanese monkeys (*Macaca fuscata*), from view point of food habits – an essay on reconstruction of the history of Japanese monkeys' distribution.] (In Japanese.) In *Morphology, evolution and primates*, ed. Y. Kato, S. Nakao & T. Umesao, pp. 189–232. Tokyo: Chuokoron-sha.

Wada, K. & Ichiki, Y. (1980). Seasonal home range use by Japanese monkeys in the snowy Shiga Heights. *Primates*, **21**, 468–83.

Wada, K. & Tokida, T. (1981). Habitat utilization by wintering Japanese monkeys (*Macaca fuscata fuscata*) in the Shiga Heights. *Primates*, **22**, 330–48.

Watanuki, Y. (1984). [Age–sex differences of feeding behavior in Japanese monkeys in northwestern part of Shimokita Peninsula.] (In Japanese.) *Annual Report of the Primate Research Institute, Kyoto University*, **14**, 73–4.

Watanuki, Y. & Nakayama, Y. (1986). [Age–sex differences of feeding behavior in Japanese monkeys in northwestern part of Shimokita Peninsula.] (In Japanese.) *Annual Report of the Primate Research Institute, Kyoto University*, **16**, 63.

Wheatley, B. P. (1982). Energetics of foraging in *Macaca fascicularis* and *Pongo pynmaeus* and a selective advantage of large body size in the orang-utan. *Primates*, **23**, 348–63.

Yotsumoto, N. (1976). The daily activity rhythm in a troop of wild Japanese monkeys. *Primates*, **17**, 183–204.

Zhao, Q., Deng, Z. & Xu, J. (1991). Natural foods and their ecological implications for *Macaca thibetana* at Mount Emei, China. *Folia Primatologica*, **57**, 1–15.

12

Population management and viability of the Gibraltar Barbary macaques

J. E. FA AND R. LIND

Introduction

Food-enhanced primate populations with historical or anthropocentric significance can be valuable centrepieces for conveying conservation messages to a wider public (Fa & Southwick, 1989). Because of their more tractable nature, these groups can also serve as important models for investigations into small population biology, behaviour and ecology (Fa & Southwick, 1989). The Barbary macaques (*Macaca sylvanus*) on the Rock of Gibraltar are one such example. These monkeys, which live in natural areas close to human habitation, have become an important attraction visited by thousands of tourists yearly. They have also been the subject of much ecological and behavioural research (see MacRoberts & MacRoberts, 1966; MacRoberts, 1970; Burton, 1972; Burton & Sawchuk, 1974, 1982, 1984; Zeller, 1980, 1986, Chapter 24, this volume; Fa, 1981, 1984a,b, 1987, 1988, 1989, 1991, 1992; Hornshaw, 1984; O'Leary & Fa, 1993).

Barbary macaques have probably been in Gibraltar since as early as AD 711, but records of their presence only appear after 1740. Fa (1981) suggested that the monkeys may have been introduced then by the British Garrison. Since 1740, the monkey population has been re-stocked three times (1813, 1860 and between 1939 and 1943). The present animals descend from the last importation, which was carried out in response to a sharp decline in numbers as a result of disease (Zeuner, 1952).

Throughout the late 1800s and early 1900s, monkey forays into built-up areas caused such problems to townsfolk (Sclater, 1900) that the British Army was given instructions in 1915 to confine the animals to the Rock's natural areas (Upper Rock) by 'whatever means' (Kenyon, 1938). The Army's brief was to perpetuate the monkey population, while ensuring minimum incursions into, and possible danger to, the adjacent human one.

By provisioning the animals in the Upper Rock, the problems of entering dwellings and stealing property, the threat of outright physical damage and the risk of spread of disease were supposedly minimised. In reality, monkey incursions continued regularly until 1975, when the feeding site for one of the troops was relocated even further away from the town and rations were increased substantially (Fa, 1984a, 1989).

The management of the Gibraltar monkeys has changed little since 1915. No scientific criteria have been used to control population size or to guide provisioning. Some efforts to implement a management plan, in consultation with the British Army and Gibraltar Government, were started in 1991 by Medambios Environmental Consultants and the Institute of Mediterranean Ecology (O'Leary, 1993a).* These were cut short (see Isola & Isola, 1993; O'Leary, 1993b) when, following the decision of the Ministry of Defence (UK) in September 1992 to relinquish all responsibility for the care of the monkeys to the Gibraltar Government, the latter then transferred the charge to a private company experienced in managing the Rock's tourist sites and beaches. The monkeys are now part of the Upper Nature Reserve tourist attraction run by Sights Management Ltd (Tilbury, 1993). This company is sub-contracted by the Gibraltar Tourism Agency (a semi-government organisation linked to tourism promotion) and it continues to feed and care for the macaques.

Despite various publications urging a reconsideration of the management of the Gibraltar macaques in order to ensure their viability, to date the relevant authorities have not acted on any proposals made to them (see O'Leary, 1993b). Indeed, more than 12 years ago a panel of over 20 expert primatologists at the International Conference on the Conservation of the Barbary Macaque in Gibraltar recorded there was a 'critical need for rational management of the Gibraltar [macaque] population if it is to survive on the Rock at all'. The conference made recommendations contained in a book by Fa (1984b). Similarly, between 1980 and 1984, the Conservation Working Party of the Primate Society of Great Britain was also involved with the Gibraltar Regiment over issues of demographic regulation and physical condition of the animals (see Chivers, 1984). The problems affecting the macaques have not altered much since then and

*In contrast to what is implied by Isola & Isola (1993), Medambios was contracted only to manage the tourist site at Apes' Den under a contract signed on April 1991. The company provided wardens and ensured traffic flow and cleanliness. The company had no jurisdiction over the control of feeding or any other activities concerning the macaques on the site. Medambios, however, initiated a scientific management plan through the Institute of Mediterranean Ecology, catalysed by the Minister of Tourism (now Minister of the Environment) to promote scientific study in Gibraltar.

revolve essentially around reconciling tourist interest with the animals' general well-being.

This chapter reviews the current situation of the Gibraltar macaques and outlines recommendations for their long-term survival. It presents an explicit action framework that will: (1) determine whether the current situation requires active intervention by estimating probability of extinction (pE) (see Maguire, Seal & Brussard, 1987); (2) evaluate risks and benefits of the existing management intervention; and (3) recommend realistic management schedules that take into account the multiple interests involved and their conflicting goals. This procedure provides a means of evaluating options in a logical and repeatable manner.

Population dynamics
Population size and composition

Between 1946 and 1991, the Gibraltar monkey population was composed of two separate groups, at Queen's Gate and at Middle Hill (O'Leary & Fa, 1993; Fa, 1984*a*). At present, six groups are recognised. Until 1980, the modal size for both troops was 15–20; the average size of the Middle Hill troop (17 animals) was slightly larger than Queen's Gate (15 animals). Queen's Gate grew to 25 monkeys by 1953 from an initial five, fell to nine in 1969, but rose to 51 animals in three groups by 1994 (Fig. 12.1). Middle Hill troop, which was formed by a splinter group of five animals from Queen's Gate in 1946, has varied in size between 12 and 22 monkeys with no significant drop in numbers. This troop had increased to 98 individuals divided into three groups by 1994.

Group fissioning was first detected in 1991 when Queen's Gate split into two groups (the Royal Anglian Way troop and the main Queen's Gate monkeys). Subsequently, in 1993 another splinter group broke away from Queen's Gate (Prince Phillip's Arch troop). Middle Hill troop divided into a main troop and a smaller group (the Farringdon's Battery troop) between 1990 and 1992. Another group (the Rock Gun troop) separated from Farringdon's Battery troop later in 1993.

Age–sex distributions of the population have remained close to normal; this is particularly true for the period from 1950 to 1980 (Burton & Sawchuk, 1984; Fa, 1984*a*). Adult sex ratios within the two sub-populations have none the less shown a high degree of variation (Fig. 12.2) due to changes in the number of adult males left in the groups. Because of the Army's perception that more than one 'leader' male in the troop would

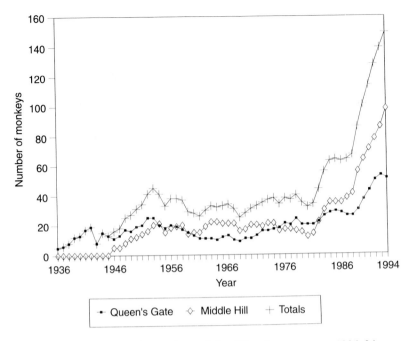

Fig. 12.1. Changes in numbers of the Gibraltar macaques 1936–94.

increase aggression, young males were exported or culled prior to reaching sexual maturity. Only a few sexually mature males were kept in the troops to replace adult males that died or those whose reproductive productivity began to decline. This changed when culling was stopped after 1980 (Fig. 12.2). Most females were retained for breeding purposes.

Breeding

Estimated birth rates, calculated as the percentage of mature females (over four years, see Fa, 1984a; Paul & Kuester, 1989; Paul, Kuester & Podzuweit, 1993) giving birth each year (Fig. 12.3), averaged 48.51% for Queen's Gate and 63.24% for Middle Hill. For each decade, mean and median birth rates in Queen's Gate and Middle Hill were highest during 1940–50 and 1950–60 but dropped during 1960–70. After 1980, birth rates have remained at similar levels for Queen's Gate (Fa, 1989) but have increased in Middle Hill.

Age of onset of breeding has varied between macaque sub-populations; at Queen's Gate average age of first breeding was 5.16 years (SD $\pm$ 1.13) but was lower for Middle Hill (4.42 $\pm$ 0.74) (Fa, 1984a). Median age of first

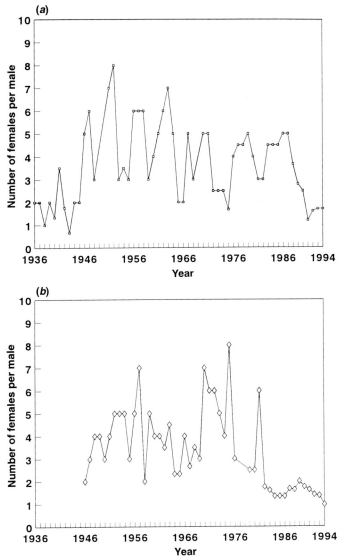

Fig. 12.2. Adult sex ratio fluctuations in both Gibraltar macaque sub-populations: (*a*) Queen's Gate; (*b*) Middle Hill.

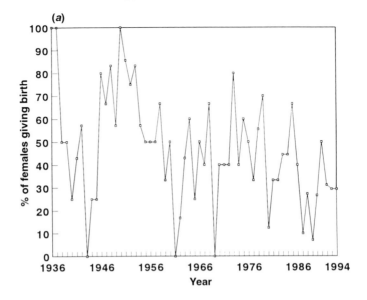

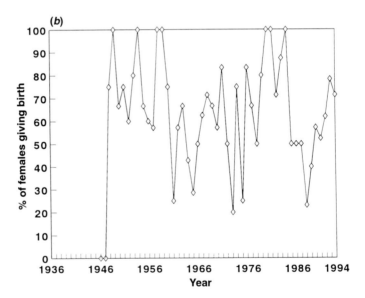

Fig. 12.3. Changes in annual birth rate for both Gibraltar macaque sub-populations: (*a*) Queen's Gate; (*b*) Middle Hill.

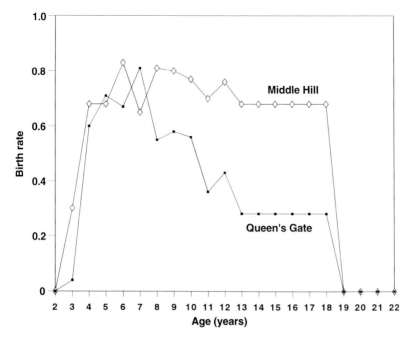

Fig. 12.4. Fecundity schedules for Barbary macaques in Gibraltar.

birth was 4.87 years for Queen's Gate and 4.49 years for Middle Hill. The range for primiparous births at Queen's Gate was 3.99–8.97 years but 3.93–6.87 years at Middle Hill. At Middle Hill and Queen's Gate, 25% and 5% of females, respectively, had their first parturition at around 4 years. The majority of females (around 50% in both sub-populations) delivered their first infant at 5 years. Fluctuations in onset of breeding age between decades were detected by Fa (1984*a*); there was an identical elevation of two years in primiparous age during the 1960–70 period.

Female fertility in Gibraltar peaked among 8–12 year old females at Middle Hill but among 4–8 year olds at Queen's Gate (Fig. 12.4). Fertility declined thereafter, dropping to zero after 19 years of age. Reproductive senescence may occur after 20 years as shown by Paul *et al.* (1993) for female Barbary macaques at Salem, Germany. Fecundity schedules are statistically different between the Gibraltar sub-populations (Fa, 1984*a*) but only Queen's Gate contrasts significantly with Salem macaques (Paul *et al.*, 1993).

Health and mortality

Systematic health screening has never been undertaken either by the Army or by the current management authorities. There has been little awareness of parasite loads despite the fact that Burton & Underwood's (1987) independent study of intestinal helminths showed that all the animals examined at Queen's Gate and a third of Middle Hill monkeys were infected with *Ascaris, Trichuris* and an unidentified hookworm. These differences between the groups were related to the prevailing hygienic conditions around their feeding areas. There is no evidence to suggest that the situation may have changed recently.

Causes of death in the Gibraltar macaques have ranged from hypothermia, bronchitis, coronary thrombosis, etc. to a variety of traumas (accidental electrocutions, injuries etc.) (see Fa, 1984a). Between 1936 and 1994, recorded mortality was above the median recorded (4%) in over 30% of all years (Fig. 12.5). Mortality rates have always been higher at Queen's Gate than at Middle Hill (Fa, 1984a). Outbreaks of disease have been a major cause of death. An epidemic of severe gastro-enteritis (Zeuner, 1952) between 1936 and 1944 killed 68% of the population at its peak in 1944. A second, smaller epidemic of ringworm in 1968 (*Trichophyton simii*) affected 35% of the animals (Clifford, Callanan & Smith, 1972) and again in 1987, viral pneumonia (O'Leary & Fa, 1993) caused the death of about 25%. All outbreaks have been associated either with inappropriate provisioning (e.g. Zeuner (1952) attributed the outbreak of 1944 to cookhouse refuse from Army barracks, while Clifford *et al.* (1972) also pointed to inadequate feeding as the cause of the ringworm outbreak) or with contagion of zoonoses from visiting tourists. The interval between reported disease peaks was 25 years, which translates into a 4% chance of occurrence.

A review of Gibraltar macaque management

General management protocols

The British Army's involvement with the monkeys was largely inspired by the legend which maintained that British sovereignty of the Rock would endure so long as the monkeys remained there; this political element ensured continuity in management for over 77 years. The Ministry of Defence (UK) deemed itself to be the animals' legal custodian (and owner) by virtue of the Governor's orders that officially included the macaques in the Army Roll (Kenyon, 1938). Management of the monkeys has taken the form of provisioning, veterinary care and population control, first at the hands of the Royal Artillery and, later, of the Gibraltar Regiment. Decisions on daily feeding and culling were made by an ape-keeper (a full-time soldier) with the approval of the Commanding Officer. Royal

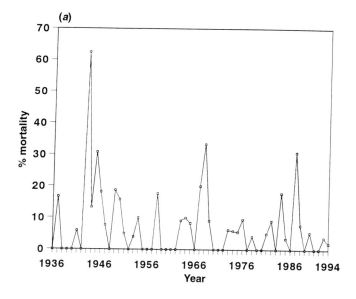

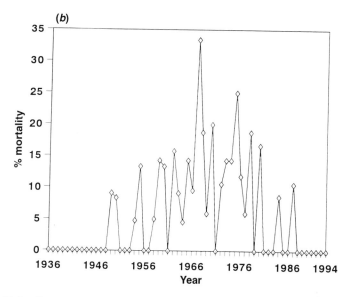

Fig. 12.5. Changes in mortality rates in both Gibraltar Barbary macaque sub-populations, 1936–94: (*a*) Queen's Gate; (*b*) Middle Hill.

Army Corps veterinarians were flown in annually for veterinary advice. Any surgical interventions were carried out at the Royal Naval Hospital. At present, a civilian veterinarian is contracted to oversee sick animals.

During the Army's period of management, all monkeys were named (usually after dignitaries) and daily censuses confirmed the birth or presumed death of any individual. However, as animals have never been marked with tags or tattoos, recognition of individuals has been visual, thus making the follow-up of any particular animal far from accurate. It has also introduced an unknown margin of error in the Army roll calls and correspondingly on the pedigree data obtained from them (see Burton & Sawchuk, 1984). The effect of this error has been mitigated, to some extent, by the continuity of personnel attending the monkeys (until the 1992 takeover there were only four Army ape-keepers, the last two holding that position for 26 and 12 years, respectively). Under the present management no attempt has yet been made to introduce a system of individual identification, and censuses are being taken only periodically.

Provisioning and tourist pressure

After more than five decades of daily provisioning by the Army the monkeys are now concentrated on the Upper Rock. As the monkeys' reputation as a tourist attraction on the Rock has grown, so the number of people visiting and feeding them has increased. Visitor feeding has been largely encouraged by public service vehicle operators (taxis, tour buses) who have realised the value of enhancing their guided tours by luring the monkeys with sweets, peanuts, etc. to approach visitors and to perform antics (e.g. sitting on shoulders for photographs). The Middle Hill animals have never had as much contact with humans as the Queen's Gate troop, although in recent years it has increased to some extent.

Provisioning levels have been rising since artificial feeding was introduced. Although the actual food volume given to the macaques had not been recorded by the authorities, annual food expenditure/monkey (once corrected for inflation) shows that the amount given has increased more than fourfold since the outset (Fa, 1984*a*, 1989). Types of food have remained similar; in a study between 1979 and 1980, Fa (1986*b*) counted 25 different types of fresh fruit and vegetables fed to the macaques. Natural foods may consist of a variety of plant species found in the Upper Rock but have been calculated to account for fewer than 1% of the food ingested (Fa, 1986*b*).

Figures for 1979–80 indicated that Queen's Gate troop was given a daily average of 18.96 ± 12.98 kg of food, while Middle Hill received more

(25.10 $\pm$ 9.50 kg) (Fa, 1986*b*). There were no monthly or seasonal differences in food volume offered. Averages of 494 kg/month for Queen's Gate and 764 kg/month for Middle Hill were calculated, giving annual totals of 5929 and 9167 kg, respectively. Expressed as calories this represented around 12 201 kcal/animal (1 kcal = 4.184 kJ) at Queen's Gate (24 fold more than the calculated energy requirement for a wild Barbary macaque, see Drucker, 1984) and 144 672 kcal at Middle Hill (289 fold more).

Apart from this food provided by the management authorities, the monkeys, especially those at Queen's Gate, receive extra food from tourists. Comparisons of the ecology and behaviour of the tourist-affected and the wilder sub-populations in 1979–80 demonstrated that significant differences were correlated with the degree of tourist exposure (Fa, 1986*b*, 1988). During this study, which was during a low tourist year because of the closed frontier with Spain, an estimated total of 42 139 people visited Queen's Gate at a rate of 15.00 $\pm$ 6.00 visitors/h (Fa, 1986*a*). Over the next decade and with the full re-opening of the frontier (in 1985), visitor numbers rose to an average of 120.09 $\pm$ 82.02 persons/h in 1987 (Fa, 1988) and 135.00 $\pm$ 25.23 persons/h in 1991 (O'Leary & Fa, 1993).

In 1991, O'Leary & Fa (1993) calculated an average of 99.6 interactions/h between people and monkeys at Queen's Gate occurring daily. Of all recorded interactions, 86% concerned tourists. Contact with people was the most common type of interaction and accounted for 30% of all macaque-initiated exchanges. Infants and juveniles spent a significantly higher proportion of all interactions engaging in contact with humans than any other class. A typical interaction would be for a monkey to be enticed, by means of food held over a tourist's head or shoulders, to sit on the person's shoulder for a photograph. This 'making friends with the apes' is a popular perk offered by taxi drivers to their passengers. Infants and juveniles are the main targets because they are more active, 'look cute', and do not alarm tourists. However, 25% of all reported 'making friends' interactions involved adult males for the most part begging or grabbing food from visitors (O'Leary & Fa, 1993).

Veterinary care

Primarily because the macaques on Gibraltar have always been considered 'wild', the authorities have never made any attempts to carry out systematic, hands-on health checks of the animals. As a general rule veterinary intervention would take place only in serious and unavoidable circumstances, usually requiring surgical operations that could be given a high publicity profile. Little attention was paid to preventing disease. This is in stark contrast to the yearly round-up, check-up and vaccination routines carried out on more than 200 Barbary macaques in the French and German Barbary macaque parks (de Turckheim & Merz, 1984).

Culling

The Army adopted a policy of culling to restrict population numbers and thereby ensure that macaques would not wander into town areas again. 'Excess' or 'problem' individuals (usually adult males) were removed. An arbitrary maximum level of 25 animals/troop was set by the British Army during World War II but this was later raised to 34. This increase was justified on the basis that regular feeding at specific sites had made the animals easier to control at higher numbers; culling ceased in 1980.

Impact of provisioning and tourist pressure
Behaviour modification

Diet has a direct bearing on an animal's use of time (see Boyd & Goodyear, 1971). A diet based largely on soluble compounds tends to release time otherwise employed in processing food. Studies of food-enhanced primate groups all concur that the animals spend less time feeding and foraging (Post & Baulu, 1978; Baulu & Redmond, 1980; Seth & Seth, 1986; Forthman Quick & Demment, 1988; Malik & Southwick, 1988; Marriott, 1988). An explanation for this is that human foods provide more energy for less effort than the same amount of wild forage. In Gibraltar, provisioned foods contain a mean of 134.08 kcal/g in comparison with 0.04 kcal/g for natural forage (Fa, 1986b). In terms of weight per food item, provisioned items weighed an average of 101.92 g compared with 0.44 g for natural foods.

There is ample evidence to show that the Gibraltar macaques' activity budgets are affected by the amount and types of food given to them. Time spent feeding is negatively correlated with access to human-derived foods. A Queen's Gate troop that received high calorie tourist-derived foods, as well as provisioned foods, was observed to spend only 7% of its overall activity feeding (H. O'Leary, unpublished data), whereas time spent foraging at Middle Hill was significantly higher (16%). Monkeys at Middle Hill foraged for natural foods to supplement their diet, an activity rarely seen at Queen's Gate. Wild Barbary macaques, in contrast to all Gibraltar troops, may spend more than half of their daytime feeding (see Deag, 1974; Drucker, 1984).

Influence on demography

The observed disparity in the annual birth rates of the two troops (see above) did not correlate with climatic variables (temperature and rainfall), provisioning levels, genetic relatedness or population densities. In contrast, Fa (1984a) pointed to a significant negative correlation between the number of people visiting the monkeys and demographic changes.

Lower breeding performance can be related to increased tourist pressure. While attacks over common food items are rare, scarce and/or nutritionally attractive foods provoke intense competition (see Wrangham, 1974). Such competition makes access to food difficult, which, in turn, gives rise to aggression. With visitors offering sweets throughout the day, the monkeys have learnt to expect these tidbits. Fa (1984a) suggested that it was the manner and level of supplemental feeding that was responsible for the fall in breeding. High visitor numbers increased tension in the troops and adversely affected breeding performance. Behavioural modifications instigated by the 'addiction to', or conditioned preference for, sweets and supplemental foods were thus likely to be responsible for breeding failures. Because the Gibraltar macaques breed for a period of no more than four months in a year, with mating concentrated in one month, and because the artificially controlled troop sex ratio has usually been one or two males to up to eight females, any distraction from the strictly seasonal mating activities would have a significant impact on mating.

Consequences for visitors

Uncontrolled visitor interactions with the monkeys have not only had deleterious effects on the animals, they have also introduced a significant potential health and safety risk to the public. This has been substantiated by Campbell (1989) and later Fa (1992) after examining hospital records of people who received medical attention for monkey bites; in a nine year period (1980–9), Fa (1992) noted 248 recorded cases, including some severe lacerations. This represents a significant medical threat to tourists given that zoonoses are clearly transmissible between humans and primates.

Viability analysis

A species' viability can be measured by evaluating population dynamics (size structure and sub-structure, density and variation in growth rate), population characteristics (morphology, physiology, and behaviour and dispersal patterns), and environmental effects (habitat quality and quantity, patterns and rates of environmental disturbance and change, and interactions with other species including man) (Soulé, 1987). Analysis of persistence of a species can be simulated using VORTEX, a Monte Carlo computer program that can predict future trends in a population (Lacy, 1993). The model considers demographic events in a population's history, with processes modelled as discrete sequential events. Events are probabilistic occurrences and are determined with the use of a random number generator. This program recreates the stochasticity often found in small, isolated populations. Stochasticity is achieved by incorporating a range of variation for each demographic variable. The program models a population's progress over a given time-span and produces summary statistics on the population. These include the mean time to extinction for those simulations reaching extinction, mean population size for populations remaining extant, and the genetic variation remaining in extant populations.

The population parameters discussed above can be used to predict future trends in the Gibraltar macaque population. Disease epidemics are the only perceived major threat to the macaque population in Gibraltar. Although it is difficult to know the extent to which the population's reproductive output has been affected by increased mortality periods, available data indicate that the effect has been only slight (Fa, 1984a). However, as population size rises and imposes restraints on available resources, stress levels are heightened and the animals become more vulnerable to disease. When this is coupled with increased contact with humans, the threat of cross-infection (especially to the monkeys) can be very real.

The known population parameters to which the population is sensitive were considered in detail and, where exact values were unavailable, a range of values was used and all scenarios considered. Table 12.1 gives the demographic values used in the simulations. A minimum value of 48 months (4 years) was used in the simulations to represent the earliest possible age for the onset of breeding. Because it is difficult to determine the age of first breeding in males, for present purposes 48 months was considered an appropriate figure, as indicated by Paul *et al.* (1993).

Female productivity for these simulations is based on the maximum number of offspring/female and the percentage of females producing offspring in a given year. Although twin births have been recorded in these populations the incidence is so low ($< 1\%$) that it was felt it could justifiably be ignored and the maximum number (litter size) was set at one offspring/female per year. The percentage of females producing offspring varied widely, with a minimum of 38.50% in Queen's Gate and a maximum of 77.60% in Middle Hill.

Table 12.1. *Results of the 'VORTEX' stochastic modelling simulations for the Gibraltar macaque population*

Simulation	Female productivity (%)	Probability of catastrophe(s)[a]	% effect of catastrophe on reproduction	% effect of catastrophe on survival	Probability of survival (%)	Mean final population size	Lambda
1	48.76	none	none	none	100	242.12 (SD 12.06)	1.036
2	48.76	0.04	0.50	0.65	98.40	174.31 (SD 73.53)	1.019
3	48.76	0.04	0.50	0.50	88.90	116.61 (SD 82.39)	1.013
4	48.76	(1)0.04	0.50	0.65	82.10	93.47 (SD 75.53)	1.003
		(2)0.04	0.50	0.65			
5	48.76	(1)0.04	0.50	0.50	44.30	57.43 (SD 60.20)	0.991
		(2)0.04	0.50	0.50			
6	55.59	0.04	0.50	0.65	99.90	205.43 (SD 56.62)	1.032
7	55.59	0.04	0.50	0.50	96.20	160.58 (SD 81.74)	1.026
8	55.59	(1)0.04	0.50	0.65	93.30	136.00 (SD 82.72)	1.016
		(2)0.04	0.50	0.65			
9	55.59	(1)0.04	0.50	0.50	61.90	74.58 (SD 73.88)	1.004
		(2)0.04	0.50	0.65			
10	62.43	0.04	0.50	0.65	100	220.48 (SD 44.01)	1.045
11	62.43	0.04	0.50	0.50	98.60	185.60 (SD 73.18)	1.038
12	62.43	(1)0.04	0.50	0.65	97.60	167.70 (SD 74.07)	1.028
		(2)0.04	0.50	0.65			
13	62.43	(1)0.04	0.50	0.50	73.40	105.24 (SD 83.62)	1.015

[a]Simulations with two probabilities of catastrophe, identified by (1) and (2) in the table, refer to simulations which were modelled to include two independent catastrophes, each with its own effect on reproduction and survival.

Demographic and environmental stochasticity

Variations in population size and structure are the result of demographic and environmental stochasticity (Goodman, 1987). Demographic stochasticity occurs in the form of random variation in the demographic parameters of the population: births, deaths and sex ratio. Environmental stochasticity results from variation in environmental factors affecting the rates of birth and death in a population. This could be the result of variation in resource availability, or of catastrophic events such as disease epidemics, fires or severe weather conditions (e.g. cyclones, hurricanes) (Gilpin, 1987; Lyles & Dobson, 1988). Such extreme levels of variation occur at random, but probabilities can often be attached to them based on the previous history of the population. In the VORTEX model, these stochastic events are seen as probabilistic occurrences determined using a Monte Carlo simulation, and are treated as discrete sequential events.

Genetic variation and inbreeding depression

Small, reproductively isolated populations (i.e. without immigration), such as the Gibraltar macaques, are expected to undergo a decrease in genetic variation through time. This then influences the probability of the population's long-term survival, because genetic variation is a requisite for evolutionary adaptation to a changing environment. However, in small groups of under 20 breeding animals the effects of drift are considerable. If the population remains small, gene frequencies will change erratically in a 'bagatelle' manner and the rarer alleles will inevitably be lost in due course. Because the probability of an allele being lost in any generation is $1/(2N)$, where N is the population size, genetic variability will decrease. The smaller the population the greater is the rate of loss.

In a normal, randomly breeding population a more realistic expression of the rate of loss of diversity is $1/(2N_e)$. Here, N_e represents the effective population size or the number of randomly breeding animals required to produce the same genetic composition as the sample population. In most cases N_e is less than N, usually only 30% of the population or less. An effective population size equal to the actual population requires that all its members breed equally and that the sex ratio of reproductively active individuals is the same.

The breeding history of the Gibraltar macaque population indicates that not all individuals have contributed to the present population (Fa, 1984a). Moreover, the founders were sub-divided into two groups. These are factors that will have accelerated the loss of diversity above the rate of loss expected for a randomly breeding population with the same number of founders. Indeed, based on pedigree information, Burton & Sawchuk (1984) estimated that the effective population size of the Gibraltar macaques was most sensitive to the number of males in a group. Their calculations showed that the effective population size progressively increased on the inclusion of more adult males. The actual populations, when these figures were calculated, were between 30 and 39 animals (mean 33.29). This gave a ratio of effective:actual population size of 0.10:0.13; a situation reflecting severe loss of heterozygosity, with potential deleterious consequences. As a corollary, for years in which paternity was unambiguous (i.e. years in which there was only one adult male capable of siring) biparental calculations indicated an effective population size ranging between 3.43 and 4.22 (Burton & Sawchuk, 1984; Fa, 1986*b*). Therefore, the current genetic structure of the population reflects the creation of a bottle-neck each time males have been removed. Allele frequencies in the population are significantly altered by reducing the male gene pool. The more often this is done, the greater the chance of eliminating altogether an allele or alleles from the gene pool. This means that: (1) the probability of close consanguineous matings in future generations increases; and (2) there is an over-representation of the alpha male's genes because the removal of other mature males allows only the resident male to sire an abnormally large number of offspring. As approximately half of these offspring will be male, the probability of this patriline dominating the population is extremely high.

The actual effect of inbreeding depression on the population is not yet known. For the purposes of our model it will be assumed that inbreeding depression has occurred and that its outcome has been to remove recessive lethals from the population. This assumption will undoubtedly lead to an underestimate of the effect of inbreeding.

Carrying capacity limits

Carrying capacity within the context of this model is considered to be the maximum number of animals that the environment can sustain. Limiting factors on the macaque population are the availability of territory and food. The area currently used by the Gibraltar macaques is approximately 26 ha: 4 ha for Queen's Gate and 22 ha for Middle Hill.

Data on Barbary macaque populations in the European parks indicate that population densities of between 15 and 26 animals/ha are possible (Paul & Kuester, 1989). However, densities at the top of this range could be associated with high mortality levels. If these densities are imposed on the known range sizes of the Gibraltar troops, populations of 60–104 animals at Queen's Gate and 330–572 at Middle Hill are to be expected. These figures represent a considerable increase from the present population size, 51 animals at Queen's Gate and 98 at Middle Hill. Associated with this potential rise in population size would be an increase in provisioning and its related costs. Therefore, the carrying capacity set should reflect both genetic requirements and economic reality.

The minimum viable population required to maintain 90% of the population's existing genetic diversity over the next 100 years can be determined using the software package CAPACITY (Ballou, 1992). This program calculates both the effective and actual population size required to maintain a specified amount of diversity over a given time-frame using generation length, rate of population growth (Lambda) and the ratio of effective to actual population (N_e/N). The minimum viable population for macaques on Gibraltar ranges from 250 to 380 animals, using values of Lambda between 1.05 and 1.25, and ratios of N_e/N of between 0.10 and 0.30. For the model, the minimum figure of 250 was chosen as a reasonable population size that could be economically provisioned. However, this population size may be ultimately unrealistic because of the conflict with the local human population that would ensue as a result of the extension of the monkey's home ranges.

Catastrophes

Disease epidemics are the only perceived major threat to the macaque population in Gibraltar. Although it is difficult to know the extent to which the population's reproductive output has been affected by increased mortality periods, available data would indicate that the effect has only been slight (Fa, 1984a). However, as population size rises and imposes restraints on available resources, stress levels are heightened and the animals may become more vulnerable to disease. When this is coupled with increased contact with humans, the threat of cross-infection (especially to the monkeys) can be very real.

The model was run using scenarios which assessed the effects of one or two disease outbreaks happening. In both scenarios, disease was considered to reduce the reproductive output of the population by 50%. The effect of disease on survival was also modelled at two levels; the first level assumed a disease epidemic resulting in 35% mortality whilst the second accounted for 50%.

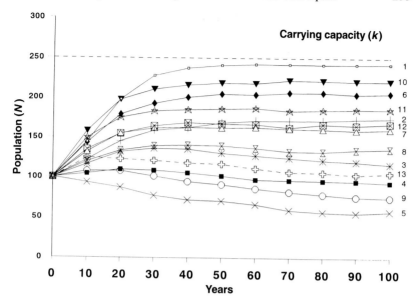

Fig. 12.6. Results of VORTEX simulations for 13 plausible scenarios (data of simulations are shown in Table 12.1).

Simulation results

Initially, simulations of 10 iterations each were run in order to assess the sensitivity of the population to the demographic variables used in the model. The results showed the population to be most affected by changes in female productivity and by the frequency and severity of disease outbreaks. Alterations to these two parameters resulted in the most dramatic changes in the model's output. Based on these results a series of 13 further simulations, each of 1000 iterations, was run. Each simulation used different combinations of female productivity and catastrophe scenarios. Table 12.1 shows the scenarios used and the results from each simulation.

Levels of female productivity below 43% resulted in the majority of the populations reaching extinction before the end of the 100 year period. This occurred even in the absence of catastrophes. When probabilities of catastrophes greater than 10% were used, the results also showed the majority of the populations, even with levels of female productivity up to 70%, reached extinction or gave a mean final population size well below carrying capacity.

These simulations show that in all instances the populations fail to reach or maintain themselves at the carrying capacity of 250 animals. Figure 12.6

illustrates the effect of each of the 13 simulations on population size over the 100 year period. In 9 of the 13 simulations, the probability of populations surviving through to the end of the model was 90% or less. Stable populations (those reaching and maintaining carrying capacity) were not obtained until female productivity rates were approximately 60% or greater and when the models included only mild effects of a catastrophe.

Disease epidemics proved to have a significant effect on the modelled populations even at low levels of probability and severity. Even with a female productivity of 60%, a 4% probability of disease with a 50% effect on reproduction and 35% or 50% effect on survival, resulted in final populations not reaching carrying capacity.

The results obtained using VORTEX suggest that the macaque population in Gibraltar is at risk from both demographic and disease catastrophes. The population shows particular vulnerability to fluctuations in female productivity and to disease epidemics. The potential for unacceptable levels of variation in productivity is real and can be appreciated from the historic census data. Only Middle Hill, of the two sub-populations, has had levels of female productivity above 60% for the majority of the census period. At Queen's Gate levels have been equal to, or less than, 60%.

Scenarios incorporating even mild effects of a disease epidemic (5%) have a significant negative effect on the population. Although historically there is no evidence to suggest a higher probability of outbreak of disease, a rise in population together with increased contact with humans may lead to just such a probability.

Conclusions

The Barbary macaque has been present in Gibraltar for at least 250 years albeit with introductions from the source country. The British Army had full responsibility for the monkeys from 1915 but this tradition has now been broken and the macaques have passed into the care of the Gibraltar Government.

The monkey population is not physically threatened by poaching, trade or destruction of habitat. However, disease has contributed heavily to mortality and inbreeding may subsequently impose further demographic constraints. The tight concentration of the troops into small home ranges, as a result of provisioning and constant tourist pressure, is likely to be responsible for breeding dysfunction, infertility and the rapid spread of disease.

After more than 50 years of provisioning, the monkeys now depend almost exclusively on provisioned foods. Natural diet accounts for a very small proportion of food weight ingested even in the non-touristic troop at

Middle Hill. The quality and quantity of the food offered are, therefore, of paramount importance. Whereas provisioning was originally intended to supplement the animals' natural diet, it now meets and even exceeds all their calorific needs. The situation of the two sub-populations on the Rock is not parallel, since Queen's Gate is subjected to incessant visitor pressure; any management of the monkeys would clearly have to take into account these differences. Yet, given the similar demographic and epidemiological history of the troops (they have suffered from peaks of relatively high deaths through disease epidemics and their fertility has fluctuated according to changes in disturbance levels from tourists) it is useful to consider them as a single population for some management purposes.

The need for scientific management

Hitherto, all decisions relating to the monkeys have been based, for the most part, on tradition and on inaccurate opinions regarding the behaviour and biology of the species. Hence, the situation where, despite the growing body of literature on the species and, particularly, on the multimale nature of Barbary macaque troops, it continued to be management policy to restrict the number of adult males in the troops. This kind of misguided intervention has had drastic consequences on the genetics of the species in Gibraltar.

Maintaining population numbers and genetic variation must be central to the long-term management of the Gibraltar macaques within the 'megazoo' concept (Foose, 1991), with much the same kind of intensive genetic and demographic control as in captivity. The Gibraltar macaques would then form part of a larger metapopulation, where migration is managed between captive and wild sub-populations. Strict record-keeping, like that followed by zoological collections, should be implemented. A studbook should be kept and published regularly, and the data made available for analysis of the population from a genetic and demographic point of view. A studbook would provide the data: (1) to determine the genetic diversity in the populations and (2) to compute the inbreeding coefficients of extant animals and of the progeny of all potential matings. This information would then be incorporated into the database at ISIS (International Species Information System at Minnesota, USA).

Before this can be done it is imperative that a method is put in place for the definite identification of individuals, their parents and their progeny. While mothers can easily be determined from pure observation, it is much more difficult without the use of genetic markers to verify paternity in multiple male groups. An annual round-up of animals, as routinely carried out in the deTurckheim parks, would allow the drawing of blood for this purpose. Although the cost of this exercise must be taken into account, there will no doubt be great interest from well-funded primatological research institutions to perform this on a regular basis.

The role of population genetics in the management and conservation of threatened species has been highlighted by various authors (Soulé & Wilcox, 1980; Frankel & Soulé, 1981; Schonewald-Cox *et al.*, 1983). The negative effects of inbreeding can be minimised by following two general policies: (1) increasing the effective population size by raising N_m, the number of adult males and (2) preventing consanguineous matings by removing one member of a close consanguineous pair from the troops. The artificial introduction of adult males from other populations (e.g. captive collections) must be looked into while natural immigration of males between troops should be allowed to continue.

Disease prevention must be given top priority in the management of the population. Veterinary surgeons who tended the Gibraltar macaques usually had little experience of primate husbandry [a problem that was further compounded by the fact that there was a high turnover of visiting surgeons], so that animal examinations have been cursory. Emphasis has been placed on emergency, clinical procedures rather than on preventive action to reduce disease outbreaks. Attention must thus be paid to gathering baseline information on the current condition of the animals with respect to obesity, parasitic infestation and pathogens. Veterinary care should consist of a regular (annual), in-depth scrutiny of the animals' epidemiological and genetic state. Macaques should be vaccinated against rabies (as performed in the deTurckheim parks) and appropriate facilities for on-site emergency or routine examination of animals must be found. At present none exists.

Causes of death have been inconsistently recorded because post-mortems were only sometimes carried out by the Army. The number of causes of death noted has risen somewhat since 1936 as a result of the greater supervision given to the animals (Fa, 1984*a*). However, the proportion of diagnosed deaths is still low; the management authorities in Gibraltar attribute this to the difficulty of finding bodies in the rugged terrain. Post-mortem examination, where possible, should be performed when a dead animal is discovered.

Provisioning and public feeding control

Provisioning has been constant, although clearly dependent on the individual ape-keeper's appraisal of what constituted a proper diet for the macaques. There has been little, if any, dietary planning in recent years.

Food is left lying on the ground and allowed to decompose on site, thereby attracting vermin (black rats and herring gulls) and increasing the threat of disease.

A major unresolved problem is that of tourist feeding. This can not be allowed to continue unabated if efforts to improve the monkey's health and promote more naturalistic behaviours (O'Leary & Fa, 1993) are to succeed. It is essential to employ personnel to supervise the animals, ensure adequate contact levels between the monkeys and visitors, and prevent negative interactions. All this would lower considerably any risk of cross-infection of diseases between humans and the monkeys.

Personnel

The role of the ape-keeper and his assistant has always been vague. The military authorities have always insisted that the ape-keeper and his assistant were only soldiers with no training in animal husbandry, and have not acknowledged that their impact on the animals has been one of management intervention. This rather *ad hoc* situation may have resulted in faults in recognition of monkeys and census-taking, as corroborated by independent workers who have spent long periods studying the monkeys (see Hornshaw, 1984). This is being perpetuated under the present system as the animal staff are untrained. Therefore, every effort should be made to provide the ape-keeper with training in animal husbandry, record-keeping, etc.

Research potential

The value of monkey groups, such as those found in Gibraltar, for scientific study is exemplified by the work emerging from the deTurckheim parks (see Paul & Kuester, Chapter 14, this volume). Scientific cooperation that would benefit the management of the animals and at the same time promote research can be achieved through carefully designed agreements involving interested parties. Management can be coordinated in such a way as to reduce conflicting goals and create a situation where data, resources and expertise are shared. The Gibraltar authorities should actively encourage universities and volunteer research groups to carry out ongoing non-invasive research on the macaques.

Tourism and environmental education

A plan to balance the touristic potential of the macaques on Gibraltar with their conservation needs was first proposed by Fa (1987). Visited by thousands of people each year, the Gibraltar macaques could become an important flagship species for environmental education. But this can become a thorny issue if monkeys are seen merely as an attraction, with little more value than performing animals. Any suggestion of controlling feeding has encountered opposition in the past from taxi drivers and tour bus operators. Public service vehicle operators must be made fully aware of how grossly detrimental such overfeeding and excessive contact with humans are to the monkeys. All interested parties should be informed of how best to work together to ensure the continuation of the monkey population in Gibraltar.

Implementation

An active conservation programme for the Gibraltar macaques must be started immediately. There must be general agreement among the interested parties that the macaques in Gibraltar need closer attention and that *ad hoc* management principles can no longer continue. If the species' long-term survival on Gibraltar is to be assured there are four basic goals that must be attained:

1. At a tourist level, visitors must be encouraged to enjoy observing the animals without touching, feeding, or teasing them. Cooperation with public service vehicle operators, rather than confrontation, is necessary and they could be encouraged to act as conservation watchdogs.
2. At a local community level, where people have grown up alongside the monkeys and seen that they have always been teased and fed, and survived, an awareness must be created of the ill-effects of poor nutrition, etc. At the level of schools, this campaign can be incorporated into more general environmental education programmes dealing with global issues as part of the Gibraltar Government's recent initiative of environmental awareness (see Tilbury, 1993).
3. Training of personnel.
4. Creation of an infrastructure for veterinary support and food storage.

Translating these goals into working policies requires the goodwill of the Gibraltar Government, and other local organisations interested in nature conservation (e.g. the Gibraltar Ornithological and Natural History Society). Political expediency has prevailed over good sense for far too long. Everyone must be made aware that a little investment now will pay off in the long term.

Acknowledgements

This chapter is dedicated to the late Sgt Alfred Holmes, in memory of his more than 38 years' tireless caring for the Gibraltar macaques. Without his attention to individual monkeys between 1954 and 1992 our record of a 250 year tradition would have been much poorer.

We are grateful to the following for advice in the development of ideas contained in this paper: Dr W. Angst, Dr F. D. Burton, Dr David Chivers and Professor Robin Dunbar. In particular, many thanks go to Monique Williamson for editorial help and encouragement of morale.

References

Ballou, J. (1992). *CAPACITY, Version 3*. Washington DC: National Zoological Park.

Baulu, J. & Redmond, D. E. (1980). Some sampling considerations in the quantitation of monkey behavior under field and captive conditions. *Primates*, **19**, 391–9.

Boyd, C. E. & Goodyear, C. P. (1971). Nutritive quality of food and ecological systems. *Archives of Hydrobiology*, **69**, 256–70.

Burton, F. D. (1972). The integration of biology and behavior in the socialization of *Macaca sylvana* of Gibraltar. In *Primate socialization*, ed. F. E. Poirier, pp. 26–62. New York: Random House.

Burton, F. D. & Sawchuk, L. A. (1974). Demography of *Macaca sylvanus* of Gibraltar. *Primates*, **23**, 271–8.

Burton, F. D. & Sawchuk, L. A. (1982). Birth intervals in *M. sylvanus* of Gibraltar. *Primates*, **23**, 140–4.

Burton, F. D. & Sawchuk, L. A. (1984). The genetic implications of effective population size for the Barbary macaque in Gibraltar. In *The Barbary macaque – a case study in conservation*, ed. J. E. Fa, pp. 307–18. New York & London: Plenum Press.

Burton, F. D. & Underwood, C. (1987). Intestinal helminths in the Barbary macaques of Gibraltar. *Canadian Journal of Zoology*, **56**, 1406–7.

Campbell, A. C. (1989). Primate bites in Gibraltar – minor casualty quirk? *Scottish Medical Journal*, **34**, 519–20.

Chivers, D. J. (1984). Foreword. In *The Barbary macaque – a case study in conservation*, ed. J. E. Fa, pp. vii–viii. New York & London: Plenum Press.

Clifford, Q. J., Callanan, C. & Smith, S. E. G. (1972). The Barbary apes of Gibraltar. *Blue Book of the Veterinary Profession*, **22**, 167–9.

Deag, J. M. (1974). A study of the social behaviour and ecology of the wild Barbary macaque (*Macaca sylvanus L.*). PhD thesis, University of Bristol.

de Turckheim, G. & Merz, E. (1984). Breeding Barbary macaques in outdoor open enclosures. In *The Barbary macaque – a case study in conservation*, ed. J. E. Fa, pp. 241–62. New York & London: Plenum Press.

Drucker, G. R. (1984). The feeding ecology of the Barbary macaque and cedar forest conservation in the Moroccan Moyen Atlas. In *The Barbary macaque – a case study in conservation*, ed. J. E. Fa, pp. 135–64. New York & London: Plenum Press.

Fa, J. E. (1981). The apes on the Rock. *Oryx*, **16**, 73–6.

Fa, J. E. (1984*a*). Structure and dynamics of the Barbary macaque population in Gibraltar. In *The Barbary macaque – a case study in conservation*, ed. J. E. Fa, pp. 263–306. New York & London: Plenum Press.

Fa, J. E. (1984*b*). Conclusions and recommendations. In *The Barbary macaque – a case study in conservation*, ed. J. E. Fa, pp. 319–34. New York & London: Plenum Press.

Fa, J. E. (1986*a*). Balancing the wild/captive equation: the case of the Barbary macaque (*Macaca sylvanus*). In *Primates – the road to self sustaining populations*, ed. K. Bernisckhe, pp. 197–211. New York: Springer Verlag.

Fa, J. E. (1986*b*). *Use of time and resources in provisioned troops of monkeys : Social behaviour, time and energy in the Barbary macaque* (Macaca sylvanus L.) *at Gibraltar*. Contributions to Primatology vol. 23. Basel: S. Karger.

Fa, J. E. (1987). A park for the Barbary macaques in Gibraltar. *Oryx*, **21**, 242–5.

Fa, J. E. (1988). Supplemental food as an extranormal stimulus in Barbary macaques (*Macaca sylvanus*) at Gibraltar – its impact on activity budgets. In *Ecology and behavior of food-enhanced primate groups*, ed. J. E. Fa & C. H. Southwick, pp. 53–78. New York: Alan R. Liss.

Fa, J. E. (1989). Influence of people on the behavior of display primates, In *Housing, care and psychological wellbeing of captive and laboratory primates*, ed. E. F. Segal, pp. 52–64. New Jersey: Noyes Publications.

Fa, J. E. (1991). Provisioning of Barbary macaques in Gibraltar. In *Primate responses to environmental change*, ed. H. Box, pp. 137–54. London: Chapman & Hall.

Fa, J. E. (1992). Visitor-directed aggression in the Gibraltar Barbary macaques. *Zoo Biology*, **11**, 43–52.

Fa, J. E. & Southwick, C. H. (1989). *Ecology and behavior of food-enhanced primate groups*. New York: Alan R. Liss.

Foose, T. J. (1989). Erstwild and megazoo. *Orion Nature Quarterly*, Spring 1989, 60–3.

Foose, T. J. (1991). Viable population strategies for reintroduction programmes. *Symposia of the Zoological Society, London*, **62**, 165–72.

Forthman Quick, D. L. & Demment, M. W. (1988). Dynamics of exploitation: differential energetic adaptations of two troops of baboons to recent human contact. In *Ecology and behavior of food-enhanced primate groups*, ed. J. E. Fa & C. H. Southwick, pp. 25–52. New York: Alan R. Liss.

Frankel, O. H. & Soulé, M. E. (1981). *Conservation and evolution*. Cambridge: Cambridge University Press.

Frisch, R. E. & MacArthur, J. W. (1974). Menstrual cycles: fatness as a determinant of minimum weight for height necessary for their maintenance or onset. *Science*, **185**, 949–51.

Gilpin, M. E. (1987). Spatial structure and population vulnerability. In *Viable populations for conservation*, ed. M. E. Soulé, pp. 125–40. Cambridge: Cambridge University Press.

Goodman, D. (1987). The demography of chance extinction. In *Viable populations for conservation*, ed. M. E. Soulé, pp. 11–34. Cambridge: Cambridge University Press.

Hornshaw, S. G. (1984). A comparison of proximity behaviour in two groups of Barbary macaques – implications for the management of the species in captivity. In *The Barbary macaque – a case study in conservation*, ed. J. E. Fa, pp. 221–40. New York & London: Plenum Press.

Isola & Isola, (1993). Barbary macaques in Gibraltar. *Oryx*, **27**, 189.

Kenyon, E. R. (1938). *Gibraltar under Moor, Spaniard and Briton*. London: Methuen.

Lacy, R. C. (1993). VORTEX: a computer simulation model for population viability analysis. *Wildlife Research*, **20**, 45–65.

Lyles, A. M. & Dobson, A. P. (1988). Dynamics of provisioned and unprovisioned primate populations. In *Ecology and behavior of food-enhanced primate groups*, ed. J. E. Fa & C. H. Southwick, pp. 167–98. New York: Alan R. Liss.

MacRoberts, M. H. (1970). The social organization of Barbary apes (*Macaca sylvana*) on Gibraltar. *American Journal of Physical Anthropology*, **33**, 83–100.

MacRoberts, M. H. & MacRoberts, B. R. (1966). The annual reproductive cycle of the Barbary ape (*Macaca sylvana*) in Gibraltar. *American Journal of Physical Anthropology*, **25**, 299–304.

Maguire, L. A., Seal, U. S. & Brussard, P. F. (1987). Managing critically endangered species: the Sumatran rhino as a case study. In *Viable populations for conservation*, ed. M. E. Soulé, pp. 141–58. Cambridge: Cambridge University Press.

Malik, I. & Southwick, C. H. (1988). Feeding behavior and activity patterns of rhesus monkeys at Tughlaqabad, India. In *Ecology and behavior of food-enhanced primate groups*, ed. J. E. Fa & C. H. Southwick, pp. 95–112. New York: Alan R. Liss.

Marriott, B. M. (1988). Time budgets of rhesus monkeys (*Macaca mulatta*) in a forest habitat in Nepal and on Cayo Santiago. In *Ecology and behavior of food-enhanced primate groups*, ed. J. E. Fa & C. H. Southwick, pp. 125–52. New York: Alan R. Liss.

O'Leary, H. (1993a). Monkey business in Gibraltar. *Oryx*, **27**, 55–7.

O'Leary, H. (1993b). Helen O'Leary replies. *Oryx*, **27**, 189–90.

O'Leary, H. & Fa, J. E. (1993). Effects of tourists on Barbary macaques at Gibraltar. *Folia Primatologica*, **61**, 77–91.

Paul, A. & Kuester, J. (1989). Life history patterns of Barbary macaques (*Macaca sylvanus*) at Affenberg Salem. In *Ecology and behavior of food-enhanced primate groups*, ed. J. E. Fa & C. H. Southwick, pp. 199–230. New York: Alan R. Liss.

Paul, A., Kuester, J. & Podzuweit, D. (1993). Reproductive senescence and terminal investment in female Barbary macaques (*Macaca sylvanus*) at Salem. *International Journal of Primatology*, **14**, 105–24.

Post, W. & Baulu, J. (1978). Time budgets of *Macaca mulatta*. *Primates*, **19**, 125–39.

Schonewald-Cox, C. M., Chambers, S. M., MacBryde, B. & Thomas, L. (1983). *Genetics and conservation*. Menlo Park, CA: Benjamin Cummings.

Sclater, P. L. (1900). Mr Sclater on *Macaca inuus*. *Transactions of the Zoological Society, London*, 20 November, 773–4.

Seth, P. K. & Seth, S. (1986). Ecology and behaviour of rhesus monkeys in India. In *Primate ecology and conservation*, ed. J. G. Else & P. C. Lee, pp. 89–104. Cambridge: Cambridge University Press.

Soulé, M. E. (1987). Where do we go from here?. In *Viable populations for conservation*, ed. M. E. Soulé, pp. 175–84. Cambridge: Cambridge University Press.

Soulé, M. E. & Wilcox, B. A. (1980). *Conservation biology: an evolutionary – ecological perspective*. Sunderland, MA: Sinauer Press.

Tilbury, D. (1993). *Working together towards an improved environment*. Gibraltar: Ministry of the Environment.

Wolfe, L. D. (1979). Sexual maturation among members of a transplanted troop of Japanese macaques (*Macaca fuscata fuscata*). *Primates*, **20**, 411–8.

Wrangham, R. (1974). Artificial feeding of chimpanzees and baboons in their natural habitat. *Animal Behaviour*, **22**, 83–93.

Zeller, A. (1980). Primate facial gestures. A study of communication. *International Journal of Human Communication*, **13**, 565–606.

Zeller, A. (1986). Comparison of component patterns in threatening and friendly gestures in *Macaca sylvanus* of Gibraltar. In *Current perspectives in primate social dynamics*, ed. D. M. Taub & F. A. King, pp. 487–504. New York: Van Nostrand Reinhold.

Zeuner, F. E. (1952). Monkeys in Europe past and present. *Oryx*, I, 265–73.

13

Etho-ecology of Tibetan macaques at Mount Emei, China

ZHAO QI-KUN

Introduction

Until 1985 most of what was known about the Tibetan macaque (*Macaca thibetana*) was based on short field survey reports or on local anecdotes (Fooden *et al.*, 1985). This situation is typical of most of the primates unique to China. Myths on the Tibetan macaques abound: these monkeys are said to sleep in caves, forage for long distances (up to 100 km), live in 'a sub-clan led by the oldest', and even bury their dead; 70–80 monkeys join to form a 'waterwheel' in order to descend from a tree branch to drink from the deep valleys, and may form a 'parabola' to cross a river (Li, 1960).

A long-term scientific project on the species was initiated in 1984 to study the behaviour and ecology of a seasonally provisioned population at Mt Emei, a tourist and Buddhist centre in China. This chapter summarises some of the main results of this research. In particular, it concentrates data gathered on:

1. Physical characteristics and growth.
2. Reproduction (including birth seasonality, male–infant caretaking, socio-sexual behaviour, male mating competition and inter-group transfer).
3. Habitat and feeding ecology.
4. Ranging behaviour and responses to the changing environment.

Study area

The study area (29° 33′ N and 103° 21′ E) is located on the north-east slope of Mt Emei at the south-west edge of the Sichuan Basin or Chengdu plain. This slope ascends gently from 500 m to 800 m above sea level but then rises sharply to 3099 m (Fig. 13.1). The terrain is highly variable, and includes rocky precipices and towering peaks with winding trails on them. The

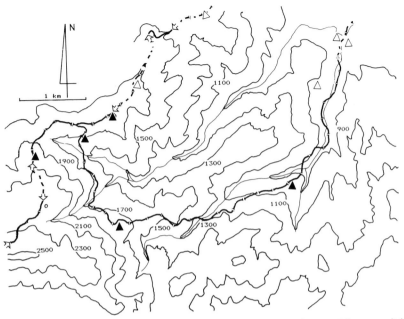

Fig. 13.1. Topography, streams, trails and temples in the study area. The summit is to the south-west; the Chengdu plain is to the north-east. The monkey-ranging sections of the trail are between any two arrows. Temples visited by monkeys are shown as black triangles.

macaque population survives here not far away from the human population, mainly on the plain, who have suffered in the past from periods of famine, and are a potential threat to the animals.

Because monkeys are linked to the Buddhist religion (Zhao & Deng, 1992), the macaques at Mt Emei have been respected by some of the people for centuries. The vegetation and landscape of the area have thus been well protected by the Buddhist temples in the past and now by the local government. Weather data (1976–85) from the Sichuan Province Meteorological Bureau show the macaque population range between 750 m and 2400 m:

1. Year-round rain but with five or more consecutive months of less than 100 mm precipitation. Precipitation, which peaks at elevations of about 2100 m (2060 mm per year), is often in the form of fog and rain, or winter snow along the upper reaches; there are on average 323 foggy days per year. There is no retarded plant growth period (Eisenberg *et al.*, 1981) in this habitat.
2. Seasonality is significant but annual variations are slight.

3. Highest temperatures typically occur in July and August. In the upper third of the macaque population range, mean monthly temperatures are $\leq 10\,^{\circ}$C for six months of the year (snow blankets for four months), and $< 17\,^{\circ}$C during the warmest months.

The vegetation changes with elevation from sub-tropical evergreen broadleaf forest below 1900 m, deciduous broadleaf and mixed forest in the 1600–2300 m belt (Fig. 13.2), to sub-alpine evergreen coniferous forest patches in the higher (2100–2400 m) altitudes. In the study area, Tibetan macaques are found in all of the three zones, not only in sub-tropical evergreen forest, like macaque groups found in the eastern-most range of the species at Mt Huang (30° 10′ N and 118° 11′ E). There, six (Xiong & Wang, 1988) and one (Wang & Xiong, 1989) wild groups were located in evergreen broadleaf forest for about 50% of the observation days. Therefore, to describe the habitat of Tibetan macaques as 'subtropical evergreen broadleaf' type, as suggested by Fooden (1982), is questionable.

Home ranges of six monkey groups in 1986 fell mostly between 1500 and 2100 m, where the main plant communities were dominated by *Chimonobambusa szechuanensis* in the shrub layer (78% cover). At 750–1500 m and > 2300 m, the shrub cover decreases to around 36% mainly due to the abundance of bamboo species in the layer being considerably reduced in the upper area and absent in the lower. Tree canopy density (60–70%) and herbs and grass cover (10%) were similar within all plant communities (Li, 1984; Zhao, Xu & Deng, 1989). Two streams and their four main tributaries flow into the deep valleys within the range (Fig. 13.1). Away from the streams and the scattered ponds found along the trail there are few other water sources in the area.

Macaque groups use trails on the mountain as well as five Buddhist temples along the trails. Paths from two bus stations, 1.5 km apart, meet at 1800 m, after which there is only one route to the summit; visitors can travel by bus to 2430 m. In recent years, the number of visitors walking to the summit, as opposed to taking buses and cable cars, has fallen, but they tend to walk down from the summit and carry food for themselves and for the monkeys. Some tourists (also food carriers) only visit the lower part of the habitat to see the monkeys and a famous temple at 1020 m on the south-eastern trail. Here, human food is more abundant at the upper (> 1700 m) and lower (≤ 1020 m) extremes of the trail than at the mid-section. However, fewer people (calculated at about one tenth of the main flow) take the north-eastern trail.

Fig. 13.2. Deciduous broadleaf and evergreen vegetation within the main range of the Tibetan macaques at Mt Emei.

Diet and feeding behaviour

From admission tickets sold in 1991, most of the one million visitors may walk through the area in a year. The active and passive feeding of the wild macaque groups by visitors has led to the monkeys developing food beg-robbing behaviours, which has resulted in serious changes in the ecology and behaviour of the macaques, and threats to the visitors themselves. Tibetan macaques at Mt Emei thus rely partly on people's food handouts (Zhao & Deng, 1992; Zhao, 1994a). However, even during the peak provisioning period, animals feed on a variety of natural foods in considerable quantity (Zhao, Deng & Xu, 1991). Samples of the natural foods eaten by the macaques were gathered in two 100 day observation periods (March–June 1986; September–December 1987). Plant food items were classified into (1) structural parts of trees, bamboos and vines; (2) reproductive parts of trees, bamboos and vines, and (3) ground items such as grasses and herbs. Plants were also grouped into those items found on the ground (low) and those that the macaques had to climb for (high).

A total of 196 plant food species of 135 genera and 72 families were identified. Plant parts eaten were significantly different between the two study periods (Kolmogorov–Simimov two group test: $*D_{max} = 0.49$; $x^2 = 34.00$, $p < 0.001$). During the first period, corresponding to seasons other than autumn the macaques fed upon 186 plant parts from 177 species, with flowers representing only 10% of the total parts. During the second period (autumn), only 44 parts belonging to 42 species were used, with reproductive parts accounting for 59% of the total. In fact, the macaques consume mainly bamboo shoots and fruits for about two months in autumn, whereas they rely on visitors' food handouts, a variety of structural parts, and some invertebrates during late spring and summer. During the rest of the year, monkeys will eat mainly mature leaves and bark. At Mt Huang, macaques are reported to eat only leaves, buds and stems in spring and summer, while fruits and seeds represent 27% of total parts eaten in autumn and winter (Xiong & Wang, 1988).

About two thirds of the vines consumed are found on trees, whereas the rest can be reached from the ground. Thus, the proportions of low and high species parts eaten were approximately equal ($z = -0.72$, $p = 0.24$). This forest-dwelling macaque species and *M. sylvanus* (Ménard & Vallet, 1988) feed on the ground and from trees, displaying a foraging strategy different from that of other macaques who use the ground only to move to tree food sources (Caldecott, 1986a). The two species also eat more leaves and fewer seeds and fruits (see Ménard & Vallet, 1988; Zhao, Deng & Xu, 1991), offering an important supplement to the diet range of macaques which had been considered as frugivores (Wheatley, 1980).

Tibetan macaques are highly selective in their choice of food, consistently going for specific parts of particular species. For example, consumption of ripe fruits can take different forms:

Spitting out the seed (e.g. *Decaisnea fragesii* and *Lindera megaphylla*) or stones *Davidia involucrata*).

Spitting out peel (e.g. *Smilax ferox*).

Spitting out both peel and seeds (e.g. *Kadsura longipedunculata*).

Noisily masticating hard seeds while eating the entire fruit (e.g. *Swida controversa*).

Eating the fruit without chewing the seeds, which are later eliminated in the faeces (e.g. *Dendrobenthamia multinervosa*).

The monkeys ate bunches of mature leaves of *C. szechuanensis* and *Nothopanax davidii*, but took a leaf-at-a-time of the mature leaves of *Lithocarpus cleistocarpus* and *Aucuba chinensis omeiensis*. They ate the bark of stems of *Sloanea omeiensis*, but consumed only the wood substance and pith of young twigs of *Rhododendron calophytum*. The bark and terminal buds of *Euaraliopsis ciliata* were favoured foods, but the fruit of the same species was never eaten.

Ranging behaviour and responses to changing environment

At Mt Emei, macaques have been fed in Buddhist temples, especially in winter, for probably hundreds of years. Buddhism, which originated in India, was introduced into China more than 1000 years ago (Chen, 1983*b*). Because Buddha is thought to have once been a monkey king (Chen, 1983*a*), monkeys have long been respected and tolerated by people influenced by the religion (Teas *et al.*, 1980; Mittermeier & Cheney, 1987).

China's economic reforms of the last decade have led to a growth in tourism. As a direct result of this, food handouts to the macaques have increased dramatically. Most visitors feed monkeys for pleasure, or as an offering to Buddha; increasingly, however, food is being used to 'pacify' the aggressive monkeys. This provisioning has made the animals dependent on food from visitors. In response to this new human–monkey interaction, the macaques have optimised/altered their foraging strategy by beg-robbing the feeder, submissive persons or those who carry bags (Fig. 13.3) (Zhao, 1994*a*). Monkeys approach the visitors as they arrive (Zhao & Deng, 1992).

Fig. 13.3. Beg-robbing a submissive bag-carrier by adult male Tibetan macaque.

However, monkeys avoid local labourers who believe 'the best defence is offence' and often hurt the animals.

Another consequence of people feeding has been the increase of trail-foraging days. The proportion of trail-foraging days in total observation days in spring and summer increased from 43% (between 15% for group F and 64% for group A) for groups A–F in 1986 (Zhao & Deng, 1988a), to 63% for groups A, B, D1 and F (61% for group F) in 1989, and then 89% for groups B, D and D1, which were radio-tracked in 1991 (for group descriptions see Table 13.1). The increments between 1986 and 1991 for the three groups were 48%, 35% and 30%, respectively. Observations in 1989 showed that as a result of intra-group competition for people's food, the foraging cycles of five high-ranging groups at above 1300 m consisted of moving up about 400 m along the trail on average per day and then returning to the sleeping site through the forest. In the same way, two or three groups from the lowest parts of the range regularly began to move down to meet visitors walking up rather than using the available transport, even though the number of people walking down is higher (Zhao, 1994a).

Inter-group competition for food resources, or for certain favourite trail sections, has led to different inter-group interactions in foraging developing at the upper range (Zhao, 1994a):

Table 13.1. *Range elevations, demographic data, and cumulative birth rates (CBR) of 11 groups of Tibetan macaques in the middle birth season from 18–23 March 1992*[a]

Elevation (m)	2400–1500				1500–1020			1020–750		
Group	A	B	D + D2	D1	C	F1	H	E	F	G
Size	19	30	21	67	13	15	9	36	30	19
Adult female	6	10	8	26	3	4	5	9	11	7
Yearling	3	3	3	9	1	1	0	5	3	1
Newborn	2	5	3	6	1	1	3	0	1	1
CBR (%)	67	71	60	35	50	33	60	0	13	17

[a]$CBR = i/(F - y)$, where i indicates infant number, F, female number, and y, yearling number.

1. Moving up then down to avoid the dominant group at a higher location.
2. Arriving after the dominant had left.
3. Stopping closely below the dominant.
4. Aggressively displacing the low-ranking groups either up or down.
5. Making a detour to avoid the dominant group (through forest) to open a new, higher section discontinuously.

Moving in the opposite direction, the pattern for the lower groups was the same. Thus, the population range extended discontinuously from 1260–2150 m (in 1986) to 750–2400 m (after 1989) along the trail. In the north-west branch, a short section at about 1500 m was used exclusively by group D in 1986, given up in 1989, and then used again by group D2 that had fissioned from D since autumn 1991.

However, among the high-groups, the most dominant group (group D) had a longer trail-foraging movement than the other three groups (Student's *t*-test: $t = 2.07$, d.f. $= 44$, $p < 0.05$), and the coefficient of variation (CV) in trail-foraging lengths increased with the decline of inter-group ranks (D (58% CV) > A (116% CV) > B (139% CV) > D1 (194% CV)). Both in relation to visitors or to food resources, the dominant group had more freedom, whereas subordinates could only react to circumstances within the range (Zhao, 1994a). In fact, the higher-ranking groups occupied the two extremes, and the small low-ranking ones remained mid-way with fewer handouts (Table 13.1).

Although the groups are becoming increasingly dependent on people's food handouts, their ranging pattern is well-regulated by foraging. When visitor numbers per day fell from thousands to hundreds to tens as the season changed from spring through to winter, three radio-tracked groups (groups B, D and D1) significantly increased their percentage of days foraging away from the trail (11%, 15% and 72%), and the proportion of the range used exclusively (19%, 31% and 54%) (Zhao, 1992).

All groups changed their trail range after 1 to 12 days. Ranging movements consisted of wandering around a densely used area and were interspersed with excursions between the areas. Sleeping sites were relatively fixed: after feeding in a certain area, a group returned to a particular sleeping site nearby, often on successive nights (Zhao & Deng, 1988a). The ranging pattern remains similar to that reported for wild *Macaca fuscata* (Wada & Tokida, 1981), *M. fascicularis* (Fittinghoff & Lindburg, 1980) and baboons (DeVore & Hall, 1965; Kummer, 1968; Altmann & Altmann, 1970). Obviously, the pattern ensures the maintenance of food resources so that no source is completely depleted under natural conditions (Clutton-Brock, 1975; Homewood, 1978).

Of the 27 sleeping sites observed between 1984 and 1986, 41% were almost entirely safe from predators, which might have been a problem in the past, 48% were relatively safe, and only 11% were vulnerable. This reflects the much reduced pressure of predation (Zhao & Deng, 1988a). The sites may be divided into five types:

1. A cluster of trees in the middle of a large cliff.
2. A group of trees on a cliff-top.
3. A ledge in the middle of a cliff, often with a protruding upper edge providing shelter from rain.
4. A group of tall trees in the forest.
5. Virtually anywhere, even as far as several meters away from the trail.

Types 1 and 3 were entirely safe; 2 and 4 were relatively safe.

Growth and body size

Dorsal fur colour is blackish with a thin, shiny, silver-sheen surface in newborns; it then becomes light yellow about 10 days after birth, and by 3.5–4.5 months it is the brown or blackish colour of the adult. Forehead hair growth differs from that of other macaques: dark hair appears on the broad, bare white forehead of infants at the end of the third month; a triangular patch forms about 30 days later with full cover developing by about 5 months (Zhao & Deng, 1988b). The female's age at first birth is about 5.5 years, and the age of onset of female adulthood is about 5 years.

These observations were made to build a base line for estimating the groups' age–sex composition in this species. This information supplements and improves that described by Fooden *et al.* (1985).

In late spring 1986, body weights were measured by using peanuts to attract the animal to stand or sit on a body scale (Fig. 13.4), thus making use of the monkeys' food beg-robbing behaviour. Sitting height was measured using 120 photographs that included a reference length. Adult body weight was 18.3 ± 2.4 kg ($n = 12$, range 14.0–21.5 kg) for males and 12.8 ± 1.8 kg ($n = 17$, range 9.7–15.0 kg) for females. Sitting height was 55.0 ± 4.8 cm ($n = 17$, range 47.1–63.2 cm) for males and 46.9 ± 5.3 cm ($n = 14$, range 37.5–52.4 cm) for females. The ponderal index, mean weight (kg) $\times$ 100/mean sitting height (cm), was 33.3 for adult males and 27.3 for adult females (Zhao & Deng, 1988*b*).

According to the literature on body weight in primates (Harvey, Martin & Clutton-Brock, 1987), Tibetan macaques and *M. sylvanus* are the largest and second largest of the macaques, respectively. When the spacing and structure of plant foods, described above, are taken into account, the large body weights are consistent with current theory. That is, within phylogenetic groups larger species tend to feed more on foliage or structural parts (Clutton-Brock & Harvey, 1977; Sailer, Gaulin & Kurland, 1985) and also terrestrial species tend to have a large body size (Clutton-Brock & Harvey, 1977); data on body weights confirmed the observations on feeding ecology of the two species (Zhao, Deng & Xu, 1991; Ménard & Vallet, 1988).

Adult males (M) and females (F) in groups above 1500 m were weighed again in late autumn (La) between 5 November and 7 December 1991, and in late winter (Lw) between 25 and 28 March 1992 (Zhao 1994*b*). Changes in mean body weight (M,F) were documented as:

Males
   (La) 19.5 ± 2.2 kg ($n = 14$, range 16.0–22.5)
   (Lw) 17.0 ± 2.7 kg ($n = 14$, range 11.5–20.0)
Females
   (La) 16.8 ± 2.1 kg ($n = 13$, range 14.0–21.5)
   (Lw) 11.4 ± 1.0 kg ($n = 15$, range 9.0–13.0)

Reported mean body weights ($M = 18.3$ kg, $F = 12.8$ kg) taken in late spring fell mid-way between the two extremes, as expected, and could be the best representative body weight for this species; as suggested by Wada & Tokida (1981) the temperate forest macaque does accumulate energy as fat in autumn and consume it in winter. With the drop in body weight in the harsh winter, weight dimorphism (M/F) increased from 1.16 to 1.49. This

Fig. 13.4. Weighing a juvenile on a body scale, using peanuts to attract the animal.

can be seen to be a result of differential parental investment by the two sexes (Trivers, 1972) if, as Pickford (1986) claims, the females have to channel more energy into the growth of their offspring, during gestation and infancy, and so lose more fat themselves under the poor foraging conditions.

Based on features of morphology and behaviour, three age-classes of sexually active males were sub-divided to clarify their age-dependent performances in sexual activity and male dispersal (Zhao, 1993; 1994c). Though essential to macaque etho-ecology, age-dependent behaviour has hitherto been all but ignored in field studies. The male age-classes are as follows:

Sub-adult males (SA): aged 5–7 years, weighed 15.4 ± 1.2 (13.0–16.5) kg ($n = 7$), fully erupted canines; able to ejaculate, tended to be peripheral, and mating usually out of sight of other males.

Young adult males (YA): over 7 years, weighed 18.2 ± 1.4 (16.5–20.0) kg ($n = 7$), very sharp canines; most active sexually, and in migration and mating competition.

Middle–old males (MO): older than the YA, weighed 19.8 ± 0.8 (18.0–21.5) kg ($n = 9$), worn canines; usually of a lower rank, much less active sexually, and lower rate of inter-group shift than the YA.

Reproduction

Birth seasonality

Surveys on birth timing were conducted for each group every 2 or 3 days in the 1986 birth season (Zhao & Deng, 1988c). Because of the uncertainty of the groups' trail foraging time, birth dates for newborns were not collected regularly. Indicators of age, such as the umbilical cord, colour of dorsal fur, stage of forehead hair growth (Zhao & Deng, 1988b) and behaviour, were used to estimate the birth date of infants born in a missing period. The dates of 32 infants were blocked into 16 periods of 14 days and processed by methods outlined by Caughley (1977) and Eisenberg *et al.* (1981).

The mean birth date for the six groups in 1986 (A to F) was 27 March (SD = 39 days), and the median birth date was 14 March. Distribution was skewed in the positive direction but still normal (Fig. 13.5); no newborn was seen in autumn. Birth timing approaches the pulse mode (SD $\leq$ 30 days) suggested by Caughley (1977). Sex ratio of newborns (F:M = 11:21) did not deviate significantly from 1:1 ($z = 1.59, p > 0.05$). The ratio of newborns *and* yearlings to adult females (65:66) was very close to 1:1. Therefore, generally speaking, each adult female gave birth every two years. These results are in keeping with findings for the majority of primate species (Altmann, Altmann & Hausfater, 1978; Eisenberg *et al.*, 1981).

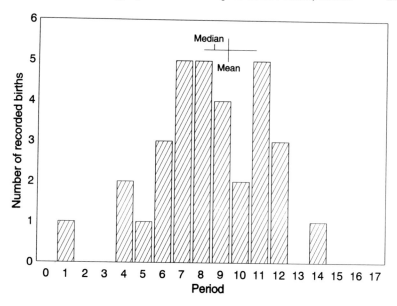

Fig. 13.5. Distribution of births in six groups of Tibetan macaques at elevations 1260–2100 m at Mt Emei in 1986. The mean birth date, the median birth date and the standard deviation are shown on the bar above the histogram. Each period code corresponds to 14 days after 10 November 1985.

Interestingly, median birth date was significantly negatively correlated with the range altitude ($r = -0.84$, $p < 0.05$). Group A's (1900 m) median birth date was three months earlier than that of group C (1300 m). This shift in births was later confirmed by a short survey on 11 groups ranging between 750 m and 2400 m from 18 to 23 March 1992. In the later study (Zhao, 1994*d*), the birth timing was estimated by the ratio of newborns observed at the survey time to those expected to be born in the whole season, or the cumulative birth rate (CBR) at the survey time, calculated using $i/(F - y)$, where i is the infant number, F is the female number and y is the yearling number. The value of $(F - y)$ yields the potential pool of females that could conceive that year (Zhao & Deng, 1988*c*). When the groups were classified according to their range elevations in the slope habitat, the CBR decreased significantly from 58% to 48% to 10% with the decline of the groups' trail-range elevations from the high, to the middle, to the low (Spearman rank-order correlation: $r = 0.83$, $p < 0.05$; critical value (two-tail, 0.05) = 0.63). In fact, 71% of infants in group B were born in winter – before 27 March in the 1992 birth season (Table 13.1).

Current hypotheses regarding reproductive regulation in primates cannot explain this phenomenon. Because of male inter-group transfer (see below), genetic distance among the groups should be small (Nozawa *et al.*, 1975); social stimuli (Gordon & Bernstein, 1973; Vandenbergh & Drickamer, 1974) appeared comparable among groups. The home ranges were, in fact, at the same latitude, so that changes in the light:dark ratio (van Horn, 1975, 1980) should be synchronised across groups. Furthermore, sunrise and sunset occurred at almost the same time, as the locations face a large plain to the east, and differences in daylight were reduced by the diffusion of light in the rainy and foggy habitat. The hypothesis that infants are produced when foraging is optimal (Small & Smith, 1986) was not borne out at this site, as newborns have more time in the snow at higher altitudes.

Only one environmental factor, altitude, can be related to reproductive regulation in the macaque population, but altitude is a compound variable, including at least rainfall, temperature and related phenological phenomena. Spring rainfall has been reported to trigger mating activity for tropical rhesus (Rawlins & Kessler, 1985), and temperature for temperate Japanese macaques (Cozzolino *et al.*, 1992). Considering that food availability depends on seasonal rainfall in the tropics, and on temperature in the temperate zone (Ricklefs, 1973), the birth timing of Tibetan macaques is very likely optimised to ensure the nutrition needed by mating parents in autumn, when their body weight considerably increased, and to offer the possibility for mothers to channel the stored nutrients to fetuses and newborns in winter and early spring. In addition, a 'bottle-neck' for infant survival, weaning, would occur in the next autumn (Zhao, 1994d).

Male–infant caretaking

In addition to the energy budget specified for reproductive activities, it is of great interest to note that Tibetan infants aged within 23 weeks were cared for 10% of their time mainly by males (Deng & Zhao, 1995). Similarly Barbary infants spent 12% of their time in contact with individuals other than their mothers in the month after the third week of age (Wilson, 1980, cited in Fa, 1986), and about 12% of observation time was spent in infant care by males (Fa, 1986). This is rarely seen in primates, and has been specified as 'intensive caretaking' by Whitten (1987).

The intensive male–infant caretaking may be related to certain ecological parameters shared by the two species in cold areas. With respect to this we know that: (1) the habitats produce few fruits and seeds, and a considerable number of food items are found on the ground, and (2) a sharper decline of female body weight relative to males appears in Tibetan macaques over

Table 13.2. *Composition of Tibetan macaque groups in the middle of the 1987 mating season[a]*

	Group					
Age–sex class	A	B	C	D	E	F
Adult female (F)	7	8	6	18	8	13
Young adult male (YA)	1	2	1	3	1	1
Middle-old male (MO)	4	4	2	6	3	3
Sub-adult male (SA)	1	1	1	4	2	2
Immature	16	21	13	43	18	22
Sex ratio$_1$	1.4	1.3	2.0	2.0	2.0	3.3
Sex ratio$_2$	1.2	1.1	1.5	1.4	1.3	2.2

[a]Sex ratio$_1$ = F/(YA + MO); sex ratio$_2$ = F/(SA + YA + MO).

winter (-32% vs -13% seen in males). Accordingly, the male caretaking may be selected to compensate for the mothers' high investment in their offspring – giving the exhausted mothers more mobility, albeit temporary, to reach food, and reducing their calorie demands for caring for the infant in the birth season when energy demands for mothers are already high.

Furthermore, the male of these two species often carried and held an infant during an interaction with another male during the birth season, forming a special, ritualised male–infant–male triadic interaction pattern. The triads in Tibetan macaques were specific to the birth season, and recognised as three types on a continuum of function changing from passive 'agonistic buffering' (4.8%) to active spatial cohesion, which resulted in a significant decline of inter-male distances ($t = 9.2$, d.f. $= 584$, $p < 0.001$). In addition positive correlations were documented between the triad initiation rate and the number of females in consort with the males in the following mating season (Spearman rank-order correlation: $r = 0.96$, critical value (one-tail, 0.05) $= 0.82$), and between the triad reception rate and the number of infants in proximity to the males in the mating season ($r = 0.90$), while the close relationship of females and infants in the birth season ($r = 0.84$) disappeared in the mating season ($r = 0.32$). Interestingly, both the initiator and recipient of triads were the YA males, the most sexually active age-class. Thus, the triad initiation appeared as a tactic of mating, and male's mating effffort and kin/sexual selection may be deeply involved in the triad of this species. These observations, however, were not in keeping with those reported for a colony of *M. sylvanus*, the other triad-species of macaques (Paul *et al.*, 1992). Possibly, the stable and rich food supply has changed the implications of triads in the colony.

Considering the paternity level was much lower in *M. sylvanus* (Taub, 1980*b*; Paul *et al.*, 1992) than that in *M. tibetana* (Zhao, 1993), but the two triad species shared similar foraging conditions, and showed comparable intensities of male–infant caretaking, the triad was very likely a by-product of male–infant caretaking. Thus, so far, the long-term arguments about the triad in *M. sylvanus* can be united to form a model of the way in which the 'male–infant caretaking' hypothesis (Taub, 1980*a*) works ultimately, and the 'regulating social relations' hypothesis (Deag, 1980) does proximately (Zhao, 1996). Therefore, the significance of ecological parameters in the expression of social behaviour, which has been focused on by few studies in primates (Fa, 1991), is addressed by these observations.

Socio-sexual behaviour

The mating season, like the birth season, was also well marked. Between August and November, typically there was a great diversity of grouping patterns with mating activities, especially in groups with a low sex ratio (F/(MO + YA + SA) ≤ 1.5; Table 13.2). Data on socio-sexual behaviour were collected in two typical sub-grouping patterns: central sub-groups (CS) in which the alpha male was together with most group members, and far-peripheral adult sub-groups (FAS) composed of only a few adult male(s) and female(s) ranging out of view of the CS for one to eight days.

Focusing on the dominant male, or the most sexually active male and his partners, 14 all-occurrence sampling (cf. Lehner, 1979) sessions were made in the CS of groups B, E, C and F (mean duration $\pm$ SD = 2.7 $\pm$ 1.5 h) and six sessions in FAS of B, E and F (2.4 $\pm$ 1.1 h). The contexts and context-dependent sexual performances of mating were (Zhao, 1993):

1. The dominant male in CS had significantly more accompanying males and consorting females than the one in FAS (2.3 vs 0.3, $p < 0.001$; and 2.8 vs 1.3, $p < 0.01$ per session, respectively).
2. Significantly more females were mated in CS (2.2 vs 1.3 in FAS, $p < 0.05$), and the dominant CS male probably mounted more often (2.0/h vs 1.2/h in FAS, $p = 0.06$) but ejaculated significantly less often (0.2/h vs 0.7/h in FAS, $p < 0.001$).
3. The copulating pair received significantly more harassment from females in CS (0.7/h, 33% of mounts) than in FAS (0/h; $p < 0.01$).
4. The mount to ejaculation ratio was 3.5 times higher in CS than in FAS (6.3 : 1.8). This can reasonably be considered to be the composite result of behavioural responses 2 and 3 of males and females to social contexts (result 1).

By comparing data within CS and FAS, results 2 and 3 can be further highlighted:

5. Females in CS showed a stronger tendency than males to engage in 'sexual harassment and replacement' (the higher ranking animal has sex by replacing the low-ranking one of the same sex during, or just before, copulation) (0.7/h vs 0/h, $p < 0.001$).
6. As a multiple mount-to-ejaculation species, mount rate is significantly higher than ejaculation rate in both CS (2.0/h : 0.2/h, $p < 0.001$) and FAS (1.2/h : 0.7/h, $p < 0.05$).

Copulatory patterns have been considered to vary systematically with genital anatomy or phylogenetic relatedness (Fooden, 1980), or with qualities of habitat (Caldecott, 1986b), and have been dichotomously classified as the single mount-to-ejaculation (SME) or the multiple mount-to-ejaculation (MME). From comparative studies on *M. mulatta*, *M. fascicularis* and *M. radiata*, and literature reviewing the genus *Macaca*, Shively *et al.* (1982) proposed that ejaculatory patterns in macaques might be more accurately thought of as being on a continuum rather than being a dichotomous system. Shively *et al.* (1982) and Caldecott (1986b) have also noted the relation between inter-male relationships ('relaxed' and 'antagonistic') and copulation patterns (SME and MME, respectively). This study offers a model for the better understanding of the connection between mount : ejaculation ratio and the socio-context in which matings occur. If the socio-context, either as a species-specific feature or an immediate situation, is important for sexual performance, Shively *et al.*'s (1982) hypothesis may be further improved: macaque copulatory patterns are on a continuum in which the location of a species/group may be changed according to certain immediate socio-context parameters.

A low-ranking male has never been seen to mate within sight of the dominant male in groups with sex ratio$_2$ ≤ 1.5, but for CS of group F (sex ratio$_2$ = 2.2, see Table 13.2), the situation was very different: a low-ranking YA made 4 mounts and 1 ejaculation with 2 females in full view of dominant males during a 2.5 h all-occurrence sampling session; one of the YA's consorting females was overtaken once by the third-ranking male. This observation implies that, to some extent, the inter-male tolerance observed in *M. radiata* as a SME species (Shively *et al.*, 1982) can also appear in a MME species, *M. thibetana*, when the immediate social context (females in 'rich supply') allows.

Although four YA males comprised fewer than 25% of the sexually active males in 4 CS, they made 84% of copulations there, and 2 YA males in different FAS, including at least 1 middle–old male, made 92% of copulations. Obviously, in Tibetan macaques, females mating with the YA are mating with the sexually most active males. Such matings also result in inbreeding avoidance (all YAs are new immigrants) and ensure a better social environment for the next offspring because the YA, even if only in FAS as the temporary or solitary male, is still the strongest competitor for the alpha position. In fact, 4 out of 6 of the alpha males and most of the dominant males in FAS in 1987, and 5 out of 6 of alpha males recognised in 1989 were YA. These observations are well in keeping with the theory of female choice in sexual activity (Trivers, 1972).

In the mating season, FAS displayed a 'space-segregation' tactic for both sexes to avoid dominants. This may be seen as the development of an 'hidden mating' tactic (Zhao, 1993, 1994c), because in the main study group 5 out of 8 females and most low-ranking males, or the losers in competition, experienced living in FAS but the middle-aged alpha male (Mq) and the most dominant female (Fh) did not.

Due to the 'space-segregation' and 'triad-initiation' tactics of mating, the relation between male rank and mating performance is likely to be bimodal in the whole Tibetan macaque group with normal sex ratio, mainly because the sexually active YA males in FAS should be low-ranked. This does not necessarily mean that male rank is a negligible factor influencing mating activity because, within each CS or FAS as a sub-unit of reproduction, the dominant male normally exclusively mated with females in sight (Fig. 13.6). This corresponds to Altmann's (1962) observation on the relation between rank and sexual activity in *M. mulatta* males in a group.

Surprisingly, in a 124 min observation on a consorting triad of the alpha male (Mj) and two females (Fo > Fn in rank) in group E, Mj mated Fn for 7 mounts with 1 ejaculation (only 1 in view of, and harassed by, Fo), and he attacked/threatened Fn on 5 occasions (2 to assist Fo; 2 as responses to Fo's returning back after a short leave). In contrast, Fo received only affiliative expression (assisting her attacking Fn, and a bout of grooming), but only 1 mount from Mj. Such segregation of males' sexual and social preferences towards females is psychologically far beyond our imaginings about non-human primate societies (Zhao, 1993).

Male mating competition and male inter-group transfer

An analysis of the socio-sexual behaviour of Tibetan macaques should show acute mating competition among males. Indeed, males had large numbers of wounds in the mating seasons of the study years (1987, 1989 and interval was 8.3 ± 6.8 days). Wound variables included number, severity

Fig. 13.6. Only the dominant male (in this case the young adult alpha male from group B) would copulate in full view of the other males in the group.

Fig. 13.7. A young immigrant (Ne) with wounds scored as I, 1[2] (lip); I, 1[4] (cheek); III, 1[3] (shoulder) using the code x, $n[s]$, where x = location (I–IV), n = number and s = severity (1–4) of wounds.

and bodily distribution (Fig. 13.7). Wounding in this species has features as follows (Zhao, 1994c):

1991). As an indicator of male mating competition, wounds were recorded in groups B, C, D, E and F at an average interval of 5 days (the longest

1. The YA males suffered significantly more wounds than did either MO or SA ($p < 0.05$), but there was no difference between MO and SA.

2. The distribution of wounds was different for MO and young males (YA and SA): they tended to have more wounds in the front than MO ($\chi^2 = 19.7$, $p < 0.001$ for wound number, $\chi^2 = 51.3$, $p < 0.0001$ for wound severity). The distribution of fresh wounds in males would indicate that the most severe fighting took place between the original alpha male and the young adult immigrant. It also points to different postures by the fighters: the newcomer seems to be more passive, tending to be bitten in the more vulnerable parts of the body. Whether the young immigrant made 'begging for mercy' gestures (Lorenz, 1966), as seen on two occasions in a newly wounded SA to the alpha male in group B (Zhao, 1994c), or whether some other behaviour underlies the interaction is not clear. If observed female choice in mating activity is taken into account, wounding in the young males could be the result of the resident alpha male's redirected aggression.

3. Log plots of wound variables (total number and severity, average number and severity for each male, and average number and severity for each wounded male) were significantly negatively correlated with the adult sex ratio, and more strongly correlated with that including SA ($r = -0.84$ to -0.97 for sex ratio$_1$; $r = -0.91$ to -0.99 for sex ratio$_2$).

4. Across the three male age-classes and body parts, the wound number correlated closely with the estimated severity score for 16 wounded males ($r = 0.997$, $p < 0.001$). Hence, number of wounds would be sufficient to assess the intensity of aggression in field studies, making it unnecessary to estimate wound severity.

As in other macaque species, the male is the dispersing sex in Tibetan macaques, but, in contrast with most other observations in macaques (Caine, 1987), the majority of male Tibetan macaques left their natal group as young adults. Based on individual recognition of males in groups (with the exception of group D, which had too many males to be recognised), 5 out of 6 YA and 5 out of 17 MO changed groups in the 1.5 years between the end of the 1986 birth season and the end of the 1987 mating season. Three middle-aged males migrated soon after they lost their alpha positions: two transferred from group A to group B (Ha in 1986 birth season, Ey at the end of 1987 mating season); and one (D1a) from group D1 to group B in 1991 mating season.

The rate of MO male inter-group transfer strongly supports the statement that a male is probably rarely present when his oldest daughter becomes reproductively mature (*M. mulatta*: Melnick *et al.*, 1984; *M. sinica*: Dittus, 1977; *Presbytis* spp.: Rudran, 1973; Hrdy, 1974; *Cercopithecus aethiops*: Henzi & Lucas, 1980). Secondary transfer (Pusey & Packer, 1987), or any further movement after natal transfer, occurs 5.1 years later (1.5 years × 17/5), which is close to the age of onset of female adulthood (5 years) in this species.

No SA transfer was observed in the 1.5 year period, but three SA immigrants observed in 1989 and 1991 were badly treated in new groups: one soon disappeared in group B, and two died of severe wounds in group F, which had had the most relaxed inter-male relations in 1986. This may explain the delay in timing of natal dispersal and may also be related to the effects of provisioning, as in some populations of Japanese monkeys (Sugiyama & Ohsawa, 1982).

In addition to the observations on wounding and mating opportunity in the three age-classes of males, it is interesting to note that after migrating from group A to an adjacent group (B), two YA immigrants were ranked an average of 2.5 positions higher, while three MO immigrants were ranked an average of 2 positions lower. On the basis of differences in benefit and cost of migration for adult males, proximate causes of inter-group transfer are likely to differ. It seems that when the males grew sexually attractive enough to unfamiliar females in other groups (Enomoto, 1974; Sugiyama, 1976), and physically strong enough to challenge the alpha position for mating, they started to transfer. However, the secondary transfer made by MO males may be motivated by the need to avoid inter-male competition, usually after they had surrendered their rank to other males or following the establishment of a new dominant male, as described by Pusey & Packer (1987).

Tibetan macaque males did not immigrate into new groups at random (Zhao, 1994c). The proportion of newly sexually matured SA and immigrants to the overall sexually active males was significantly negatively correlated with the change in sex ratio$_2$ ($r = -0.97$, $p < 0.01$), but not significantly with the change in sex ratio$_1$ ($r = -0.84$, $p > 0.05$). There was a clear tendency to transfer into adjacent groups, as observed in Japanese macaques (Sugiyama, 1976) and rhesus (Melnick, Pearl & Richard, 1984), and into groups that had previously received males from their former group as in *M. mulatta* (Drickamer & Vessey, 1973; Meikle & Vessey, 1981; Colvin, 1983) and *M. fascicularis* (van Noordwijk & van Schaik, 1985). For example, all five adult males in group B at the end of 1989 had emigrated from group A, an adjacent group; another immigrant (the new alpha male) in 1990 came from group D, another adjacent group.

These results on the rate and direction of inter-group transfers and the wounding rate as indicators of the degree of male mating competition show that the SA, as an age-class, is an active element in the mating system, as reported for *M. mulatta* (McMillan, 1982), *M. fuscata* (Takahata, 1982) and *M. sylvanus* (Paul, 1989). Therefore, all males that are equal to, or larger than, adult females in body size can be treated as 'adult males'; this would make it simpler to determine adult sex ratios in field studies. During the 1.5 year study period, both adult sex ratios and their variation were obviously reduced ($n = 6$ groups) mainly by an increase in male numbers in groups with higher sex ratios.

	June 1986	December 1987
Sex ratio$_1$	3.1 ± 1.9	2.1 ± 0.8
Sex ratio$_2$	2.8 ± 2.1	1.5 ± 0.4

This was a result of male migration and also of improved conservation in recent years (Zhao & Deng, 1988*d*; Zhao, 1994*c*).

Non-random male shifts (Drickamer & Vessey, 1973) and random shifts (Koford, 1966; Melnick *et al.*, 1984), in terms of the relation between adult sex ratio and shift direction, have been documented in rhesus monkeys. According to the findings of the present study, the differences in the reports may be traced to the population status, i.e. whether the population is stable or disturbed but recovering. In the latter case, as reported here, previous poaching for meat and bones targeted mainly large males and led to greater diversity of adult sex ratios in groups according to their proximity to human habitation. Hence, changes in sex ratios can be used to estimate the hunting pressure on the populations of large primates.

Concluding remarks

Taking advantage of a habitat with a great difference in elevation and a slight annual variation in climate, and of macaques living in a rich social context and in semi-commensalism with people, this long-term study has been able to provide conclusive scientific information on the behaviour and ecology, especially the significance of ecological parameters on the expression of social behaviours, and the age- and socio-context dependence of sexual performances, of a hitherto little-known macaque. Because of the effect of visitors feeding these macaques, the upper and lower limits of the population range has been extended considerably in recent years. Despite the fact that the macaques have persisted on Mt Emei for centuries because of both the variable terrain and the protection by the Buddhist culture, we now have to face the emerging problem in relations between the wild macaques and humans under improved economic conditions. As part of a famous mountain with great aesthetic, cultural, and economical value, and also as an important resource for the study of the species, the macaques should be conserved by mitigating the impact of visitors.

Acknowledgements

The studies were made possible mainly by grants from the Winner–Gren Foundation for Anthropological Research (no. 4739), the National Geographic Society (no. 4022-89) and the National Natural Science Foundation of China (no. 389701147). I thank Drs J. Erwin, G. A. Doyle, R. D. Martin, T. Hasegawa, and D. G. Lindburg for their valuable comments and/or careful editorial work on the original manuscripts. Special thanks are given to Dr J. E. Fa for his great contribution to the expression of this review, to Dr H. J. Stewart for careful editorial work on the final manuscript, to Professors J.-H. Pan, L. T. Nash and J. Fooden for their help and scientific stimulation to initiate the study, and to Deng, Z.-Y. for her cooperation. I am indebted to the Administrative Commission of Mt Emei, and the Xianfeng Temple for their understanding and warm support.

References

Altmann, J., Altmann, S. A. & Hausfater, G. (1978). Primate infant's effect on mother's future reproduction. *Science*, **201**, 1028–30.

Altmann, S. A. (1962). A field study of the sociobiology of the rhesus monkey, *Macaca mulatta*. *Annals of the New York Academy of Sciences*, **102**, 338–435.

Altmann, S. A. & Altmann, J. (1970). *Baboon ecology: African research*. Chicago: University of Chicago Press.

Caine, N. G. (1987). Behavior during puberty and adolescence. In *Comparative primate biology*, vol. 2A: *Behavior, conservation and ecology*, ed. G. Mitchell & J. Erwin, pp. 327–62. New York: Alan R. Liss.

Caldecott, J. O. (1986a). An ecological and behavioural study of the pig-tailed macaque. *Contributions to Primatology*, vol. 21. Basel: S. Karger.

Caldecott, J. O. (1986b). Mating patterns, societies and ecogeography of macaques. *Animal Behaviour*, **34**, 208–20.

Caughley, G. (1977). *Analysis of vertebrate population*. New York: John Wiley & Sons.

Chen, Q.-X. (translator) (1983a). [Origin and development of Buddhistic art.] (In Chinese.) *Wonders of the world's museums*, vol. 3, ed. K. Muchida, pp. 37–56. Taibei: Publisher.

Chen, Q.-X. (1983*b*). [Survey on eight ruins of Buddha.] (In Chinese.) *Wonders of the world's museums*, vol. 3, ed. K. Muchida, pp. 177–9. Taibei: Publisher.

Clutton-Brock, T. H. (1975). Ranging behaviour of red colobus (*Colobus badius tophrosceles*) in the Gombe National Park. *Animal Behaviour*, **23**, 706–22.

Clutton-Brock, T. H. & Harvey, P. H. (1977). Species differences in feeding and ranging behaviour in primates. In *Primate ecology*, ed. T. H. Clutton-Brock, pp. 557–84. London: Academic Press.

Colvin, J. (1983). Influences of the social situation on male emigration. In *Primate social relationships: an integrated approach*, ed. R. A. Hinde, pp. 160–71. Oxford: Blackwell Scientific Publications.

Cozzolino, R., Cordischi, C., Aureli, F. & Scucchi, S. (1992). Environmental temperature and reproductive seasonality in Japanese macaques (*Macaca fuscata*). *Primates*, **33**, 329–36.

Deag, J. M. (1980). Interactions between males and unweaned Barbary macaques: testing the agonistic buffering hypothesis. *Behaviour*, **75**, 54–81.

Deng, Z.-Y. & Zhao, Q.-K. (1995). Alloparenting for newborns of Tibetan macaques. (In Chinese, English abstract.) *ACTA Anthropologica Sinica*, in press.

DeVore, I. & Hall, K. R. L. (1965). Baboon ecology. In *Primate behavior: field studies of monkeys and apes*, ed. I. DeVore, pp. 20–52. New York: Holt, Rinehart & Winston.

Dittus, W. (1977). The social regulation of population density and age-sex distribution in the toque monkey. *Behaviour*, **63**, 281–321.

Drickamer, L. C. & Vessey, S. H. (1973). Group changing in free-ranging male rhesus monkeys. *Primates*, **14**, 359–68.

Eisenberg, J. F., Dittus, W. P. J., Fleming, T. H., Green, K., Struhsaker, T., Thorington, R. W., Jr, Altman, N. H., Brackett, B. G., Goy, R. W., Marriott, B. M., New, A. E. & Senner, J. W. (1981). *Techniques for the study of primate population ecology*. Washington, DC: National Academy of Sciences.

Enomoto, T. (1974). The sexual behavior of Japanese monkeys. *Journal of Human Evolution*, **3**, 351–72.

Fa, J. E. (1986). *Use of time and resources in provisioned troops of monkeys: social behaviour, time and energy in the Barbary macaque* (Macaca sylvanus L.) *at Gibraltar*. Contributions to Primatology, vol. 23. Basel: S. Karger.

Fa, J. E. (1991). Provisioning of Barbary macaques on the Rock of Gibraltar. In *Primate responses to environmental change*, ed. H. O. Box, pp. 137–54. London: Chapman & Hall.

Fittinghoff, N. A., Jr & Lindburg, D. G. (1980). Riverine refuging in East Bornean *Macaca fascicularis*. In *The macaques: studies in ecology, behavior and evolution*, ed. D. G. Lindburg, pp. 182–214. New York: Van Nostrand Reinhold.

Fooden, J. (1980). Classification and distribution of living macaques. In *The macaques: studies in ecology, behavior and evolution*, ed. D. G. Lindburg, pp. 1–9. New York: Van Nostrand Reinhold.

Fooden, J. (1982). Ecogeographic segregation of macaque species. *Primates*, **28**, 574–9.

Fooden, J., Quan, G.-Q., Wang, Z.-R. & Wang, Y.-X. (1985). The stumptail macaques of China. *American Journal of Primatology*, **8**, 11–30.

Gordon, T. P. & Bernstein, I. S. (1973). Seasonal variation in sexual behavior in all-male rhesus troops. *American Journal of Physical Anthropology*, **38**, 221–7.

Harvey, P. H., Martin, R. D., & Clutton-Brock, T. H. (1987). Life history in comparative perspective. In *Primate societies*, ed. B. B. Smuts, D. L. Cheney,

R. M. Seyfarth, R. W. Wrangham & T. Struhsaker, pp. 181–96. Chicago: University of Chicago Press.

Henzi, S. P. & Lucas, J. W. (1980). Observations on inter-troop movement of adult vervet monkeys (*Cercopithecus aethiops*). *Folia Primatologica*, **33**, 220–35.

Homewood, K. M. (1978). Feeding strategy of the tana mangabey (*Cercocebus galeritus galeritus*; Mammalia: Primates). *Journal of Zoology*, **186**, 375–91.

Hrdy, S. B. (1974). Male–male competition and infanticide among the langurs (*Presbytis entellus*) of Abu, Rajasthan. *Folia Primatologica*, **22**, 19–58.

Koford, C. B. (1966). Population changes in rhesus monkeys: Cayo Santiago, 1960–64. *Tulane Studies in Zoology*, **13**, 1–7.

Kummer, H. (1968). *Social organization of Hamadryas baboons*. Chicago: University of Chicago Press.

Lehner, P. N. (1979). *Handbook of ethological methods*. New York: Garland STPM Press.

Li, X.-G. (1984). [The preliminary investigation on the vertical distribution of the forest vegetation on Emei mountain, Sichuan Province.] (In Chinese, English abstract.) *Acta Phytoecologica et Geobotanica Sinica*, **1**, 52–66.

Li, Y.-L. (1960). [The Tibet macaque (*M. thibetana*) of Emei Shan.] (In Chinese.) *Chinese Journal of Zoology*, **4**, 202–4.

Lorenz, K. (1966). *On aggression*. London: Methuen.

McMillan, C. A. (1982). Male age and mating success among rhesus macaques. *International Journal of Primatology*, **3**, 312.

Meikle, D. B. & Vessey, S. H. (1981). Nepotism among rhesus monkey brothers. *Nature*, **294**, 160–1.

Melnick, D. J., Pearl, M. C. & Richard, A. F. (1984). Male migration and inbreeding avoidance in wild rhesus monkeys. *American Journal of Primatology*, **7**, 229–43.

Ménard, N. & Vallet, D. (1988). Disponibilités et utilisation des ressources par le magot (*Macaca sylvanus*) dans différents milieux en Algérie. *Revue d'Ecologie (Terre Vie)*, **43**, 201–50.

Mittermeier, R. A. & Cheney, D. L. (1987). Conservation of primates and their habitats. In *Primate societies*, ed. B. B. Smuts, D. L. Cheney, R. M. Seyfarth, R. W. Wrangham & T. Struhsaker, pp. 477–90. Chicago: University of Chicago Press.

Nozawa, K., Shotake, T., Ohkura, Y., Kitajima, M. & Tanabe, Y. (1975). Genetic variations within and between troops of *Macaca fuscata fuscata*. In *Contemporary primatology*, ed. S. Kondo, A. Ehara & M. Kawai, pp. 75–89. Basel: S. Karger.

Paul, A. (1989). Determinants of male mating success in a large group of Barbary macaques (*Macaca sylvanus*) at Salem (FRG). *American Journal of Primatology*, **30**, 461–76.

Paul, A., Kuester, J. & Amemann, J. (1992). Male–infant–male interactions in Barbary macaques (*Macaca sylvanus*): testing functional hypotheses. In *Abstracts of the XIIth Congress of the International Primatological Society*, p. 221. Strasbourg: Charles River.

Pickford, M. (1986). Sex differences in higher primates: a summary statement. In *Sexual dimorphism in living and fossil primates*, ed. M. Pickford & B. Chiarelli, pp. 191–9. Giugno, Firenze: Il Sedicesimo.

Pusey, A. E. & Packer, C. (1987). Dispersal and philopatry. In *Primate societies*, ed. B. B. Smuts, D. L. Cheney, R. M. Seyfarth, R. W. Wrangham & T. Struhsaker, pp. 250–6. Chicago: University of Chicago Press.

Rawlins, R. G. & Kessler, M. J. (1985). Climate and seasonal reproduction in the Cayo Santiago macaques. *American Journal of Primatology*, **9**, 87–9.

Ricklefs, R. E. (1973). *Ecology*. New York: Chiron Press.

Rudran, R. (1973). Adult male replacement in one-male troops of purple-faced langurs (*Presbytis senex*) and its effect on the population structure. *Folia Primatologica*, **19**, 166–92.

Sailer, L. D., Gaulin, S. J. C. & Kurland, J. A. (1985). Measuring the relationship between dietary quality and body size in primates. *Primates*, **26**, 14–27.

Shively, C., Clarke, S., King, N., Schapiro, S. & Mitchell, G. (1982). Patterns of sexual behavior in male macaques. *American Journal of Primatology*, **2**, 373–84.

Small, M. F. & Smith, D. G. (1986). The influence of birth timing upon infant growth and survival in captive rhesus macaques (*Macaca mulatta*). *International Journal of Primatology*, **7**, 289–304.

Sugiyama, Y. (1976). Life history of male Japanese monkeys. In *Advances in the study of behavior*, vol. 7, ed. J. S. Rosenblatt, R. A. Hinde, E. Shaw & C. Beer, pp. 255–84. New York: Academic Press.

Sugiyama, Y. & Ohsawa, H. (1982). Population dynamics of Japanese monkeys with special reference to the artificial feeding. *Folia Primatologica*, **39**, 238–63.

Takahata, Y. (1982). The socio-sexual behaviour of Japanese monkeys. *Journal of Comparative Ethology*, **59**, 89–108.

Taub, D. M. (1980*a*). Testing the 'agonistic buffering' hypothesis. *Behavioral Ecology and Sociobiology*, **6**, 187–97.

Taub, D. M. (1980*b*). Female choice and mating strategies among wild Barbary macaques (*Macaca sylvanus* L.). In *The macaques: studies in ecology, behavior and evolution*, ed. D. G. Lindburg, pp. 287–344. New York: Van Nostrand Reinhold.

Teas, J., Richie, T., Taylor, H. & Southwick, C. (1980). Population patterns and behavioral ecology of rhesus monkeys (*Macaca mulatta*) in Nepal. In *The macaques: studies in ecology, behavior and evolution*, ed. D. G. Lindburg, pp. 247–62. New York: Van Nostrand Reinhold.

Trivers, R. L. (1972). Parental investment and sexual selection. In *Sexual selection and the descent of man, 1871–1971*, ed. B. Campbell, pp. 136–79. Chicago: Aldine.

Van Horn, R. N. (1975). Primate breeding season: photoperiod regulation in captive *Lemur catta*. *Folia Primatologica*, **24**, 203–20.

Van Horn, R. N. (1980). Seasonal reproductive patterns in primates. *Progress of Reproductive Biology*, **5**, 181–221.

Van Noordwijk, M. & Van Schaik, C. P. (1985). Male migration and rank acquisition in wild long-tailed macaques, *Macaca fascicularis*. *Animal Behaviour*, **33**, 849–61.

Vandenbergh, J. G. & Drickamer, L. C. (1974). Reproductive coordination among free-ranging rhesus monkeys. *Physiology and Behavior*, **13**, 373–6.

Wada, K. & Tokida, E. (1981). Habitat utilization by wintering Japanese monkeys (*Macaca fuscata fuscata*) in the Shiga Heights. *Primates*, **22**, 330–48.

Wang, Q.-S. and Xiong, C.-P. (1989). [The study of seasonal home range of Tibetan monkey's Yulinkeng troop in Huangshan mountain.] (In Chinese, English abstract.) *Acta Theriologica Sinica*, **9**, 239–46.

Wheatley, B. P. (1980). Feeding and ranging of East Bornean *Macaca fascicularis*. In *The macaques: studies in ecology, behavior and evolution*, ed. D. G. Lindburg, pp. 215–46. New York: Van Nostrand Reinhold.

Whitten, P. L. (1987). Infants and adult males. In *Primate societies*, ed. B. B. Smuts, D. L. Cheney, R. M. Seyfarth, R. W. Wrangham & T. Struhsaker, pp. 343–57. Chicago: University of Chicago Press.

Xiong, C.-P. & Wang, Q.-S. (1988). [Seasonal habitat used by Tibetan monkeys.] (In Chinese, English abstract.) *Acta Theriologica Sinica*, **8**, 176–83.

Zhao, Q.-K. (1992). Range use and its seasonal change in commensal Tibetan macaques at Mt Emei, China. I. *Abstracts of the XIVth Congress of the International Primatological Society*, Strasbourg, pp. 147–8.

Zhao, Q.-K. (1993). Sexual behavior of Tibetan macaques at Mt. Emei. *Primates*, **4**, 431–44.

Zhao, Q.-K. (1994a). A study on semi-commensalism of Tibetan macaques at Mt. Emei, China. *Revue d'Ecologie (Terre Vie)*, **49**, 259–71.

Zhao, Q.-K. (1994b). Seasonal changes in body weight of *Macaca thibetana* at Mt. Emei, China. *American Journal of Primatology*, **32**, 223–6.

Zhao, Q.-K. (1994c). Mating competition and intergroup transfer of males in Tibetan macaques (*Macaca thibetana*) at Mt. Emei, China. *Primates*, **35**, 57–68.

Zhao, Q.-K. (1994d). Birth timing shift with altitude and its ecological complication for *Macaca thibetana*. *Ecologia Montana*, **3**(1–2), 24–6.

Zhao, Q.-K. (1996). Male–infant–male interactions in Tibetan macaques. *Primates*, **37**(2), in press.

Zhao, Q.-K. & Deng, Z.-Y. (1988a). Ranging behavior of *Macaca thibetana* at Mt. Emei, China. *International Journal of Primatology*, **9**, 37–47.

Zhao, Q.-K. & Deng, Z.-Y. (1988b). *Macaca thibetana* at Mt. Emei, China. I. A cross-sectional study of growth and development. *American Journal of Primatology*, **16**, 251–60.

Zhao, Q.-K. & Deng, Z.-Y. (1988c). *Macaca thibetana*, at Mt. Emei, China. II. Birth seasonality. *American Journal of Primatology*, **16**, 261–8.

Zhao, Q.-K. & Deng, Z.-Y. (1988d). *Macaca thibetana* at Mt. Emei, China. III. Group composition. *American Journal of Primatology*, **16**, 269–73.

Zhao, Q.-K. & Deng, Z.-Y. (1992). Dramatic consequences of food handouts to *Macaca thibetana* at Mt. Emei, China. *Folia Primatologica*, **58**, 24–31.

Zhao, Q.-K., Deng, Z.-Y. & Xu, J.-M. (1991). Natural foods and their ecological implications for *Macaca thibetana* at Mt. Emei, China. *Folia Primatologica*, **57**, 1–15.

Zhao, Q.-K., Xu, J.-M. & Deng, Z.-Y. (1989). Climate, vegetation and topography of the slope habitat of *Macaca thibetana* at Mt. Emei, China. *Zoological Research*, **10** (Suppl.), 91–9.

III
Mating and social systems

14

Differential reproduction in male and female Barbary macaques

A. PAUL AND J. KUESTER

Introduction

From an evolutionary standpoint, differential reproduction is the most important aspect of life because it is the central focus of natural selection (Alcock, 1989). Not surprisingly then, systematic research on differential reproduction in non-human primates dates back to the early days of modern primatological field studies (Carpenter, 1942). Our understanding of factors determining individual reproductive success has since grown (Fedigan, 1983; Harcourt, 1987; Silk, 1987; Clutton-Brock, 1988; Abbott, 1991). Nevertheless, the relationship between key variables, such as dominance and reproductive success, is still a source of major controversy (Bernstein, 1976; Fedigan, 1983; Gray, 1985; Bercovitch, 1986; Dunbar, 1988; McMillan, 1989; Cowlishaw & Dunbar, 1991). There are several reasons for this. Firstly, primates are long-lived animals and few studies have been able to cover extended periods of their lives. Lifetime reproductive success is important to natural selection, and such data are only just accumulating (see Fedigan *et al.*, 1986, for a rare example). Secondly, male reproductive success is notoriously difficult to measure. Field research has relied heavily on the assumption that male mating success is a sufficiently reliable indicator of male reproductive success. This has been repeatedly questioned by studies that have incorporated genetic paternity data (Curie-Cohen *et al.*, 1983; Stern & Smith, 1984; Shively & Smith, 1985). Moreover, studies designed to test the association between rank and true reproductive success in male macaques have yielded inconsistent results (e.g. Smith, 1981; Stern & Smith, 1984). However, the study of male reproductive success has only recently 'come of age'. The discovery of hypervariable DNA sequences in the eukaryotic genome by Jeffreys and his colleagues (Jeffreys, Wilson & Thein, 1985*a,b*; Ali, Müller & Epplen, 1986), and the application of 'fingerprint' techniques to behavioural studies have stimulated a significant amount of new research (e.g. Martin, Dixson & Wickings, 1992). Undoubtedly, our understanding of factors affecting male reproductive success will improve in the very near future. Thirdly, primates may use alternative reproductive strategies with a more or less balanced reproductive output (Dunbar, 1983; see Noë & Sluijter, 1990, for an excellent example in male savanna baboons). Finally, ecological and demographic factors have important influences on the mode of competition and its reproductive consequences (for female savanna baboons, see Altmann, Hausfater & Altmann (1988) and Smuts & Nicholson (1989); for female vervet monkeys, see Whitten (1983), Fairbanks & McGuire (1984) and Cheney *et al.* (1988)). The data on reproductive success in male and female Barbary macaques that we present in this chapter are, therefore, necessarily only a small piece of a complex puzzle, but hopefully a helpful one.

Study population

Data on the reproduction of free-ranging Barbary macaques were collected during an 11 year study. The animals live in a 14.5 ha outdoor enclosure near Salem in south-west Germany. The park was founded in 1976 and stocked with 164 monkeys from two French enclosures (see de Turckheim & Merz, 1984, for details on history and management). The population consisted of several multimale, multifemale groups of various sizes (see Paul & Kuester, 1988, for group composition, population growth, etc.). The monkeys were and are provisioned daily, but food is widely distributed and cannot be effectively monopolised.

Complete demographic records of the population are available for the period from 1978 until the end of 1988. Additional short-term surveys were conducted thereafter. With few exceptions, newborn infants were detected during the morning after delivery. All infants that were found dead at this time were classified as 'stillbirths'. For management purposes, several groups were removed from the area during the period from 1984 to 1986, and during the last two years of our study (1987 and 1988) large numbers of females were implanted with contraceptives to reduce population growth.

The ages of all animals born in France and Germany were known, ages of wild-born monkeys were estimated (E. Merz, personal communication). Rank relations were based on agonistic displays and approach/retreat interactions during dyadic encounters. For analytical purposes the female hierarchy was divided into a high-ranking third, a mid-ranking third and a low-ranking third. The terms 'high-born', 'mid-born' and 'low-born' refer to the maternal rank of an individual at birth.

The complete data set on differential female reproduction consists of 961 reproductive years of 207 females. Components of females reproductive success (RS), i.e. longevity, fecundity (birth rate: number of infants born per female reproductive year) and offspring survival (proportion of infants surviving to one year of age), and their relative contribution to the variance of female RS were calculated according to the method of Brown (1988). Ideally, such analyses should consider only females whose entire breeding life is known (see Fedigan *et al.*, 1986). Unfortunately, our sample does not allow this, but the relative importance of components may become apparent if only parts of the reproductive life span are known (cf. Cheney *et al.*, 1988).

Oligonucleotide fingerprinting was used to determine paternity and male reproductive success (Ali *et al.*, 1986; Arnemann *et al.*, 1989). DNA was extracted from 1–4 ml of blood and digested with three different restriction enzymes (*Alu*I, *Hae*III, and *Hinf*I). Fragments were separated electrophoretically, hybridised with six synthetic radiolabelled oligonucleotide probes ($(GTG)_5$, $(GACA)_4$, $(GATA)_5$, $(GA)_8$, $(GT)_8$, $(GGAT)_4$) and the result was autoradiographed. Banding patterns of 15 enzyme/probe combinations per individual were analysed (for further details see Kuester, Paul & Arnemann, 1992). The present analysis of male reproductive success is restricted to data from one large group (C) during four successive years (breeding seasons 1984/85–1987/88). Sexually mature males, 16–33 in number, were present in the group during this period, and 74 infants were available for paternity analysis. Two fetuses and eight infants that did not survive their first year were unavailable for testing. Therefore, male reproductive success refers to offspring surviving to at least one year of age.

Factors affecting breeding success of females

Components of reproductive success

The vast majority, about 88%, of females born in this population reached sexual maturity (Paul & Kuester, 1988). Tables 14.1 and 14.2 summarise components of reproductive success (RS) of 182 breeding females. Most of the overall variance in female RS (61%) can be explained by differences in breeding life span, i.e. the number of reproductive years we were able to follow. Both differences in fecundity and offspring survival to one year of age were relatively unimportant. If only females that contributed at least five (potentially) reproductive years are considered ($n = 94$), breeding life span still accounted for 56% of the overall variance; however, fecundity

Table 14.1. *Mean and variance of components of reproductive success and their products in female Barbary macaques*

Component	Original		Standardised variance
	Mean	Variance	
L	5.06	7.315	0.286
F	0.76	0.040	0.069
S	0.87	0.063	0.083
LF	3.99	6.564	0.413
LS	4.55	7.314	0.353
FS	0.67	0.064	0.143
LFS	3.59	6.012	0.466

$n = 182$ breeding females.
L indicates breeding life span; F, fecundity; S, proportion of offspring surviving to their first year.

Table 14.2. *Percentage contribution of the components of reproductive success to overall variance in reproductive success among female Barbary macaques*

Component	L	F	S
L	61.4		
F	12.5	14.7	
S	−3.3	−1.8	17.7
LFS	−1.2		

Proportion of breeders, 89.5%
Overall variance (OV), 6.671%
OV due to non-breeders, 16.5%
OV due to breeders, 83.5%

For definition see Table 14.1.

(37%) was more important than offspring survival (14%). The covariance between fecundity and offspring survival was −6.1% in this sample.

Rank

The relationship between female dominance rank, fertility and offspring survival was examined in three social groups (Table 14.3). Although high-ranking females slightly out-reproduced low-ranking ones in all three groups, a consistent positive relationship between female dominance and

Table 14.3. *Female rank and reproductive success (RS, infants surviving to one year per reproductive year (RY)) in three social groups. RYs of females >25 years omitted*

Group	Rank	RY (n)	Births (n)	Birth rate	Infant survival rate	RS
A2	High	43	38	0.884	0.912	0.795
	Middle	38	29	0.763	0.920	0.676
	Low	35	30	0.857	0.931	0.794
B	High	132	98	0.742	0.892	0.654
	Middle	156	122	0.782	0.882	0.686
	Low	149	99	0.664	0.874	0.572
C	High	70	63	0.900	0.889	0.800
	Middle	62	53	0.855	0.906	0.774
	Low	73	51	0.699	0.920	0.644

Group A2: 1980–85, 27 females. Group B: 1978–88, 97 females. Group C: 1978–88, 48 females.

fertility was detected only in group C (χ^2 high vs low $= 8.96$, $P = 0.003$). The relationship was inconsistent and weak at best in the other groups. Infant survival to at least one year of age was not affected by maternal rank in any of the three groups.

If the data from all three groups were pooled, high-ranking females produced significantly more surviving infants per reproductive year (171/236) than low-ranking females (157/252; $\chi^2 = 5.70$, $P = 0.016$). This difference diminished when offspring survival to breeding age was taken into account. The analysis, based on age-specific mortality rates, revealed that more than 80% of all offspring reached the age of 4 years (84.2% high-borns, 85.5% middle-borns, and 86.8% low-borns). The product of fecundity and offspring survival to age 4 was 0.684 for high-ranking females, 0.681 for mid-ranking females, and 0.608 for low-ranking females. Translated into actual number of surviving offspring per reproductive year, this difference suggests only a slight reproductive advantage for high-ranking females over low-ranking ones ($\chi^2 = 2.99$, $P = 0.08$).

Age

In many animals (Clutton-Brock, 1988) female reproductive success increases early in life, and declines towards the end of the life span, the analysis resulting in an inverted U-shaped curve. This is exactly the pattern

A. Paul and J. Kuester

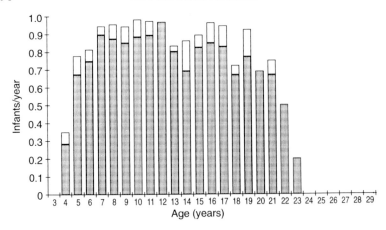

Fig. 14.1. Female reproductive success as a function of age. Open bars refer to the number of infants born per reproductive year, shaded sections refer to infants surviving to at least one year of age.

observed in the Salem Barbary macaque females (Fig. 14.1). Fecundity in this population was strongly age-dependent, with a significant increase in fertility rates during the first years of reproductive life, a significant decrease during the second half of the life span, and a post-reproductive life span of up to eight years (Paul, Knester & Podzuweit, 1993). No consistent relationship between maternal age and offspring survival was found, although first offsprings were at a significantly higher risk than those born later (see below).

The interplay of age, dominance and breeding success

Several studies suggest that the effect of rank on reproductive success is not independent of age. Especially, differences in the timing of sexual maturation, or first successful conception, appear to be common (Abbott, 1991). It seems, therefore, useful to have a closer look at the determinants and implications of first births and at the relationship between age, dominance and reproduction in general.

Timing of maturation: determinants

Barbary macaque females are capable of conceiving their first offspring at the age of approximately 3.5 years, but more frequently at the age of about 4.5 years (Burton & Sawchuk, 1974; Fa, 1984; Paul & Kuester, 1988). At Salem, 55 out of 157 females (35%) had their first parturition at the age of 4

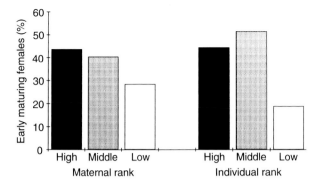

Fig. 14.2. Percentage of early maturing females in relation to maternal rank and individual rank at first parturition.

years. The majority of females (61%) delivered their first offspring at the age of 5 years, and only a small minority of about 4% delayed their first reproduction until they were 6 or even 7 years old.

In a previous analysis, Paul & Kuester (1988) found a significant relationship between maternal rank and age at first reproduction. This relationship still holds true if we divide the female hierarchy into only two rank classes: 25 out of 50 high-born females delivered their first offspring at the age of 4 years, whereas barely a quarter of all low-born females (12/49) were able to reproduce at this early age ($\chi^2 = 6.88$, $P = 0.009$). With three female rank classes, the significance of this difference disappears (χ^2 high vs low $= 1.48$, $P = 0.221$; Fig. 14.2). This is partly due to the fact that several of the highest-born females of one group suffered from severe inter-group competition (see below). A slightly different picture emerges if individual instead of maternal rank is taken into consideration (Fig. 14.2). In low-ranking females first parturition was significantly more often delayed than in high- and mid-ranking females (χ^2 high vs low $= 7.04$, $P = 0.008$). The difference between the samples was mainly due to the fact that, in large groups, many daughters of mid-ranking females did not reach their matriarchal rank at the age of 4 years (cf. Paul & Kuester, 1987a). Most 4-year-old mid-born non-mothers were actually still low-ranking, while those with babies often had reached their final rank. It seems plausible that especially for mid-born females an early attainment of rank and an early onset of reproduction are both controlled by a third factor, namely early attainment of a critical body weight. Body weight is certainly as critical for early onset of reproduction in high-born females, but rank attainment may be facilitated by the existence of powerful allies.

Table 14.4. *Birth rates of 4 and ≥ 5 year old females and their relation to social stress induced by inter-group competition (number of females per age class in brackets)*

Group	Female age (years)	Birth rates 1982–86	1987–88
Group B	4	0.50 (36)	0.00 (13)
Group C	4	0.53 (17)	0.38 (13)
Group B	≥ 5	0.87 (214)	1.00 (18)
Group C	≥ 5	0.91 (100)	0.96 (27)

Another factor that appeared to affect the timing of sexual maturation was social stress induced by severe inter-group competition. As has been previously observed in other groups (Paul & Kuester, 1988), one of our study groups (B) came under social pressure from spring 1986 onwards, when it was almost constantly harassed by other groups. Until this period, the proportion of B-group females delivering their first baby at the age of 4 years did not differ from females belonging to other large groups (Table 14.4). In 1987 and 1988, however, the birth rate of 4 year old B-group females dropped to zero, while it remained at the normal level in C-group females. During this period, the difference between groups B and C was significant ($\chi^2 = 4.76$, $P < 0.05$), as was the difference in group B between the two periods ($\chi^2 = 8.75$, $P < 0.01$). The fecundity of females older than 4 years was not affected (Table 14.4). This, and the fact that nearly all members of group B weighed about 1 kg less than their age-mates in other groups (unpublished data from Affenberg Salem), indicate that the reproductive delay was not caused by harassment-induced non-ovulatory cycles or abortions. Instead, due to social stress and/or reduced feeding time these young females did not reach the critical body weight and body fat to conceive at the age of 3.5 years.

The costs of early motherhood

Barbary macaque females, like other primates (Watts & Gavan, 1982), reach adult weight only after 6–7 years of age. Among young, adolescent mothers the demands of maternal growth are expected to be in conflict with the costs of fetal growth and lactation (e.g. Lancaster, 1986; Altmann, 1987). Thus, the advantage of an early parturition in terms of Darwinian

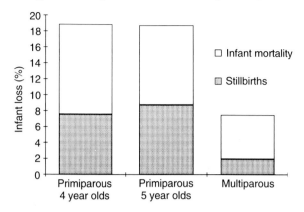

Fig. 14.3. Offspring mortality of primiparous and multiparous females.

fitness may be limited, since an early onset of reproduction also has its extra costs.

In fact, primiparous females suffered from higher rates of stillbirths ($\chi^2 = 11.93$, $P < 0.01$) as well as from higher mortality rates among live-born infants ($\chi^2 = 5.42$, $P < 0.05$) than did multiparous females (Fig. 14.3). A comparison of first- and second-born infants of 92 mothers revealed that first-born infants were at a significantly higher risk than second-born infants (sign test, $n = 15$, $x = 1$, $P < 0.001$). Age appeared to influence infant mortality of low-born primiparae only, where the few early maturing females were less successful than the later maturing ones (3/6 vs 11/15, n.s.). Infant loss was also higher for low-born 4-year old mothers than for their high-born age-mates (3/6 vs 2/14), but the difference did not reach statistical significance ($P > 0.05$, Fisher test). Among the later maturing females, low-born mothers were as successful in rearing their first infants as high-born mothers (11/15 vs 7/9).

Mothers who raised their first infant often failed to become pregnant during the next mating season. Among primiparous females who raised an infant, 34 out of 78 (43.6%) skipped one birth season before becoming pregnant again compared with only 1 (4 year old) out of 16 females (6.3%) whose first infant had died. This difference was highly significant ($\chi^2 = 7.92$, $P = 0.005$), suggesting that the costs of lactation are indeed high for primiparous females. Early maturing (4 year old) females skipped one birth season significantly more often than did later maturing females (60.0% vs 30.2%, $\chi^2 = 6.95$, $P = 0.008$). Low- and mid-born females delayed the next conception more frequently than did high-born females (66.7% and 70.0%, respectively, vs 40.0%), but this rank-related difference again did not reach statistical significance ($\chi^2 = 2.58$, $P = 0.104$). Notably, however, none of the four 5 year old high-born mothers skipped the next birth season, while four of the eight mid- and low-born 5 year olds did so.

Dominance and reproductive senescence

As mentioned above, fertility significantly declined with advancing age. Is this process to any extent affected by dominance status? To answer this question, we partitioned females into four age classes (Fig. 14.4): young females (aged 4–6 years), prime-aged females (7–12 years), middle-aged females (13–19 years), and old females (20–25 years).

While differences in fecundity within all four age classes did not reach statistical significance, differences between age classes were apparent. Fecundity started to decline from prime to middle age, and this effect was much stronger in low-ranking ($\chi^2 = 6.52$, $P = 0.01$) and middle-ranking females ($\chi^2 = 2.47$, $P = 0.11$) than in high-ranking females ($\chi^2 = 0.17$, $P > 0.60$). These data suggest that low-ranking females age more rapidly, or, more precisely, that the cost of rearing infants increases more rapidly with age for low-ranking females.

Determinants of male reproductive success

Rank

Paternity determination was possible for 69 out of 74 infants, yielding a success rate of 93%. For the remaining five infants, two males remained as possible fathers, and in these cases half of the respective offspring was assigned to each of the males. Individual reproductive success refers to the number of offspring (T) for each male divided by the total number of offspring (ΣT, $n = 74$; Smith, 1981).

Overall, male RS was highly correlated with the male's average rank position during the study period ($r_s = 0.710$, $P = 0.001$, $n = 36$; see Fig. 14.5). While this association seems convincing at first glance, two cautionary notes have to be sounded. First, if all breeding seasons are analysed separately, a significantly positive correlation between male rank and RS was found during the first three years, but not in the last year (Table 14.5). Second, no correlation between rank and RS was found in any breeding season when only adult males (aged at least 7 years at the time of conception of their offspring) were considered (Table 14.5).

Observational data have suggested that low-ranking males have very

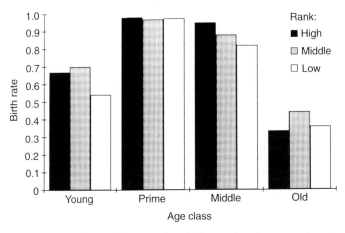

Fig. 14.4. Fertility differences in relation to female age and rank.

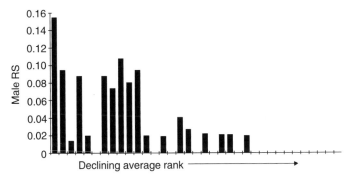

Fig. 14.5. Male rank and overall reproductive success (RS) during four breeding seasons. As rank relations were not completely stable during the entire study period, average rank positions are indicated.

little chance to sire offspring with high-ranking females. To examine this hypothesis, we compared the maternal rank of infants sired by the three highest-ranking males during each breeding season with that sired by the other males (Table 14.6). There was no statistically significant relation between maternal rank and paternal rank in this sample ($\chi^2 = 0.94$, d.f. = 2, n.s.). At least in this group, access to high-ranking females was not restricted to high-ranking males.

Age

Male age was strongly correlated with reproductive success in this sample (Fig. 14.6). Statistically significant correlations ($r_s = 0.555$ to 0.753, $P = 0.000$

A. Paul and J. Kuester

Table 14.5. *Spearman's rank order correlations (r_s) between male reproductive success and agonistic rank*

Breeding season	All sexually mature males	Adult males only
1984/85	$r_s = 0.552$, $n = 16$, $P = 0.030$	$r_s = 0.593$, $n = 8$, n.s.
1985/86	$r_s = 0.467$, $n = 23$, $P = 0.027$	$r_s = -0.399$, $n = 10$, n.s.
1986/87	$r_s = 0.720$, $n = 33$, $P = 0.001$	$r_s = 0.050$, $n = 11$, n.s.
1987/88	$r_s = 0.202$, $n = 31$, $P =$ n.s.	$r_s = -0.251$, $n = 17$, n.s.

Table 14.6. *Distribution of the number of offspring by the ranks of fathers and mothers*

	Male rank		
	1–3	4–33	Total offspring
Female rank			
High	7.5	15.5	23
Middle	7	20	27
Low	4	18	22
Total offspring	18.5	53.5	72

to 0.027) were found in all but the last ($r_s = 0.151$, $P = 0.415$) breeding seasons. Overall, adult males were much more successful than sub-adult males (Mann-Whitney $U = 749.5$, $n = 46$, $n = 57$, $P = 0.000$). Young adult males (7.5–9.5 years of age at the time of conception) were not less successful, statistically, than older males, but clearly more successful than sub-adult males ($U = 304.5$, $P = 0.005$).

Other factors

Dominance rank and age are certainly not the only factors that may affect male RS. The effect of length of residence has not yet been examined quantitatively. However, due to the fact that most males migrate as sub-adults (Paul & Kuester, 1985, 1988), immigrants are, on average, less successful during their first years than older residents. For older males after secondary transfer, the number of similarly aged competitors appears to be the most important constraint for successful reproduction (Paul & Kuester, unpublished data).

Natal transfer was often delayed in large-sized groups. There was no

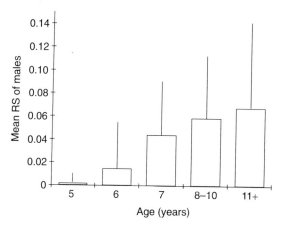

Fig. 14.6. Male age and reproductive success (RS). The age of males at the infants' birth is indicated; males between 5 and 7 years old were sub-adult at the time of the infants' conception.

indication, however, that the reproduction of natal males was suppressed. Males breeding in their natal groups were, on average, as successful as their age-mates breeding in non-natal groups (Paul, Kuester & Arnemann, 1992).

Finally, maternal rank affected male RS. Males born to high-ranking mothers were found to reproduce earlier, and to sire significantly more offspring during the first four years of their reproductive career than males born to low-ranking mothers. The available data further indicate that this advantage persists after natal dispersal (Paul *et al.*, 1992).

Discussion

Female reproductive success

As might be expected in a population with high fertility and survival rates (Paul & Kuester, 1988), differences in the breeding life span accounted for most of the variance in female RS. Fedigan *et al.* (1986) found the same relationship in their analysis of the components of life time RS in a cohort of provisioned Japanese macaque females. Among free-ranging vervet monkeys, however, offspring survival accounted for most of the variance in female RS (Cheney *et al.*, 1988). The vervet data are probably more representative for unprovisioned populations (for free-ranging toque macaques, cf. Dittus, 1979), but precise data for larger macaque species are still unavailable. The main difference between the Barbary macaque data presented here and the

Japanese macaque data provided by Fedigan *et al.* (1986) concerns the covariation between offspring survival and fecundity. Fedigan and her colleagues found a strong negative covariation (-18.4%), indicative of relatively high costs of reproduction. This covariation in our study sample, although negative too, was relatively negligible. Apparently, the fecundity of adolescent and ageing mothers only was constrained by lactation. This suggests that under harsher living conditions, a stronger negative covariance between fecundity and infant survival would emerge (cf. Deag, 1984; Mehlman, 1989).

Rank-related variance in female RS was clearly neither large nor consistent across groups. According to Fedigan *et al.* (1986), this is not surprising in a population which is not food-limited, because competitive hierarchies affect female RS through differential access to resources. Even in food-limited populations, however, variance in female RS may be low (Altmann *et al.*, 1988; Cheney *et al.*, 1988), while in several provisioned groups high-ranking females significantly out-reproduced low-ranking females (Fairbanks & McGuire, 1984; Meikle, Tilford & Vessey, 1984; and references in Fedigan, 1983). In fact, Fairbanks & McGuire (1984) argued that the strongest rank effects were found in provisioned, growing populations, because high-ranking females might be able to respond more rapidly with increased fecundity under such conditions than could low-ranking ones (cf. Sugiyama & Oshawa, 1982). Apparently, both arguments have been somewhat overstated. Variance in female RS certainly depends on environmental conditions (e.g. Whitten, 1983), but the crucial point is the distribution of food, not its abundance (Harcourt, 1987).

Our data suggest that dominance affects fecundity at the onset of the reproductive life and during the period of reproductive senescence. Earlier onset of sexual maturation among the members of high-ranking matrilines have been found in several other studies (reviewed in Harcourt, 1987; Abbott, 1991), and it seems that this life stage is most sensitive to reproductive failure. Even if low-ranking females conceive at an early age, their offspring have a comparatively low survival prospect, and even if the offspring survives, the mother's next conception is significantly delayed. Rank-related differences in single measures were generally small and statistically not significant in our study, but the complete set of data suggests that high-born daughters tend to conceive their first offspring earlier than do low-born daughters, and they tend to cope better with the burden of motherhood at a young age.

Although not directly addressed, our data strongly suggest that nutritional state is crucial for successful reproduction. In humans, infant nutrition is the single most important cause of the timing of menarche (Wu, 1988; for non-human primates, see also Altmann, 1986). Delayed first conception of females in the harassed group was clearly caused by undernutrition, and it seems reasonable that differential nutritional states are also the cause for rank-related differences in the timing of first births. Our data also indicate that both the high lactational effort and the demands of maternal growth influence the period of suppressed fertility after the first birth (for further discussion, see Wilson, Walker & Gordon, 1983; Wilson *et al.*, 1988; Gomendio, 1989). Since aging females generally lose weight, it seems plausible that these factors may also have some influence on their declining fecundity.

In the Salem Barbary macaque colony, a considerable proportion of females reach old age (Paul & Kuester, 1988). An examination of the reproductive history of individual females revealed that fecundity began to decline during the second decade of life, dropping to zero in the first half of the third decade. A post-reproductive life period of five or more years seems to be common in this colony. While survival to old age is certainly enhanced by favourable living conditions, a considerable body of evidence suggests that in many Old World primate populations, female fecundity declines with advancing age, although certainly not in all populations (e.g. the Koshima macaques, Watanabe, Mori & Kawai, 1992). The data presented here show that the costs of reproduction with advancing age are not independent of social status and lead to an earlier onset of reproductive senescence in low-ranking females.

The fact that even in provisioned populations the consequences of differential nutrition may become obvious underlines the crucial importance of food resources for the evolution of female social strategies (Wrangham, 1980). Inter-individual differences in female lifetime RS due to intra-group competition are apparently small in this colony, and probably in many other populations as well; but even small differences may have biologically significant effects if they persist over generations (Smith & Smith, 1988). Our data also support the prediction of Wrangham's (1980) model that a female's RS depends, at least partly, on her group's ability to compete successfully with other groups. Strong inter-group competition has been observed in this population (Paul & Kuester, 1988), although neither the causes for its occurrence in a provisioned population nor its dynamics are well understood. Obviously, overt inter-group competition can be observed more easily and can have more severe consequences under conditions of restricted space (see also Merz, 1976). But space, and especially access to favourable habitats, is not unlimited in the wild, and deleterious effects of inter-group competition on female RS have also been documented in wild populations (Dittus, 1986, 1987; Cheney & Seyfarth, 1987).

Male reproductive success

Variance in reproductive success during this 4-year period was clearly much higher among males than among females. Male reproductive success varied from 0 to 5 offspring/year. During the first two breeding seasons, with no human interference in female fecundity, variance in male RS was 8.8 to 12.3 fold higher than variance in female RS. But even during the last two breeding seasons, where many adult females were not allowed to reproduce, variance in male RS was still 2.1 to 3.8 fold higher. Of course, these data refer to short-term variation in reproductive success, mainly reflecting age-dependent differences in male reproductive performance. But even males that were fully adult from the beginning of the study period showed great differences in reproductive success (1 to 11.5 offspring over the entire 4 year study period). Using a larger sample, Kuester, Paul & Arnemann (1995) found that variance in adult male RS was about 3 to 4 fold higher than that of females, a value quite similar to earlier findings (for free ranging *Macaca fuscata* [estimated], see Fedigan *et al.*, 1986; for free ranging *M. mulatta* [estimated], see Meikle *et al.*, 1984; for a much higher difference in captive *M. mulatta* [confirmed], see also Smith, 1986). While all these data come from provisioned populations, observations in wild populations of baboons and vervet monkeys appear to confirm these results (Altmann *et al.*, 1988; Cheney *et al.*, 1988).

While the available evidence indicates that variance in male RS in polygamous species indeed exceeds that of females, as sexual selection theory predicts (Bateman, 1948; Trivers, 1972), the relationship between male rank and reproductive success is still the subject of major controversy. The genetic paternity study in the rhesus monkey colony of the Californian Primate Research Center by Smith (1981) yielded a close association between male dominance rank and reproductive success, but some years later this strong correlation disappeared (Stern & Smith, 1984). Shively & Smith (1985) found even a negative correlation between dominance and RS among the long-tailed macaque males of the same facility. The authors of a paternity study on captive Barbary macaques concluded that male rank was strongly correlated with 'Darwinian fitness' (Witt, Schmidt & Schmitt, 1981). The two adult males of this group, however, frequently changed ranks and produced apparently similar numbers of offspring.

Several other recent genetic paternity studies found more convincing evidence for an association between male rank and RS. Curie-Cohen *et al.* (1983) found a positive relationship between male dominance and RS in a group of captive rhesus monkeys. The alpha male of this group, however, although sexually the most active, did not sire the highest number of infants. In relatively small groups of wild rhesus monkeys (Melnick, 1987), long-tailed macaques (de Ruiter *et al.*, 1992), and red howlers (Pope, 1990), alpha males sired most or all of the infants. In a group of captive Japanese macaques, male rank was significantly correlated with RS (Inoue *et al.*, 1990). This was also the case in the present study, but the correlation vanished when sub-adult males were excluded from the analysis. Vanishing correlations due to the exclusion of sub-adult males have also been observed in studies on male mating success, leading Bercovitch (1986) and McMillan (1989) to the conclusion that, hitherto, virtually no study convincingly demonstrated an association between male rank and RS. Clearly, if male RS is simply a function of age-dependent differences in resource-holding potential, inter-individual differences in lifetime RS would be diminished. A recent analysis of a large number of studies by Cowlishaw & Dunbar (1991) suggested, however, that this is not the case. Rather, high-ranking males experience increasing difficulty in monopolising females in oestrus as the number of male competitors in the group increases. It may not be coincidental, therefore, that in the present study the highest correlation between rank and RS among adult males was found when the number of adult males was lowest (see also de Ruiter *et al.*, 1992).

A number of behavioural studies suggested that males generally prefer high-ranking females as mates. Because high-ranking males attempt to monopolise high-ranking females, low-ranking males have little chance to gain access to high-ranking females (Paul, 1989, and references therein). In terms of sexual selection, this scenario sounds plausible from both the male's and the female's point of view. But the currently available evidence from studies with known patterns is certainly poor. While the data of Witt *et al.* (1981) appear to support the idea that high-ranking males can effectively monopolise high-ranking females (but see Fedigan, 1983, for a critique of this study), neither the present study, nor the studies by Small & Smith (1982) on captive rhesus monkeys, and Shively & Smith (1985) and de Ruiter *et al.* (1992) on long-tailed macaques were able to prove a relationship between maternal and paternal rank. This discrepancy appears to be due partly to the fact that females are by no means passive resources handled by males, but often quite active sexually, mating promiscuously with many different males on a given day (e.g. for *M. sylvanus*, Taub, 1980; Small, 1990; for *M. fascicularis*, de Ruiter *et al.*, 1992). This should not imply that 'female choice' overrides male competition. Evidence that females can or do prefer certain males over others is quite limited (Small, 1989). But undoubtedly, low-ranking males have better chances to inseminate high-ranking females if the female's behaviour hampers effective monopolisation by any male.

Are Barbary macaques an egalitarian species (*sensu* Vehrencamp, 1983), in comparison with other members of the genus? In the light of the data presented here, and those available from other species, they are certainly not. Clearly, however, there are differences between species in 'dominance style' (*sensu* de Waal, 1989; e.g. de Waal & Luttrell, 1989) and other behavioural syndromes (Thierry, 1990). For example, in Barbary macaques the male hierarchy appears to be less stable or consistent than in rhesus or Japanese macaques (cf. Taub, 1980; Witt *et al.*, 1981; Paul, 1989; and, e.g., Chapais, 1983; Inoue *et al.*, 1990). This may suggest that rank-related differences in male lifetime RS are less pronounced in Barbary macaques than in other, more despotic species. This should not imply, however, that male competition for mates is, for whatever reason, weak or absent in Barbary macaques, as is sometimes suggested (Taub, 1980; Small, 1990). In fact, there is strong evidence for intense male competition during the breeding season, including fights causing dangerous wounds in both large and small groups (Witt *et al.*, 1981; Paul, 1989; Kuester & Paul, 1992).

Alternative reproductive tactics

The discussion concerning the association between male rank and reproductive success will certainly continue, as more and more 'hard' data are accumulated. But clearly, striving for rank is only one tactic for maximising reproductive fitness, although for male macaques probably the most successful one. If individuals are not able to achieve high rank, they are expected to use alternative tactics (Dunbar, 1983).

The low resource holding potential of sub-adult Barbary macaque males does not keep adult males away from their mating partners. Their 'sphere of influence' is not respected, simply because strong power asymmetries eliminate any 'tolerance' of adult males. The alternative tactic sub-adult males use, therefore, is 'sneaking' copulations (Kuester & Paul, 1989). This tactic is widespread in the animal kingdom, but probably only practicable in species where males are capable of achieving ejaculation within a few seconds. We would hardly expect this tactic in macaque species with a multiple mount-to-ejaculation pattern. After a successful copulation, the ability of sub-adult males to defend their investment is restricted, but not completely lacking. Every attempt of other, higher-ranking males to copulate with the sub-adult's female partner is constantly harassed, until the next male achieves ejaculation (Kuester & Paul, 1989). Five year olds are more consistent and successful than four year olds in this respect, and these differences are reflected by the paternity data.

Intrinsic power is certainly highest among young adult males, due to superior physical strength and canine condition. The reproductive success of these males was, on average, higher than expected from behavioural observations (Paul, 1989, and unpublished data). Nevertheless, these males in their physical prime were not the most successful ones, apparently because older males cooperatively peripheralised young adult competitors during the breeding season (Paul, 1989). Perhaps not coincidentally, the oldest male of the group, aged more than 20 years, achieved the second highest reproductive success despite his rather low rank in the adult male hierarchy. Coalitions are formed by older males beyond their physical prime and younger, intrinsically more powerful males are targeted. In this respect, the system resembles that of savanna baboons, although Barbary macaques rarely form coalitions as a tactic for directly gaining access to a female in consort with another male, as is widespread among baboons (see e.g. Packer, 1977; Bercovitch, 1988; Noë, 1989).

The adaptive significance of alternative reproductive decisions is often difficult to assess. Several studies suggested that primate mothers are able to bias the sex ratio of their offspring towards the sex with the highest expected fitness return (for a recent overview, see van Schaik & Hrdy, 1991). Unfortunately, strong evidence that such sex ratio trends are adaptive is limited, because we know little about offspring reproductive success in relation to the maternal condition. Previous research in the Salem Barbary macaque colony showed that high-ranking mothers skewed the sex ratio of their offspring in favour of their sons, while low-ranking females produced comparatively more daughters (Paul & Thommen, 1984; Paul & Kuester, 1987b). Data on maternal rank and male mating success suggested that high-born males were reproductively more successful than low-born males, while maternal rank had little influence on the reproductive success of daughters (Paul & Kuester, 1990). The paternity data now available confirm this observation (for details see Paul, Kuester & Arnemann, 1992; also for captive rhesus monkeys see Smith & Smith, 1988). The hypothesis that the advantage of a high maternal rank vanishes after natal dispersal (the 'poor little rich boy' phenomenon, Berard, 1989) was not supported by the data. Our data are, therefore, in line with the predictions of Trivers & Willard (1973). Under different ecological conditions, however, other reproductive decisions may be adaptive (for the 'local resource competition hypothesis' see Silk, 1983, and for an integrative interpretation of the available data van Schaik & Hrdy, 1991). No doubt, we will learn much more about alternative reproductive decisions and their consequences in the near future.

Acknowledgements

We are grateful to Ellen Merz and Gilbert de Turckheim for permission to study their Barbary macaques for such a long time and to take blood samples. Walter Angst and the staff at the 'Affenberg Salem' greatly facilitated our work. We thank all students who participated in the project and shared information about the animals, the Institutes of Human Genetics and Legal Medicine, and the Zentrales Isotopenlabor, University of Goettingen, for providing facilities. We also thank Mauvis Gore, Eckart Voland and Paul Winkler for helpful comments on the manuscript. Special thanks go to Joachim Arnemann for introducing us to the oligonucleotide fingerprint technique and to the late Christian Vogel, without whose support and encouragement this project would not have been possible. This study was supported financially by the Deutsche Forschungsgemeinschaft (grants An 131/1-6, Vo 124/15-1, Vo 124/18-1).

References

Abbott, D. H. (1991). The social control of fertility. In *Primate responses to environmental change*, ed. H. O. Box, pp. 75–89. London: Chapman & Hall.

Alcock, J. (1989). *Animal behavior. An evolutionary approach*, 4th edn. Sunderland, MA: Sinauer Press.

Ali, S., Müller, C. R. & Epplen, J. T. (1986). DNA finger printing by oligonucleotide probes specific for simple repeats. *Human Genetics*, **74**, 239–43.

Altmann, J. (1986). Adolescent pregnancies in non-human primates: an ecological and developmental perspective. In *School-age pregnancy and parenthood. Biosocial dimensions*, ed. J. B. Lancaster & B. A. Hamburg, pp. 247–62. New York: Aldine.

Altmann, J. (1987). Life span aspects of reproduction and parental care in anthropoid primates. In *Parenting across the life span. Biosocial dimensions*, ed. J. B. Lancaster, J. Altmann, A. S. Rossi & L. R. Sherrod, pp. 15–29. New York: Aldine.

Altmann, J., Hausfater, G. & Altmann, S. (1988). Determinants of reproductive success in savannah baboons, *Papio cynacephalus*. In *Reproductive success. Studies of individual variation in contrasting breeding systems*. ed. T. H. Clutton-Brock, pp. 403–18. Chicago: University of Chicago Press.

Arnemann, J., Schmidtke, J., Epplen, J. T., Kuhn, H. J. & Kaumanns, W. (1989). DNA fingerprinting for paternity and maternity in 'group O' rhesus monkeys at the German Primate Center. Results from a pilot study. *Puerto Rico Health Science Journal*, **8**, 181–4.

Bateman, A. J. (1948). Intra-sexual selection in *Drosophila. Heredity*, **2**, 349–68.

Berard, J. D. (1989). Life histories of male Cayo Santiago macaques. *Puerto Rico Health Science Journal*, **8**, 61–4.

Bercovitch, F. B. (1986). Male rank and reproductive activity in savanna baboons. *International Journal of Primatology*, **7**, 533–50.

Bercovitch, F. B. (1988). Coalitions, cooperation and reproductive tactics among adult male baboons. *Animal Behaviour*, **36**, 1198–209.

Bernstein, I. S. (1976). Dominance, aggression and reproduction in primate societies. *Journal of Theoretical Biology*, **60**, 459–72.

Brown, D. (1988). Components of lifetime reproductive success. In *Reproductive success. Studies of individual variation in contrasting breeding systems*, ed. T. H. Clutton-Brock, pp. 439–53. Chicago: University of Chicago Press.

Burton, F. D. & Sawchuk, L. A. (1974). Demography of *Macaca sylvanus* of Gibraltar. *Primates*, **15**, 271–8.

Carpenter, C. R. (1942). Sexual behavior of free ranging rhesus monkeys (*Macaca mulatta*). II. Periodicity of estrus, homosexual, autoerotic and noncomformist behavior. *Journal of Comparative Psychology*, **33**, 143–62.

Chapais, B. (1983). Reproductive activity in relation to male dominance and the likelihood of ovulation in rhesus monkeys. *Behavioral Ecology and Sociobiology*, **12**, 215–28.

Cheney, D. L. & Seyfarth, R. M. (1987). The influence of intergroup competition on the survival and reproduction of female vervet monkeys. *Behavioral Ecology and Sociobiology*, **21**, 375–86.

Cheney, D. L., Seyfarth, R. M., Andelman, S. J. & Lee, P. C. (1988). Reproductive success in vervet monkeys. In *Reproductive success. Studies of individual variation in contrasting breeding systems*, ed. T. H. Clutton-Brock, pp. 384–402. Chicago: University of Chicago Press.

Clutton-Brock, T. H. (ed.) (1988). *Reproductive success. Studies of individual variation in contrasting breeding systems*. Chicago: University of Chicago Press.

Cowlishaw, G. & Dunbar, R. I. M. (1991). Dominance rank and mating success in male primates. *Animal Behaviour*, **41**, 1045–56.

Curie-Cohen, M., Yoshihara, D., Luttrell, L., Benforado, K., MacCluer, J. W. & Stone, W. H. (1983). The effects of dominance on mating behavior and paternity in a captive troop of rhesus monkeys (*Macaca mulatta*). *American Journal of Primatology*, **5**, 127–38.

Deag, J. M. (1984). Demography of the Barbary macaque at Ain Kahla in the Moroccan Moyen Atlas. In *The Barbary macaque. A case study in conservation*, ed. J. E. Fa, pp. 113–33. New York: Plenum Press.

de Ruiter, J. R., Scheffrahn, W., Trommelen, G. J. J. M., Uitterlinden, A. G., Reber, C., Martin, R. D. & van Hooff, J. A. R. A. M. (1992). Male social rank and reproductive success in wild long-tailed macaques: paternity exclusions by blood protein analysis and DNA-fingerprinting. In *Paternity in primates: tests and theories*, ed. R. D. Martin, A. F. Dixson & E. J. Wickings, pp. 175–91. Basel: S. Karger.

de Turckheim, G. & Merz, E. (1984). Breeding Barbary macaques in outdoor open enclosures. In *The Barbary macaques. A case study in conservation*, ed. J. E. Fa, pp. 241–61. New York: Plenum Press.

de Waal, F. B. M. (1989). Dominance 'style' and primate social organization. In

Comparative socioecology. The behavioural ecology of humans and other mammals, ed. V. Standen & R. A. Foley, pp. 243–64. Oxford: Blackwell Scientific Publications.

de Waal, F. B. M. & Luttrell, L. M. (1989). Towards a comparative socioecology of the genus *Macaca*: different dominance styles in rhesus and stumptail monkeys. *American Journal of Primatology*, **19**, 83–109.

Dittus, W. P. J. (1979). The evolution of behaviors regulating density and age-specific sex ratios in a primate population. *Behaviour*, **69**, 265–302.

Dittus, W. P. J. (1986). Sex differences in fitness following a group take-over among toque macaques: testing models of social evolution. *Behavioral Ecology and Sociobiology*, **19**, 257–66.

Dittus, W. P. J. (1987). Group fusion among wild toque macaques: an extreme case of inter-group resource competition. *Behaviour*, **100**, 247–91.

Dunbar, R. I. M. (1983). Life history tactics and alternative strategies of reproduction. In *Mate choice*, ed. P. P. G. Bateson, pp. 423–33. Cambridge: Cambridge University Press.

Dunbar, R. I. M. (1988). *Primate social systems*. London: Croom Helm.

Fa, J. E. (1984). Structure and dynamics of the Barbary macaque population in Gibraltar. In *The Barbary macaque. A case study in conservation*, ed. J. E. Fa, pp. 263–306. New York: Plenum Press.

Fairbanks, L. A. & McGuire, M. T. (1984). Determinants of fecundity and reproductive success in captive vervet monkeys. *American Journal of Primatology*, **7**, 27–38.

Fedigan, L. M. (1983). Dominance and reproductive success in primates. *Yearbook of Physical Anthropology*, **26**, 85–123.

Fedigan, L. M., Fedigan, L., Gouzoules, S., Gouzoules, H. & Koyama, N. (1986). Lifetime reproductive success in female Japanese macaques. *Folia Primatologica*, **47**, 143–57.

Gomendio, M. (1989). Differences in fertility and suckling patterns between primiparous and multiparous rhesus mothers (*Macaca mulatta*). *Journal of Reproduction and Fertility*, **87**, 529–42.

Gray, J. P. (1985). *Primate sociobiology*. New Haven, CO: HRAF Press.

Harcourt, A. H. (1987). Dominance and fertility among female primates. *Journal of Zoology*, **213**, 471–87.

Inoue, M., Takenaka, A., Tanaka, S., Kominami, R. & Takenaka, O. (1990). Paternity discrimination in a Japanese macaque group by DNA fingerprinting. *Primates*, **31**, 563–70.

Jeffreys, A. J., Wilson, V. & Thein, S. L. (1985a). Hypervariable 'minisatellite' regions in human DNA. *Nature*, **314**, 67–73.

Jeffreys, A. J., Wilson, V. & Thein, S. L. (1985b). Individual specific 'fingerprints' of human DNA. *Nature*, **316**, 76–9.

Kuester, J. & Paul, A. (1989). Reproductive strategies of subadult Barbary macaque males at Affenberg Salem. In *Sociobiology of reproductive strategies in animals and humans*, ed. O. A. E. Rasa, C. Vogel & E. Voland, pp. 93–109. London: Chapman & Hall.

Kuester, J. & Paul, A. (1992). Influence of male competition and female mate choice on male mating success in Barbary macaques (*Macaca sylvanus*). *Behaviour*, **120**, 192–217.

Kuester, J., Paul, A. & Arnemann, J. (1992). Paternity determination by oligonucleotide DNA 'fingerprinting' in Barbary macaques (*Macaca sylvanus*). In *Paternity in primates: genetic tests and theories*, ed. R. D. Martin, A. F.

Dixson & E.J. Wickings, pp. 141–54. Basel: S. Karger.

Kuester, J., Paul, A. & Arnemann, J. (1995). Age, sex and individual differences of reproductive success in Barbary macaques (*Macaca sylvanus*). *Primates*, **36**, in press.

Lancaster, J.B. (1986). Human adolescence and reproduction: an evolutionary perspective. In *School-age pregnancy and parenthood. Biosocial dimensions*, ed. J.B. Lancaster & B.A. Hamburg, pp. 17–37. New York: Aldine.

Martin, R.D., Dixson, A.F. & Wickings, E.J. (eds.) (1992). *Paternity in primates: tests and theories*. Basel: S. Karger.

McMillan, C.A. (1989). Male age, dominance and mating success among rhesus macaques. *American Journal of Physical Anthropology*, **80**, 83–9.

Mehlman, P.T. (1989). Comparative density, demography, and ranging behavior of Barbary macaques (*Macaca sylvanus*) in marginal and prime conifer habitats. *International Journal of Primatology*, **10**, 269–92.

Meikle, D.B., Tilford, B.L. & Vessey, S.H. (1984). Dominance rank, secondary sex ratio and reproduction of offspring in polygynous primates. *American Naturalist*, **124**, 173–88.

Melnick, D.J. (1987). The genetic consequences of primate social organization: a review of macaques, baboons and vervet monkeys. *Genetica*, **73**, 117–35.

Merz, E. (1976). Beziehungen zwischen Gruppen von Berberaffen (*Macaca sylvana*) auf La Montagne des Singes. *Zeitschrift des Kölner Zoo*, **19**, 59–67.

Noë, R. (1989). Coalition formation among male baboons. PhD thesis, University of Utrecht.

Noë, R. & Sluijter, A.A. (1990). Reproductive tactics of male savanna baboons. *Behaviour*, **113**, 117–70.

Packer, C. (1977). Reciprocal altruism in *Papio anubis*. *Nature*, **265**, 441–3.

Paul, A. (1989). Determinants of male mating success in a large group of Barbary macaques (*Macaca sylvanus*) at Affenberg Salem. *Primates*, **30**, 461–76.

Paul, A. & Kuester, J. (1985). Intergroup transfer and incest avoidance in semifree-ranging Barbary macaques (*Macaca sylvanus*) at Salem (FRG). *American Journal of Primatology*, **8**, 317–22.

Paul, A. & Kuester, J. (1987a). Dominance, kinship and reproductive value in female Barbary macaques (*Macaca sylvanus*) at Affenberg Salem. *Behavioral Ecology and Sociobiology*, **21**, 323–31.

Paul, A. & Kuester, J. (1987b). Sex ratio adjustment in a seasonally breeding primate species: evidence from the Barbary macaque population at Affenberg Salem. *Ethology*, **74**, 117–32.

Paul, A. & Kuester, J. (1988). Life history patterns of semifree-ranging Barbary macaques (*Macaca sylvanus*) at Affenberg Salem (FRG). In *Ecology and behavior of food-enhanced primate groups*, ed. J.E. Fa & C.H. Southwick, pp. 199–228. New York: Alan R. Liss.

Paul, A. & Kuester, J. (1990). Adaptive significance of sex ratio adjustment in semifree-ranging Barbary macaques (*Macaca sylvanus*) at Salem. *Behavioral Ecology and Sociobiology*, **27**, 287–93.

Paul, A., Kuester, J. & Arnemann, J. (1992). Maternal rank affects reproductive success of male Barbary macaques (*Macaca sylvanus*): evidence from DNA fingerprinting. *Behavioral Ecology and Sociobiology*, **30**, 337–41.

Paul, A., Kuester, J. & Podzuweit, D. (1993). Reproductive senescence and terminal investment in female Barbary macaques (*Macaca sylvanus*) at Salem. *International Journal of Primatology*, **14**, 105–24.

Paul, A. & Thommen, D. (1984). Timing of birth, female reproductive success and infant sex ratio in semifree-ranging Barbary macaques (*Macaca sylvanus*). *Folia Primatologica*, **42**, 2–16.

Pope, T. R. (1990). The reproductive consequences of male cooperation in the red howler monkey: paternity exclusion in multi-male and single-male troops using genetic markers. *Behavioral Ecology and Sociobiology*, **27**, 439–46.

Shively, C. & Smith, D. G. (1985). Social status and reproductive success of male *Macaca fascicularis*. *American Journal of Primatology*, **9**, 129–35.

Silk, J. B. (1983). Local resource competition and facultative adjustment of sex ratios in relation to competitive abilities. *American Naturalist*, **121**, 56–66.

Silk, J. B. (1987). Social behavior in evolutionary perspective. In *Primate societies*, ed. B. B. Smuts, D. L. Cheney, R. M. Seyfarth, R. W. Wrangham & T. T. Struhsaker, pp. 318–29. Chicago: University of Chicago Press.

Small, M. F. (1989). Female choice in nonhuman primates. *Yearbook of Physical Anthropology*, **32**, 103–27.

Small, M. F. (1990). Promiscuity in Barbary macaques (*Macaca sylvanus*). *American Journal of Primatology*, **20**, 267–82.

Small, M. F. & Smith, D. G. (1982). The relationship between maternal and paternal rank in rhesus macaques (*Macaca mulatta*). *Animal Behaviour*, **30**, 626–7.

Smith, D. G. (1981). The association between rank and reproductive success of male rhesus monkeys. *American Journal of Primatology*, **1**, 83–90.

Smith, D. G. (1986). Incidence and consequences of inbreeding in three captive groups of rhesus macaques (*Macaca mulatta*). In *Primates. The road of self-sustaining populations*, ed. K. Benirschke, pp. 867–74. New York: Springer-Verlag.

Smith, D. G. & Smith, S. (1988). Parental rank and reproductive success of natal rhesus males. *Animal Behaviour*, **36**, 554–62.

Smuts, B. & Nicholson, N. (1989). Reproduction in wild female olive baboons. *American Journal of Primatology*, **19**, 229–46.

Stern, B. R. & Smith, D. G. (1984). Sexual behaviour and paternity in three captive groups of rhesus monkeys (*Macaca mulatta*). *Animal Behaviour*, **32**, 23–32.

Sugiyama, Y. & Oshawa, H. (1982). Population dynamics of Japanese monkeys with special reference to the effect of artificial feeding. *Folia Primatologica*, **39**, 238–63.

Taub, D. M. (1980). Female choice and mating strategies among wild Barbary macaques (*Macaca sylvanus* L.). In *The macaques. Studies in ecology, behavior and evolution*, ed. D. G. Lindburg, pp. 287–344. New York: Van Nostrand Reinhold.

Thierry, B. (1990). Feedback loop between kinship and dominance: the macaque model. *Journal of Theoretical Biology*, **145**, 511–21.

Trivers, R. L. (1972). Parental investment and sexual selection. In *Sexual selection and the descent of man*, ed. B. Campbell, pp. 136–79. Chicago: Aldine.

Trivers, R. L. & Willard, D. E. (1973). Natural selection of parental ability to vary the sex ratio of offspring. *Science*, **179**, 90–2.

van Schaik, C. P. & Hrdy, S. B. (1991). Intensity of local competition shapes the relationship between maternal rank and sex ratios at birth in cercopithecine primates. *American Naturalist*, **138**, 1555–62.

Vehrencamp, S. (1983). A model for the evolution of despotic versus egalitarian societies. *Animal Behaviour*, **31**, 667–82.

Watanabe, K., Mori, A. & Kawai, M. (1992). Characteristic features of the reproduction of Koshima monkeys, *Macaca fuscata fuscata*: a summary of thirty-four years of observation. *Primates*, **33**, 1–32.

Watts, E. S. & Gavan, J. A. (1982). Postnatal growth of non-human primates: the problem of the adolescent spurt. *Human Biology*, **54**, 53–70.

Whitten, P. L. (1983). Diet and dominance in female vervet monkeys (*Cercopithecus aethiops*). *American Journal of Primatology*, **5**, 139–59.

Wilson, M. E., Walker, M. L. & Gordon, T. P. (1983). Consequences of first pregnancy in rhesus monkeys. *American Journal of Physical Anthropology*, **61**, 103–10.

Wilson, M. E., Walker, M. L., Pope, N. S. & Gordon, T. P. (1988). Prolonged lactational infertility in adolescent rhesus monkeys. *Biology of Reproduction*, **38**, 163–74.

Witt, R., Schmidt, C. & Schmitt, J. (1981). Social rank and Darwinian fitness in a multimale group of Barbary macaques (*Macaca sylvana* Linnaeus, 1785). *Folia Primatologica*, **36**, 201–11.

Wrangham, R. W. (1980). An ecological model of female-bonded primate groups. *Behaviour*, **75**, 262–300.

Wu, F. C. W. (1988). The biology of puberty. In *Natural human fertility: social and biological mechanisms*, ed. P. Diggory & M. Potts, pp. 89–101. London: Macmillan.

15

Reproductive biology of captive lion-tailed macaques

D. G. LINDBURG AND N. C. HARVEY

Introduction

Lion-tailed macaques (*Macaca silenus* Linnaeus, 1758) are found only in rain forests of the Western Ghats of southern India. Never numerous, they are today reduced to a population estimated at less than 5000 individuals (Kumar, 1987). Their existing range consists of a narrow strip of forest extending from about 14° 55′ N to near the southern tip of the sub-continent at 8° 25′ N (Fooden, 1975). Although the north–south dimension of the range covers a distance of about 850 km, the forested strip at its widest rarely exceeds more than 30 km. In many places the range is discontinuous, resulting in demes that are today genetically isolated from one another.

Lion-tailed macaques are among the smaller of the macaques, with captive weights averaging 8.9 kg for males and 6.1 kg for females (Harvey, Clarke & Lindburg, 1991). The pelage is predominantly black, with a tuft of guard hairs at the tip of the tail from which their common name derives. Both sexes possess a prominent, silvery mane or ruff (Fig. 15.1) that first begins to appear when infants are about 3–5 months of age. The species appears to belong to a single, monomorphic population (Fa, 1989), but no systematic morphological or genetic investigations across its geographical range have yet been carried out. Within the genus *Macaca*, lion-tails are often placed with *M. nemestrina*, *M. sylvanus*, and the Sulawesi macaques on an evolutionary branch (the *silenus–sylvanus* sub-group) believed to have retained certain phylogenetically primitive features of the genital morphology (Fooden, 1976).

The lion-tail has been described by Green & Minkowski (1977, p. 291) as 'an obligate rain forest dweller', and Kurup & Kumar (1993, p. 28) portrayed their adaptation to relatively uniform tropical wet evergreen forests as 'unique among macaques'. Although often regarded as the most

318

Fig. 15.1. Adult female and infant lion-tailed macaque, showing facial ruff and dark facial pigmentation as traits of adults, but not of infants.

arboreal of the macaques (less than 1% of time terrestrial in the study by Kurup & Kumar, 1993), a recent summary of the literature on this point suggests that, in the proportion of day-ranging spent arboreally, lion-tails are similar to other closely related macaques in spending about 90% of the time above ground (Caldecott, 1986a). In two lion-tail groups carefully monitored by Kurup & Kumar (1993), the middle story from 5 m to 20 m was used 66% of the time. Dietarily, lion-tails appear to be similar to other *Macaca* in consumption of flowers, buds, leaves, fruits, and insects, but with a tendency towards frugivory and insectivory (Johnson, 1980; Kumar, 1987; Kurup & Kumar, 1993). Demographic studies have led some to conclude that a single adult male per group is typical (e.g. Hohmann & Herzog, 1985), but recent census data suggest that one-male groups may occur mainly in impoverished habitats such as cardamom plantations, where the numbers of females are also reduced (Johnson, 1985). For eight groups in rain forest habitat, censused over four to five years, Kumar (1987) reported a mean group size of 20.8, with an average of 1.6 adult males (>6 years), 1.5 sub-adult males, 8.1 adult females, and 9.6 immatures per group. Although several of the groups censused by Kumar were increasing in size, the ratio of adult males : adult females averaged 1 : 5.7 over the course of his

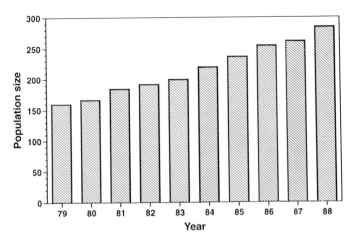

Fig. 15.2. Increase in size of the North American zoo population of lion-tails from 1979 through 1988.

study. Apparently, although no detailed information is as yet available, males are the dispersing sex (Kumar, 1987).

Perhaps because of their relative rareness and possibly their remoteness, wild-living lion-tails have not yet been widely investigated scientifically. Field work to date consists primarily of census and demographic information, and ecological studies (Sugiyama, 1968; Green & Minkowski, 1977; Kurup, 1978; Johnson, 1980, 1985; Ali, 1985; Joseph, 1985; Karanth, 1985, 1992; Kumar, 1987; Ramachandran, 1990). Anatomical studies of museum specimens have been reported by Kurup (1964, 1966) and Fooden (1975); Kumar & Kurup (1981, 1985a,b) have provided first reports on sexual, maternal, and inter-group behavior in wild populations, and activity budgets and feeding have also recently been described (Kurup & Kumar, 1993).

Zoos have been instrumental in focusing attention on this macaque, having sponsored four international symposia since 1982. In addition, an international studbook has been compiled, resulting in useful reproductive information on the captive population (Gledhill, 1990; Lindburg & Gledhill, 1992). Wild-caught lion-tails were last imported to North American zoos in 1968; therefore over 90% of the captive population alive today is captive born (Lindburg & Forney, 1992). Improved husbandry and more systematic breeding, beginning in the 1970s, led to a doubling of the North American population between 1979 and 1988 (Fig. 15.2). The most recent update of the studbook (Gledhill, 1992) revealed a known captive population of 542 individuals, found primarily in North America, Europe, and Japan.

Although the first captive lion-tails on record appeared in a European zoo in 1826 (M. Jones, personal communication), slight attention has traditionally been given to the study of captive individuals. Here, we present partial results from observations initiated in 1979 on a colony maintained in outdoor facilities at the San Diego (California) Zoo and Wild Animal Park. The data are derived from colony records and from detailed investigations of reproductive behavior and physiology. Information from studbook records for the North American population, covering the period from 1899 through 1982, is included where appropriate.

Materials and methods

The original subjects for this study were 11 adult females (> 6 years of age), and numerous captive-born males ranging in age from a few months to over 20 years. In addition, data are reported from nine females subsequently born into the colony and followed to a maximum age of approximately > 8 years. An initial group was assembled through acquisitions from other zoological institutions, and is here identified as the San Diego (SD) group. In 1983 a breeding group of 15, having been socially housed since 1963, and including three founders that were probably wild-caught, was obtained from the Centre d'Acclimatization de Monaco and transported by air to the San Diego Zoo. This second group, identified hereafter as the Monaco (MN) group, may be unique among captive lion-tails in having overlapping generations of related individuals, and a time-depth of about 30 years at present. Since a founder male was still actively breeding at the time of acquisition, and other adults assumed to be his offspring were present in the group, he and three other males of breeding age were immediately replaced by a single, unrelated adult male. Altogether, five different males were rotated into the group between 1983 and 1992. Although birth dates were known for the animals derived from the Monaco Centre, pedigrees of those born prior to 1983 were reconstructed by DNA fingerprint analysis (Morin & Ryder, 1991). The age and gender composition of the entire study population is presented in Table 15.1.

The initial phase (I) of study (1979–83) focused on studies of the menstrual cycle (Shideler *et al.*, 1983, 1985; Lasley, Czekala & Lindburg, 1985) and its behavioral correlates (Lindburg, Shideler & Fitch, 1985). Procedures for urine collection and the assays used in physiological monitoring of female cycles have been described by Shideler *et al.* (1983,

Table 15.1. *Status of individual lion-tails from time of acquisition to end of 1988, and offspring produced to end of 1992. An asterisk (*) indicates estimated birth dates and ages at acquisition*

Studbook no.	Born	Provenance	Age at acquisiton (years)	Reproductive status at acquisition	Number of offspring produced
Females					
557	1963*	Wild	20*	Cycling	0
650	1963*	Unknown	20*	Pregnant	1
651	1963*	Unknown	20*	Cycling	0
673	1965*	Unknown	18*	Cycling	0
767	1967	Unknown	12	Cycling	0
970	1969	Captive	10	Cycling	4
1013	1972	Captive	11	Cycling	8
1088	1973	Captive	6	Cycling	5
1336	1977	Captive	6	Cycling	2
1341	1977	Captive	6	Cycling	5
1344	1977	Captive	6	Pregnant	4
1367	1978	Captive	1	Immature	2
1598	1982	Captive	0	Immature	3
1620	1982	Captive	0	Immature	1
1770	1985	Captive	0	Immature	3
1881	1986	Captive	0	Immature	0
1890	1986	Captive	0	Immature	0
1900	1986	Captive	0	Immature	3
1952	1987	Captive	0	Immature	1
2006	1988	Captive	0	Immature	0
2019	1988	Captive	0	Immature	0
Males					
553	1961*	Unknown	22*	Proven	5
655	1963*	Wild	17*	Unproven	2
778	1967	Captive	14	Unproven	0
969	1969*	Captive	10*	Unproven	1
984	1968*	Unknown	14*	Proven	0
1062	1973*	Captive	10	Unproven	0
1068	1973	Captive	10	Unproven	15
1120	1974	Captive	5	Unproven	8
1200	1975	Captive	7	Unproven	7
1422	1979	Captive	4	Immature	0
1458	1979	Captive	4	Immature	1
1514	1980	Captive	3	Immature	2
1519	1980	Captive	0	Immature	4
1530	1981	Captive	5	Unproven	0
1577	1981	Captive	2	Immature	0
1580	1981	Captive	2	Immature	0
1590	1981	Captive	2	Immature	0
1607	1982	Captive	0	Immature	0
1691	1983	Captive	0	Immature	0
1695	1983	Captive	0	Immature	0
1712	1984	Captive	0	Immature	0
1748	1984	Captive	0	Immature	0
1879	1986	Captive	0	Immature	0
1928	1987	Captive	<1	Immature	0
2009	1988	Captive	0	Immature	0
2010	1988	Captive	0	Immature	0
2029	1988	Captive	0	Immature	0

1985). Behavioral data collected during phase I were based on 1 h sessions, with males introduced daily into the pens of individually housed females (for details of methodology, see Lindburg *et al.*, 1985).

With the addition of group MN to the colony in 1983, the emphasis shifted to studies of behavior in a group-living context and such periodic physiological monitoring of cycles with urine as could be accomplished under these conditions. At this time, also, a checklist was developed for keepers to use in tracking reproductive events in females of breeding age. On a daily basis, staff recorded degree of sexual skin swelling (0 = flat, 3 = maximum tumescence) and noted the occurrence of menstruation, mating, and proceptive calling. In contrast to some other macaques, vaginal plugs or other signs of coagulum were rarely visible. Menstruation was determined primarily from blood residues on structures and by visual inspection of the labia of unrestrained females. For data analysis, cycles in excess of 55 days were discarded in those instances where females underwent successive episodes of calling, tumescence–detumescence and mating, on the assumption there was a failure to detect intervening menses between successive cycles. The behavioral data reported for phase II were collected from February to July 1989, by a single observer who recorded all occurrences of selected events in focal animals (Altmann, 1974). All sessions were conducted for three days weekly, between 9:00 and 11:00 hours.

Phase III began in mid-1989, when group MN ($n = 13$ at the time) was transferred to a $\frac{1}{3}$ ha corral at the San Diego Wild Animal Park, where it remains under observation to the present time. Simultaneously, a portion of group SD ($n = 7$) was moved to a newly constructed exhibit in the zoo grounds, and ceased at that time to be a part of the research effort. The MN group has continued breeding, whereas the adult male in the SD group was vasectomized to preclude the production of unwanted offspring.

During phases I and II, all subjects were housed in an off-exhibit, open-air (chain-link) facility at the San Diego Zoo. Up to three adjoining pens, each measuring approximately 5 m × 4 m × 4 m, were available for each of the two social groups. Additional adult males were housed singly in nearby pens measuring 2 m × 2 m × 1.5 m. Other details of housing and care have been previously described (Lindburg *et al.*, 1985).

The breeding corral occupied by group MN from 1989 onward is circular in shape, with walls of corrugated metal about 4 m in height, and sloped inward 15° from vertical (Fig. 15.3). The substrate is grass, and several protected trees provide shade during the summer months. Three metal sheds, open on the two sides facing the observation deck and equipped with heated perches placed about a meter above ground, afford protection from

Fig. 15.3. The corral habitat occupied by group MN during phase III observations. Note observation deck at left side of photo.

rain and cool winter temperatures. A swimming pool and dead tree tops, positioned vertically, provide additional enrichment opportunities.

At the time of its transfer to the corral, group MN consisted of a single adult male, a sub-adult male, five sexually mature females, and six immatures of various ages. At the beginning of 1994, two sub-adult males were vasectomized to preclude inbreeding. Behavioral data during phase III were recorded from an observation platform stationed at the perimeter of the corral, raised above ground to permit unobstructed viewing of all parts of the habitat (Fig. 15.3). The day-time observations reported here were collected by a single observer between 8:00 and 10:00 hours, Monday through Friday, from May through October 1991. Data on sexual activities at night were collected by the same observer during 55 all-night sessions (17:30 to 06:30) between April 1993, and March 1994. Except for sexual skin ratings and occasional detection of menses, no physiological data were available for group MN after its transfer to the corral facility.

Results

Sexual maturation of females

Information on sexual development is available for 11 females born into the San Diego population (Table 15.2). Data are incomplete for females withheld from breeding ($n = 2$) or who did not become pregnant ($n = 1$). Invariably, the first indication of puberty was slight swelling of the sexual skin, typically for periods of short duration initially, and beginning at a mean age of 2.8 years. Menarche was detected at an average of 159 days after first swellings ($n = 7$), but one female was exceptional in conceiving during her first episode of tumescence. More commonly, young females exhibited numerous swelling episodes (range from 2 to 24) of variable length and variable tumescence between the first swelling and the first conception. In adolescent females the first bouts of proceptive calling occurred virtually simultaneously with first tumescence in three females, but began an average of about 156 days after onset of swelling in five others. Using age from menarche to first conception as the criterion, adolescent infertility averaged 118.8 ± 27.2 (SEM) days in five individuals. The average age for first parturition, at 1430.8 ± 71.3 days (3.9 years), compares with a North American zoo population average of 4.9 years (Lindburg, Lyles & Czekala, 1989), a difference that is probably attributable to the fact that, with rare exceptions, young females in the San Diego population were continuously available to breeding males from before onset of puberty.

Menstrual cycles

Menstrual cycle data for this chapter were derived from females' daily log sheets during the period of mid-1983 to mid-1989 (phase II), while housed in quarters where close inspection was possible. Visually detected menses ranged from 1 to 8 days in sexually mature females, with a mean duration of 1.9 ± 0.96 days in 146 cycles. From onset of menstruation to the day before the next menstruation, 149 cycles from 12 multiparae averaged 32.0 ± 0.4 days in length (Fig. 15.4). Of this total, 38 cycles monitored by urinalysis yielded a mean length of 32.5 ± 6.2 days, confirming the reliability of using overt criteria such as menses in establishing onset and length of the menstrual cycle.

The tumescent phase of the sexual skin in 136 cycles ranged from 3 to 37 days, with a mean length of 12.4 ± 0.4 days. First indications of tumescence were seen an average of 4.2 ± 0.3 days after onset of menstruation, gradually rising to the phase of maximum swelling (range $1-18$, $\bar{x} = 6.9 \pm 3.7$

Table 15.2. *Age (in days) of first occurrence of reproductive events in maturing lion-tailed females*

Reproductive event	No. of females	Range (days)	Mean (days)	SEM
Sexual skin	11	878–1255	1039.0	35.3
Menarche	7	930–1357	1140.3	65.2
Calling	8	921–1333	1120.5	52.9
Conception	8	1074–1556	1270.6	60.8
Birth	7	1228–1728	1430.8	71.3
Adolescent sterility*	5	52– 199	118.8	27.2

'*' represents the interval (in days) from first menarche to first conception

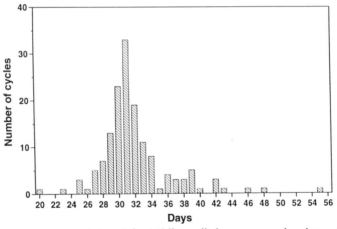

Fig. 15.4. Menstrual cycle length for 149 lion-tailed macaque cycles, determined by visual detection of menses.

days). Urinary analyses were used to determine the relationship of the mid-cycle estrogen peak to onset of sexual skin detumescence. Detumescence occurred from one day prior to the estrogen peak to six days following, with day $+2$ being the mode (40 cycles from seven females, Fig. 15.5). Since it has been shown in lion-tails (Shideler *et al.*, 1983; Lasley *et al.*, 1985) and other macaques (e.g. Monfort *et al.*, 1986) that ovulation normally occurs from 24 to 48 h after the mid-cycle peak in urinary estrogen, these data indicate that careful tracking of the sexual skin can provide an alternative and reliable basis for predicting conception, and thus parturition. The majority of the data on the timing of other events in this study derive from

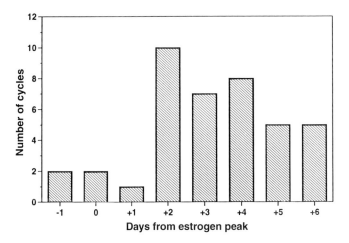

Fig. 15.5. Deflation of the sexual skin in relation to the mid-cycle estrogen peak for 40 cycles monitored by urinary assay.

the use of menstruation and sexual skin deflation as the biologically observable end points.

In females recovering from lactational amenorrhea, initial swelling cycles (analyzed separately from the above) averaged 81 days in length, but gradually declined in length to the species' mean after three to five cycles (Clarke, Harvey & Lindburg, 1993). It was hypothesized that an increased rate of copulation during these extended swelling phases would reduce male inseminations of competing females, or, alternatively, function to attract extra-group males by virtue of increasing females' visual conspicuousness beyond the normal confines of the group.

The endocrine aspects of the menstrual cycle, based on females from this colony, have been described by Shideler *et al.* (1983, 1985), and Lasley *et al.* (1985).

The annual reproductive cycle

In the San Diego population, situated at $32°\ 40'$ N, and kept out doors all year, births ($n = 42$) occurred in all months of the year. This distribution parallels that of the larger North American zoo population, for which 305 births have been shown to be randomly distributed throughout the year (Lindburg *et al.*, 1989). A plot (Fig. 15.6) of the month of menstrual onset for the 149 cycles described earlier failed to show any evidence of a seasonal amenorrhea in captive lion-tails, in agreement with the data from births.

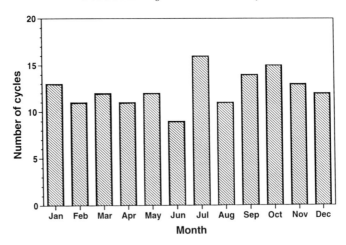

Fig. 15.6. Month of onset of visually detected menstrual bleeding in 149 cycles.

Although the species is not strictly seasonal in captivity, data from the San Diego population suggests a measure of reproductive coordination among females of a group, apparently based on social facilitation (Clarke, Harvey & Lindburg, 1992).

Sexual behavior

In our initial studies of behavior in relation to cycle state (Lindburg *et al.*, 1985), a pair-test procedure was used to elucidate sexual interactions and the copulatory sequence (defined as a series of mounts with insertion and thrusting), using females from whom daily urine samples were simultaneously collected. Under these conditions, a male newly introduced to a female for a 1 h test usually responded with the 'len' expression (Bobbitt, Jensen & Gordon, 1964; and see p. 336) followed by approach, movement of the female's tail to one side as she presented, a moment of intent visual inspection of the sexual skin, and copulation. In 89.3% of mount sequences, two or more mounts with insertion and thrusting occurred before ejaculation, leading to the tentative conclusion that the lion-tailed macaque is best described as a multiple-mount ejaculator (terminology of Fooden, 1980). It was noted that 87% of the mounts in a copulation sequence were initiated by the male, although females showed a marked increase in proceptive behavior as they approached mid-cycle. Also, in four swelling cycles aligned by the day of the mid-cycle estrogen peak, copulation rarely occurred during the luteal phase (Lindburg *et al.*, 1985).

Additional information is now available on copulations and associated behavior. In a group-living situation with a single adult male continuously present (phase II), lion-tails living in outdoor pens at the San Diego Zoo began to copulate on average 2.3 ± 4.2 days (mode = day 4) after the first day of menstruation (59 cycles, 10 females), thus following closely the onset of sexual skin swelling. Copulations ceased, on average, 1.1 ± 0.3 days before the day of sexual skin detumescence and, as in the pair-test situation, were rarely seen during the luteal phase of the cycle. Mean length of the copulatory phase was 10.6 ± 0.5 days.

The initial conclusion that the lion-tailed macaque is a multi-mount ejaculator was confirmed by data from an additional nine males, all observed in a group-living context. *Ad libitum* notes for these individuals indicate that ejaculation after a single mount was rarely seen. For 21 complete copulation sequences occurring during 1 h observation sessions, males averaged 9.1 ± 1.2 mounts to ejaculation (range from 1 to 25). Mean duration of 16 copulation bouts was 33.3 ± 5.2 min.

In the group-living situation of phase II, male and female typically separated from one another after each mount in a copulatory sequence (88.9% of all mounts, $n = 205$). Males were responsible for the majority of physical separations (89.6%; $\chi^2 = 113.9$, 1 d.f., $P < 0.001$). However, the next mount in the bout was significantly more often initiated by females than by males (56.4%; $\chi^2 = 21.97$, 6 d.f., $P < 0.001$), in contrast to the pair-test situation. Females' initiation of mounts was typically by approaching or by present/approach overtures. In addition, females were seen to call to males, and to 'parade' in the area in front of them with tail slightly elevated and frequent *over-the-shoulder* visual monitoring. Less common were *body-tweaks* and *tail-yanks* from females as they ran by. Males initiated mounts by approaching, often preceded by or in conjunction with the 'len' expression. They briefly inspected the females' genitalia before 67.3% of all copulatory mounts, and were significantly more likely to do so in mounts initiated by females ($\chi^2 = 6.62$, 1 d.f., $P < 0.01$). The female copulatory vocalization, usually of seven or eight syllables, was recorded during 74.3% of all mounts with copulating females, usually just before penis withdrawal. Males were significantly more likely than females to walk away after mounts during which the female vocalized ($\chi^2 = 98.5$, 1 d.f., $P < 0.001$). Some males uttered a soft squeak during virtually every episode of thrusting, while others were consistently silent. Harassment of copulations was exceedingly rare, having been recorded in *ad libitum* notes (never during scheduled observations) on only two occasions.

A vocalization very similar to the copulation call was uttered by cycling females at times other than during copulation (Lindburg, 1990). Once begun, calling continued on a daily basis for a period averaging 13.8 ± 5.0 days in length. This period fell almost exclusively between menstruation and onset of detumescence. Calling was accompanied by visual targeting of males, particularly those that were visible in nearby pens, and was sometimes accompanied by manual stimulation of a female's own vulvar region. Although these calls infrequently stimulated copulation by the resident male, they produced strong behavioral responses such as the 'len' expression, agitation, and sometimes masturbation, in males in adjoining pens. Calling is now known to be widely prevalent in the captive population, and has become useful as an unambiguous indication that females are in the follicular stage of the menstrual cycle.

During the corral phase of observation (phase III), which began in 1989, two different adult males were serially introduced to group MN. Breeding females at the onset of this phase were older than in the previous observation periods, and some of the younger females had become sexually mature. However, the major difference during this phase was a > 136 fold increase in the space available to the group. Under these conditions, it was determined that proximity, grooming, and other affiliative behaviors occurred between members of the same matriline more often than with other group members (Birky, 1993). Night-time studies in progress indicate that sleeping clusters are also formed on the basis of kinship. However, the two adult males that were serially introduced into this group showed a strong preference for areas (e.g. perch structures) away from females and young. Consequently, the spatial separation that occurred between mounts in a copulation sequence in the smaller pens at the San Diego Zoo was even more pronounced in the large corral structure.

Mounts in a copulation sequence during this observational phase (III) were preceded by one or more of four different activities: approaching from a distance, allogrooming (nearly all instances were of female as groomer), presenting by the female, and 'lens' by the male. Summarized briefly, a female in estrus would typically leave her matrilineal associates and approach the male on the periphery. She usually presented to the male who, in turn, pulled her tail aside and looked intently at the sexual skin before mounting. If the male ignored her present, the female would groom him briefly, then present a second time. Following mounts, the female either groomed the male or left the area, or the male walked away. This pattern was repeated until ejaculation occurred.

Approaches preceded 74.2% of all copulatory mounts, and females accounted for 60.7% of these (Fig. 15.7), suggesting that females put more

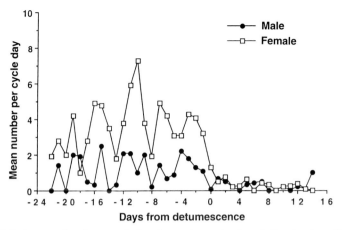

Fig. 15.7. Approaches for copulation by each gender in relation to day of sexual skin detumescence.

effort than males into initiating copulations ($\chi^2 = 5.52$, 1 d.f., $P < 0.02$). For three females scored in phases II and III, the ratio of female initiated copulations to the total received was significantly higher in the spatious (corral) environment of phase III ($t = 6.3$, 2 d.f., $P = 0.024$), though not when primiparae were included. In the corral, females also accounted for a greater, but not significant, proportion of departures after a mount (54.9%, $\chi^2 = 0.98$, 1 d.f., $P > 0.30$) than did males. However, distance comparisons (Fig. 15.8) indicate that females lingered in the general vicinity of the male when the sexual skin was swollen more so than when it was flat ($t = 3.66$, 9 d.f., $P < 0.01$). This conclusion is reinforced by the observation that females were the groomers in 51 out of 52 copulations that were followed by grooming.

Proceptive calling by swelling or swollen females was evident in the corral environment, but at a lower rate than in phase II, when males could be seen in nearby cages. The absence of outsider males during this phase, coupled with the lowered rate of proceptive calling, adds support to the hypothesis that this call is primarily a mechanism for attracting non-group males for mating (Lindburg, 1990).

Night-time observations still in progress suggest a remarkable flexibility in the mating tactics of male lion-tails. In the absence of females in estrus the one male observed to date maintained his peripheral relationship to the female matrilines by sleeping alone in a separate structure. However, on 10 out of 12 nights when sleeping huddles contained swollen females, he positioned himself directly beneath or within a meter or two of the huddle

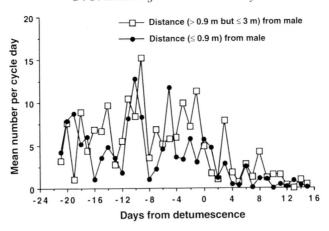

Fig. 15.8. Proximity of estrous females to adult male in relation to day of sexual skin detumescence, sampled at one minute intervals.

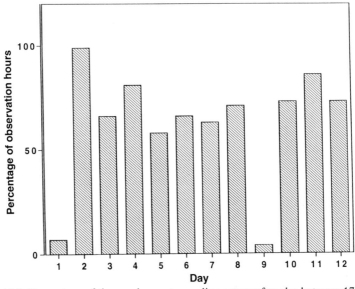

Fig. 15.9. Percentage of time male spent guarding estrous females between 17:30 to 06:30 hours on 12 different nights.

from 63% to 99% of the night-time hours (Fig. 15.9). Copulations from approximately midnight to just before dawn (Fig. 15.10) were initiated almost exclusively by the male, and resulted in frequent disturbance and relocation of sleeping huddles. With the pre-dawn arousal of the group

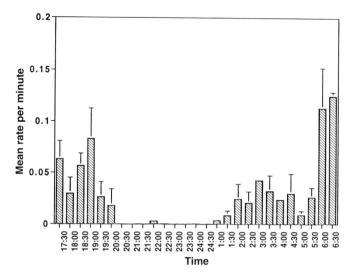

Fig. 15.10. Copulatory mounts during 55 nights of observation.

from sleep, the daytime pattern in which females initiated most copulations resumed.

Inter-male relations

We have reported elsewhere that attempts at forming an all-male group of *M. silenus* had to be abandoned because of high and persistent levels of inter-male aggression, whereas a similar attempt with *M. fascicularis* was successful (Clarke, Czekala & Lindburg, 1995). In addition, by the end of the period from which data for this chapter were derived, young males in our corral group had attained the ages of 5, 6, and 8 years, and were becoming sexually active. However, as a result of aggressive responses from the adult male, these maturing males spatially distanced themselves from him, and their copulations have been confined to the earliest and presumed infertile stages of females' cycles. The composite picture which is emerging is of a fairly high level of inter-male intolerance in *M. silenus*.

Discussion

The lion-tailed macaque displays a constellation of phenotypic traits which, in their totality, distinguishes it from other macaques. A most readily observed difference is in the pelage, consisting of a body fur that is exteme in its blackness, coupled with a prominent silvery ruff that is present in both genders and appears during development at about the time natal coat changes occur in other macaques. We know of no hypotheses that have been put forward to account for these phenotypic contrasts to other *Macaca*, but can rule out sexual selection, since adults are monomorphic. Theories based on predator avoidance also seem unlikely, given the conspicuousness conveyed by the ruff. The appearance of the ruff at 3–5 months of age coincides in time with the change from the pink infantile face to the darkly pigmented face of the adult, perhaps signalling loss of immunity by dependent infants from adult aggression, as has been suggested for other macaques. However, the functional significance of the ruff itself remains unknown.

Certain aspects of lion-tail life history, sexual behavior, and physiology are more likely to be affected by captive living than others. For example, age at first birth is estimated to be 6.6 years ($n = 5$) for females in wild populations (Kumar, 1987), compared with 3.9 years in the San Diego colony (this chapter) and 4.9 years for the entire North American zoo population (Lindburg *et al.*, 1989). An earlier onset of reproduction in captivity may be attributable to better nutritional conditions, as has been reported for numerous other primates (see, e.g. Mori, 1979; Loy, 1988).

Some macaques are known to be highly seasonal in reproduction, with mating seasons occurring in the fall and births in the spring (reviewed by Lindburg, 1987). Kumar & Kurup (1985a) observed sexual swellings and copulations in wild lion-tails in all months of the year except March and April, but with peaks in June and July. On the basis of nine births during their study and three reported by Sugiyama (1968), they suggested two annual peaks in conception (June–July and December), possibly related to major and minor monsoon rains. Since captive lion-tails are kept primarily in zoos and are, therefore, outdoors for portions of each year, it is unlikely that any seasonal tendencies would be completely obscured by the effects of captivity. Although not yet investigated in any detail, the existing evidence from menstrual cycles and from the timing of births in captive populations provides no indication of temporal clustering. The absence of clearly defined mating and birth seasons may, therefore, be typical of all but *M. sylvanus* in the *silenus–sylvanus* sub-group of macaques.

Data from the San Diego population indicate that the menstrual cycle in lion-tails is similar in duration to that of other macaques (Hrdy & Whitten, 1987), and physiological similarity is also evident (Shideler *et al.*, 1983, 1985). Like other macaques in the *silenus–sylvanus* sub-group, lion-tail females have a pronounced phase of sexual skin tumescence and emit a staccato call during most copulations. Lion-tails mate nearly exclusively in the follicular portion of the cycle, even when housed in small quarters.

Those macaques grouped by Fooden (1976) into the *silenus–sylvanus* group have the most pronounced edema of the sexual skin. In *M. nigra* and *M. nemestrina*, the swelling is extreme, causing a funnel shape that appears to aid males in penile insertion (Dixson, 1983). Swelling rarely reaches such extreme proportions in *M. silenus*, and in fact there appears to be a tendency for swellings in this species to become less prominent as females age. However, male lion-tails clearly observe swollen females from a distance and, as noted earlier, will often approach and visually inspect or possibly even sniff the genitalia before mounting. This pattern of male response may be an indication that the sexual skin functions primarily as a distance cue, but that close range inspection is important to confirm attractivity, as suggested by Dixson (1983).

Clutton-Brock & Harvey (1976) suggested that, in addition to its importance in attracting the attention of males for mating, sexual swelling might represent a female tactic for stimulating competition among males as it occurs primarily in species having a multi-male composition. However, Burt (1992) claimed that swellings need not be so extreme if excitation of males is the objective, and suggests alternatively that the conspicuousness of this feature may be an indication of its importance as a distance cue in communicating females' sexual state. Usually, those who regard the sexual skin as a distance signal consider its targets to be primarily those males residing in the females' natal group. That this may not always be the case is evident in Packer's (1979) report on baboon males (*Papio anubis*) who not only survey a new group before transferring in, but usually enter groups having more cycling females than their own. In addition, an hypothesized loss of sexual swellings in other species of macaques in order to reduce the risk of hybridization (Dixson, 1983) makes sense only if it is first thought of as a distance signal that would attract out-group males.

Although the sexual skin is often discussed as functioning in combination with olfactory and tactile modalities (Dixson, 1983), its relationship to vocal behavior is infrequently considered. Our work with lion-tails strongly implicates their non-copulatory calls during the period of sexual skin tumescence as a proceptive behavior. This signal unequivocally causes males, particularly those living outside her group, to visually focus on the swollen female. The time course of this call, the behavioral responses by males to it, and its abolishment through estrogenic manipulation, strongly support the conclusion that it is used by females to solicit matings. Being acoustically similar to copulation calls which, in rain forest habitat, can be heard at a distance of 60–70 m (Kumar, 1987), we have suggested (Lindburg, 1990) that proceptive calls might be directed primarily at out-group males in a habitat where the animals' visual range is limited. The discovery that this behavior declined during phase III of our study, i.e. after relocation to a facility where out-group males were no longer present, reinforces this view. We find it interesting, furthermore, that inter-group contacts in a wild population were more frequent when swollen females were present (Kumar & Kurup, 1985*a*), raising the possibility that in dense forest habitat its occurrence in the context of inter-group encounters could be mistaken for an intra-group signal if observers relied only on auditory detection to record it. The possibility that proceptive and copulation calls have different informational content awaits further behavioral study in combination with sonographic analysis.

Although proceptive calling was originally thought to be unique to *M. silenus*, it now appears that *M. tonkeana* and possibly also *M. nemestrina* have a similar call (Masataka & Thierry, 1993). The failure to routinely document this behavior in wild groups might be attributable to a combination of poor observation conditions and infrequent observations at times when it is most likely to be noticed, i.e. when extra-group males are in the vicinity. Another macaque that is said to use vocalizations proceptively is *M. fuscata* (Oda & Masataka, 1992). However, this species' call seems to be acoustically different from the copulation-type call of other macaques, and may be limited to the solicitation of in-group males.

The 'len' expression, also called facial pucker, flehmen, jaw thrust, or protruded-lips face, and initially thought to be unique to *M. nemestrina* (Redican, 1975; Caldecott, 1986*a*), is commonly seen in the lion-tailed macaque, including infants as young as 4 days of age (D. G. Lindburg, personal observation). Male lion-tails often use this expression when initiating mounts in a copulation sequence (Lindburg *et al.*, 1985). In a comparative study of captive groups of *M. radiata* and *M. nemestrina*, Nadler & Rosenblum (1973) reported its occurrence prior to copulation in the latter.

Although we are not prepared to fully characterize lion-tail consortships on the basis of patterns observed in one-male captive groups, some tentative conclusions about this relationship can be advanced. Assuming the inter-sexual interactions observed in the more spacious environment of phase III are the least compromised by captive conditions, the consortship may be said to consist of a loose series of short-term affiliations largely controlled by females. Swollen females avidly solicited males on the group periphery by a combination of calling, approaching, presenting, grooming, and occasional yanks and tweaks of the males' anatomy. The data further suggest that males approached females for mounting often after giving the 'len', but at times did so in response to the females' overtures. Following mounts in a copulatory series, females, furthermore, elected to move off a short distance from the male at about the same frequency as they groomed him. Although male guarding of females in estrus was seen during the night hours, we found no evidence of the sustained proximity and highly coordinated travel, or the intense, exaggerated, and easily elicited threats, head-bobs, body slaps, etc., which have been described for other consorting macaques such as *M. mulatta* (Carpenter, 1942; Lindburg, 1983). The absence of competition from other adult males undoubtedly affected the patterns seen in the San Diego population, but it is equally plausible that the contrast reflects a real difference between species in the nature of the consort relationship.

During a brief survey of wild populations, Sugiyama (1968) noted 'mating couples' at distances of 100–200 m from their group. Kumar & Kurup (1985a) also mention peripheralized consort pairs, at times marked by females' persistent following of males, and including entry into feeding trees together. At the same time, females in their study are described as using invitational gestures such as approach, dashing past, presenting, and calling, and as primarily responsible for initiating and sustaining consortships. Females were also able to reject solicitations from males for post-mount grooming, electing instead to move a short distance away and sit. Although characterized as having exclusive relationships of long duration, many features in common with our captive groups are evident.

Caldecott (1986b) included lion-tails with *M. nemestrina*, *M. mulatta*, and *M. fuscata* in a 'group 1' classification of Asian macaques that show common adaptations to a predominantly dipterocarp forest where food is chronically in short supply. Insofar as our results confirm that lion-tails are multi-mount copulators, exhibit low inter-male tolerance, and have a mating system largely controlled by females, his model is validated. However, the nature of the consort relationship, to date based on observations in both wild and captive groups having only a single, fully adult male, awaits further investigation.

Acknowledgements

We thank Drs Susan Shideler, Bill Lasley, Nancy Czekala and Lonnie Kasman for their assistance with endocrine work during the years of our study, and Helena Fitch-Snyder and Dr Susan Clarke for their participation with us in behavioral studies. Bonnie Fox, Densie Fitzgerald, Jill Mellon, Cathie Wertis, Peter Balcaen, Lisa Gouse, Dee Conkel, Teresa Everett, Karen Barnes, Paula Augustus, Gail Thurston and Bob Cisneros maintained daily checklists and other colony records in their role as animal technicians. We also thank Patricia Pedrozo, Ernest Hartt, and Katherine DeFalco for their substantial contributions as observers over the years of this study.

References

Ali, R. (1985). An overview of the status and distribution of the lion-tailed macaque. In *The lion-tailed macaque: status and conservation*, ed. P. G. Heltne, pp. 13–25. New York: Alan R. Liss.

Altmann, J. (1974). Observational study of behaviour: sampling methods. *Behaviour*, **49**, 227–67.

Birky, W. A. (1993). Female-female social relationships in a captive group of lion-tailed macaques (*Macaca silenus*). MSc thesis, California State University.

Bobbitt, R. A., Jensen, G. D. & Gordon, B. N. (1964). Behavioral elements (taxonomy) for observing mother–infant–peer interaction in *Macaca nemestrina*. *Primates*, **5**, 71–80.

Burt, A. (1992). 'Concealed ovulation' and sexual signals in primates. *Folia Primatologica*, **58**, 1–6.

Caldecott, J. O. (1986a). *An ecological and behavioural study of the pig-tailed macaque*. Contributions to Primatology, vol. 21. Basel: S. Karger.

Caldecott, J. O. (1986b). Mating patterns, societies and the ecogeography of macaques. *Animal Behaviour*, **34**, 208–20.

Carpenter, C. R. (1942). Sexual behavior of free ranging rhesus monkeys (*Macaca mulatta*). *Journal of Comparative Psychology*, **33**, 113–42.

Clarke, A. S., Czekala, N. M. & Lindburg, D. G. (1995). Behavioral and adrenocortical responses of male cynomolgus and lion-tailed macaques to social stimulation and group formation. *Primates*, **36**, 41–56.

Clarke, A. S., Harvey, N. C. & Lindburg, D. G. (1992). Reproductive coordination in a nonseasonally breeding primate species, *Macaca silenus*. *Ethology*, **91**, 46–58.

Clarke, A. S., Harvey, N. C. & Lindburg, D. G. (1993). Extended postpregnancy estrous cycles in female lion-tailed macaques. *American Journal of Primatology*, **31**, 275–85.

Clutton-Brock, T. H. & Harvey, P. H. (1976). Evolutionary rules and primate societies. In *Growing points in ethology*, ed. P. P. G. Bateson & R. A. Hinde, pp. 195–237. Cambridge: Cambridge University Press.

Dixson, A. F. (1983). Observations on the evolution and behavioral significance of the 'sexual skin' in female primates. In *Advances in the study of behavior*, vol. 13, ed. J. S. Rosenblatt, R. A. Hinde, C. Beer & M. C. Busnel, pp. 63–106. New York: Academic Press.

Fa, J. E. (1989). The genus *Macaca*: a review of taxonomy and evolution. *Mammal Review*, **19**, 45–81.

Fooden, J. (1975). Taxonomy and evolution of liontail and pigtail macaques (Primates: Cercopithecidae). *Fieldiana Zoology*, **67**, 1–169.

Fooden, J. (1976). Provisional classification and key to living species of macaques (Primates: *Macaca*). *Folia Primatologica*, **25**, 225–36.

Fooden, J. (1980). Classification and distribution of living macaques (*Macaca Lacépède*, 1799). In *The macaques: studies in ecology, behavior and evolution*, ed. D. G. Lindburg, pp. 1–9. New York: Van Nostrand Reinhold.

Gledhill, L. (1990). *Lion-tailed macaque international studbook*. Seattle, WA: Woodland Park Zoo.

Gledhill, L. (1992). *Supplement, lion-tailed macaque international studbook*, Seattle, WA: Woodland Park Zoo.

Green, S. & Minkowski, K. (1977). The lion-tailed monkey and its south Indian rain forest habitat. In *Primate conservation*, ed. His Serene Highness Prince Rainier III of Monaco & G. H. Bourne, pp. 289–337. New York: Academic Press.

Harvey, N. C., Clarke, A. S. & Lindburg, D. G. (1991). Morphometric data for adult lion-tailed macaques (*Macaca silenus*). *American Journal of Physical Anthropology*, **85**, 233–6.

Hohmann, G. M. & Herzog, M. O. (1985). Vocal communication in lion-tailed macaques (*Macaca silenus*). *Folia Primatologica*, **45**, 148–78.

Hrdy, S. B. & Whitten, P. L. (1987). Patterning of sexual activity. In *Primate societies*, ed. B. Smuts, D. Cheney, R. Seyfarth, W. Wrangham & T. Struhsaker, pp. 370–84. Chicago: University of Chicago Press.

Johnson, J. M. (1980). The status, ecology and behaviour of lion-tailed macaque (*Macaca silenus*). *Journal of the Bombay Natural History Society*, **75**, 1017–26.

Johnson, J. M. (1985). Lion-tailed macaque behavior in the wild. In *The lion-tailed macaque: status and conservation*, ed. P. G. Heltne, pp. 41–63. New York: Alan R. Liss.

Joseph, K. J. (1985). *Macaca silenus*, the lion-tailed macaque: its status and habitat management in Kerala. In *The lion-tailed macaque: status and conservation*, ed. P. G. Heltne, pp. 27–39. New York: Alan R. Liss.

Karanth, U. K. (1985). Ecological status of the lion-tailed macaque and its rain forest habitats in Karnataka, India. *Primate Conservation*, **6**, 73–84.

Karanth, U. K. (1992). Conservation prospects for lion-tailed macaques in Karnataka, India. *Zoo Biology*, **11**, 33–41.

Kumar, A. (1987). The ecology and population dynamics of the lion-tailed macaque (*Macaca silenus*) in South India. PhD thesis, University of Cambridge.

Kumar, A. & Kurup, G. U. (1981). Infant development in the lion-tailed macaque, *Macaca silenus* (Linnaeus): the first eight weeks. *Primates*, **22**, 512–22.

Kumar, A. & Kurup, G. U. (1985a). Sexual behavior of the lion-tailed macaque (*Macaca silenus*). In *The lion-tailed macaque: status and conservation*, ed. P. G. Heltne, pp. 109–30. New York: Alan R. Liss.

Kumar, A. & Kurup, G. U. (1985b). Inter-troup interactions in the lion-tailed macaques, *Macaca silenus*. In *The lion-tailed macaque: status and conservation*, ed. P. G. Heltne, pp. 91–107. New York: Alan R. Liss.

Kurup, G. U. (1964). On the cranial characters of *Macaca silenus* (Linnaeus) Primates: Cercopithecidae. *Journal of the Bombay Natural History Society*, **60**, 246–9.

Kurup, G. U. (1966). A comparative study of the cranial characters of *Presbytis* and *Macaca* based on Indian species (Primates: Cercopithecidae). *Mammalia*, **30**, 64–81.

Kurup, G. U. (1978). Distribution, habitat and survey of the liontailed macaque, *Macaca silenus* (Linnaeus). *Journal of the Bombay Natural History Society*, **75**, 321–40.

Kurup, G. U. & Kumar, A. (1993). Time budget and activity patterns of the lion-tailed macaque (*Macaca silenus*). *International Journal of Primatology*, **14**, 27–39.

Lasley, B. L., Czekala, N. M. & Lindburg, D. G. (1985). Urinary estrogen profile in the lion-tailed macaque. In *The lion-tailed macaque: status and conservation*, ed. P. G. Heltne, pp. 91–107. New York: Alan R. Liss.

Lindburg, D. G. (1983). Mating behavior and estrus in the Indian rhesus monkey. In *Perspectives in primate biology*, ed. P. K. Seth, pp. 45–61. New Delhi: Today & Tomorrow's Printers and Publishers.

Lindburg, D. G. (1987). Seasonality of reproduction in primates. In *Comparative primate biology*, vol. 2B: *Behavior, cognition, and motivation*, ed. G. Mitchell & J. Erwin, pp. 167–218. New York: Alan R. Liss.

Lindburg, D. G. (1990). Proceptive calling by female lion-tailed macaques. *Zoo Biology*, **9**, 437–46.

Lindburg, D. G. & Forney, K. A. (1992). Long-term studies of captive lion-tailed macaques. *Primate Report*, **32**, 133–42.

Lindburg, D. G. & Gledhill, L. (1992). Captive breeding and conservation of lion-tailed macaques. *Endangered Species Update*, **10**, 1–4.

Lindburg, D. G., Lyles, A. M. & Czekala, N. M. (1989). Status and reproductive potential of lion-tailed macaques in captivity. *Zoo Biology Supplement*, **1**, 5–16.

Lindburg, D. G., Shideler, S. & Fitch, H. (1985). Sexual behavior in relation to time of ovulation in the lion-tailed macaque. In *The lion-tailed macaque: status and conservation*, ed. P. G. Heltne, pp. 131–48. New York: Alan R. Liss.

Loy, J. (1988). Effects of supplementary feeding on maturation and fertility in primate groups. In *Ecology and behavior of food-enhanced primate groups*, ed. J. E. Fa & C. H. Southwick, pp. 153–66. New York: Alan R. Liss.

Masataka, N. & Thierry, B. (1993). Vocal communication of Tonkean macaques in confined environments. *Primates*, **34**, 169–80.

Monfort, S. L., Jayaraman, S., Shideler, S. E., Lasley, B. L. & Hendrickx, A. G. (1986). Monitoring ovulation and implantation in the Cynomolgus macaque (*Macaca fascicularis*) through evaluations of urinary estrone conjugates and progesterone metabolites: a technique for the routine valuation of reproductive parameters. *Journal of Medical Primatology*, **15**, 17–26.

Mori, A. (1979). Analysis of population changes by measurement of body weight in the Koshima troop of Japanese monkeys. *Primates*, **20**, 371–98.

Morin, P. A. & Ryder, O. A. (1991). Founder contribution and pedigree inference in a captive breeding colony of lion-tailed macaques, using mitochondrial DNA and DNA fingerprint analysis. *Zoo Biology*, **10**, 341–52.

Nadler, R. D. & Rosenblum, L. A. (1973). Sexual behavior of male pig-tail macaques in the laboratory. *Brain, Behavior and Evolution*, **7**, 18–33.

Oda, R. & Masataka, N. (1992). Functional significance of female Japanese macaque copulatory calls. *Folia Primatologica*, **58**, 146–9.

Packer, C. (1979). Inter-troop transfer and inbreeding avoidance in *Papio anubis*. *Animal Behaviour*, **27**, 1–36.

Ramachandran, K. K. (1990). Feeding and ranging patterns of lion-tailed macaques in Silent Valley National Park. In *Ecological studies and long-term monitoring of biological processes in Silent Valley National Park*, Government of India, Final Report to Ministry of Environment and Forests, pp. 109–32. Kerala: Kerala Forest Research Institute.

Redican, W. K. (1975). Facial expressions in nonhuman primates. In *Primate* behavior: developments in field and laboratory research*, vol. 4, ed. L. A. Rosenblum, pp. 103–94. New York: Academic Press.

Shideler, S. E., Czekala, N. M., Kasman, L. H., Lindburg, D. G. & Lasley, B. L. (1983). Monitoring ovulation and implantation in the lion-tailed macaque (*Macaca silenus*) through urinary estrone conjugate evaluations. *Biology of Reproduction*, **29**, 905–11.

Shideler, S. E., Mitchell, W. R., Lindburg, D. G. & Lasley, B. L. (1985). Monitoring luteal function in the lion-tailed macaque (*Macaca silenus*) through urinary progesterone metabolite measurements. *Zoo Biology*, **4**, 65–73.

Sugiyama, Y. (1968). The ecology of the lion-tailed macaque [*Macaca silenus (Linnaeus)*] – A pilot study. *Journal of the Bombay Natural History Society*, **65**, 283–92.

16

Sexual behaviour and mating system of the wild pig-tailed macaque in West Sumatra

T. OI

Introduction

Macaque species have the widest distribution among non-human primates, ranging from the tropical to cool temperate zones. The great variation in macaque behaviour and social structure might be related to differences in their habitats (Shively et al., 1982; Thierry, 1985; Caldecott, 1986a). One example of this is the relationship between environmental and mating seasonality (Vandenbergh & Vessey, 1968; van Schaik & van Noordwijk, 1985; Lindburg, 1987). Mating seasonality can be thought of as the temporal distribution pattern of females in oestrus. As females in oestrus can be considered to be a limited resource for males (e.g. Emlen & Oring, 1977; Wrangham, 1979, 1980; Berenstain & Wade, 1983; Ridley, 1986; Newton, 1988), mating seasonality will have important effects on competition between males for access to mates.

In this chapter, first I describe the sexual behaviour of wild pig-tailed macaques (pig-tails). The sexual behaviour and mating system of pig-tails have been studied most frequently under laboratory conditions (e.g. Kuehn et al., 1965; Tokuda, Simons & Jensen, 1968). Studies on the sexual behaviour of wild pig-tails are few in number and are based on fragmentary observations (Bernstein, 1967; Caldecott, 1986b). Perhaps because social behaviour is flexible and may be influenced by external factors, results from the laboratory and those from the natural habitat do not always agree (e.g. Yamagiwa, 1986). Therefore, results from the laboratory need to be verified and reinterpreted in the light of results from studies in the field.

Second, I analyse the relationships between the pattern of oestrus in females and mate monopolisation by males in the pig-tail mating system. Finally, I present a model which explains the variation in the mating systems of macaques in terms of mate monopolisation by males from an evolutionary view point, based on the above-mentioned theory that the

Table 16.1. *Sizes and age–sex composition of troops*

Troop name	AdM	AdlM	AdF	AdlF	J	I	Total
Troop A	3	5	22	2	27	15	74
Troop B	3	4	14	1	18	9	49
Troop C	4	4	26	2	34	11	81

AdM, adult male; AdlM, adolescent male; AdF, adult female; AdlF, adolescent female; J, juvenile; I, infant.

temporal distribution of females in oestrus can determine male mating strategy.

Materials and methods

Three troops of wild pig-tails were studied in West Sumatra province, Indonesia, during three periods: from January 1985 to March 1985, from June 1985 to March 1986, and from July 1986 to February 1987. A detailed description of the study area is given by Oi (1990a). From 22 January to 2 March 1985, troop A was partially provisioned, and from July 1986 to February 1987, all three troops, including troop A, were provisioned and observed regularly at two baiting sites.

In this chapter, data obtained on 158 observation days, from July 1986 to February 1987, are analysed. During this period, troops A, B, and C visited one baiting site on 104 occasions (2109 min), on 102 occasions (4534 min), and on 19 occasions (794 min), respectively. Troop A stayed at the site for 20 min on average (range: 2–49 min), troop B for 44 min (range: 3–102 min), and troop C for 42 min (range: 15–66 min). Table 16.1 shows the age–sex composition of the three troops. Adult males (fully mature, estimated age of over 9 years), adolescent males (mature, estimated age of 5–9 years), adult females (parous, estimated age of over 5 years), and adolescent females (nulliparous, estimated age of 3.5–6 years) were the subjects for analysis (for detailed definition, see Oi, 1990a).

The baiting site was square (20 m × 20 m) and the animals were observed from a hide. Five to 15 kg of unshelled peanuts (*Arachis hypogaea*) were distributed at the site before observations. All adult and adolescent animals in troops A and B were identified visually. As troop C was observed much less frequently, only certain individuals could be recognised. For analyses that required complete individual identification, only data for troops A and

B were used. The male dominance order was determined from the outcome of dyadic agonistic interactions (Oi, 1990*b*).

The oestrous condition of each mature female was monitored and all births were recorded. All sexual behaviour, including sexual solicitation, copulation, interference in copulation, and rejection of copulation, was recorded. Females in oestrus were easily distinguished by the conspicuous swelling of their sexual skin (Fooden, 1975). Bullock, Paris & Goy (1972) reported that ovulation occurred between the day of peak swelling and the first day of detumescence of the sexual skin. To examine the seasonality of oestrus, females in oestrus were monitored from July 1986 to February 1987. For the period from March to June 1986, the minimum number of females in oestrus per month was calculated by counting back 171 days, the median gestation length (range: 154–230 days: Hadidian & Bernstein, 1979), from the possible birth dates.

Results

Copulatory behaviour

Copulatory behaviour could begin with solicitation by either a male or a female. A male typically solicited a female in oestrus by approaching her while puckering (lowering his chest, with protruding mouth shut, and retracting his ears) and putting his hand on her rump. A female solicited a male by approaching him from behind, passing by his nose, and finally presenting her genitalia 0.5–1.5 m in front of him. In this posture the female often looked back over her shoulder at the male.

A mount with intromission usually followed sexual solicitation. The mean length of a mount was 9.0 ± 3.0 s ($n = 92$), and 14 ± 5.5 pelvic thrusts ($n = 220$) occurred in a single mount. Ejaculation could not be recognised. Near the end of a mount, the female turned her face back towards the male and grasped his thigh or scrotum, though this 'reaching-back behaviour' is not always accompanied by ejaculation (Bullock *et al.*, 1972). The male sometimes emitted a short scream, 'kya', with a grimace, during the mount. In 246 of the 249 mounts observed, when the male dismounted the female rushed away from him, emitting the sounds 'ga-ga-ga'.

Although ejaculation could not be recognised, the results of the present study strongly suggest that the pig-tail studied was a single-mount ejaculator. According to Nadler & Rosenblum (1973), pig-tail males required 10 min to achieve ejaculation after a series of mounts in the laboratory. Therefore, I defined the termination of a mounting series as either: (1) discontinuation of mounts for more than 10 min, or (2) solicitation towards or copulation with other individuals.

A total of 194 mounting series, consisting of 249 sexual mounts, were observed in troops A, B and C. Within the mounting series, 162 (83.5%) were single mounts, only 20 (10.3%) were double mounts, eight (4.1%) were triple mounts, none consisted of four mounts, two (1.0%) of five mounts, one (0.5%) of six mounts, and one (0.5%) of seven mounts. Between mounts, the animals often fed. In addition, during tracking of the troops before provisioning, two single-mount copulations were observed. Therefore, I have treated one mount as one copulation in the analysis of sexual interactions since most mounting series consisted of only single mounts.

Oestrous cycle of pig-tail females

Swelling of the sexual skin occurred in a series of marked sequential changes. Initially, the skin around the anus turned red, then it became swollen, and, at its most extreme, the swelling extended to the caudal base. Two young females in troop A, one in troop B, and two in troop C showed the pubertal form of swelling (Dixson, 1983). All females who appeared to have reached 3.5–4 years of age showed swelling of the sexual skin.

The two successive first days of maximal swelling were separated by 39 ± 17 days ($n = 10$), and 10 ± 3.6 days ($n = 10$) were required to reach the maximal swelling from the beginning of swelling (tumescent phase). Maximal swelling continued for 13 ± 6.1 days ($n = 20$) (full swelling phase), and then gradually deflated over 11 ± 5.8 days ($n = 16$) (detumescent phase). The complete absence of swelling was sometimes observed between the detumescent phase and the next tumescent phase. Sexual solicitation by both males and females, and copulation, occurred most frequently during the full swelling phase (Table 16.2). Females also became red around the anus for 43 ± 18 days ($n = 7$) before parturition, but did not show any swelling of the sexual skin. Sexual solicitation and copulation were not observed during this period before parturition.

It took 170 ± 28 days ($n = 11$) for a female to resume post-partum sexual swelling, and 170 ± 9.6 days ($n = 5$) for a fertile female to give birth after full swelling ceased. Out of a total of 20 cases of consecutive oestrous cyclings, two of these had cycled once or twice before cycling ceased, possibly indicating conception, but the remaining 18 animals had one or more cycles prior to conception. However, the actual number could not be ascertained since the initiation or termination of a series of oestrous cyclings was outside the observation period. Nevertheless, two or more

Table 16.2. *Phase of sexual swelling and incidence of sexual behaviour*

Sexual behaviour	Phase of sexual swelling				
	Tumescent	Full swelling	Detumescent	Undiscerned	Total
Sexual solicitation by males	14 (13)	85 (79)	8 (7)	1 (1)	108 (100)
Sexual solicitation by females	43 (15)	205 (74)	24 (9)	6 (2)	278 (100)
Copulation	27 (11)	196 (79)	21 (8)	5 (2)	249 (100)

The numbers in parentheses indicate percentage incidence of each behaviour.

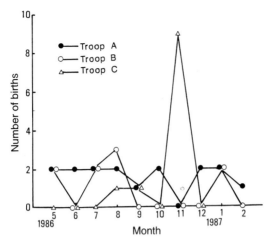

Fig. 16.1. Distribution of live births from May 1986 to February 1987. Seven births from May through to the middle of July were estimated by the developmental stage of body size and behaviour of the babies.

oestrous cycles were observed before cessation in 11 cases. This result implies that at least 65% of individuals either failed to conceive during the first oestrous cycle or showed post-conception swelling.

Seasonality of oestrus and births

There were 24 births in troops A and B between May 1986 and February 1987. Sixteen (73%) of the 22 adult females in troop A, and eight (57%) of the 14 adult females in troop B, gave birth. Births occurred throughout the period of observation (Fig. 16.1) and were not concentrated in any particular months. Data for troop C were obtained between May and December 1986. Eleven (42%) of 26 adult females gave birth, and 82% of these were in November.

Females in oestrous were present almost throughout the year, with an increase towards the end of 1986 in each troop (Fig. 16.2). The number of females in oestrous in troop A reached its maximum, seven (29% of mature females) in November and January, that in troop B reached its maximum, four (27% of mature females) in November, and that in troop C reached its maximum, nine (32% of mature females) in May, with a second peak in November.

During observation, 14 females showed signs of oestrus in troop A, and seven females did so in troop B. However, only a few females were in oestrus

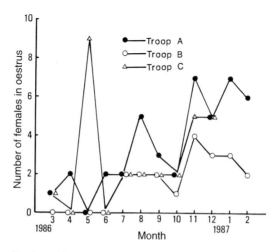

Fig. 16.2. Distribution of females in oestrus from March 1986 to February 1987. The numbers from March 1986 to July 1986 were estimated (see text).

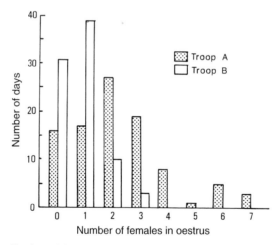

Fig. 16.3. Distribution of the numbers of females in oestrus on each observation day.

simultaneously. The mean number of females in oestrus on a daily basis was 2.3 ± 1.1 (mode: 2, $n = 96$) in troop A, and 0.83 ± 0.78 (mode: 1, $n = 81$) in troop B (Fig. 16.3).

Sexual solicitation and copulation

In troop A, there were 23 solicitations by males, 57 solicitations by females and 66 copulations. In troop B, there were 79 solicitations by males, 196 solicitations by females and 157 copulations. In troop C, there were six solicitations by males, 25 solicitations by females and 26 copulations.

The larger number of solicitations by females in all groups might be due to the fact that while the higher-ranking male could solicit at any time any female in oestrus, the other males were inhibited from soliciting females in oestrus in the presence of the highest-ranking male. For example, when the alpha male was present, solicitation by other males occurred infrequently (0.037 solicitations/h; this rate is calculated from the number of solicitations by the males divided by the total number of hours of observation for each male). In the absence of the alpha male, solicitations by other males increased (0.54 solicitations/h). By contrast, females in oestrus solicited any male, even in the presence of the alpha male (0.38 solicitations/h). The solicited males ignored the females or moved around to escape from such females. However, the rate of sexual solicitation by females also increased, to 3.4 solicitations/h, in the absence of the alpha male.

Only high-ranking males performed sexual solicitations. In troop A, only the three top-ranking adult males solicited females in estrous, while in troop B, the two top-ranking adult males and the fourth-ranking adolescent male did so. The rate of sexual solicitation by each male, given by the number of solicitations divided by the total nmber of hours of observation for each male, is significantly correlated with the dominance rank of males in troop A ($\tau = -0.51$, $P < 0.05$, two-tailed), and not so in troop B ($\tau = -0.81$, $P > 0.05$ two-tailed).

All adult males and three adolescent males were sexually solicited by females in troop A, and the rate of such sexual solicitation for each male (abbreviated to FSR), given by the number of solicitations divided by the total number of hours of observation for each male, is significantly correlated with the dominance rank of the males ($\tau = -0.78$, $P < 0.01$, two-tailed). In troop B, all males except for one adolescent (SR) were sexually solicited by females. A solitary male (BK) was also solicited by a female in estrous (Kn), who had temporarily separated herself from the other troop members. The FSR for the second-ranking male (1.7 solicitations/h) surpassed that for the first-ranking male (1.3 solicitations/h) in troop B, and the correlation between FSR and the dominance rank of males is not significant ($\tau = -0.52$, $P > 0.05$, two-tailed).

Females seemed to seek more mating partners than did males. In 46 cases (81%), a male solicited a single female in one observation day, and in 11 cases (19%) a male solicited two females in one observation day. The males who solicited two females were the alpha (eight cases), and the beta (three

Table 16.3. *Mating partner*

(a) *Troop A*

Male	Female																								Total
	Cw	Ef	Mr	Mi	Dn	Mo	Em	Mm	Up	Uu	Hr	Ww	Rt	Ii	La	Li	Pi	Me	Ma	Bt	Ys	Ad	AF	YF	
NU(AdM)	8	1	2	0	0	0	0	0	0	1	0	0	2	2	0	1	1	0	0	8	0	0	3	2	31
NS(AdM)	3	2	5	0	0	0	0	0	0	0	0	0	1	0	0	0	0	0	0	3	0	0	0	1	15
NR(AdM)	2	0	0	0	0	0	1	0	0	0	0	0	0	1	0	0	0	0	0	0	0	0	0	4	8
DO(AdM)	0	0	1	0	0	0	0	1	0	0	0	0	0	0	0	0	0	0	0	0	0	0	0	0	2
KR(AdM)	0	0	0	0	0	0	0	0	0	0	0	0	0	0	0	0	0	0	0	0	0	0	0	0	0
KM(AdlM)	0	0	0	0	0	0	0	0	0	0	0	0	0	0	0	0	0	0	0	1	0	0	1	0	2
IK(AdlM)	0	0	0	0	0	0	0	0	0	0	0	0	0	0	0	0	0	0	0	0	0	0	0	0	0
KA(AdlM)	0	0	0	0	0	0	0	0	0	0	0	0	0	1	0	0	0	0	0	0	0	0	0	0	1
AK(AdlM)	0	0	0	0	0	0	0	0	0	0	0	0	0	0	0	0	0	0	0	0	0	0	0	0	0
IN(AdlM)	0	0	0	0	0	0	0	0	0	0	0	0	0	0	0	0	0	0	0	0	0	0	0	0	0
MN(AdlM)	0	0	0	0	0	0	0	0	0	0	0	0	0	0	0	0	0	0	0	0	0	0	0	0	0
UI(AdlM)	0	0	0	0	0	0	0	0	0	0	0	0	1	1	0	0	0	0	0	0	0	0	0	5	7
Total	13	3	8	0	0	0	1	1	0	1	0	0	4	5	0	1	1	0	0	12	0	0	3	13	66

(b) *Troop B*

Male	Female															Total
	St	Hn	Kg	Ks	Ab	Kn	Tr	Yt	Po	Na	Ni	Ng	Tk	Gi	Mg	
AM(AdM)	1	6	0	0	4	27	0	0	0	13	6	0	29	0	0	86
RS(AdM)	0	1	0	0	3	15	0	0	0	16	7	0	14	0	0	56
ID(AdM)	0	2	0	0	0	2	0	0	0	0	4	0	0	0	0	8
JN(AdlM)	0	0	0	0	0	0	0	0	0	0	0	0	1	0	0	1
SR(AdlM)	0	0	3	0	0	0	0	0	0	0	0	0	0	0	0	3
HG(AdlM)	0	0	0	0	0	0	0	0	0	0	0	0	0	0	0	0
PT(AdlM)	0	0	0	0	0	0	0	0	0	0	0	0	0	0	0	0
BK(AdlM)	0	0	0	0	0	3	0	0	0	0	0	0	0	0	0	3
Total	1	9	3	0	7	47	0	0	0	29	17	0	44	0	0	157

AF, unidentified adult females; YF, unidentified adolescent females; UI, unidentified individuals. DO, KR and MN, males that temporarily stayed with the troop. BK, a solitary male. Individuals are arranged from top to bottom or from left to right almost according to their dominance rank order. Underlined females had swollen sexual skin during the study period.

cases). Females solicited a single male on one observation day in 62 cases (70%), two males in 15 cases, three males in nine cases, and four males in three cases. In all, a female solicited more than one male in any one observation day in 27 cases (30%).

In troop A and B, 50% and 75% respectively of females in oestrus copulated with more than one male, and all adult males copulated with more than one female (Table 16.3). Fixed mating relationships between a particular male and female were not observed. The rate of copulation for each male, given by the number of copulations divided by the total number of hours of observation for each male, is significantly correlated with dominance rank of males (troop A: $r = -0.78$; troop B: $r = -0.81$, $P < 0.01$, two-tailed). In troop A, all adult troop males (NU, NS, NR), two adolescent males (KM, KA), and an adult males (DO), who was a temporary member of troop A, and all adult males (AM, RS, ID), two adolescent males (JN, SR), and one solitary male (BK) in troop B, were observed to copulate (Table 16.3). The two top-ranking males performed 70% of copulations in troop A and 90% of copulations in troop B.

The rates of copulation for each male dominance rank changed according to the number of females in oestrus (Fig. 16.4). There seemed to be a limit to the number of females in oestrus, above which the ability of the first-ranking male to monopolise females decreased. When the number of females was one or two, the copulation rate for the first-ranking male was much greater than those of other males and was maximal when two females were in oestrus. However, when the number of females was equal to three, the copulation rate for the first-ranking male decreased (Cochran–Cox test, $P < 0.001$, two-tailed), and that for the second-ranking male increased. The difference between the two top-ranking males became non-significant (Student's t-test, $P > 0.05$, two-tailed). However, the copulation rates for the first- and second-ranking males were still 5–49 fold larger than those for the other males. When the number of females was more than three, the rate for the third-ranking male reached a maximum. In this case, the rates for the three top-ranking categories were almost equal (but were still significantly different) (Student's t-test, $P < 0.05$, two-tailed), and the rates were only 4–10 fold larger than those for the fourth-ranking male and those below him.

Duration of mating relationships

As the duration of any set of observations was limited (20–42 min on average), it is not known how long a mating relationship may have continued. However, as the frequency of social interactions between a male

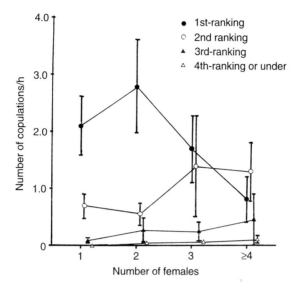

Fig. 16.4. Rates of copulation for each male dominance rank according to the number of females in oestrus. Data from troop A and troop B have been combined. Vertical intervals indicate standard errors.

and a female increased when the female became oestrous, the time during which they were in proximity to each other might also increase. A total of 130 grooming bouts (Dunbar, 1976) were observed in troops A, B and C. Thirty-six bouts (28%) involved heterosexual pairs, of which 75% occurred when the female's sexual skin was tumescent, and 22% occurred with pregnant females.

Interference in mating interactions

In all three troops, the higher-ranking males, especially the alpha male, interfered with the mating interactions of other males. The interference could be directed against either a female or a male. The object of the interference may be subtly influenced by male–male relationships.

Mating interactions are defined as sexual solicitations, irrespective of whether copulations occur. In a total of 95 mating interactions in troop A, one case of interference was observed; the alpha male attacked the beta male, who was copulating with a female in oestrus.

Interference occurred in 12 cases of 319 mating interactions observed in troop B. In 11 cases, the alpha male interfered with a pair composed of the beta male and a female in oestrus. In most cases, the beta male just moved away from the female as soon as the alpha male interfered with the pair. A more severe encounter took place on one occasion when the alpha male found the beta male copulating; the alpha male approached the pair and presented to the female, then he chased her and the beta male joined the attack.

In troop C, in a total of 32 cases of mating interactions, one case of interference was observed. The former alpha male attacked a female in oestrus after she had copulated with the current alpha male.

Although these interferences were effective in separating the mating pair at the time, it is uncertain whether they were also effective in the long term because of the short duration of daily observations. In only two cases was a post-interference interaction among the three participants observed: in one case, the interrupted pair copulated 27 min after the interference had occurred; in another case, the interferer copulated with the target female 22 min after the interference.

Rejection of copulation by females

Females in oestrus were able to reject solicitations by males, either by running away or by sitting down in order to prevent copulation. In troop A, the alpha male was rejected by three females in four out of 35 attempts at copulatory mounting, and in troop B, the alpha male was rejected by four females in eight out of 93 attempts, and the beta male was rejected by two females in two out of 57 attempts at copulatory mounting. In troop C, no rejection was observed in 26 attempts at copulatory mounting. All but one of the pairs in which a rejection had taken place were observed to copulate on other occasions.

Discussion

Sexual behaviour

Fooden (1980) classified *Macaca* into four species groups largely on the basis of the morphology of the sexual organs. Furthermore, he suggested that these morphological characteristics are related to mount–ejaculation patterns: males of some species ejaculate after repeated mounts (multi-mount ejaculators), while others ejaculate after a single mount (single-mount ejaculators). Subsequent studies have supported Fooden's hypothesis to some degree (Shively *et al.*, 1982; Caldecott, 1986a). There have been changes, however, in the classification to two macaque species in terms of their mount–ejaculation patterns. The Barbary macaque, which was previously considered to be a multi-mount ejaculator, was shown to be a single-mount ejaculator (Taub, 1982), and the crab-eating macaque, previously considered to be a multi-mount ejaculator, was found to perform both types of copulation (Shively *et al.*, 1982). This changing perception suggests that there is still much to be learned about the mating patterns of male macaques.

Laboratory studies indicated that pig-tails are multi-mount ejaculators (Tokuda *et al.*, 1968; Nadler & Rosenblum, 1973). Only Bernstein (1967) and Caldecott (1986*b*) have reported on mount–ejaculation patterns observed in the field. Bernstein mentioned that copulation occurred via multi-mount ejaculation but also reported that '113 potentially reproductive mountings were observed, many in a single copulatory sequence. Many mountings, however, were isolated and probably indicative of tension produced by provisioning and observers rather than reproductive behaviour.' Caldecott, after observing only 17 mountings, concluded that copulation was composed of a series of mounts with an inter-mount interval ranging from 1–20 min. However, he stated that he did not observe a complete set of mountings, from the onset to the cessation of copulation, because of poor visibility. Since he could not discern when ejaculation occurred, Caldecott's results are not conclusive.

The present data, obtained in both the provisioned and the unprovisioned situation, suggested that each copulation was composed of a single mount. If the mounts that were observed in the provisioned situation were performed as part of a series, the inter-mount interval was more than 10 min, much longer than the average (3 min) reported by Tokuda *et al.* (1968). Furthermore, the behaviour of females, whereby they separated from males after each mounting with emission of characteristic vocalisations, gave the strong impression that a single-mount ejaculation or, at least, a completed interaction had occurred.

Observations in the laboratory have shown that the sexual behaviour of females is most frequent (Goldfoot, 1971, cited in Caldecott, 1986*b*), and the ejaculation rate highest (Bullock *et al.*, 1972), when swelling is maximal. Although the ejaculation rate could not be determined in the present study, the frequency of sexual behaviours by both males and females reached a maximum that coincided with maximum swelling.

As the mean duration of gestation and of post-partum anoestrus was 170 days, the period of anoestrus after conception continued for 340 days on average. Therefore, females recovered behavioural receptivity in 340 days, and so if the first oestrus was fertile a female could give birth every year, as reported by Hadidian & Bernstein (1979) from their laboratory study. However, since it was possible that more than half of the females only conceived after multiple cycles in the present study (1.7 cycles in the laboratory, as reported by Kuehn, Jensen & Morill, 1965), the birth interval might be more than a year under natural conditions.

Mating seasonality

Parturition was observed throughout the study period. Furthermore, direct observation and the distribution of births showed that oestrus occurred throughout the year, with a peak from November to January when the heaviest rainfall of the year was recorded (Oi, 1990a). The duration of daylight did not fluctuate enough during the year to produce marked seasonality in this study area because it was at a latitude of around 1° S. The possibility that the oestrous condition of females was influenced by rainfall could not be eliminated. Pig-tails in Malaysia also show an annual pattern in the occurrence of oestrus, with peaks just after the relatively dry season when high productivity of fruits can be expected (Caldecott, 1986b). Similarly, a peak in conceptions among lion-tailed macaques (*Macaca silenus*) in India coincides with the onset of the monsoon (Kumar & Kurup, 1985b). In seasonal environments, where levels of available food change dramatically, monkeys may adjust conception or lactation to the times when the richest supply of food is available (van Schaik & van Noordwijk, 1985).

Mate monopolisation by males

As most individuals of both sexes copulated with multiple partners, the mating system of the study troops can be described as polygamous. Caldecott (1986b) and Robertson (1986) suggested that a harem-type breeding unit was possibly the breeding system of pig-tails. In their model, copulation would occur exclusively between a certain male and some defined females. The present data do not support this model.

Pig-tail males compete for mating partners. Sexual access by subordinate males to the females in oestrus was restricted by the presence of dominant males. The dominant males interfered with mating interactions by the subordinate males. As a result, two top-ranking males in each troop performed most of the copulations.

The monopoly of copulation by the dominant males was achieved because there were usually only a few females in oestrus at one time. The smaller the number of females in oestrus, the more conspicuous the mate monopolisation by the dominant males became. By contrast, as the number of females in oestrus increased, the rates of copulation by the subordinate males and non-troop males increased. This finding agrees with the priority of access model, first deduced from the observations on the mating of rhesus macaques (Altmann, 1962; Suarez & Ackerman, 1971; reviewed in Dunbar, 1988).

Female choice counteracted this monopolistic tendency of the dominant males. Females solicited even the lower-ranking males and tended to copulate with many males. Furthermore, on some occasions they rejected attempts by top-ranking males to copulate. In troop B, more sexual solicitations by females were directed at the beta male than at the alpha, implying that females sometimes preferred the subordinate male. It could be said that female choice drove the tendency towards promiscuity.

In the small area of the baiting sites where individuals crowded together, males could easily monitor the condition of females in oestrus, and their positions in relation to other males. Thus, dominant males could easily maintain exclusive access to females in oestrus. In the forest, bushes and undergrowth provide cover where subordinate males can copulate surreptitiously, thereby weakening the dominant male's ability to monopolise females in oestrus. However, since there are only a few females in oestrus with conspicuous sexual signs at any given time in a troop, a few dominant males could probably herd them. Therefore, the tendency towards monopolisation of copulation with the fertile females by a few males is highly probable, even outside the baiting site.

Evolution of mating systems among macaques

Based on the implications of the relationship between mate monopolisation by males and the number of females in oestrus in the pig-tail mating system, I constructed a model of the evolution of mating systems among macaques.

Two different types of promiscuity in macaques – a hypothesis

Macaque species have the widest distribution among non-human primates, extending from non-seasonal to seasonal environments. Species living in a seasonal habitat have a shorter breeding season, during which mature females exhibit oestrous synchrony. Therefore, the number of females in oestrus at any given time during the breeding season is expected to be larger in a seasonal habitat, compared with a reproductive unit of the same size in a non-seasonal habitat (Ridley, 1986).

The findings of a relationship between the number of females in oestrus and the tendency towards mate monopolisation by males for pig-tailed macaques suggests that in the non-seasonal environment, a clear tendency towards mate monopolisation by a few males and smaller numbers of males in a troop could be expected. The mating system should be more promiscuous in species that live in seasonal habitats. Furthermore, such seasonality might also result in a greater proportion of males in a troop.

Ridley (1986) described the significant relationship between the length of the breeding period and the nature of the reproductive units, i.e. multi-male or single-male, in primates. However, he chose only seasonal species of macaques for his analysis.

Mating seasonality and troop adult sex ratios

Figure 16.5 illustrates data for eight species of macaques with regard to breeding seasonality, and intra-troop adult sex ratios. Data for examination of troop sex ratios were from the studies of populations in which multiple troops were contiguously distributed, because in isolated troops the lack of natural migration tends to create unnatural distortions in the sex ratio (Takasaki & Masui, 1984).

I used the monthly distribution of the number of births as an indicator of seasonal mating because it may also be a good indicator of the temporal distribution of fertile oestrous, where males may actually compete with each other (Ridley, 1986). As breeding seasonality is a relative term, species with more than 70% of the number of births concentrated within three months were considered to be seasonal breeders, and all others were considered to be year-round breeders. *M. nemestrina* exhibited 33% concentration (present study); *M. silenus*, 50% concentration (Kumar & Kurup, 1985*b*); North Sumatran *M. fascicularis*, 61% concentration (van Schaik & van Noordwijk, 1985); *M. fuscata*, 93% concentration (Kawai, Asuma & Yoshiba, 1967); North Indian *M. mulatta*, 90% concentration (Lindburg, 1971); *M. radiata*, 93% concentration (Rahaman & Parthasarathy, 1969); and *M. sinica*, 71% concentration (Dittus, 1977). In the case of *M. sylvanus*, Mehlman (1986) stated 'the study population had a discrete birth season each year (between 14 April and 15 June)', which means 100% concentration. Thus, *M. nemestrina*, *M. silenus* and North Sumatran *M. fascicularis* were considered to be year-round breeders, and *M. fuscata*, North Indian *M. mulatta*, *M. sylvanus*, *M. radiata* and *M. sinica* were considered to be seasonal breeders.

The number of females:male in a troop of year-round breeders was higher than that for seasonal breeders (Mann–Whitney U-test: $z = 4.1$,

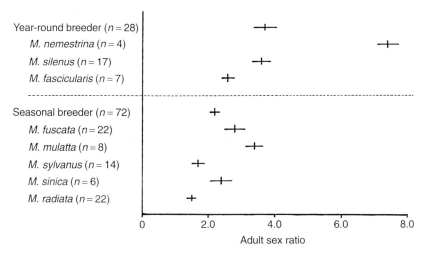

Fig. 16.5. Breeding seasonality and intra-troop adult sex ratios in eight species of macaques. Data for adult sex ratios were from Oi (1990*a*) and Caldecott (1986*b*) for *M. nemestrina*; Sugiyama (1968), Johnson (1985), and Kumar & Kurup (1985*a*) for *M. silenus*; van Schaik *et al.* (1983) and van Schaik & van Noordwijk (1988) for *M. fascicularis*; Yamada (1966), Izawa (1972, 1978), Yotsumoto (1976), Watanabe (1978), Maruhashi (1980, 1982), Furuichi (1983) and Oi (1988) for *M. fuscata*; Lindburg (1971) for *M. mulatta*; Deag & Crook (1971), Whiten & Rumsey (1973), Taub (1980) and Mehlman (1986) for *M. sylvanus*; Simonds (1965), Sugiyama (1971) and Koyama & Shekar (1981) for *M. radiata*; Dittus (1975, 1977, 1979) for *M. sinica*. Horizontal bars indicate standard errors.

$n_1 = 72$, $n_2 = 28$, $P < 0.01$, two-tailed). Even if *M. fascicularis*, for which the value was near the boundary, is included among seasonal breeders the result is the same ($z = 5.1$, $n_1 = 79$, $n_2 = 21$, $P < 0.01$, two-tailed).

Most troops of *M. silenus*, which was a year-round breeder and formed small troops ($\bar{x} = 11$, SD $= 6.6$, $n = 16$), had a single-male–multifemale composition (Johnson, 1985; Kumar & Kurup, 1985*a*). It appears that, in *Macaca*, not only seasonally restricted oestrus in females but also large numbers of females (Altmann, 1990) may provide suitable conditions for the evolution of troops with a multimale configuration, and to increase the number of troop males.

Mating seasonality and promiscuity

Wild troops of seasonal breeders tended to demonstrate no correlation between social rank and mating activity of males, whereas confined troops of seasonal breeders, and both wild and confined troops of year-round breeders did demonstrate a correlation between the two parameters (Table

T. Oi

Table 16.4. *Breeding seasonality and promiscuity in macaque societies*

Species	Breeding seasonality	Correlation between social rank and mating activity of males	
		Wild troops	Confined troops
M. nemestrina	Year-round	Yes[a]	Yes[b]
M. silenus (North Sumatran)	Year-round	Many one-male troops[c]	?
M. fascicularis	Year-round	Yes[d]	Yes[e]
M. fuscata (North Indian)	Restricted	No[f]	Yes[g]
M. mulatta	Restricted	No[h]*	Yes[i]
M. sinica	Restricted	Males are competitive when acquiring mating partners[j]	?
M. radiata	Restricted	No[k]	No[e]
M. sylvanus	Restricted	No[l]	Yes[m]

[a]This study; [b]Tokuda *et al.* (1968); [c]Johnson (1985) and Kumar & Kurup (1985a); [d]van Noordwijk (1985); [e]Shively *et al.* (1982); [f]Enomoto (1978) and Takahata (1982); [g]Hanby, Robertson & Phoenix (1971) and Modahl & Eaton (1977); [h]Lindburg (1983); [i]Smith (1981); [j]Dittus (1975, 1979); [k]Simonds (1965); [l]Taub (1980); [m]Witt, Schmidt & Schmitt (1981).
*Kendall rank correlation coefficient was calculated with significance level at 0.05 (two-tailed test), from data in Table 3, p. 52, in Lindburg (1983).

16.4). Wild troops of year-round breeders, such as *M. nemestrina* and North Sumatran *M. fascicularis*, show a tendency towards mate monopolisation by a few top-ranking males. The single-male composition of *M. silenus* is indicative of mate monopolisation by a single male. The tendency towards mate monopolisation by pig-tail males (Tokuda *et al.*, 1968) and long-tail males (Shively *et al.*, 1982) was also reported from a laboratory study. By contrast, in some wild troops of seasonal breeders such as *M. fuscata* (Enomoto, 1978; Takahata, 1982), North Indian *M. mulatta* (Lindburg, 1983), *M. radiata* (Simonds, 1965), and *M. sylvanus* (Taub, 1980), there was no correlation between the dominance rank of males and frequency of copulation. However, even such seasonally breeding species show a tendency towards male monopolisation by dominant males, and correlations between dominance rank and frequency of copulation were reported for confined troops (Hanby *et al.*, 1971; Modahl & Eaton, 1977, for *M. fuscata*; Smith, 1981, for *M. mulatta*; Witt *et al.*, 1981, and Fa, 1986, for *M. sylvanus*). In confined troops, dominant males can easily perceive the behaviour of females, and can easily control the behaviour of subordinate males.

Exclusive and non-exclusive mating strategy

Taub (1980) reported that, in a wild troop of *M. sylvanus*, the frequency of copulation was not correlated with male dominance rank. Both males and females changed their mating partners frequently, and consortships were short (17 min on average). This mating system was sustained by the mate choice of females, and the apparent unwillingness of males to interfere with the mating of other males (Taub, 1980; Small, 1990). *Macaca radiata* might have the same pattern, in that males appeared to be inhibited from interfering with the mating of any other male (Rahaman & Parthasarathy, 1969), and the dominance rank of males was not correlated with the number of copulations (Shively *et al.*, 1982). Furthermore, *M. radiata* and *M. sylvanus* had other similar behavioural characteristics, for example tolerance between males was high, males copulate with single-mount ejaculation, length of consortships was short, etc. (Shively *et al.*, 1982; Caldecott, 1986a). These similarities imply that they have the same mating system. The most striking similarity is that both of them are seasonal breeders.

Macaca fuscata, *M. mulatta*, *M. fascicularis*, and *M. nemestrina* males are explicitly competitive, in that dominant males interfere with the mating of subordinate males in any situation. Their mating relationships should be called a superficial promiscuous state, because of the inability of dominant males to exert mate monopolisation at times when too many females are in oestrous. In that superficial promiscuous state, if dominant males continue a monopolistic strategy with interference of the matings of other males, the rates of copulation by the dominant males would decrease, as suggested above with the example from pig-tails. Takahata (1982) and Huffman (1987) suggested that interference by the dominant males was not effective in preventing female mate choice in *M. fuscata*. The decrease in the rate of copulation might be ascribed to excessive time spent guarding many females or interfering with the mating of others. In such cases, it would be adaptive to stop such ineffective guarding and interference with copulation. *Macaca sylvanus* males practise a non-exclusive mating strategy of this sort, by abandoning any interference when ineffective, but they can also employ an exclusive mating strategy in situations where mate monopoly is possible, as reported in confined *M. sylvanus* (Witt *et al.*, 1981). Males of *M. fuscata* do not practise such a non-exclusive mating strategy; this may be due to phylogenetic inertia, as argued below.

Phylogeny

Fooden (1980), Shively *et al.*, (1982) and Caldecott (1986*a*) recognised several behavioural differences between phylogenetic species groups of *Macaca* (Fooden, 1980; Cronin, Cann & Sarich, 1980; Delson, 1980; Melnick & Kidd, 1985). It might be expected that each phylogenetic species group would demonstrate one or both of the two strategies, exclusive and non-exclusive mating. Only the *sylvanus* group (*M. sylvanus*) and the *sinica* group (*M. sinica, M. radiata, M. assamensis, M. thibetana*) include species in which males have a non-exclusive mating strategy. The *silenus* group (*M. silenus, M. nemestrina, M. pagensis*, and Sulawesi macaques), and the *fascicularis* group (*M. fascicularis, M. mulatta, M. cyclopis, M. fuscata*) include the species with an exclusive strategy. In the *sinica* group, only *M. radiata* is reported to have a non-exclusive mating strategy, whereas *M. sinica* (Dittus, 1975, 1979) and *M. thibetana* (C. Xiong, personal communication) are explicitly competitive over mates. This inconsistency in mating strategy, in terms of mate interference in the *sinica* group, might be ascribed to other factors which affect mate monopoly by males, for example the intra-troop spatial dispersion of females. *Macaca thibetana* females demonstrate cohesive spacing (C. Xiong, personal communication).

A scenario of behavioural evolution

A scenario whereby each species group came to have one or other mating strategy can be deduced from the process of geographical distribution of each species group as proposed by some researchers (Fooden, 1980; Delson, 1980; Eudey, 1980) (Fig. 16.6). If the multimale troop of the macaques is derived from a single-male troop by the addition of extra males (Eisenburgh, Muckenhirn & Rudran, 1972), the proto-macaques that formed multimale troops should appear in a seasonal environment, where females come into oestrus simultaneously in a single reproductive unit; in such a unit group, several males could coexist. The dry areas, where rainy and dry seasons differ dramatically, and the high latitude areas represent clearly seasonal environments. Such areas are found around the Mediterranean Sea where the oldest fossil macaque was excavated. The mating relationships of this proto-macaque should involve superficial promiscuity, with males being exclusively competitive, assuming that this behavioural tendency was an inheritance from the ancestors of the proto-macaques that dwelt in a tropical, non-seasonal environment and formed single-male troops.

Species groups descended from the proto-macaque that intruded into

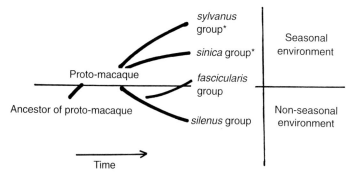

Fig. 16.6. Possible process of distribution into seasonal and non-seasonal environments of each species group of the genus *Macaca*. Males of the species groups with asterisks have a non-exclusive mating strategy, while males of the other species groups do not. The present distribution of the *fascicularis* group extends into both environments.

tropical Asia may have retained the tendency for males to be exclusively competitive for mates in the non-seasonal environments as in the proto-*silenus* and the extant *silenus* group. The proto-*fascicularis* group speciated in tropical Asia and spread recently into higher latitude zones in a relatively short time (Fooden, 1980; Delson, 1980; Eudey, 1980). The males of this species group existing today may have retained the tendency to be exclusively competitive for mates even at higher latitudes in seasonal environments, because their intrusion into the new habitat occurred rapidly and relatively recently (phylogenetic inertia).

Meanwhile, *M. sylvanus* (and possibly *M. radiata*) has been living in a seasonal environment and has developed a promiscuous mating system in which males have a non-exclusive mating strategy. The oldest fossil macaques are regarded as sub-species of *M. sylvanus* and have been excavated from Pliocene strata around the Mediterranean Sea (Delson, 1980). The climate of Western Europe has been a warm temperate or cool temperate climate at least since the Pliocene era (Dorf, 1964). The faunal assemblage excavated in Europe, including fossil macaques, suggests that the 'macaque originally may have adapted to seasonal or deciduous or even tree savanna' (Eudey, 1980). In the process of adaption to such seasonal environments, the males of this species may have developed a non-exclusive mating strategy whereby they gain access to mates without wasteful sexual interference from other males.

Acknowledgements

I thank the Indonesian Institute of Science (LIPI), the Directorate of Forest Protection and Nature Conservation (PHPA), Universitas Andalas in Padang, and Universitas Nasional in Jakarta. I also thank Dr S. Kawamura, Dr O. Takenaka of Kyota University and Dr Amsir Bakar for their support for the field study. I express my great appreciation to Drs T. Kano, Y. Takahata, K. Watanabe, Y. Sugiyama, and T. Nishida of Kyota University who gave useful advice in revising the manuscript. The study was financed by a Kajima Foundation Research Grant, and by the Grants-in-Aid for Scientific Research from the Japanese Ministry of Science, Education and Culture (60041043, 61041044, 63790234).

References

Altmann, J. (1990). Primate males go where the females are. *Animal Behaviour*, **39**, 193–5.

Altmann, S. A. (1962). A field study of the sociobiology of rhesus monkeys, *Macaca mulatta*. *Annals of the New York Academy of Sciences*, **102**, 338–435.

Berenstain, L. & Wade, T. D. (1983). Intrasexual selection and male mating strategies in baboons and macaques. *International Journal of Primatology*, **4**, 201–35.

Bernstein, I. S. (1967). A field study of the pigtail monkey (*Macaca nemestrina*). *Primates*, **8**, 217–28.

Bullock, D. W., Paris, C. A. & Goy, R. W. (1972). Sexual behaviour, swelling of the sex skin and plasma progesterone in the pigtail macaque. *Journal of Reproduction and Fertility*, **31**, 225–36.

Caldecott, J. O. (1986a). Mating patterns, societies and the ecogeography of macaques. *Animal Behaviour*, **34**, 208–20.

Caldecott, J. O. (1986b). *An ecological and behavioural study of the pig-tailed macaque*. Contributions to Primatology, vol. 21. Basel: S. Karger.

Cronin, J. E., Cann, R. & Sarich, V. M. (1980). Molecular evolution and systematics of the genus *Macaca*. In *The macaques: studies in ecology, behavior and evolution*, ed. D. G. Lindburg, pp. 31–51. New York: Van Nostrand Reinhold.

Deag, J. M. & Crook, J. H. (1971). Social behaviour and 'agonistic buffering' in the wild Barbary macaque, *Macaca sylvana* L. *Folia Primatologica*, **15**, 183–200.

Delson, E. (1980). Fossil macaques, phyletic relationships and a scenario of deployment. In *The macaques: studies in ecology, behavior and evolution*, ed. D. G. Lindburg, pp. 10–30. New York: Van Nostrand Reinhold.

Dittus, W. P. J. (1975). Population dynamics of the toque monkey, *Macaca sinica*. in *Socioecology and psychology of primates*, ed. R. H. Tuttle, pp. 125–51. The Hague: Mouton.

Dittus, W. P. J. (1977). The social regulation of population density and age–sex distribution in the toque monkey. *Behaviour*, **63**, 281–322.

Dittus, W. P. J. (1979). The evolution of behaviors regulating density and age-specific sex ratios in a primate population. *Behavior*, **69**, 265–302.

Dixson, A. F. (1983). Observations on the evolution and behavioral significance of 'sexual skin' in female primates. In *Advances in the study of behavior*, vol. 13, ed. J. S. Rosenblatt, R. A. Hinde, C. Beer & M. C. Busnel, pp. 63–106. New York: Academic Press.

Dorf, E. (1964). The use of fossil plants in palaeoclimatic interpretations. In *Problems in palaeoclimatology*, ed. A. E. Nairin, pp. 13–31. London: Wiley-Interscience.

Dunbar, R. I. M. (1976). Some aspects of research design and their implications in the observational study of behaviour. *Behaviour*, **58**, 78–98.

Dunbar, R. I. M. (1988). *Primate social systems*. London & Sydney: Croom Helm.

Eisenberg, J. F., Muckenhirn, N. A. & Rudran, R. (1972). The relation between ecology and social structure in primates. *Science*, **176**, 863–74.

Emlen, S. T. & Oring, L. W. (1977). Ecology, sexual selection, and the evolution of mating systems. *Science*, **197**, 215–23.

Enomoto, T. (1978). On social preference in sexual behavior of Japanese monkeys (*Macaca fuscata*). *Journal of Human Evolution*, **7**, 283–93.

Eudey, A. A. (1980). Pleistocene glacial phenomena and the evolution of Asian macaques. In *The macaques: studies in ecology, behavior and evolution*, ed. D. G. Lindburg, pp. 52–83. New York: Van Nostrand Reinhold.

Fa, J. E. (1986). *Use of time and resources in provisioned troops of monkeys: social behaviour, time and energy in the Barbary macaques* (Macaca sylvanus L.) *at Gibraltar*. Contributions to Primatology, vol. 23. Basel: S. Karger.

Fooden, J. (1975). Taxonomy and evolution of liontail and pigtail macaques (Primates: Cercopithecidae). *Fieldiana Zoology*, **67**, 1–169.

Fooden, J. (1980). Classification and distribution of living macaques (*Macaca* Lacepede, 1799). In *The macaques: studies in ecology, behavior and evolution*, ed. D. G. Lindburg, pp. 1–9. New York: Van Nostrand Reinhold.

Furuichi, T. (1983). [Dominant–subordinate relationships in the social life of Japanese macaques.] (In Japanese.) *Iden*, **371**, 3–9.

Hadidian, J. & Bernstein, I. S. (1979). Female reproductive cycles and birth data from an Old World monkey colony. *Primates*, **20**, 429–42.

Hanby, J. P., Robertson, L. T. & Phoenix, C. H. (1971). The sexual behaviour of a confined troop of Japanese macaques. *Folia Primatologica*, **16**, 123–43.

Huffman, M. A. (1987). Consort intrusion and female mate choice in Japanese macaques (*Macaca fuscata*). *Ethology*, **75**, 221–34.

Izawa, K. (1972). Japanese monkeys living in the Okoppe Basin of the Shimokita Peninsula: the second report of the winter follow-up survey after the aerial spraying of herbicide. *Primates*, **13**, 201–12.

Izawa, K. (1978). [Ecological study of Japanese snow monkey in Hakusan National Park – troop movement and intertroop relationships in snowy season II.] (In Japanese.) *Annual Report of Hakusan Nature Conservation Center*, **4**, 93–109.

Johnson, T. J. M. (1985). Lion-tailed macaque behavior in the wild. In *The lion-tailed macaque: status and conservation*, Monographs in Primatology, vol. 7, ed. P. G. Heltne, pp. 41–63. New York: Alan R. Liss.

Kawai, M., Azuma, S. & Yoshiba, K. (1967). Ecological studies of reproduction in Japanese monkeys (*Macaca fuscata*). I. Problems of the birth season. *Primates*, **8**, 35–74.

Koyama, N. & Shekar, P. B. (1981). Geographic distribution of the rhesus and the bonnet monkeys in west central India. *Journal of the Bombay Natural History Society*, **78**, 240–55.

Kuehn, R. E., Jensen, G. D. & Morill, R. K. (1965). Breeding *Macaca nemestrina*: a program of birth engineering. *Folia Primatologica*, **3**, 251–62.

Kumar, A. & Kurup, G. U. (1985a). Inter-troop interactions in the lion-tailed macaques, *Macaca silenus*. In *The lion-tailed macaques: status and conservation*, Monograph in Primatology, vol. 7, ed. P.G. Heltne, pp. 91–107. New York: Alan R. Liss.

Kumar, A. & Kurup, G.U. (1985*b*). Sexual behavior of the lion-tailed macaque, *Macaca silenus*. In *The lion-tailed macaques: status and conservation*, Monographs in Primatology, vol. 7, ed. P.G. Heltne, pp. 109–30. New York: Alan R. Liss.

Lindburg, D.G. (1971). The rhesus monkeys in North India: an ecological and behavioral study. In *Primate behavior; developments in field and laboratory research*, vol. 2, ed. L.A. Rosenblum, pp. 1–106. New York: Academic Press.

Lindburg, D.G. (1983). Mating behavior and estrus in the Indian rhesus monkey. In *Perspectives in primate biology*, ed. K. Seth, pp. 45–61. New Delhi: Today & Tomorrow's Printers and Publishers.

Lindburg, D.G. (1987). Seasonality of reproduction in primates. In *Comparative primate biology*, vol. 2, part B, *Behavior, cognition, and motivation*, ed. G. Mitchell & J. Erwin, pp. 167–218. New York: Alan R. Liss.

Maruhashi, T. (1980). Feeding behaviour and diet of the Japanese monkey (*Macaca fuscata yakui*) in Yakushima Island, Japan. *Primates*, **21**, 141–60.

Maruhashi, T. (1982). An ecological study of troop fission of Japanese monkey (*Macaca fuscata yakui*) in Yakushima Island, Japan. *Primates*, **23**, 317–37.

Mehlman, P. (1986). Male intergroup mobility in a wild population of the Barbary macaque (*Macaca sylvanus*), Ghomaran Rif Mountains, Morocco. *American Journal of Primatology*, **10**, 67–81.

Melnick, D.J. & Kidd, K.K. (1985). Genetic and evolutionary relationships among Asian macaques. *International Journal of Primatology*, **6**, 123–60.

Modahl, K.B. & Eaton, G.G. (1977). Display behaviour in a confined group of Japanese macaques (*Macaca fuscata*). *Animal Behaviour*, **25**, 525–35.

Nadler, R.D. & Rosenblum, L.A. (1973). Sexual behavior during successive ejaculations in bonnet and pigtail macaques. *American Journal of Physical Anthropology*, **38**, 217–20.

Newton, P.N. (1988). The variable social organization of hanuman langur (*Presbytis entellus*), infanticide, and the monopolization of females. *International Journal of Primatology*, **9**, 59–77.

Oi, T. (1988). Sociological study on the troop fission of wild Japanese monkeys (*Macaca fuscata yakui*) on Yakushima Island. *Primates*, **29**, 1–19.

Oi, T. (1990*a*) Population organization of wild pig-tailed macaques (*Macaca nemestrina nemestrina*) in West Sumatra. *Primates*, **31**, 15–31.

Oi, T. (1990*b*). Patterns of dominance and affiliation in wild pig-tailed macaques (*Macaca nemestrina nemestrina*) in West Sumatra. *International Journal of Primatology*, **11**, 339–56.

Rahaman, H. & Parthasarathy, M.D. (1969). Studies on the social behaviour of bonnet monkeys. *Primates*, **10**, 149–62.

Ridley, M. (1986). The number of males in a primate troop. *Animal Behaviour*, **34**, 1848–58.

Robertson, J.M.Y. (1986). On the evolution of pig-tailed macaque societies. PhD thesis, University of Cambridge.

Shively, C., Clarke, S., King, N., Schapiro, S. & Mitchell, G. (1982). Patterns of sexual behavior in male macaques. *American Journal of Primatology*, **2**, 373–84.

Simonds, P.E. (1965). The bonnet macaque in south India. In *Primate behavior: field studies of monkeys and apes*, ed. I. De Vore, pp. 176–96. New York:

Holt, Rinehart & Winston.

Small, M. F. (1990). Promiscuity in Barbary macaques (*Macaca sylvanus*). *American Journal of Primatology*, **20**, 267–82.

Smith, D. G. (1981). The association between rank and reproductive success of male rhesus monkeys. *American Journal of Primatology*, **1**, 83–90.

Suarez, B. & Ackerman, D. R. (1971). Social dominance and reproductive behavior in male rhesus monkeys. *American Journal of Physical Anthropology*, **35**, 219–22.

Sugiyama, Y. (1968). The ecology of the lion-tailed macaque [*Macaca silenus* (Linnaeus)] – a pilot study. *Journal of the Bombay Natural History Society*, **65**, 283–92.

Sugiyama, Y. (1971). Characteristics of the social life of bonnet macaques (*Macaca radiata*). *Primates*, **12**, 247–66.

Takahata, Y. (1982). The socio-sexual behavior of Japanese monkeys. *Zeischrift für Tierpsychologie*, **59**, 89–108.

Takasaki, H. & Masui, K. (1984). Troop composition data of wild Japanese macaques reviewed by multivariate methods. *Primates*, **25**, 308–18.

Taub, D. M. (1980). Female choice and mating strategies among wild Barbary macaques (*Macaca sylvanus* L.). In *The macaques: studies in ecology, behavior and evolution*, ed. D. G. Lindburg, pp. 287–344. New York: Van Nostrand Reinhold.

Taub, D. M. (1982). Sexual behavior of wild Barbary macaque males (*Macaca sylvanus*). *American Journal of Primatology*, **2**, 109–13.

Thierry, B. (1985). Patterns of agonistic interactions in three species of macaques (*Macaca mulatta, M. fascicularis, M. tonkeana*). *Aggressive Behaviour*, **11**, 223–33.

Tokuda, K., Simons, R. C. & Jensen, G. D. (1968). Sexual behavior in a captive group of pigtailed monkeys (*Macaca nemestrina*). *Primates*, **9**, 283–94.

van Noordwijk, M. A. (1985). Sexual behaviour of Sumatran long-tailed macaques (*Macaca fascicularis*). *Zeischrift für Tierpsychologie*, **70**, 277–96.

van Schaik, C. P. & van Noordwijk, M. A. (1985). Interannual variability in fruit abundance and the reproductive seasonality in Sumatran long-tailed macaques (*Macaca fascicularis*). *Journal of Zoology, London*, **206**, 533–49.

van Schaik, C. P. & van Noordwijk, M. A. (1988). Scramble and contest in feeding competition among female long-tailed macaques (*Macaca fascicularis*). *Behaviour*, **105**, 77–98.

van Schaik, C. P., van Noordwijk, M. A., de Boer, R. J. & den Tonkelaar, I. (1983). The effect of group size on time budgets and social behaviour in wild long-tailed macaques (*Macaca fascicularis*). *Behavioral Ecology and Sociobiology*, **13**, 173–81.

Vandenbergh, J. G. & Vessey, S. (1968). Seasonal breeding of free-ranging rhesus monkeys and related ecological factors. *Journal of Reproduction and Fertility*, **14**, 71–9.

Watanabe, K. (1978). Some social alterations in the early periods following the commencement of provisioning in Japanese monkeys (*Macaca fuscata*). *Japanese Journal of Ecology*, **28**, 35–41.

Whiten, A. & Rumsey, T. J. (1973). 'Agonistic buffering' in the wild Barbary macaque, *Macaca sylvana* L. *Primates*, **14**, 421–5.

Witt, R., Schmidt, C. & Schmitt, J. (1981). Social rank and Darwinian fitness in a multimale group of Barbary macaques (*Macaca sylvanus* Linnaeus, 1758). *Folia Primatologica*, **36**, 201–11.

Wrangham, R.W. (1979). On the evolution of ape social systems. *Social Science Information*, **18**, 335–68.

Wrangham, R.W. (1980). An ecological model of female-bonded primate groups. *Behaviour*, **75**, 262–300.

Yamada, M. (1966). Five natural troops of Japanese monkeys in Shodoshima Island. I. – Distribution and social organization. *Primates*, **7**, 315–62.

Yamagiwa, J. (1986). [Social structure of Yaku monkeys and male mating strategy.] (In Japanese.) In *Japanese monkeys on Yaku-Island*, ed. T. Maruhashi, J. Yamagiwa & T. Furuichi, pp. 60–125. Tokyo: Tokai University Press.

Yotsumoto, N. (1976). The daily activity in a troop of wild Japanese monkey. *Primates*, **17**, 183–204.

17

Determinants of reproductive seasonality in the Arashiyama West Japanese macaques

L. M. FEDIGAN AND L. GRIFFIN

Introduction

Reproductive seasonality has been well documented in Japanese macaques (*Macaca fuscata*) (Tokuda, 1961–2; Kawai, Azuma & Yoshiba, 1967; Nigi, 1976; Eaton *et al.*, 1987; Gouzoules, Gouzoules & Fedigan, 1981). A number of possible influences on the timing of annual cycles have been identified, for example photoperiod, temperature, rainfall patterns, endogenous circannual rhythms, social synchrony and social drift. General reviews of primate seasonal reproduction (Lancaster & Lee, 1965; Van Horn, 1980; Lindburg, 1987) all refer in some detail to the data on Japanese macaques, which show that the exact timing of birth seasons, and by inference mating seasons, in different populations varies along a latitudinal gradient. The further south a monkey group is located in Japan, the later in the calender year its mean dates of conception and parturition. Each group has a characteristic timing that is relatively consistent over the sampled years (Kawai *et al.*, 1967). One inference that may be drawn from this relationship between reproductive timing and latitude is that various seasonal, environmental factors that also vary along a latitudinal cline (e.g. photoperiod and, to a lesser extent, daily minimum/maximum temperatures and rainfall patterns) may act as necessary proximate cues to modulate endocrine and reproductive functions. A second inference, however, which may be drawn from these data showing that regional populations in Japan characteristically conceive and deliver young at different mean dates, is that seasonal environmental factors have exerted their effects over evolutionary time by altering gene frequencies in the separated populations.

This chapter compares the seasonal timing of births in the Arashiyama population in Japan (at latitude 35° 00′ N) during a 15-year period to that of the descendant Arashiyama West population in Texas (at latitude 28° 05′ N) over a subsequent 15-year period. The fundamental question addressed is whether the birth season (and inferentially, the breeding season) has shifted in time in the Texas habitat. The translocation of the Arashiyama West group from Japan to Texas and the ongoing data collection at both sites offer a unique opportunity to examine the effects of extrinsic and intrinsic factors on seasonal reproduction. No other primate population has been studied over many annual breeding cycles in its native habitat and then moved as an entire group to a new location that differs extensively in the environmental cues for reproductive seasonality.

A distinction between the proximate and ultimate effects of environmental factors is made in most descriptions of seasonality in animals (e.g. Baker, 1938; Sadleir, 1969; Jones, 1981; Bronson, 1989). Ultimate effects refer to environmental factors (such as food availability) that exert selective pressure to restrict reproductive activities to a particular time of the year, whereas proximate effects refer to the 'fine tuning' of reproductive seasonality within a narrower time frame, primarily through the immediate impact of extrinsic factors on physiological processes. In species, such as monkeys, that exhibit long periods of gestation and lactation, the proximate cues will be separate in time, and different in nature, from the 'optimal season factors' that select for annual cyclicity.

In addition, Gwinner (1981, 1986) has noted that some species may evolve emancipation, or 'uncoupling' from all extrinsic effects by incorporating seasonal timing into an endogenous circannual rhythm. The presence of an internal circannual 'clock' has been demonstrated through experimental research on many vertebrate species (see references in Gwinner, 1986). Furthermore, some captive studies of rhesus macaques (e.g. Michael & Bonsall, 1977; Michael & Zumpe 1978; Wickings & Nieschlag, 1980) have found that male monkeys maintain seasonality of reproductive functions over several years even under constant and controlled laboratory conditions. Finally, Nozaki, Oshima & Mori (1990) concluded that individually caged Japanese monkeys (both intact and chronically estradiol-treated ovariectomized females) exhibited striking circannual variations in endocrine levels, appropriate to the breeding and non-breeding seasons.

Of the several environmental factors that have been suggested to modulate annual patterns, photoperiod has received the most empirical support (e.g. Sadleir, 1969; Immelman, 1973; Herbert, 1977; Hoffman, 1981; Gwinner, 1981, 1986). Seasonal change in daylength is the meteorological variable that is most correlated with latitude and is the most consistent, and thus reliable, from year to year. Observational research on mammals that occur over wide geographical areas, such as hamsters, deer and lagomorphs, has often found that the timing of breeding seasons follows a latitudinal gradient, and thus a photoperiod cline. In addition, a large body of experimental research has demonstrated that alterations of light/dark cycles in the laboratory can affect the endocrine status of a variety of vertebrates (Sadleir, 1969; Hoffman, 1981; Bronson, 1989). However, the relationship between light and reproductive cycles is a complex one and varies from species to species. Several hypotheses, such as the 'hourglass' system and the 'Bunning model', have been proposed to explain how photoperiod might act as a proximate cue for the onset of mating. It is sometimes argued that monkeys inhabiting the tropics would experience very little seasonal change in daylength. However, many organisms exhibit a very precise 'critical photoperiod' and one interpretation is that they are sensitive, and can respond, to very small changes in daylength. Furthermore, the 'Bunning' model, proposes that neither changes in daylength nor the changing ratio of light to dark are the cueing factors: rather the critical photoperiod occurs when light coincides with a particular phase of the intrinsic rhythm of sensitivity in the organism.

In terms of primates, support for the hypothesis that photoperiod regulates annual breeding seasons comes from the six-month reversal found in rhesus macaques living in zoos in the southern hemisphere (e.g. Bielert & Vandenbergh, 1981). Also, Van Horn (1980) reanalyzed data from Kawai *et al.* (1967) on Japanese macaques, as well as data from a variety of sources on lemurs and rhesus monkeys, to argue that photoperiod cycles are the primary regulatory factor ('zeitgeber') in the reproductive seasonality of anthropoids. He further concluded that Japanese macaques respond to a critical daylength of slightly less than 12 hours.

In contrast, Rawlins & Kessler (1985) argued that changes in photoperiod at the latitude of Cayo Santiago (18° N) are not sufficient to bring about seasonality in the rhesus monkeys found there, and reiterated that two groups, located at the same latitude but on different islands, exhibit different timing. These two researchers analyzed eight years of rainfall and temperature data and concluded that Koford's (1965) original hypothesis was correct. Koford postulated (and see also Vandenbergh & Vessey, 1968) that the onset of spring rains triggers the onset of the mating season in the Cayo Santiago rhesus monkeys. Building on this postulate, Rawlins & Kessler (1985) suggested that photoperiod sets the temporal limits of annual seasonal reproduction, while the onset of annual spring rains regulates reproductive activity *within* that range.

Kawai *et al.* (1967) also considered a number of environmental variables correlated with latitude in their survey of Japanese monkey groups, and concluded that the relationship between extrinsic factors and reproductive timing was not simple. However, they did note that conceptions did not begin until the daily maximum temperatures dropped below 30 °C in all of the 25 groups studied across Japan. They also grouped 25 populations into three patterns of reproductive timing: northerly, middle range and southerly group patterns.

We suggest that if environmental factors act primarily as proximate cues for the onset of spermatogenesis, ovulatory cycles and conception, then the breeding and birth seasons should have shifted in Texas as compared to the Japanese pattern in accordance with the altered environmental cues (i.e. to a more southerly pattern). Conversely, if the environmental factors that vary according to latitude in Japan have acted primarily to select for stable, endogenous circannual rhythms in geographically and genetically separated populations of monkeys, then the timing of reproductive seasonality in the Arashiyama West group should have remained consistent with the pattern of origin, even after the monkeys were moved to a new habitat.

Gouzoules *et al.* (1981) previously compared the timing of the birth seasons in Arashiyama, Japan to birth dates from six of the first years in Texas, and found that the onset and termination of the seasons remained unchanged in the new habitat, although the distribution of births was altered. Their finding of a characteristic and stable group-timing for reproductive seasonality, which was maintained in the population even as the habitat changed, led the authors to suggest that the regional differences seen among Japanese monkey groups in terms of the birth season are due to a phenomenon akin to 'social drift', enhanced by the genetic separation known to exist between groups. Some researchers have argued that when seasonal animals maintain their circannual rhythms under new conditions it may be due simply to 'refractoriness' (e.g. Sadleir, 1969; Hoffman, 1981; Jones, 1981), and that experimental studies which transport seasonal organisms to new conditions must be conducted over many breeding cycles. The present paper extends Gouzoules *et al.* (1981) by comparing 15 years of data from Texas with 15 years from Japan, and by examining daily rainfall, temperature and photoperiod patterns in relation to various aspects of reproductive timing in Texas.

Methods

The study subjects were the breeding age females in the Arashiyama and descendant Arashiyama West (AW) populations of Japanese macaques that have been studied by Japanese and Western primatologists from 1954 to the present (for details, see Fedigan & Asquith, 1991). This population, which numbered 34 on first contact, fissioned into two daughter groups in 1966; these were named Arashiyama A and B. Six years later, in 1972, one of these two daughter groups (Arashiyama A) was translocated almost in its entirety (150 out of 158 group members) to a large ranch in southern Texas and renamed the Arashiyama West group. Most of the breeding females in the early years in Texas had previously given birth in Japan, and thus constitute a cohort that was sampled, and then translocated and sampled again. However, over the years, younger females who were born in Texas reached reproductive maturity and became part of the sample.

For the past 38 years, with the exception of 1977, the exact birth dates and identity of the mothers of virtually all infants born into the Arashiyama groups have been recorded. This paper analyzes 30 of these birth seasons, those from 1957–71 in Japan, and 1974–89 in Texas. The first three years of data from Japan were deleted due to small sample sizes and missing data points. In the Texas data set, the 1972 birth season was deleted because the monkeys arrived from Japan already pregnant, 1973 was deleted due to an unusually small sample size of births ($n = 7$), and 1977 was not included because exact birth dates were not available. Birth dates for 416 infants born in Japan and 790 infants born in Texas were used in the analyses.

The timing of a birth season has at least six components: the dates of onset, termination, mean birth and median birth, the length of the season and the frequency distribution of birth dates. Since the median birth date (and inferred median conception date) is the least affected by outlying observations, we focused primarily on it. Earlier studies of seasonally breeding macaques, such as that of Drickamer (1974), have shown that any female's exact date of parturition in a given year may be affected by her recent reproductive history (e.g. when she last gave birth, and whether the infant survived). On the assumption that the median conception/birth date is the measure of the group's reproductive timing least affected by such individual life history variables, we use it to best-represent the 'core timing' of the season. (See also Lindburg's (1987) comments on the preferability of the median over the mean or onset dates.) Conception dates were calculated by counting back 173 days from birth dates (Nigi, 1976), and all dates are presented as consecutive days in the Julian calendar (1–365). Descriptive data on all six components are presented; standard parametric and non-parametric tests (analyses of variance, *t*-tests, Kolmogorov–Smirnov tests, regression analyses), from the SPSS PC package, were used to compare the results from Japan and Texas. Daily maximum and minimum temperatures and daily rainfall patterns between 1974 and 1989 in Laredo and San Antonio, Texas, were obtained from the National Climatic Data Center. Daylength was calculated by Dr D. Hube of the University of Alberta, by interpolating from data provided in the *Observers' handbook*, Royal Astronomical Society of Canada, and checked against sunrise/sunset charts provided by the US Naval Observatory.

Results

Components of reproductive timing in the Arashiyama monkeys in Japan and Texas

To control for the possibility that the timing of the birth season was more variable in one location than another, F_{max} tests were run. When the variances were homogeneous ($p > 0.05$) a pooled variance t-test was carried out; when variances were heterogeneous ($p < 0.05$) a separate variance t-test was used. Only one of the variables, mean birth date, proved to be heteroscedastic.

Table 17.1 displays descriptive data for five components of the birth season: dates of mean and median births, and onset, termination and length of the season. It can be seen that the mean birth date shifted slightly forward in Texas. Thus, day 148 (28 May) was the mean birth date for the 15 years in Japan as compared to day 143 (23 May) in Texas. However, the difference was not significant ($t = 1.74$; $df = 22.27$; $p = 0.095$).

The median birth date was nearly the same in both locations: day 143 in Japan, day 142 in Texas ($t = 0.53$; $df = 28$; $p = 0.602$). Table 17.1 also shows that the mean date for the *onset* of the birth season in Japan (day 113, 20 April) shifted forward by 20 days to the mean onset date in Texas (day 91, 31 March), which was a significant difference ($t = 4.23$; $df = 28$; $p = 0.000$), but the termination dates were similar (day 206, 25 July, in Japan, and day 207, 26 July in Texas; $t = -0.24$; $df = 28$; $p = 0.810$).

The mean length of the birth season was significantly longer in Texas, having increased from 92 days in Japan to 117 in Texas ($t = -2.39$; $df = 28$; $p = 0.024$). However, there were, on average, nearly twice as many infants born in a Texas season (52.7) as compared to a Japanese birth season (27.7), and regression analysis showed that the length of the season was related to the number of infants born ($r^2 = 0.57$; $F_{1,28} = 13.31$; $p = 0.001$). Furthermore, an analysis of co-variance demonstrates that the relationship between the length of the season and the number of births was not significantly different in the two locations ($F_{1,27} = 0.15$; $p = 0.703$).

Table 17.1. *Timing of the birth seasons*

Site	Year	Mean	Median	Onset	Termination	Length	No. births
Japan	57	152.0[a]	145	121	201	80	13
Japan	58	147.7	146	121	183	62	11
Japan	59	147.2	145	118	215	97	15
Japan	60	153.0	149	127	177	50	22
Japan	61	152.0	149	130	179	49	16
Japan	62	154.9	150	126	192	60	21
Japan	63	146.2	142	108	212	104	25
Japan	64	142.9	142	99	215	116	27
Japan	65	141.4	135	101	252	151	35
Japan	66	149.3	143	114	203	89	40
Japan	67	159.3	149	120	211	91	38
Japan	68	148.8	144	118	210	92	32
Japan	69	142.0	137	85	224	139	45
Japan	70	143.1	133	105	198	93	28
Japan	71	137.9	136	107	218	111	48
Mean		147.9	143	113.3	206	92.3	27.7
Texas	74	172.9	172	137	205	68	31
Texas	75	144.4	139	84	247	163	22
Texas	76	133.9	129	99	190	91	35
Texas	78	135.0	131	93	201	108	43
Texas	79	147.2	150	100	200	100	43
Texas	80	130.6	129	67	196	129	40
Texas	81	131.5	129	93	180	87	56
Texas	82	137.1	136	81	183	102	52
Texas	83	139.2	143	88	193	105	57
Texas	84	147.9	149	71	233	162	79
Texas	85	144.3	146	96	224	128	57
Texas	86	140.6	145	73	194	121	78
Texas	87	145.7	144	99	244	145	70
Texas	88	143.1	142	87	204	117	46
Texas	89	145.5	145	94	223	129	86
Mean		142.6	142	90.8	207.8	117	53

[a]Dates are expressed as days in the Julian calendar (1–365).

Thus, the seasons were the 'same' length in Japan and Texas when adjusted for the number of births.

There was also a significant relationship between the onset date and the number of births per season, when all 30 years are considered ($r^2 = 0.423$; $F_{1,28} = 20.59$; $p = 0.0001$), but there was no significant relationship between the termination date and the number of infants born. This indicated that, as more infants were born, the likelihood increased of an

Table 17.2. *Cumulative proportions of births in different intervals of the season*

Period	Month	Japan	Texas	Difference
Early	March	0.00	0.10	−0.10
Middle	March	0.00	0.40	−0.40
Late	March	0.20	1.20	−1.00
Early	April	0.50	4.80	−4.30
Middle	April	2.90	11.40	−8.50
Late	April	11.80	22.50	−10.70[a]
Early	May	27.20	35.60	−8.40
Middle	May	43.80	47.20	−3.40
Late	May	65.40	64.90	0.50
Early	June	76.90	78.00	−1.10
Middle	June	83.90	86.30	−2.40
Late	June	90.40	91.60	−1.20
Early	July	94.70	96.20	−1.50
Middle	July	96.90	98.30	−1.40
Late	July	98.60	99.10	−0.50
Early	August	99.50	99.40	0.10
Middle	August	99.80	99.60	0.20
Late	August	99.80	99.70	0.10
Early	September	100.00	100.00	0.00

[a]Largest unsigned difference $(D) = 10.70$.

earlier birth date and thus of a longer season. However, an analysis of co-variance showed that the relationship between the onset dates and the number of births was marginally significantly different in the two locations $(F_{1,27} = 3.87; p = 0.059)$. Thus, even when adjusted for the number of births per season, the onset of reproduction could be said to be somewhat earlier in Texas than Japan.

The sixth component of seasonal timing, the distribution of birth dates, was compared for Japan and Texas with a Kolmogorov–Smirnov test (Sokal & Rohlf, 1981). Birth dates were divided into 19 intervals, three time periods per month (early, middle and late), from early March to early September, and the cumulative proportions of births in each consecutive time period were calculated. Table 17.2 shows that there was a significant difference (D) between the distribution of birth dates in Japan and Texas $(D = 10.7, p < 0.01)$.

To further examine the comparative distributions, the 416 births in Japan and 790 in Texas were plotted as percentages per month of the season (Fig. 17.1). Again, each month was divided into early, middle and late periods. From Fig. 17.1 it can be seen that the peak period of births in both Japan and Texas was in late May. However, in Japan, the late May birth peak was greater than the one in Texas (53.6% of all births occurred in May in Japan, compared with 42.4% in Texas). Also, a smaller percentage of births occurred in April in Japan (11.5%) than in Texas (21.3%). The percentages of infants born in June and July were quite similar in both locations (June: 25% vs 26.7%; July: 8.2% vs 7.5%), and there were very few births in March, August or September in either location. Thus, the significant difference that was found in the *overall* distributions of birth dates in Japan and Texas (Table 17.2) can be attributed primarily to differences occurring in the months of April and May. Although similar proportions of births had taken place in both locations by the median birth dates of 23 May (Japan) and 22 May (Texas), a higher proportion of births were distributed in May in Japan and April in Texas. This may indicate a slight shift forward into April of the birth season in Texas, a shift forward that was also suggested by the somewhat earlier onset and mean birth dates in Texas as compared to Japan.

Comparison of the Arashiyama monkeys with other populations of Japanese macaques

Overall, the two distributions of birth dates in Fig. 17.1 look very similar. However, the Kolmogorov–Smirnov test is highly sensitive to sample size and it is, perhaps, not surprising that significant differences were found between the 416 birth dates from Japan and 790 from Texas. A more revealing question might be how the patterns of Arashiyama and AW birth dates compare to the timing of the birth season in other populations of Japanese macaques. Kawai *et al.* (1967) presented birth data for 25 groups in Japan, located at latitudes ranging from 31° N to 41° N. From these data, they concluded that each group has a characteristic timing of its birth season that is more or less fixed. Some of these 25 groups were sampled for fewer than 4 birth seasons or 10 birth dates, however, and several groups had been translocated from their original sites. Thus, for his subsequent analysis, Van Horn (1980) chose only those 11 of the 25 groups that were adequately sampled and still living in their original locations. With this sample, he was able to demonstrate a highly significant correlation between latitude and mean birth dates.

Following Van Horn, we also selected birth data from these 11 groups and plotted the mean conception date of the groups as a function of latitude (Fig. 17.2). As reported previously by Van Horn (1980), we found that there is a significant linear relationship between the latitudinal location of these Japanese macaque groups and the timing of their birth seasons ($r^2 = 0.79$,

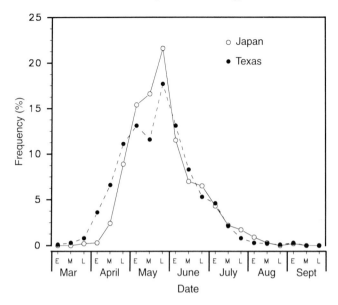

Fig. 17.1. Distribution of birth dates in Japan (1957–71, $n = 416$) and Texas (1974–76, 1978–89, $n = 790$), plotted as percentages per month of the season. Each month is divided into early (E), middle (M) and late (L) periods.

$F_{1,9} = 33.7$, $p = 0.0003$). Figure 17.2 also displays the mean birth date from AW according to its Texas location at 28° N; this group falls far off the regression line for Japanese macaque groups. However, if plotted according to its 'site of origin', rather than transfer site, the AW group falls appropriately on the regression line.

Environmental factors and the timing of reproductive events in Texas

It has already been demonstrated that the study group did not shift to a later breeding season when translocated to a more southerly location (as predicted by the hypothesis that photoperiod, or some aspect of the correlation of seasonal light/dark ratios with latitude, is a proximate determinant of reproductive timing), therefore comparative daylength data at the two sites are presented here simply for descriptive purposes. Table 17.3 shows daylight hours in Kyoto, Japan and Laredo, Texas, calculated as the interval between sunrise and sunset. From this it can be seen that, during the longest day of the year, 21 June, daylength in Kyoto is 34 minutes longer than in south Texas, and during the shortest day of the year, 21 December, daylength in Kyoto is 25 minutes shorter than in south

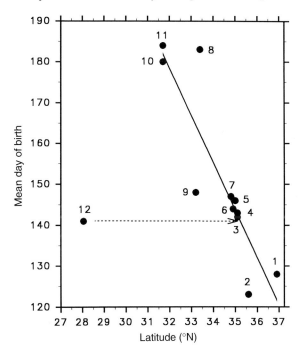

Fig. 17.2. Mean birth date of the population as a function of latitude, for 12 groups of Japanese macaques. Arashiyama West population is group no. 12, which is plotted according to its current location at approximately 28° 05′ N in Texas, and also according to its site of origin at 35° 00′ N, near Kyoto, Japan.

Texas. For arguments that photoperiod may regulate the timing of conception in Japanese macaques, the crucial time period would presumably occur between the fall equinox of 23 September (when daylength is equal to approximately 12 hours at all latitudes) and the late December median conception dates of the Arashiyama monkeys. During this time period, daylength is decreasing by a mean of approximately 2 min per day (rate of 0.3%) in Kyoto, and a mean of less than 1 min per day (rate of 0.1%) in south Texas.

Apart from photoperiod, the two climatic factors that have been identified as possibly regulating the reproductive seasonality of Japanese monkeys (and other seasonal macaques) are rainfall and temperature patterns. Gouzoules *et al.* (1981) presented a comparative climatograph, which showed that rainfall and temperature patterns vary considerably between Kyoto and south Texas. For example, in Japan, June and July are the most rainy months, whereas in south Texas, rainfall since 1974 has occurred mainly in May–June and September–October (National Climatic

Table 17.3. *Monthly changes in daylength at the different Arashiyama sites*

Date	Daylight hours[a]		Difference (min)
	Kyoto	Laredo	
1 January	9:51	10:22	−31
1 February	10:28	10:51	−23
1 March	11:23	11:32	−9
1 April	12:32	12:27	5
1 May	13:35	13:14	21
1 June	14:19	13:49	30
21[b] June	14:31	13:56	35
1 July	14:29	13:55	34
1 August	13:56	13:28	28
1 September	12:55	12:46	9
23[c] September	12:00	12:00	0
1 October	11:51	11:53	−2
1 November	10:46	11:05	−19
1 December	10:00	10:30	−30
21[d] December	9:48	10:13	−25

[a]Daylength calculated as the interval between sunrise and sunset, which depends on latitudes, but not longitude. No corrections for refraction or altitude have been made. (Calculations made by Dr D. Hube of the University of Alberta, by interpolating data from the *Observer's handbook*, Royal Astronomical Society of Canada.)
[b]21 June is the summer solstice and the longest day of the year in the northern hemisphere.
[c]23 September is the autumnal equinox and daylength is 12 hours in all parts of the world.
[d]21 December is the winter solstice and the shortest day of the year in the northern hemisphere.

Data Center data). To examine the possibility that relatively heavy rainfall in the months immediately preceding the mating season might 'trigger' the requisite endocrine changes, we focused on the pattern of rainfall in August, September and October in Texas (mean onset of conceptions in Texas is 10 October). We also calculated the time period when maximum temperatures dropped below 30 °C.

Table 17.4 summarizes daily rainfall and temperature data for Laredo, between 1974 and 1979, and similar climatic data for San Antonio, Texas, between 1980 and 1989. It should be noted that weather patterns in San Antonio may be slightly wetter and cooler than those prevailing in Dilley (approximately 156 km to the south), but San Antonio is the closest weather station for which daily records are available from the National Climatic Data Center. Also presented in Table 17.4 are the relevant

Table 17.4. *Climatic variables and conception in Texas*

Year	Rain onset[a]	Rain total[b]	Temp. <30°C[c]	1st Con. interval[d]	1st Con. date[e]	Onset interval[f]	Median/con. interval[g]	Median/con. date[h]
74	1	21.79	8	12	330	12	15	365
75	4	9.85	11	7	277	11	12	332
76	7	10.12	7	8	292	9	11	322
78	4	10.32	9	8	286	9	11	324
79	0	1.15	10	8	293	10	13	343
80	1	8.78	9	4	260	8	11	322
81	2	12.38	9	8	286	8	11	322
82	7	4.26	8	7	274	10	12	329
83	1	7.50	6	7	281	9	13	336
84	2	10.04	6	5	264	10	13	342
85	6	9.16	6	8	289	9	13	339
86	3	11.27	8	6	264	9	13	338
87	4	3.01	8	8	292	9	13	337
88	2	3.43	9	7	280	9	13	331
89	7	7.83	8	8	287	10	13	338

Data from 1974–79 are for Laredo, and from 1980–89 are for San Antonio.

[a]First interval in which 1.5 inches fell. Each month was divided into 3 intervals, beginning with the month of August.

[b]Total accumulated rainfall in August, September and October, in inches.

[c]First interval in which daily maximum temperatures (temp.) dropped to <30 °C.

[d]Interval during which the first conception (con.) took place.

[e]First conception date (Julian calendar date out of 365).

[f]First interval of the season by which 10% of the conceptions had occurred.

[g]Interval during which the median conception date occurred.

[h]Median conception date (Julian calendar day).

conception dates for the Arashiyama monkeys in Texas. Regression analyses were run comparing three climatic variables (rainfall onset, total rainfall (August, September and October), and maximum temperatures $< 30\,°C$) to three aspects of the timing of the mating season (first conception, first 10% of conceptions, and median conception date). None of the nine regressions resulted in significant relationships (or even approached significant relationships) between these climatic variables and the timing of the breeding season. Unlike the conclusion of Kawai *et al.* (1967), who found that conceptions did not begin until the temperature dropped below $30\,°C$, we found that the onset of conception in Texas occurred during periods when daily maximums were still regularly reaching $33\,°C$.

Discussion

Many discussions of annual reproductive cycles in primate populations do not identify which components of the timing of reproductive seasons are at issue. In our presentation, six aspects of the timing of reproductive seasonality were distinguished, and only one of these was shown to vary significantly between the Arashiyama groups in Japan and Texas; that is, the distribution of birth dates. A frequency plot of the distribution of birth dates showed that proportionally more infants were born in April in Texas and in early and mid-May in Japan, but by late May equal proportions had been born in both locations.

The median and terminal birth dates were virtually identical in the two locations and the mean birth dates were similar. Differences between Japan and Texas in the length and onset of the birth seasons were accounted for largely by the increasing number of infants born per season over the years of the study. A comparison of AW with other groups of Japanese monkeys showed that the timing of the AW season was consistent with their original location (Kyoto, $35°\,N$) and was different from that of groups found in other parts of Japan.

We interpret these results to indicate that the AW monkeys have maintained a stable, group-specific, circannual rhythm of seasonal reproduction over many years under environmental conditions different from those found in their native habitat. Thus, it seems unlikely that the environmental factors traditionally linked to seasonality are currently acting as proximate cues to regulate endocrine and reproductive processes in our study population.

However, since proximate effects are often said to 'fine-tune' the season, we also tested to see whether daily changes in rainfall and temperature might be related to the variation that does exist in the onset and median dates of conception in Texas; no relationships were found. According to the latitudinal cline found in Japan, if proximate cues were 'fine-tuning' the seasons, we would have expected AW to move toward later dates in the calender year. However, if we have any evidence for a shift at all (and this is a matter of interpretation of marginally significant differences), they could be said to have shifted toward an earlier season.

Further evidence for the stability of reproductive timing in the AW monkeys can be obtained from an examination of the dates of 21 births that have occurred in a small subgroup of AW monkeys that was transported to Montreal, Canada in 1985. The median birth date in this subgroup, living at 45° 30′ N, is day 140, 20 May, and the mean birth date is day 143, 23 May (calculated from data provided by Dr B. Chapais). These values are virtually identical to the characteristic timing of the Arashiyama population from which the Montreal monkeys originate and, therefore, different from the timing of Japanese macaque groups that occur at more northerly latitudes in Japan. Nozaki *et al.* (1990) found that artificial manipulations of photoperiod in the laboratory, or chronic implantation of melatonin, had no effects on the timing of ovulatory cyclicity, which also suggests that photoperiod is not acting as a proximate effect.

If the timing of AW mating and birth seasons is not reliant on the proximate environmental cues that have been traditionally proposed, but remains consistent over many years in a new habitat, what *does* regulate the timing of reproductive seasonality in this group? We will suggest and examine a number of possible answers to this question, many of which are not mutually exclusive. Firstly, a characteristic circannual rhythm may persist in the AW group simply as an example of a very lengthy refractory period (i.e. a period when the environmental stimuli that normally regulate reproduction are not effective). However, refractory periods in laboratory tests are normally very brief, and then the animals adjust their timing to the current environmental cues. Even for long-lived animals such as macaques, 15 years would be a long time to maintain an annual pattern that is not modulated by either intrinsic rhythms or extrinsic cues. In any case, the monitoring of the birth season in Texas continues, so this hypothesis could be further tested in the future.

Secondly, it might be argued that the proximate cues, such as photoperiod changes, in south Texas are not sufficiently different from the conditions near Kyoto to bring about a change in reproductive timing. This seems unlikely, given the latitudinal gradient of reproductive timing in the different Japanese macaque groups that was demonstrated by Kawai *et al.*, (1967) and Van Horn (1980). There is not a great deal of difference between the latitudinal locations in Japan of the 11 groups included in the analyses ($36°\,90' - 31°\,70' = 5°\,20'$ of difference), but the mean birth dates of the most northerly and southerly groups differ by approximately 60 days. These different mean dates of birth and their significant correlation with latitude would indicate that Japanese monkeys should be capable of responding to very small or precise differences in photoperiod, if that is the mechanism involved. The AW monkeys were moved $7°$ south of their original habitat and yet their mean birth date differs by only five days from that of the Arashiyama group in Japan. The timing of AW reproduction continues to fit clearly into the 'middle range' group identified by Kawai *et al.* (1967), even though the south Texas habitat exhibits quite different temperature and rainfall patterns.

That being said, it is still possible that the AW monkeys rely on some proximate cues (such as magnetic fields) that may occur similarly in Texas and Japan, but have not yet been considered here. As one example, if the monkeys had some way of 'measuring' the number of days subsequent to the autumnal equinox, this proximate cue would remain the same anywhere in the world. Perhaps all groups in Japan are ultimately selected to mate in the fall and give birth in the spring, but each group has its own 'critical day', or critical period of days, for the onset of spermatogenesis and ovulatory cycling, as calculated from the fall equinox. However, no such mechanism has been described for any other species, and one wonders why the characteristic 'critical days' for each group would fall along a latitudinal gradient.

Another proximate factor that has remained similar in the new habitat, and has not yet been discussed, is the social environment. For a number of species, social stimuli, especially pheromones, exchanged between females have been shown to be the proximate cause of reproductive synchrony (reviewed by Ims, 1990; see also Herndon, 1983). The AW group was translocated intact from Japan to Texas, and that may have played a role in the maintenance of their characteristic reproductive timing. It is possible to envision that one individual's reproductive status may have a stimulatory influence on her fellow group members, and thus enhance reproductive synchrony in the group. It is less easy to envision the mechanism that would continue to couple this synchronized group of females to some precise calendar date in the year. In a group with such consistent circannual rhythms, such as the Arashiyama monkeys, who is the primary keeper of the annual clock? If social stimuli alone were responsible for maintaining reproductive synchrony in a new habitat, we might expect that females of the group would coordinate the timing of their ovulatory cycles but that the mating season itself would become uncoupled from its temporal location in the calendar. Most discussions of social cues (e.g. Lindburg, 1987; Bronson, 1989; Ims, 1990) seem to accept that either environmental or intrinsic factors set the timing of the reproductive seasons, and then social cues act to 'cluster' the conceptions or births within that season.

It is also important to note here two constraining characteristics of the AW group that may have an impact on reproductive seasonality. The first characteristic is that the AW group is provisioned (and indeed, almost all the groups in the Kawai *et al.* (1967) study have been provisioned since the early 1950s.) Such food supplementation may have largely ameliorated the selective pressure of seasonal food availability over the past 40 years. It could be argued that if the monkeys in Texas were forced to rely more extensively on native resources, then the mating and the birth seasons would shift in accordance with local patterns of food availability. The second characteristic is that the AW group has not experienced the immigration of new males since its translocation to Texas, although it has exhibited high rates of male emigration. If the timing of mating seasons in Japan remains consistent over the years, even as males transfer groups, one would assume that the philopatric females are the 'keepers of the circannual rhythm' and that males adjust to the timing of the group into which they have immigrated. However, the real situation may be more complicated, and perhaps the mating season in Texas would shift with the introduction of new breeding males.

Finally, we would like to suggest the possibility that reproductive seasonality in Japanese macaques relies on an endogenous circannual clock that has become emancipated from extrinsic factors, and is set to initiate reproductive events differently in each of the genetically distinct groups in Japan. It is possible that the precise timing found in the different groups is the result of genetic and/or social drift (Gouzoules *et al.*, 1981) and, thus, has no necessary functional explanation. However, the strong correlation between reproductive timing and latitude in Japan seems unlikely to be entirely fortuitous or due to chance differences between populations, especially as such correlations are found in other vertebrate species. We have argued that photoperiod, in spite of its relationship to latitude, is unlikely to be operating as a proximate cue for seasonality in our study group. But photoperiod may well have been involved in the evolutionary selection for circannual clocks with different 'critical days' in different genetic populations of Japanese monkeys. Once the clock was 'set' it would continue to follow the same temporal pattern as long as the group remains socially and genetically intact.

From the results of the 'Arashiyama translocation experiment', and from the evidence on seasonal rhythms in other animals, we suggest that reproductive seasonality in this group relies primarily on an endogenous circannual clock that has become emancipated from the traditionally examined climatological cues, but which may be affected by the social environment and by other proximate cues that have yet not been identified.

Acknowledgements

It is not possible to thank by name all the people who have contributed to the 30 years of birth data analyzed in this chapter, but we would like to acknowledge gratefully the cooperative work of the many Arashiyama researchers who have participated in the project since 1954. We would also like to thank the numerous people who have given of their time, energy, and money to maintain the monkey groups, both in Texas and Japan, especially Mr N. Asaba, Mrs. E. J. Dryden Jr, and Mrs P. Burns Dailey. Dr J. F. Addicott helped extensively with the statistical analyses, B. Chapais provided birth dates for the AW monkeys located in Montreal, and C. Allen, D. Becker, H. Gouzoules, D. Hube and F. LaRose also contributed to improvements in the manuscript. L. M. Fedigan's research is supported by Operating Grant no. A7723 from the Natural Sciences and Engineering Research Council of Canada (NSERCC).

References

Baker, J. R. (1938). The evolution of breeding seasons. In *Evolution: essays on aspects of evolutionary biology*, ed. G. R. deBeer, pp. 161–77. London: Oxford University Press.

Bielert, C. & Vandenbergh, J. G. (1981). Seasonal influences on births and male sex skin coloration in rhesus monkeys (*Macaca mulatta*) in the southern hemisphere. *Journal of Reproduction and Fertility*, **62**, 229–33.

Bronson, F. H. (1989). *Mammalian reproductive biology*. Chicago: University of Chicago Press.

Drickamer, L. C. (1974). A ten-year summary of reproductive data for free-ranging *Macaca mulatta*. *Folia Primatologica*, **21**, 61–80.

Eaton, G. G., Rostal, D. C., Glick, B. B. & Senner, J. W. (1987). Seasonal behavior in a confined troop of Japanese macaques (*Macaca fuscata*). In *Progress in biometeorology*, vol. 5: *Seasonal effects on reproduction, infection and psychoses*, ed. T. Miura, pp. 29–40. The Hague: SPB Publishing.

Fedigan, L. M. & Asquith, P. J. (eds.) (1991). *The monkeys of Arashiyama: thirty-five years of research in Japan and the West*. Albany: State University of New York Press.

Gouzoules, H., Gouzoules, S. & Fedigan, L. (1981). Japanese monkey group translocation: effects on seasonal breeding. *International Journal of Primatology*, **2**, 323–34.

Gwinner, E. (1981). Circannual systems. In *Handbook of behavioral neurobiology*, vol. 4: *biological rhythms*, ed. J. Aschoff, pp. 391–410. New York: Plenum Press.

Gwinner, E. (1986). *Zoophysiology*, vol. 18: *circannual rhythms: endogenous annual clocks in the organization of seasonal processes*. Berlin: Springer-Verlag.

Herbert, J. (1977). External factors and ovarian activity in mammals. In *The ovary*, 2nd edn, vol. 2: *physiology*, ed. S. Zuckerman & B.J. Weir, pp. 457–505. New York: Academic Press.

Herndon, J.G. (1983). Seasonal breeding in rhesus monkeys: influence of the behavioral environment. *American Journal of Primatology*, **5**, 197–204.

Hoffman, K. (1981). Photoperiodism in vertebrates. In *Handbook of behavioral neurobiology*, vol. 4: *biological rhythms*, ed. J. Aschoff, pp. 449–73. New York: Plenum Press.

Immelman, K. (1973). Role of the environment in reproduction as a source of 'predictive' information. In *Breeding biology of birds*, ed. D.S. Farner, pp. 121–47. Washington, DC: National Academy of Sciences.

Ims, R.A. (1990). The ecology and evolution of reproductive synchrony. *Tree*, **5**, 135–40.

Jones, R.E. (1981). Mechanisms controlling seasonal ovarian quiescence. In *Dynamics of ovarian function*, ed. N.B. Schwartz & M. Hanzicker-Dunn, pp. 205–34. New York: Raven Press.

Kawai, M., Azuma, S. & Yoshiba, K. (1967). Ecological studies of reproduction in Japanese monkeys (*Macaca fuscata*). I. Problems of the birth season. *Primates*, **8**, 35–74.

Koford, C.B. (1965). Population dynamics of rhesus monkeys on Cayo Santiago. In *Primate behavior: field studies of monkeys and apes*, ed. I. DeVore, pp. 160–74. New York: Holt, Rinehart, & Winston.

Lancaster, J.B. & Lee, R.B. (1965). The annual reproductive cycle in monkeys and apes. In *Primate behavior: field studies of monkeys and apes*, ed. I. DeVore, pp. 484–513. New York: Holt, Rinehart, & Winston.

Lindburg, D.G. (1987). Seasonality of reproduction in primates. In *Comparative primate biology*, vol. 2B: *behavior, cognition, and motivation*, ed. G. Mitchell & J. Erwin, pp. 167–218. New York: Alan R. Liss.

Michael, R.D. & Bonsall, R.W. (1977). A 3-year study of an annual rhythm in plasma androgen levels in male rhesus monkeys (*Macaca mulatta*) in a constant laboratory environment. *Journal of Reproduction and Fertility*, **49**, 129–31.

Michael, R.P. & Zumpe, D. (1978). Annual cycles of aggression and plasma testosterone in captive male rhesus monkeys. *Psychoneuroendocrinology*, **3**, 217–20.

Nigi, H. (1976). Some aspects related to conception of the Japanese monkey (*Macaca fuscata*). *Primates*, **17**, 81–7.

Nozaki, M., Oshima, K. & Mori, Y. (1990). Environmental and internal factors affecting seasonal breeding of the Japanese monkey. In *Primatology today. Proceedings of the XIIIth Congress of the International Primatology Society in Nagoya and Kyoto*, ed. A. Ehara, T. Kimura, O. Takenaka & M. Iwamoto. Amsterdam: Elsevier Science Publishers.

Rawlins, R.G. & Kessler, M.J. (1985). Climate and seasonal reproduction in the

Cayo Santiago macaques. *American Journal of Primatology*, **9**, 87–99.

Sadleir, R. M. F. S. (1969). *The ecology of reproduction in wild and domestic animals.* London: Methuen.

Sokal, R. R. & Rohlf, F. J. (1981). *Biometry: the principles and practice of statistics in biological research*, 2nd edn. New York: W. H. Freeman.

Tokuda, K. (1961–2). A study on the sexual behavior in a Japanese monkey troop. *Primates*, **3**, 1–40.

Van Horn, R. N. (1980). Seasonal reproductive patterns in primates. *Progress in Reproductive Biology*, **5**, 181–221.

Vandenbergh, J. G. & Vessey, S. (1968). Seasonal breeding of free-ranging rhesus monkeys and related ecological factors. *Journal of Reproduction and Fertility*, **15**, 71–9.

Wickings, E. J. & Nieschlag, E. (1980). Seasonality in endocrine and endocrine testicular function of the adult rhesus monkey (*Macaca mulatta*) maintained in a controlled laboratory environment. *International Journal of Andrology*, **3**, 87–104.

18

Behavior of mixed species groups of macaques

F. D. BURTON AND L. CHAN

Introduction

Primate polyspecific associations are widespread and variable. They have been described among closely related *Cercopithecus* (Struhsaker, 1978, 1981; Gautier-Hion, Quris & Gautier, 1983; Cords, 1987) and *Cebus* spp. (Terborgh, 1983, 1990; Buchanan-Smith, 1990). Mixed-species or inter-taxa interactions have also been reported among species of different genera or families (MacKinnon & MacKinnon, 1978), such as: the association of patas (*Erythrocebus patas*), olive baboons (*Papio anubis*) and vervets (*Cercopithecus aethiops*) (Haddow, 1952; Hall, 1962, 1963, 1965); olive colobus (*Procolobus verus*) and Diana guenons (*Cercopithecus diana*); and chimpanzee (*Pan troglodytes*) and red colobus (*Colobus badius*) (Goodall, 1986). These associations vary in terms of duration, range, ecological relations, biological exchange and social interactions. Assumptions that inter-species contacts would necessarily be aggressive due to competitive exclusion have not been validated by early reports. Indeed, as early as the mid-1960s, Hall (1962) observed that 'multi-species aggregations' did not show aggression, at least between baboons, patas and vervet monkeys. He attributed this to the fact that 'groups which encounter other species groups regularly in their habitat know very well tolerance limits to be observed on such occasions.'

As polyspecific relationships occur frequently, it is unlikely that they simply occur by chance. Recent socio-ecological hypotheses concerning the causes of polyspecific association share an emphasis on the cost and benefit of living with another species group as the primary concern (Waser, 1982; Richard, 1985; Gautier-Hion, 1988; Struhsaker, Butynski & Lwanga, 1988; Boinski, 1989; Norcork, 1990; Terborgh, 1990). While ecological relationships between interacting species have received much attention, less is known about social and biological interactions between individuals of species groups.

It is also important to note that terminologies used by researchers to describe inter-species interaction have not been consistently applied. The term 'polyspecific association' has been used in a variety of contexts. Without clarification, this term does not exclude commensal, parasitic, and predatory relationships between species. Clearly, relationships that result in hybridization should be named differently from other ecological associations and opportunistic encounters. We suggest a clarification in terminology for contacts between non-human primates based on interaction and duration, with specific referents for each term. This terminology considers the quintessential definition of species as a distinct reproductive unit, and highlights the behavioral mechanisms facilitating cross-species breeding:

1. **Inter-specific** or heterospecific **associations** are sympatric, but individuals are not interspersed; home ranges overlap.
2. **Mixed-species** occur where two or more species may be interspersed but do not physically interact (such as between langur and rhesus, or langur and bonnet as described by Sugiyama (1967).
3. **Polyspecific associations** are those where species are interspersed and behave like members of a group: grooming, playing, feeding etc., for varying, but limited lengths of time, occasionally mating (Struhsaker & Gartlan, 1970; Struhsaker, 1975).
4. **Hybrid groups** are polyspecific associations that are cohesive, stable units, characterized by recurrent genetic exchange.

This chapter reports on sustained relationships between several macaque species in Kowloon, Hong Kong. Observations of these monkeys raise important methodological and conceptual problems about current views of inter-species relations. These patterns of association within and between macaque groups are far more complex than previously assumed, and they do not appear to be motivated solely by ecological concerns. More importantly, at present the ecological 'norm' for many macaque species has not yet been established. The Kowloon macaques therefore represent a natural test case of the flexibility and adaptability of the most widespread and abundant genus of non-human primates.

Study area and methods

Kowloon is part of Hong Kong, a British Colony and Protectorate until 1997. Kowloon (and the New Territories, which comprise 957 km²) lies to the north, across the water from Hong Kong Island, contiguous with

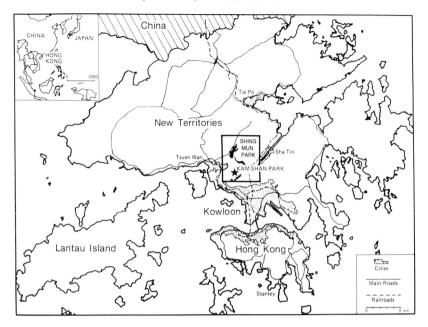

Fig. 18.1. The study area in relation to its geographical location.

Guandong Province in mainland China (Fig. 18.1). The area is sub-tropical, and prior to human settlement about 1000 years ago it was covered with thick sub-tropical forest in which tigers and elephants roamed. Annual rainfall averages 2246 mm, and falls mostly during the summer monsoons (May–September). Temperature ranges from 10 °C in January–March and soars to 32 °C in June–September. Cyclones occur between July and September and typhoons throughout mid-summer.

Human occupation has radically altered the region by reducing forests and their varied wildlife. Today, there is still a variety of fish, water and land snakes, birds, and invertebrates. However, mammals, other than monkeys, have declined, although wild boar (*Sus scrofa*) and leopard (*Panthera pardus*) have been sighted recently in the Shing Mun forests.

Completed in 1992, this chapter reviews data gathered in 1984, 1986 and 1988; a total of 1000 observation hours. A breakdown of animals during the 1984 survey and counts for 1986 and 1988 are shown in Table 18.1. The monkeys are habituated to human presence, as they receive donations of food. The censuses were based on direct counts of all visible monkeys at a given location. As the animals travel around, and as group locations are not static, duplicated counts of animals can occur. To circumvent this problem,

View looking west over part of the range showing Kowloon Reservoir.

teams of two to five researchers were assigned to different locations and at specified times noted number and composition of animals in that area. The count is asymptotically the number that were present, as corrections were made by repeated counts at intervals over the day's study period at a given location. Marker monkeys, individuals that were easily recognized, were noted. Up to three marker monkeys were entered into the database so that group identity, relationships between groups, and membership of groups could be ascertained.

The Kowloon mixed-species macaque group

The monkey population

Southwick & Southwick (1983) identified three species (rhesus, *Macaca mulatta*; Japanese macaque, *M. fuscata*; long-tailed macaque, *M. fascicularis*) in their survey of the monkeys of Kowloon and New Territories. Their breakdown of species' numbers and hybrids is shown in Table 18.2. During

Table 18.1. *Group size of the monkeys of Kowloon (1984–88)*

	Year		
Group	1984	1986	1988
M. thibetana	2	2	2
M. fascicularis	9	5	4
M. mulatta, large	60	70	90
M. mulatta, small	20	30	20
M. fuscata-like	10	20	25
M. nemestrina		1	1
Hybrids			
thibetana × mulatta		1	2
mulatta × fuscata	10	20	40
mulatta × fascicularis	25	25	30
Total	136	174	214

Table 18.2. *Population size of the monkeys of Kowloon (according to Southwick & Southwick, 1983)*

Group	*M. mulatta*	*M. fascicularis*	*M. fuscata*	*M. fascicularis × mulatta*	Total
Kam Shan entrance	40	19	0	9	68
Eagle's Nest tree	23	5	0	4	32
Lower Tai Po Road	2	5	5	1	13
Total	65	29	5	14	113

our study period (1984–8), in addition to those species reported by Southwick and Southwick (1983), two others, the Tibetan macaque (*M. thibetana*) and the pig-tailed macaque (*M. nemestrina*), were also present. Only two adult female Tibetan macaques, and two hybrid offspring, a male and a younger female were present during the review period.

Fluctuations of the monkey population in Kowloon and New Territories follow closely the history of human activity. The monkeys can be traced continuously through the twentieth century. The current predominant type, the rhesus monkey, is, according to early naturalists' reports, indigenous to the area but there are no reliable observations until relatively recently, and reports are contradictory. Herklots (1951) mentioned monkeys at Tai Tam Reservoir on Hong Kong Island in 1947 and occasional sightings of monkeys are reported throughout the 1950s (*Hong Kong Annual Reports*, 1952–4). As described in Herklots (1951), animals seen on Hong Kong Island may have been an indigenous species, while those in the New Territories were probably introduced during World War I (Herklots, 1951). Reports are vague, and through the 1950s rhesus reports decrease until during the mid-1960s the report is that rhesus 'no longer occur here' (*Hong Kong Annual Report*, 1967). Six years later, however, there are 'breeding groups' in the Kowloon Reservoir area (*Hong Kong Annual Report*, 1973, published 1974). Sightings of rhesus monkeys during the 1970s do not record group size. It is feasible, however, for one pair of adult monkeys, released in the mid-1960s, and having maximal success, to have formed a 'well established group' a decade later. Alternatively, it is equally reasonable to assume a continuous presence.

The long-tailed or crab-eating macaque is the common monkey of South-East Asia. At the time that reports of the disappearance of the rhesus were being filed, long-tailed monkeys were recorded apparently for the first time. Romer (1966) also noted their presence as a 'breeding group', although exact numbers are not given. He safely placed their escape or release 'at some time in the past'. Other scholars assume the monkeys were released in the 1920s (Department of Agriculture and Fisheries, published 1976), or after World War I (*Hong Kong Annual Report*, 1973, published 1974). From the early 1970s, long-tails were found in association with rhesus, that is in 'their old haunts' along the Tai Po Road, to the Kowloon Reservoir and onto Piper's Hill (*Hong Kong Annual Report*, 1973, published 1974). Lance (1976) remarked that the flourishing colony of *M. fascicularis* in the Kowloon Reservoir area had recently been joined by rhesus monkeys that 'have only successfully established themselves in the same area in the past few years'. Lance added that the two species seemed 'to live fairly amicably side by side in the same restricted habitat'.

The two adult female *M. thibetana* known during the study period were released in the mid-1960s. Two adult males were killed in 1984, one in a traffic accident. No written documentation on these monkeys or of the presence of *M. fuscata* has been found. The latter could have been introduced during the Second Sino–Japanese conflict, but it is not known whether for pets or for military purposes.

Conservation legislation enacted in 1954 (Wild Birds and Wild Mammals Protection Ordinance) and extended to restrict importation and possession of animals and birds in the mid-1960s accounts for the release of some pet and entertainment macaques. Hong Kong was a major transshipment center from Asia to the rest of the world. Records from 1960 show 9600 monkeys on the 'inspected and certified' list (*Hong Kong Annual Reports*, 1960–1), some of which will have escaped and taken refuge in the forests. Another source of monkeys was those purchased in order to be released as an act of veneration by Buddhists. When monkeys like *M. thibetana* were set free, it was probably along the Tai Po Road, where monkeys were already known to frequent the area. By 1986 there were only nine recognizable long-tailed monkeys. Hybridization with *M. mulatta* and 'phenotypic masking' (Struhsaker *et al.*, 1988) has made rhesus the predominant morphotype.

Hybridization levels

There has been extensive hybridization, especially between the long-tailed and rhesus monkeys. Evaluation of hybrids is difficult. In some species, especially rhesus, fur color is a poor species diagnostic, since the range is from dun-grey-white to dark brown, and changes seasonally. Redness of face and hind quarters as well as tail length, shape and location of ear pinna on the head, and glabrous or hairy faces are amongst other variable traits. Even though tail length has been used as an indicator of hybridization (Bernstein, 1966) the variability of this character makes it a poor diagnostic measure.

Estimation of parentage is further complicated by generation. The term 'hybrid' is used to describe the F1. In Kowloon, interbreeding of species probably began in the 1960s and reckoning a monkey generation at five years, some members of the population will have been F4 or F5 by the beginning of the study period. Relatively few 'pure' types are then left to identify. In addition, 'dominant' traits for each species are not adequately known, and swamping of the phenotype by some traits would obscure parentage. Struhsaker *et al.* (1988) pointed out that in cercopithecine hybrids in Kibale forest phenotypic masking obliterates expression of hybridization. This may account for the apparent vast numbers of 'rhesus' now in the population, but until an appropriate sample can be physically measured, weighed, and, above all, blood-typed, parentage continues to be based on estimation.

'Typical' forms, that is those most closely resembling a species morphotype, are recognizable amidst intermediate ones. There are, at times, some of each represented in all observed groups. At other times, however, sub-groups of a predominant type are found such as the long-tail and Tibetan members

Pure *fascicularis* (left) and hybrid *fascicularis* × *mulatta* (right).

of the group that frequent the Caldecott Road area, or the whitish furred, red-faced monkeys primarily seen along the bottom of the KowLoon Hills Management Centre Road (KHMC), near the beginning of the MacLehose Trail. The mixed group in Shing Mun, which remains elusive, has been heard making *fuscata*-type infant coos. Males identified and videoed near this group include one three-legged *fuscata* hybrid, and three *mulatta* types. Our video collection of more than 30 h of tape permits examination of these animals in their varying locations.

Group composition and sizes

In 1984, the central area of the monkeys' range contained more than 100 animals of the rhesus type, including hybrids. This area encompasses the Byewash Reservoir, across the Water Works Access Road (WWAR), north to the Kowloon Reservoir, and across the Tai Po Road, onto the Fitness Trail, Eagle's Nest Hill and the MacLehose Trail, down past the KHMC post and onto Piper's Hill (Fig. 18.2). Nine long-tails, Tibetan macaques

Unknown *mulatta* cross.

and 10 *fuscata*-like monkeys were also found (these may be the *M. fuscata*
that Southwick & Southwick (1983) described, or the lighter fur, redder-faced
form of rhesus well known in China, especially in Hainan). The smaller
group of *M. fascicularis* and *M. thibetana* was integrated with the large one,
although it tended to frequent a portion of the terrain at the bottom of Tai
Po Road. They were often seen, however, with members of the rhesus group
at locations within Kam Shan and across the Tai Po Road at Piper's Hill
and from the entrance to the KHMC to the post itself.

Group sizes were best counted when there were large troop movements
to sleeping sites, or 'parades'. Numbers reached slightly over 100 but would
fragment into smaller units of between 20 and 50 animals. The larger
aggregate was indeed cohesive, since marker individuals were seen in
several of the smaller groups as well as the 'parades'.

Group sizes changed considerably over the study period. There were
solitary monkeys at Shing Mun in 1984 but by 1986 there was a group
there, although too elusive to count. *Fuscata*-like infant coos, and the
scouting of *mulatta* males, suggest a mixed group. Unlike the situation with
long-tails, the rhesus population appears to be growing, with an adult

Group picture: 'golden' colour *mulatta* in foreground; *fascicularis* × *thibetana* (?) behind.

female to adult male sex ratio of 3 : 1 to 5 : 1, depending on the sub-group, with juveniles and sub-adults of both sexes well represented.

Range and grouping patterns

When 'parades' formed, monkeys moved quickly on the ground, crossing the reserve in the open from part of Kam Shan Park, across Tai Po Road, and up towards Eagle's Nest or the MacLehose Trail. Until the aerial overpass was completed in 1991, the monkeys crossed the roadway. Despite their caution and timing, they were not always fast enough to avoid cars and trucks; in 1984 at least two males died this way.

The order of progression in 'parades' varied, but often it was adult females who led, followed by mothers with infants, with males interspersed. There were all-male groups, usually of four or five males, and almost invariably this form of sub-group moved in the van of the larger aggregate. These aggregates tended to form towards evening and to move rapidly along the KHMC Road to the stairs behind the post, and up past the Wardens' dormitories onto the forested hill. Then, near the helicopter pad they could swing back onto the foothills of Eagle's Nest Hill (Fig. 18.2).

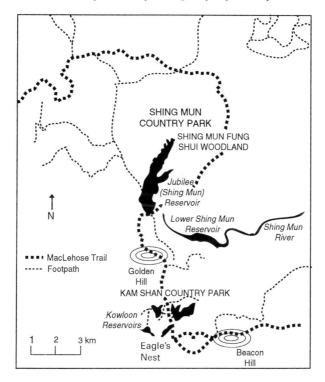

Fig. 18.2. The study area.

By 1986, grouping patterns changed and there were no 'parades'. The *fuscata*-like monkeys tended to congregate along the bottom of the KHMC Road. Distinct groupings of rhesus were observed, one at the bottom of the WWAR, the other closer to the police station on the hill before the Kowloon Reservoir. The *thibetana*-long-tail group were frequently found inter-mixed with the rhesus group centered at the KHMC post, as well as down by Black's College. The two rhesus-type groups along the WWAR each had a 'golden'-furred individual: one a juvenile male, the other an infant female. Their ranges overlapped and extended from the Kowloon Reservoir to the Byewash Reservoir.

Issues about polyspecific groupings

Although many macaque species live in sympatry, they seldom form polyspecific groups (Fooden, 1982). Researchers have noted that where closely related species live in sympatry, they tend to segregate ecologically as well as socially (Crockett & Wilson, 1980; Fooden, 1982; Rodman, 1991). Eudey (1979, 1980) reported possible, but unsubstantiated, polyspecific interactions between several macaque species (*M. nemestrina, M. assamensis, M. arctoides, M. mulatta,* and *M. fascicularis*) and noted the possibility of hybridization between *M. mulatta* and *M. fascicularis*. She did not find apparent differences in habitat use. Gautier-Hion (1973) observed a similar situation among cercopithecines in Gabon where not only *Cercopithecus pogonias* and *Cercopithecus nictitans* occurred regularly in the same sleeping sites, but no differences in vertical stratification were found. A review of her data (Gautier-Hion, 1988) also suggested that polyspecific associations permitted a wider use of the vertical dimension, especially for the smaller bodied cercopithecine members of trispecific associations.

Richard (1985) reviewed the literature on primate polyspecific associations and identified five common 'reasons' for their existence:

1. To detect and avoid or dilute predators.
2. To increase the ability to locate food without increasing competition.
3. To keep track of resources that have already been depleted, thereby minimizing the number of food patches they must visit by maximizing the return provided by each one. There is a gain in feeding efficiency.
4. Large home range species use small home range species to guide them to local resources. This results in social commensalism if competition increases, social parasitism occurs.
5. Rates of insect capture increase for one species.

Because predation is not likely to be an important factor in the formation of polyspecific groupings of macaques in Kowloon, it is interesting to ask whether increased feeding efficiency is a primary motivation for inter-species association. In an earlier study, Southwick & Southwick (1983) suggested other factors that might contribute to polyspecific groupings in Kowloon including (1) release of captive animals, (2) donations of food, and (3) environmental disturbances. They considered feeding by people and the resulting competition as the primary factor promoting polyspecific grouping, since individuals from a minority species would benefit from feeding competition by joining a larger extra-specific group. However, although (1) and (2) may have brought the animals into regional proximity, they do not explain their interactions.

It may be obvious that animals tend to aggregate at sites where food is more abundant and easily accessible. It is, however, difficult to see why the animals would form long-term relationships when competition for human handouts only occupies a small, albeit important, portion of their daily activities. Temporary aggregations are not the same as true social units. Clearly, the motivations for associations are diverse.

Option (3) proposed by Southwick & Southwick presumes that because polyspecific interactions are rare, they therefore occur primarily in disturbed habitat. Yet nearly two decades ago, Struhsaker (1975) noted in reference to cercopithecines that polyspecific associations are common in the West African forests that are undisturbed. However, breakdown of ecological barriers, due to the development of agriculture and the attendant increase in secondary forest, may account for polyspecific encounters (Gartlan & Struhsaker, 1972; Dunbar & Dunbar, 1974). While human interference can destroy barriers, it does not, however, *cause* association. The animals must do so. In other words, opportunities for inter-species encounter may increase because of anthropogenic factors, but sustained social interaction between species groups can occur only when the animals are willing to do so. In this light, Gautier-Hion (1988) asked whether polyspecific associations are merely an elaborate strategy for optimizing resources of the environment and defending against predators, or if this 'peculiar life style' affected guenon speciation through hybridization. The same question can be posed for the Kowloon macaques.

Habitat disturbance, hybridization and polyspecific life

The relationship between polyspecific life, hybridization and disturbed habitat is an important one. Hybrids are commonly regarded as products of abnormal situations and 'disturbed habitat' is commonly cited as the primary cause of hybridization. The assumption is that under 'natural' conditions, reproductive isolating mechanisms would operate to maintain the integrity of the species.

'Disturbance' is a term with considerable history in ecology, and engendering equal debate. The traditional view of 'disturbance' is 'uncommon, irregular events' that 'move communities away from static, near equilibrium conditions' (Sousa, 1984). Because equilibrium conditions virtually never exist, the notion of 'disturbance' has been redefined to refer to situations where there is 'punctuated killing', or where individuals or colonies have been displaced or damaged so that new individuals or colonies have an opportunity to become established in that area (Sousa, 1984).

Ecological disturbances are often intensified by human agency, as when suppression of forest fires permits such a build up that, when inevitably the fire occurs, it is catastrophic. Human activity generally intensifies consequences by altering the course, but not the fact of disturbance.

The effects of human disturbance on primate distribution and abundance have received considerable attention. Bishop *et al.* (1981), for example, identified four variables, each scaled from 1–4, to assess the degree of disturbance. These are (1) interference with habitat and (2) predators, and (3) amount of habituation to or (4) harassment from humans. Their classification was an attempt to facilitate discussion about non-human primates that live near or under human influence. The implicit assumption is that the state of 'nature' in which the primates evolved has been distorted. Their study also emphasizes the difficult question of what constitutes 'natural' and 'disturbed' habitat, as well as the difficulty of characterizing the animals' habitat.

It is worth remembering that for many cercopithecines, the habitats that species live in today are those that have been heavily modified by human agency. Modification renders habitat a junction zone or 'ecotone' (Odum, 1959), wherein the community contains many organisms of overlapping communities. In addition, organisms that are characteristic of, and often restricted to, the ecotone are also found. Consequently, the number of species and the population density of some of the species are greater in the ecotone than in the communities flanking it. The tendency for increased variety and density at community junctions is known as the edge effect. The influence such environments have on species' relations, hybridization, and speciation was noted by Paterson (1976). More recently, Richard, Goldstein & Dewar (1989) expanded on this broad observation and pointed out that many macaque species exemplify varying abilities in accommodating close association with human settlements, and may over prosper. Indeed, macaques and humans are inextricably bound in polyspecific relationships in most parts of the world. Wherever humans are or have been, they have imposed, sought, or accommodated such relationships. Therefore, the process of speciation, and the subsequent distribution and abundance of many macaques cannot be explained independently from their relationship with human activities.

Reproductive patterns and biological exchange between species

Since the biological species concept developed by Dobzhansky (1937) and Mayr (1942) emphasized species as groups of inter-breeding populations that are reproductively isolated from other such groups, the diagnosis of species has been sought in terms of those characters that confer reproductive isolation. Events such as gamete formation, gestation period, timing of oestrus, courtship behavior, etc. would confer species integrity.

The genus *Macaca* consists of seasonal and non-seasonal breeders. Both *M. mulatta* and *M. fuscata* are well-known seasonal breeders, the former having mating periods largely between October and February and birth seasons between March and July (Lindburg, 1987), while the latter exhibits autumn mating and spring births (Kawai, Azuma & Yoshiba, 1967; Migi, 1976). *Macaca fascicularis*, however, shows a year-round occurrence of births but with a definite peak in the middle of the year (Kavanagh & Laursen, 1984). During the study period in Kowloon, slightly more than half the rhesus-type females had 1 year birth intervals; they carried a new infant while their 1 year olds trailed close by. About a third had a 2 year old nearby but no infant present, and the inter-birth intervals of the remainder were indeterminate. This flexibility in the reproductive pattern permits rapid adaptation to ecological (including human) effects and biological and behavioral demands, and is common to other species (e.g. Crockett & Eisenberg, 1987; Melnick & Pearl, 1987; Struhsaker & Leland, 1987). Similarly, while the birth season peak is in June and July, births range from mid-winter to early autumn. Infant sex ratio for the years 1986 and 1988, however, was in favor of males, by a factor of about 4:1. Rate of growth since 1984 has been accelerating. A ratio in favor of males curbs adult population size in that not only do less males survive to adulthood, but also peripheralization of males into all-male groups removes potential breeders (Lyles & Dobson, 1988), while creating a 'reservoir' against exigencies (Carpenter, 1964). In addition, there are solitary males such as the locally named 'silver monkey', who used to frequent the Fitness Trail in 1984. Some of these males eventually integrate with other sub-groups.

The labile pattern of reproductive events in Kowloon is in accord with the genus *Macaca* in general (Dunbar, 1987, 1988; Bercovitch & Goy, 1990). In addition, inter-birth intervals in macaques can be responsive to the level of sustenance (Burton & Sawchuk, 1974). Initiation of mating may be affected by light (van Horn, 1980) or food, or both (Richard, 1985), and may include a considerable amount of social influence (Burton & Sawchuk, 1974; Wallis, King & Roth-Meyer, 1986; Dunbar, 1988). Birth peaks can be bimodal, or extend over six months. While birth season can be bounded in some species, mating, even though infrequent, often occurs over much of the year (Lancaster & Lee, 1965; Hrdy & Whitten, 1987). Therefore, the probability of overlap in reproductive schedules is high. In so far as social events (e.g. grooming) facilitate reproductive events, as has long been known in spermatogenesis (e.g. Koford, 1963), mating between members of different groups in Kowloon appears to have been reasonably assured.

Species recognition and mate choice

Although various aspects of sexual behavior in macaques have been studied extensively, there is little information on how sexual behavior may function as isolating mechanisms despite the feeling that 'genetics . . . may influence mate choice' (Struhsaker *et al.*, 1988). Because primates are highly visual animals, it is generally assumed that the visual appearance of a species plays a crucial role in species recognition and mate choice.

Primate taxonomists (e.g. Kingdon, 1980; Napier and Napier, 1985) have sought to identify characters that would 'prevent hybridization'. It is commonly assumed that in diurnal animals, with well developed color vision, the color of the coat is an important character enabling individuals to recognize their own. Behavior, linked to imprinting of facial markings, is considered an important adjunct in maintaining species integrity. Nevertheless, hybridizations do take place in the wild (Aldrich-Blake, 1968), and in captivity (e.g. Gitzen, 1963; Gautier-Hion, 1973). Similarly, Bernstein & Gordon (1980) considered that since hybridizations occurred rather easily under captive conditions, this is evidence that in 'natural' situations behavioral barriers operate as barriers to gene flow within polyspecific groupings.

Implicitly there must be a mechanism by which *bona fide* species come to recognize and mate with each other. Macaque coat color ranges from light greyish to black, and macaques vary considerably in body size, tail length, crown hair pattern and facial pattern. Whether these differences relate to specific-mate recognition systems remains to be seen.

More recently, Fujita (1987) has attempted to study 'species preference' experimentally by allowing individual macaques to select pictures of other macaques they prefer to see. The premise of the study was that mate recognition functions as a form of pre-mating isolating mechanism. He concluded that adult monkeys do prefer pictures of their own species, but whether this translates into preference to mate with conspecifics remains to be demonstrated. While Fujita (1990) was also interested in when and how species preference developed and whether it was 'natural' or 'acquired', he did not address the question of why individuals should prefer to mate with conspecifics.

Yoshikubo (1987) hypothesized that macaques attain reproductive isolation from closely related species by actively choosing their mate from their own gene pool, and labelled this the 'psychological reproductive isolating mechanism'. He distinguished this from regular behavioral reproductive isolating mechanisms in that these are reflexive and rigid, as are often observed in species-specific courtship behaviors of many non-primates. The psychological mechanism, however, refers to an active learning process of the animal (Fujita, 1990). Struhsaker *et al.* (1988) noted that social attachments and group specific identity in guenons develop from primary relations, so that early association may influence mate choice.

Richard (1985) noted that olive baboons at the species' border, where they occupy a range of habitats, are behaviorally adapted to local conditions. The major characteristics of their social system, however, remain the same. In contrast, five hybrid groups of known genetic make-up (as reflected in their phenotype), and occupying a similar habitat, differed in their social systems. She concluded that it was possible, given an 'appropriate social environment', for the social possessiveness of hamadryas males and, although less pronounced, of hybrid males, to be learned behavior.

In Kowloon, the process of hybridization may have been initiated by habitat destruction. Although the primary forest was continually destroyed, forest patches remained; in particular, the Fung Shui Lam (or sacred groves) have existed for hundreds of years (Nicholls, 1978). During the war years, the forests around the reservoirs were destroyed but groves of sacred trees were cared for and remained standing. These became important refuges for wildlife, and may have served as the meeting ground for the various macaque species.

When closely related species live in proximity, associations between members of different species groups, or individuals, will have ranged from affiliative to aggressive. Different communication systems will not have impeded association, and behaviors such as playing and grooming will have decreased distance. A scarcity of mates in one of two closely related species occurring in small, isolated populations within the forest refuge could lead to a high incidence of hybridization and extensive inter-specific interaction. In Kowloon the inter-specific infant care given by *M. thibetana* females to infants of *M. fascicularis* (Burton & Chan, 1987) may facilitate inter-specific mating, thus fostering hybridization between polyspecific species.

Minimally then, there are two stages in the process of hybridization: the first is recognition of signals and/or overlap of signals of the interacting species; the second is socialization within a complex behavioral environment where repertoires mix. This process would be greatly facilitated by cross-species infant care. Because primate groups live by their traditions, whatever 'fixed-action patterns', that is phylogenetic characters, they have may be insufficient as barriers between species (Struhsaker, 1970).

In his study of behavior as a taxonomic device, Kummer (1970) identified tradition as a possible mechanism of the transfer over generations of action

Macaca thibetana 'babysitting' *M. fascicularis* infant.

patterns that could even extend from one species to another if the two are sympatric. The distinct patterns of *Papio anubis* and *Papio hamadryas* in following behavior do not preclude hybridization as *anubis* learn not to flee from the *hamadryas* neck bite (Kummer, Goetz & Angst, 1970). Therefore, resemblance of behavioral characters may not be a consequence of either common descent or convergence.

Phylogenetic constraint can, therefore, under certain conditions be overcome by proximity. While 'ethospecies', which are morphologically similar, are kept separate by their behavior (Eibl-Eibesfeldt, 1975), inter-specific communication is readily observable: the threat postures of a lizard are comprehensible to a bird, etc. Amongst primates, generally, there is a considerable similarity in communication systems (Shirek-Ellefson, 1972). Where there is comprehension or overlap of behavioral repertoires, interaction readily develops (van Hooff, 1967). Recognition of the communication repertoire is requisite. Backcrossing to parental 'types' increases the range of intelligibility as hybrids easily communicate with both parental forms. With time, local traditions develop (Kummer, 1970; Burton, 1984; Gautier-Hion, 1988). A group's traditions define the set of conventions by which it operates. These include foraging, socialization, hierarchies and role behavior (Burton, 1992). When infants are socialized by members of another species, as in Kowloon, this tendency is reinforced.

Much remains to be learned about the process of species recognition and mate choice, particularly where species interact on a daily basis. We feel that it is more parsimonious to document the conditions and the ontogeny of the social traditions than to invoke ecological models of polyspecific relationships predicated on cost/benefit analysis. Past events provide the conditions and context for inter-species relations. The successful formation of social tradition underlies stable species interaction. Our study, therefore, has emphasized the need to understand associations between species in historical terms.

Acknowledgements

We thank Mr S. P. Lau, Conservator of Forest, Kowloon, and Mr C. L. Wong, Manager of the Kowloon Hill Management Center, Ministry of Fisheries and Agriculture, for making the research possible. The Center of Field Research and the University of Toronto, through its Humanities and Social Sciences Research Grants, provided funding. In addition, the OGS program of the Government of Ontario and the University of Toronto Open Fellowship supported, in part, L. K. W. Chan.

The final revision of this chapter was submitted in July 1992.

References

Aldrich-Blake, F. P. G. (1968). A fertile hybrid between two Cercopithecus spp. in the Budongo forest, Uganda. *Folia Primatologica*, **9**, 15–21.
Altmann, S. A. & Altmann, J. (1970). *Baboon ecology*. Chicago: University of Chicago Press.
Bercovitch, F. B. & Goy, R. W. (1990). The socioendocrinology of reproductive development and reproductive success in macaques. *Socioendocrinology of primates*, ed. C. Worthman, pp. 59–83. New York: John Wiley & Sons.
Bernstein, I. S. (1966). Naturally occurring primate hybrid. *Science*, **154**, 1559–60.
Bernstein, I. S. (1967). Intertaxa interactions in a Malayan primate community. *Folia Primatologica*, **7**, 198–207.
Bernstein, I. S. & Gordon, T. P. (1980). Mixed taxa introductions, hybrids and macaque systematics. In *The macaques: studies in ecology, behavior and*

evolution, ed. D. G. Lindburg, pp. 125–147. New York: Van Nostrand Reinhold.

Bishop, N., Blaffer Hrdy, S., Teas, J. & Moore, J. (1981). Measures of human influence in habitats of South Asian monkeys. *International Journal of Primatology*, **2**, 153–67.

Boinski, S. (1989). Why don't *Saimiri oerstedii* and *Cebus capucinus* form mixed-species groups? *International Journal of Primatology*, **10**, 103–14.

Buchanan-Smith, H. (1990). Polyspecific association of two tamarin species, *Saguinus labiatus* and *Saguinus fuscicollis*, in Bolivia. *American Journal of Primatology*, **22**, 205–14.

Burton, F. D. (1984). Inferences of cognitive abilities in Old World monkeys. *Semiotica*, **50**, 69–81.

Burton, F. D. (1992). The social group as information unit: cognitive behaviour, cultural processes. In *Social processes and mental abilities in non-human primates*, ed. F. D. Burton, pp. 31–60. Lewiston, NY: Edwin Mellen.

Burton, F. D. & Chan, L. K. W. (1987). Notes on the care of long-tail macaque (*Macaca fascicularis*) infants by stump-tail (*M. thibetana*) females. *Canadian Journal of Zoology*, **65**, 752–5.

Burton, F. D. & Sawchuk, L. A. (1974). Demography of *Macaca sylvanus* of Gibraltar. *Primates*, **15**, 271–8.

Carpenter, C. (1964). *Naturalistic behavior of non-human primates*. University Park, PN: Pennsylvania State University Press.

Castro, R. & Soini, P. (1978). Field studies on *Saguinus mystax* and other callitrichids in Amazonian Peru. In *The biology and conservation of the Callitrichidae*, ed. D. G. Kleinman, pp. 73–8. Washington, DC: Smithsonian Institution Press.

Chan, L. K. W. (1992). Problems with socioecological explanations of primate social diversity. In *Social processes and mental abilities in non-human primates*, ed. F. D. Burton, pp. 1–30. Lewiston, NY: Edwin Mellen.

Cords, M. (1987). *Mixed-species association of* Cercopithecus *monkeys in the Kakamega Forest, Kenya*. Berkeley & Los Angeles: University of California Press.

Cords, M. (1990). Mixed-species association of East African guenons: general patterns or specific examples? *American Journal of Primatology*, **21**, 101–14.

Crockett, C. M. & Eisenberg, J. F. (1987). Variations in group size and demography. In *Primate societies*, ed. B. B. Smuts, D. L. Cheney, R. M. Seyfarth, R. W. Wrangham & T. T. Struhsaker, pp. 54–68. Chicago: University of Chicago Press.

Crockett, C. M. & Wilson, W. L. (1980). The ecological separation of *Macaca nemestrina* and *M. fascicularis* in Sumatra. In *The macaques: studies in ecology, behavior and evolution*, ed. D. G. Lindburg, pp. 148–81. New York: Van Nostrand Reinhold.

Dobzhansky, T. (1937). *Genetics and the origin of species*. New York: Columbia University Press.

Dunbar, R. I. M. (1987). Demography and reproduction. In *Primate societies*, ed. B. B. Smuts, D. L. Cheney, R. M. Seyfarth, R. W. Wrangham & T. T. Struhsaker, pp. 240–9. Chicago: University of Chicago Press.

Dunbar, R. I. M. (1988). *Primate social systems*. New York: Comstock Publishing.

Dunbar, R. I. M. & Dunbar, E. P. (1974). Ecological relations and niche separation between sympatric terrestrial primates. *Ethiopia Folia Primatologica*, **21**, 36–60.

Eibl-Eibesfeldt, I. (1975). *Ethology: the biology of behavior*, 2nd edn. New York: Holt, Rinehart & Winston.

Eisenberg, J. F. (1979). Economy and society: some correlations and hypotheses for the neotropical primates. In *Primate ecology and human origins*, ed. I. S. Bernstein & E. O. Smith, pp. 215–62. New York: Garland.

Eudey, A. A. (1979). Differentiation and dispersal of macaques (*Macaca* spp.) in Asia. PhD thesis, University of California, Davis.

Eudey, A. A. (1980). Pleistocene glacial phenomena and the evolution of Asian macaques. In *The macaques: studies in ecology, behavior and evolution*, ed. D. G. Lindburg, pp. 52–83. New York: Van Nostrand Reinhold.

Fa, J. & Southwick, C. (1988). Introduction. In *Ecology and behavior of food-enhanced primate groups*, ed. J. Fa & C. Southwick, pp. xv–xvii. New York: Alan R. Liss.

Fittinghoff, N. A. & Lindburg, D. G. (1980). Riverine refuging in East Bornean *Macaca fascicularis*. In *The macaques: studies in ecology, behavior and evolution*, ed. D. G. Lindburg, pp. 182–214. New York: Van Nostrand Reinhold.

Fooden, J. (1982). Ecogeographic segregation of macaque species. *Primates*, **23**, 574–9.

Fujita, K. (1987). Species recognition by five macaque monkeys. *Primates*, **28**, 353–66.

Fujita, K. (1990). Species preference by infant macaques with controlled social experience. *International Journal of Primatology*, **11**, 553–73.

Gartlan, J. S. & Struhsaker, T. T. (1972). Polyspecific associations and niche separation of rain-forest Anthropoids in Cameroon, West Africa. *Journal of Zoology* (London), **168**, 221–66.

Gautier-Hion, A. (1973). Social and ecological features of Talapoin monkey (sic) – comparisons with sympatric cercopithecines. In *Comparative ecology and behavior of primates*, ed. R. P. Michael & J. H. Crook, pp. 148–69. New York: Academic Press.

Gautier-Hion, A. (1988). Polyspecific associations among forest guenons: ecological, behavioural and evolutionary aspects. In *A primate radiation: evolutionary biology of the African guenons*, ed. A. Gautier-Hion, F. Bourlière, J.-P. Gautier & J. Kingdon, pp. 452–76. Cambridge: Cambridge University Press.

Gautier-Hion, A., Quris, R. & Gautier, J.-P. (1983). Monospecific vs. polyspecific life: a comparative study of foraging and antipredatory tactics in a community of *Cercopithecus* monkeys. *Behavioral Ecology and Sociobiology*, **12**, 325–35.

Gitzen, A. (1963). Croisement accidentel entre deus espaces de Cercopitheques, *C. mona m.* et *C. mitis dogetti*. *Zoologifche Medelingen*, **39**, 522–5.

Goodall, J. (1965). Chimpanzees of the Gombe Stream Reserve. In *Primate behavior*, ed. I. DeVore, pp. 425–73. New York: Holt, Rinehart & Winston.

Goodall, J. (1986). *The chimpanzees of Gombe: patterns of behavior*. Harvard, MA: Belnap.

Gordon, T. P. & Bernstein, I. S. (1973). Seasonal variation in sexual behavior of all-male rhesus troops. *American Journal of Physical Anthropology*, **38**, 221–6.

Haddow, A. J. (1952). Field and laboratory studies of an African monkey, *Cercopithecus ascanius schmidti*. *Proceedings of the Zoological Society, London*, **122**, 297–394.

Hall, K. R. L. (1962). Numerical data, maintenance activities and locomotion of the wild Chacma baboon, *Papio ursinus*. *Proceedings of the Zoological Society, London*, **139**, 283–327.

Hall, K. R. L. (1963). Variations in the ecology of the Chacma baboon, *Papio ursinus*. *Symposium of the Zoological Society, London*, **10**, 1–28.

Hall, K. R. L. (1965). Behaviour and ecology of the wild Patas, *Erythrocebus patas*, in Uganda. *Journal of the Zoological Society, London*, **148**, 15–87.

Herklots, G. A. C. (1951). *The Hong Kong countryside*. Hong Kong: South China Morning Post, Ltd.

Hladik, C. M. (1979). Ecology, diet, and social patterning in Old and New World primates. In *Primate ecology: problem-oriented field studies*, ed. R. W. Sussman, pp. 513–42. New York: John Wiley & Sons.

Hrdy, S. B. & Whitten, P. L. (1987). Patterning of sexual activity. In *Primate societies*, ed. B. B. Smuts, D. L. Cheney, R. M. Seyfarth, R. W. Wrangham, & T. T. Struhsaker, pp. 370–84. Chicago: University of Chicago Press.

Jay, P. (1965). The common langur of north India. In *Primate behaviour: field studies of monkeys and apes*, ed. I. DeVore, pp. 197–249. New York: Holt, Rinehart & Winston.

Kavanagh, M. & Laursen, E. (1984). Breeding seasonality among long-tailed macaques, *Macaca fascicularis*, in penisular Malaysia. *International Journal of Primatology*, **5**, 17–30.

Kawai, M., Azuma, S. & Yoshiba, K. (1967). Ecological studies of reproduction in Japanese monkeys (*M. fuscata*). I. Problems of the birth season. *Primates*, **8**, 35–74.

Kingdon, J. S. (1980). The role of visual signals and face patterns in African forest monkeys (guenons) of the genus *Cercopithecus*. *Transactions of the Zoological Society, London*, **35**, 425–75.

Klein, L. L. & Klein, D. J. (1973). Observations of two types of neotropical primate intertaxa associations. *American Journal of Physical Anthropology*, **38**, 649–53.

Koford, C. B. (1963). Rank of mothers and sons in bands of rhesus monkeys. *Science*, **141**, 356–7.

Kummer, H. (1970). Behavioral characters in Primate taxonomy. In *Old World monkeys: evolution, systematics and behaviour*, ed. J. R. Napier & P. H. Napier, pp. 25–36. New York: Academic Press.

Kummer, H., Goetz, W. & Angst, W. (1970). Cross-species modification of social behaviour in baboons. In *Old World monkeys: evolution, systematics and behavior*, ed. J. R. Napier & P. H. Napier, pp. 351–63. New York: Academic Press.

Lancaster, J. B. & Lee, R. B. (1965). The annual reproductive cycle in monkeys and apes. In *Primate behavior: field studies of monkeys and apes*, ed. I. DeVore, pp. 486–513. New York: Holt, Rinehart & Winston.

Lance, V. A. (1976). The land vertebrates of Hong Kong. In *The fauna of Hong Kong*, ed. B. Lofts, pp. 6–22. Hong Kong: Royal Asiatic Society.

Lindburg, D. G. (1987). Seasonality of reproduction in primates. In *Comparative primate biology*, vol. 2B, *Behavior, cognition, and motivation*, ed. G. Mitchell & J. Erwin, pp. 167–218. New York: Alan R. Liss.

Lyles, A. M. & Dobson, A. P. (1988). Dynamics of provisioned and unprovisioned primate populations. In *Ecology and behavior of food-enhanced primate groups*, ed. J. E. Fa & C. H. Southwick, pp. 167–98. New York: Alan R. Liss.

MacKinnon, J. R. & MacKinnon, K. S. (1978). Comparative feeding ecology of six sympatric primates in West Malaysia. In *Recent advances in primatology*, ed. D. J. Chivers & J. Herbert, pp. 305–21. London: Academic Press.

Mayr, E. (1942). *Systematics and the origin of species*. New York: Columbia University Press.

Melnick, D. J. & Pearl, M. C. (1987). Cercopithecines in multimale groups: genetic diversity and population structure. In *Primate societies*, ed. B. B. Smuts, D. L. Cheney, R. M. Seyfarth, R. W. Wrangham & T. T. Struhsaker, pp. 121–34. Chicago: University of Chicago Press.

Migi, H. (1976). Some aspects related to conception of the Japanese monkey (*M. fuscata*). *Primates*, **21**, 230–40.

Napier, J. R. & Napier, P. H. (1985). *The natural history of the primates.* Cambridge, MA: MIT Press.

Nicholls, D. (1978). Some aspects of the vegetation of Hong Kong with special reference to the Fung Shui Wood. Department of Geography: Leicester University.

Norcork, M. A. (1990). Mechanisms promoting stability in mixed *Saguinus mystax* and *S. fuscicollis* troops. *American Journal of Primatology*, **21**, 159–70.

Odum, E. P. (1959). *Fundamentals of ecology.* Philadelphia: Saunders.

Paterson, J. D. (1976). Variations in ecology and adaptation of Ugandan baboons, *Papio cynocephalus anubis*: with special reference to forest environments and analog models for early hominids. PhD thesis, University of Toronto.

Richard, A. F. (1985). *Primates in nature.* New York: W. H. Freeman.

Richard, A. F., Goldstein, S. J. & Dewar, R. E. (1989). Weed macaques: the evolutionary implications of macaque feeding ecology. *International Journal of Primatology*, **10**, 569–97.

Rodman, P. S. (1991). Structural differentiation of microhabitats of sympatric *Macaca fascicularis* and *M. nemestrina* in East Kalimantan, Indonesia. *International Journal of Primatology*, **12**, 357–75.

Romer, J. D. (1966). Notes: long-tailed macaques in Hong Kong. *Memoirs of the Hong Kong Natural History Society*, **7**, 16.

Shirek-Ellefson, J. (1972). Social communication in some Old World monkeys and gibbons. In *Primate patterns*, ed. P. Dolhinow, pp. 297–311. New York: Holt, Rinehart & Winston.

Sousa, W. P. (1984). The role of disturbance in natural communities. *Annual Review of Ecology and Systematics*, **15**, 353–91.

Southwick, C. & Southwick, K. (1983). Polyspecific groups of macaques on the Kowloon Penisula, New Territories, Hong Kong. *American Journal of Primatology*, **5**, 17–24.

Struhsaker, T. T. (1970). Phylogenetic implications of some vocalizations of *Cercopithecus* monkeys. In *Old World monkeys: evolution, systematics and behavior*, ed. J. R. Napier & P. H. Napier, pp. 365–403. New York: Academic Press.

Struhsaker, T. T. (1975). *The red colobus monkey.* Chicago: University of Chicago Press.

Struhsaker, T. T. (1978). Food habits of five monkey species in the Kibale Forest, Uganda. In *Recent advances in primatology*, vol. 11, ed. D. J. Chivers & J. Herbert, pp. 225–48. New York: Academic Press.

Struhsaker, T. T. (1981). Polyspecific associations among tropical rain forest primates. *Zeitschrift für Tierpsychologie*, **57**, 268–304.

Struhsaker, T. T., Butynski, T. & Lwanga, J. (1988). Hybridization between redtail (*Cercopithecus ascanius schmidti*) and blue (*C. mitis stuhlmanni*) monkeys in the Kibale Forest, Uganda. In *A primate radiation: evolutionary biology of the African guenons*, ed. A. Gautier-Hion, F. Bourlière, J.-P. Gautier & J. Kingdon, pp. 477–97. Cambridge: Cambridge University Press.

Struhsaker, T. T. & Gartlan, J. S. (1970). Observations on the behavior and ecology of the patas monkey (*Erythrocebus patas*) in the Waza Reserve, Cameroon. *Journal of Zoology, London,* **161**, 49–63.

Struhsaker, T. T. & Leland, L. (1987). Colobines: infanticide by adult males. In *Primate societies,* ed. B. B. Smuts, D. L. Cheney, R. M. Seyfarth, R. W. Wrangham & T. T. Struhsaker, pp. 83–97. Chicago: University of Chicago Press.

Strum, S. & Southwick, C. (1986). Translocation of primates. In *Primates: the road to self-sustaining populations,* ed. K. Benirschke, pp. 949–58. New York: Springer-Verlag.

Sugawara, K. (1979). Sociological study of a wild group of hybrid baboons between *Papio anubis* and *P. hamadryas,* in the Awash Valley, Ethiopia. *Primates,* **20**, 21–56.

Sugiyama, Y. (1967). Social organization of hanuman langurs. *Social communication among primates,* ed. S. A. Altmann, pp. 207–19. Chicago: University of Chicago Press.

Terborgh, J. (1983). *Five New World primates.* Princeton: Princeton University Press.

Terborgh, J. (1990). Mixed flocks and polyspecific associations: costs and benefits of mixed groups of birds and monkeys. *American Journal of Primatology,* **21**, 87–100.

van Hooff, J. A. R. A. M. (1967). The facial displays of the Catarrhine monkeys and apes. In *Primate ethology,* ed. D. Morris, pp. 7–68. Chicago: Aldine.

van Horn, R. N. (1980). Seasonal reproductive patterns in primates. In *Seasonal reproduction in higher vertebrates,* ed. R. J. Reiter & B. K. Follett, pp. 181–221. Basel: S. Karger.

Wallis, J., King, B. & Roth-Meyer, C. (1986). The effect of female proximity and social interaction on the menstrual cycle of crab-eating monkeys (*Macaca fascicularis*). *Primates,* **27**, 83–94.

Waser, P. M. (1982). Primate polyspecific associations: do they occur by chance? *Animal Behaviour,* **30**, 1–8.

Waser, P. M. (1987). Interactions among primate species. In *Primate societies,* ed. B. B. Smuts, D. L. Cheney, R. M. Seyfarth, R. W. Wrangham, & T. T. Struhsaker, pp. 210–26. Chicago: University of Chicago Press.

Yoshikubo, S. (1987). A possible reproductive isolation through a species discrimination learning in genus *Macaca. Primate Research,* **3**, 43–7.

19

The population genetic consequences of macaque social organisation and behaviour

D. J. MELNICK AND G. A. HOELZER

Introduction

Among the non-human primates, no genus has been more thoroughly studied than the macaques. This chapter presents some of that research, particularly as it relates to the relationship between behaviour of individuals, the organisation of social groups, and the distribution of genetic variation within and among macaque populations. Our review is not comprehensive as an annotated bibliography of macaque behavioural ecology and genetics could fill the present volume. Instead, we describe the salient features of both macaque social organisation and population genetic structure, and attempt to develop causal links between the two.

Previous papers (Melnick, 1987, 1988) examined the effects of behaviour (i.e. patterns of dispersal, mating, and social group fission) on the genetic make-up of a number of macaque and other cercopithecine species. This review differs from previous reports in three ways. First, we focus almost exclusively on macaques, though some comparisons with other primate and non-primate mammals are made. Second, we incorporate the results of studies that have been completed in the past five years. Third, we extend our analyses to include the population genetic structure of variation in the mitochondrial genome (mtDNA).

It is the aim of this chapter to review the effect of macaque social organisation on the distribution of variation in both the nuclear and mitochondrial genomes. We also discuss the implications of these effects for the molecular and organismal evolution of macaques.

Macaque social organisation

Macaque species exhibit a multimale/multifemale social structure (Melnick & Pearl, 1987) commonly found in many other primate and non-primate mammals (Greenwood, 1980). For the purposes of this chapter, the important features of macaque social structure are:

1. The members of a population are sub-divided into groups containing several adult males, two to three times as many adult females, and an equal number of juvenile offspring.
2. Within each group there exist social dominance hierarchies that determine access to certain limited resources.
3. New groups are formed by the division of pre-existing ones, usually along matrilineal lines.
4. Male macaques almost always migrate from their natal group before reaching sexual maturity, while females ordinarily remain in their natal group throughout their life.

In the absence of selection (see below), the single most important factor determining the genetic structure of primate populations is dispersal. While on the face of it this appears to discount the effects of random genetic drift, one can easily see that if dispersal is ubiquitous the population, or 'neighbourhood' (Wright, 1969), will be large enough to inhibit genetic divergence (Hartl & Clark, 1989). Hence, dispersal will have a significant impact on the relatedness of individuals within a social group, the level and direction of gene flow, the effective population size, and the distribution of genetic variation at all levels of animal organisation, from the individual to the species.

Among most mammal species, males are much more likely to leave their natal group or area than females (Greenwood, 1980). In many higher primates, and in particular among macaques, this form of gender-biased dispersal is taken to an extreme (Clutton-Brock, 1989) with virtually all males dispersing from, and all females remaining in, their natal social group (Lindburg, 1969; Sade, 1972; Pearl, 1982; Pusey & Packer, 1987; for exceptions see Moore, 1984). Several different evolutionary models for this type of asymmetrical dispersal have been developed (Moore, 1984; Pusey & Packer, 1987), including those based on inbreeding avoidance and resource competition (Moore & Ali, 1984; Clutton-Brock, 1989). While much effort has been put into examining these models, it is doubtful whether any single explanation will hold as a rule. Moreover, our analysis of the effects of macaque social organisation on population genetic structure does not require a solution to this problem, only the realisation that extant dispersal patterns are a contingent fact of macaque population dynamics (Melnick & Pearl, 1987).

The extreme gender-biased dispersal pattern exhibited by macaques may cause the distribution of nuclear and mitochondrial genetic variation to differ dramatically because the two genomes are inherited in different ways. Half the nuclear genome is inherited from each parent, whereas the mitochondrial genome is inherited in a clonal fashion from the mother (Hutchison *et al.*, 1974; Brown, 1983; Honeycutt & Wheeler, 1989). Consequently, mitochondrial gene flow should be greatly reduced relative to nuclear gene flow. Additionally, the single-sex transmission pattern should, as in the case of surnames, result in a sorting process (Avise, Neigel & Arnold, 1984) that inevitably leads to local homogeneity when female migration is limited.

The data

Macaque species

Genetic data exist for all 19 extant species of *Macaca* (see Fooden & Lanyon, 1989; Melnick, Hoelzer & Honeycutt, 1992; and references therein). However, there have been very few population genetic analyses of macaques. Additionally, it is difficult to make comparisons between species if their biogeography differs to such an extent that the effects of behaviour become conflated with the effects of geographical distribution. Therefore, to judge properly the effects of macaque behaviour on population genetic structure one must constrain comparisons even further. Finally, we wish to compare the effects of macaque social organisation on variation in the nuclear and mitochondrial genomes; however, studies of macaque mtDNA at the population level are extremely limited. To accommodate all of these considerations we found it necessary to narrow our focus to five macaque species: the rhesus monkey (*M. mulatta*), the long-tailed macaque (*M. fascicularis*), the Japanese macaque (*M. fuscata*), the toque macaque (*M. sinica*) and the pig-tailed macaque (*M. nemestrina*). The geographical ranges of these species, and the sources of genetic data used in our analyses can be found in Fig. 19.1 and Table 19.1.

The nuclear genome

In this review, an array of studies of electrophoretic variation in blood protein loci are used to examine the population structure of nuclear variants. The shortcomings of this approach lie primarily in the unknown action of natural selection on these loci, which may result in broad

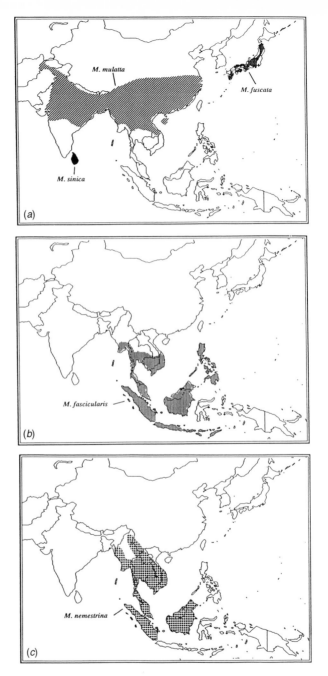

Fig. 19.1. The ranges of the five macaque species described in this chapter are indicated in these maps of southern and eastern Asia: (a) the rhesus macaque (M. mulatta – hatched), the toque monkey (M. sinica – solid black), and the Japanese macaque (M. fuscata – white spots on a black background); (b) the long-tailed macaque (M. fascicularis – vertical lines); (c) the pig-tailed macaque (M. nemestrina – checkered).

Table 19.1. *Sources of data on macaque species*

Species	Population	Sample size	No. of loci	Sources
A. Nuclear genome				
M. mulatta	Pakistan	32–221	25–37	Shotake, 1979; Melnick *et al.*, 1984*a,b*; Melnick, Jolly & Kidd, 1986; D.J.M., unpublished data.
	India	214–240	25	Darga, 1975; Shotake, 1979
	Thailand	31–59	25	Darga, 1975; Shotake, 1979
	China	60–76	25	Nozawa *et al.*, 1977; Shotake, 1979
M. fascicularis	Thailand	119–197	22	Weiss & Goodman, 1972; Darga, 1975
	Malaysia	32–263	29	Weiss & Goodman, 1972; Darga, 1975; Nozawa *et al.*, 1977; Shotake, 1979
	Sumatra	118	33	Kawamoto *et al.*, 1984
	Java	91–113	33	Darga, 1975; Kawamoto *et al.*, 1984
	Bali	136	33	Kawamoto *et al.*, 1984
	Lombok	33	33	Kawamoto *et al.*, 1984
	Sumbawa	78	33	Kawamoto *et al.*, 1984
	Philippines	21–99	29	Weiss & Goodman, 1972; Darga, 1975; Nozawa *et al.*, 1977; Shotake, 1979
M. sinica	Sri Lanka	256	32	Shotake *et al.*, 1991
M. fuscata	Kyushu	538	32	Nozawa *et al.*, 1982
	Shikoku	80	32	Nozawa *et al.*, 1982
	Honshu	845	32	Nozawa *et al.*, 1982
	Awajishima	87	32	Nozawa *et al.*, 1982
	Shodoshima	96	32	Nozawa *et al.*, 1982
M. nemestrina	Thailand	11–92	25	Darga, 1975; Nozawa *et al.*, 1977; Shotake, 1979
	Malaysia	151–314	25	Anderson & Giblett, 1975; Darga, 1975
	Sumatra	36–203	25	Anderson & Giblett, 1975; Darga, 1975
	Mentawi	7	33	Kawamoto *et al.*, 1985
B. Mitochondrial genome				
M. mulatta	Pakistan	3		Melnick *et al.*, 1993
	India	4		Melnick *et al.*, 1993
	Burma	4		Melnick *et al.*, 1993
	China	7		Melnick *et al.*, 1993
M. fascicularis	Malaysia	48		Harihara *et al.*, 1988
	Indonesia	5		Harihara *et al.*, 1988
	Philippines	48		Harihara *et al.*, 1988
M. sinica	Sri Lanka	21		Hoelzer *et al.*, 1994
M. fuscata	Honshu	8		Hayasaka *et al.*, 1988
	Shodoshima	1		Hayasaka *et al.*, 1988
	Yakushima	1		Hayasaka *et al.*, 1988
M. nemestrina	China	2		Rosenblum, Hoelzer & Melnick, 1992
	Thailand	2		Rosenblum *et al.*, 1992
	Malaysia	1		Williams, 1990
	Sumatra	2		Williams, 1990
	Borneo	3		Rosenblum *et al.*, 1992
	Siberut	3		Williams, 1990
	Pagai	1		Williams, 1990

similarities between populations that do not regularly exchange genes or share a recent common ancestor (Ehrlich & Raven, 1969). However, our analyses indicate that selection is unlikely to be an important factor (Melnick, 1988; see below), and therefore, the distribution of variation at these effectively neutral loci should accurately reflect the effects of dispersal, mating and group fission.

The data used in the analyses reported here come from a detailed study of one specific rhesus population (Melnick, Jolly & Kidd, 1984a, 1986), and numerous other studies of social groups, local populations and regions in all five species (see Table 19.1). The specific allele frequencies at the loci used are described elsewhere (Melnick, 1988; Fooden & Lanyon, 1989), as is the application of Nei's (1973) gene diversity analysis to these data (Melnick, 1988). The details of other analytical methods employed in this report can be found in the references cited in the relevant sections below.

Selective neutrality of gene products

It has been commonly assumed that the genetic relationships between primate species (Melnick & Kidd, 1985), the genetic structure of primate species (Nozawa *et al.*, 1982; Kawamoto, Ischak & Supriatna, 1984; Melnick *et al.*, 1986), and the genetic affinities of social groups within primate populations (Melnick *et al.*, 1984a) are largely the products of processes other than selection (i.e. mutation, genetic drift and gene flow). However, recent evidence from studies of captive rhesus indicates that some commonly assayed polymorphic protein loci exhibit allele frequencies that may have been shaped by natural selection (Smith & Ferrel, 1980; Smith & Small, 1982; Smith & Rolfs, 1984; Smith & Ahlfors, 1984).

Since the appearance of the neutral mutation hypothesis (Kimura, 1968), numerous methods have been employed to compare empirically determined distributions of allele frequencies, or locus-specific heterozygosities, with theoretical distributions of these variables derived from assumptions of selective neutrality (see Nei, 1987). To examine the potential influence of selection on the genetic structure of macaque species two tests of the null hypothesis of selective neutrality were employed. The first test, developed by Nei, Fuerst & Chakraborty (1976) and further elaborated by Fuerst *et al.* (1977), is a statistical comparison between a theoretical distribution of single-locus heterozygosity, assuming an infinite alleles model of selective neutrality, and the empirical distributions in each macaque population examined. The second test, incorporating the work of Ewens (1972) and

Watterson (1977, 1978), provides a means for calculating, in each macaque population, the probability of observed single-locus homozygosities under selective neutrality for each locus. These can then be aggregated by species and compared with a random expectation of the distribution of these probabilities (Slatkin, 1982).

The results of the two tests applied here (for details, see Melnick, 1988) suggest that, while there may be some evidence of selection at a few loci in a limited number of populations, most loci are selectively neutral. These results fit well with the neutral theory of molecular evolution (Kimura, 1983) that allows for a small proportion of overdominant or selectively advantageous mutations. However, in general, the distribution of allele frequencies (and their derivatives, homozygosity and heterozygosity) at the 32 loci examined and, therefore, the genetic structure of the species described here, is best explained as the product of evolutionary forces other than selection. It should be noted that these tests do not constitute an exhaustive evaluation of the absence or presence of selection in primates. They are meant solely as an estimate of the relative importance of this evolutionary force in determining macaque population genetic structure.

The mitochondrial genome

As we suggested, macaque mtDNA data are limited. The most detailed intra-specific analyses of variation in the mitochondrial genome have been carried out on the rhesus (Melnick & Hoelzer, 1992; Melnick *et al.*, 1993) and Japanese (Hayasaka *et al.*, 1986, 1988) macaques, where, respectively, individuals from five regions (Pakistan, India, Burma, South-West China, South-East China) and three islands (Honshu, Shodoshima, Yakushima) have been surveyed by restriction enzyme analysis, using 15 restriction endonucleases (*Ava*I, *Bam*HI, *Bgl*II, *Cla*I, *Dra*I, *Eco*RI, *Eco*RV, *Hae*II, *Hinc*II, *Hind*III, *Kpn*I, *Pst*I, *Sst*I and *Xba*I).

The specific conditions of DNA extraction, restriction enzyme digestion and electrophoresis, and DNA transfer and visualisation are detailed in the original research reports (for additional information on molecular methods of mtDNA research, see also Hillis & Moritz, 1990, and Melnick *et al.*, 1992). In the rhesus, ten unique multienzyme haplotypes were identified, each composed of 40 to 47 restriction fragments. Four unique haplotypes, comprising a similar number of fragments, were uncovered among the Japanese macaques surveyed. The restriction sites demarcating each fragment were located on a linear mtDNA map using single and double digestions, and the logic of parsimony (Dowling, Moritz & Palmer, 1990).

Restriction site data were used to estimate within-population diversity and between-population differences employing the methods of Nei & Tajima (1983) and the computer program MAXLIKE (courtesy of M. Nei and L. Jin). The gene diversity analysis developed for mtDNA data is analogous to the methods previously described for nuclear genetic variation and, therefore, can be directly compared (Nei, 1982).

Published and unpublished data on mtDNA variation in the other species of macaques (Hayasaka *et al.*, 1988; Harihara *et al.*, 1988; Rosenblum *et al.*, 1992; Hoelzer *et al.*, 1992, 1994; A. K. Williams, M. V. Ashley, R. Tenaza & D. J. Melnick, unpublished data) included in this review were also used to shed light on the degree to which the distribution of mtDNA variation in the rhesus and Japanese macaques can be found in macaque species with similar dispersal patterns. Some of these data sets consist of restriction fragment profiles, rather than mapped restriction sites, and were therefore analysed using the methods of Nei & Li (1979).

Population genetic structure

The nuclear genome

Rhesus macaque

As reported elsewhere (Melnick, 1988; Melnick & Hoelzer, 1992), the distribution of nuclear genetic variation in the rhesus monkey, at all levels of population sub-division, is extremely homogeneous (Table 19.2). Nearly 96% of the nuclear gene diversity found in the local rhesus monkey population of Dunga Gali, Pakistan, can be found in any one of its constituent social groups. Approximately 99% of the diversity found throughout the rhesus range in the Pakistan region was found in both the local populations surveyed. Furthermore, across the entire species range, which extends several thousand kilometres, as much as 91% of the total species diversity could be apportioned to individual differences within the same region.

Long-tailed macaque

The population structure of long-tailed macaques is similar to that found among the rhesus at several levels of sub-division, but differs in one important way. In local populations about 96% of the existing diversity can be attributed to differences between individuals in the same social group. Similarly, the diversity within each region can be largely apportioned to

Table 19.2. *Nuclear gene diversity in four macaque monkey species*

	H_S/H_L	H_L/H_P	H_L/H_T	H_P/H_T
Widely distributed species				
M. mulatta	0.96	0.99	0.90	0.91
M. fascicularis	0.96	0.93	0.55	0.67
M. nemestrina				0.73
Narrowly distributed species				
M. sinica	0.97		0.97	
M. fuscata	0.99	0.93	0.76	0.82

H_S, H_L, H_P and H_T, average gene diversity in a social group, a local population, a regional population, and the species, respectively.

differences within local populations. However, if one looks across the species' range, which covers many islands, only 67% of the total gene diversity is found in any single region.

Pig-tailed macaque

The data for pig-tailed macaques is not nearly as detailed as it is for other species. Nevertheless, in this species, which like the long-tailed macaque has both continental and insular populations, one can estimate the average proportion of total species gene diversity found in any of the regions surveyed (Thailand, Malay peninsula, Sumatra, Mentawi Islands). Taking all regional populations into account, average regional gene diversity is only 73% of that found in the entire species. This value is roughly comparable to that which we found in the long-tailed macaque.

Toque macaque

The toque monkey has a very limited geographical range that is largely continuous (Fooden, 1979). Given the degree to which nuclear genes appear to be homogeneously spread across large, continuous areas, such as the range of the rhesus monkey, one might expect a fairly uniform distribution of nuclear gene diversity across the island of Sri Lanka. Using data presented in Shotake & Santiapillai (1982) and Shotake, Nozawa & Santiapillai (1991), we examined this prediction. As expected, most genetic variants can be found across the island in roughly equal frequencies. Hence, all comparable gene diversity parameters exceed 90% (Table 19.2) and on average over 86% of total toque monkey gene diversity (or 89% if one corrects for the unevenness of the data – see Melnick, 1987) can be attributed to differences within any small social group (Shotake *et al.*, 1991), matching a similarly calculated value in the rhesus monkey ($H_S/H_T \times 100 = 86.5\%$, where H_S and H_T are the average gene diversity in a social group and the species, respectively).

Japanese macaque

The Japanese macaque, like the toque macaque, is spread over a relatively small geographical area. However, unlike the toque monkey its range is sub-divided across an archipelago (Fig. 19.1). As in the case of the other macaque species, nuclear gene diversity is quite uniformly arrayed across the continuous parts of the species' range. This uniformity begins to decline, however, when one compares populations between islands. Here we find that only 76% to 82% (depending upon which measure one uses) of the total species diversity can be found in each local or regional population, respectively. These values are comparable to the other insular species examined (i.e. the long-tailed and pig-tailed macaques), but are 10% to 20% below what we find in those species with a continuous species range.

Summary

A comparison of the five macaque species presented here suggests that: (1) the average social group contains most of its local population's genetic diversity; (2) most of the genetic diversity of the average regional population can be found in its constituent local populations; and (3) the proportion of species diversity found in the average regional population (or its constituent local populations and social groups) varies substantially among species, depending upon the degree to which a species' geographical range is fragmented by water barriers.

The mitochondrial genome

Rhesus macaque

Estimated mtDNA sequence divergence (0.2%–4.5%) among rhesus monkeys (Melnick *et al.*, 1993) is much higher than that found in most species of primates (Melnick *et al.*, 1992) or other mammals (Wilson *et al.*, 1985). Apportioning this diversity to within and between population differences reveals two characteristics of mtDNA population structure that are markedly different from our findings for the nuclear genome. First, there is very little sequence variation within any regional population (an average of 0.23%). Second, the sequence differences between populations (an average

Table 19.3. *Estimated mitochondrial DNA percentage sequence divergence within and between macaque populations*

Species	Within population	Between population
M. mulatta	0.23 (0–0.8)	2.45 (0–4.5)
M. fascicularis	0.38 (0.25–0.54)	1.59 (0.68–2.07)
M. sinica	3.10	
M. fuscata	0.0	1.74 (0–2.4)
M. nemestrina	0.15 (0–0.3)	3.51 (1.0–5.3)

of 2.45%) are more than an order of magnitude larger than the differences found within populations. Hence, mtDNA diversity analysis (Nei, 1982) indicates that approximately 9% of the total species diversity can be apportioned to within population differences, whereas 91% of the species diversity consists of between population differences. These statistics are the exact opposite of what one finds in the nuclear genome (Table 19.2 vs 19.3).

Long-tailed macaque

Paralleling what we found in the rhesus monkeys, Harihara *et al.* (1988) estimated the level of intra-populational diversity in long-tailed macaques at 0.38. Similarly, average sequence divergence between populations of *M. fascicularis* was about four times as large as the differences within populations (Harihara *et al.*, 1988). Because it is likely that Harihara *et al.*'s 'Indonesian' sample contains individuals from several of that country's islands (what we would define as separate regions), it was not considered a single regional population and was not included in this analysis. In addition, sequence divergence was measured using restriction fragment length polymorphisms (RFLPs) from only five restriction enzymes. It has been our experience that as the number of restriction enzymes increases the level of sequence divergence estimated increases and this does not level off until you exceed about ten enzymes. Hence, the absolute diversity values reported here are probably underestimates of the true levels of divergence, particularly for inter-populational comparisons, and the conclusions one can draw from these data are limited.

Pig-tailed macaque

Because of sample size constraints, intra-populational diversity could be estimated in only a few cases. Nevertheless, in those cases it ranged from 0.0% to 0.3%. Divergence between haplotypes in different populations ranged from 1.03% to 5.27%, with an average of 3.51% (A. K. Williams,

M. V. Ashley, R. Tenaza & D. J. Melnick, unpublished data). Hence, differences between populations were about 10 fold higher than those within populations, roughly what we have found in the other macaque species. The absolute values of inter-populational divergence are again very high.

Toque macaque

The only data on toque macaque mtDNA come from a study of the Polonnaruwa, Sri Lanka, population (Hoelzer, Hoelzer & Melnick, 1992; Hoelzer *et al.*, 1994). In this population, two multienzyme haplotypes have been identified and they differ from one another by more than 3.0%. Single social groups contained only one of the two haplotypes whereas neighbouring groups tended to share the same haplotype, such that groups in the south-west portion of the population exhibited one haplotype and all other groups shared the second. A distinct margin exists between the areas dominated by the two types. The level of difference between these two haplotypes, occurring as they do in a single local population and among other groups with overlapping home ranges that regularly exchange males, has not been observed in other species; however, this may be a result of insufficient sampling.

Japanese macaque

Four Japanese macaque populations on three separate islands were examined for mtDNA sequence diversity by Hayasaka *et al.* (1986, 1988). Following the emerging pattern among macaque populations, though more extreme than most, intra-population diversity was calculated as 0.0%. Inter-population diversity, however, ranged from 0.0% to 2.4% and averaged 1.74% (data from Hayasaka *et al.*, 1988, reanalysed in Melnick *et al.*, 1993). These values are well within the range seen for the other macaque species (Table 19.3).

Summary

These results demonstrate that macaque social groups and local populations contain highly homogeneous mitochondrial genomes. The one exception to this pattern, the Polonnaruwa population of the toque monkey, revealed a boundary between two areas containing different mtDNA haplotypes. This suggests that the homogeneity found in other populations may simply reflect their location on one side of a haplotype boundary that exists outside the sampling area. In contrast, large differences are typically found between haplotypes from different regions, whether or not they are separated by barriers formed by water. In the absence of barriers to dispersal, the differences seen between regions may be the result of one or more haplotype boundaries, such as the one observed at Polonnaruwa.

Contrasting population structures

It is clear from our analysis of five macaque species that the population structures of nuclear and mitochondrial genetic variation are very different. With the exception of major water barriers, nuclear genetic variation in all macaques studied is quite uniformly distributed throughout a species' range. Thus, most of a species' diversity can be found in any single regional population and differences among populations are surprisingly low. In contrast, almost all of a species' mtDNA diversity is distributed as among population differences and there is relatively little within-group or within-population diversity.

This lack of congruence indicates that, at least in some cases, mtDNA data should not be used as a substitute for nuclear data when one is assessing the overall genetic structure of a species (Melnick *et al.*, 1992). Among macaques, not only is the population structure of mitochondrial genetic variation a poor reflection of nuclear population genetic structure, it is a positively misleading one. Large mtDNA differences between populations could, in general, be associated with either large or small nuclear genetic differences. In the macaques examined, the latter case is more prevalent.

Dispersal and gene flow

Given that male macaques virtually always disperse from their natal group and females almost never do, nuclear genes flow almost entirely through the transfer, and subsequent mating success, of males from one group to another. The rate and pattern of male migration between a population's social groups, and in and out of the population, should affect (in combination with specific patterns of mating) the distribution of nuclear genetic variation and the level of inbreeding in macaques. Female sedentism, however, should have its most profound impact on the distribution of maternally inherited mitochondrial genetic differences.

The nuclear gene

The ubiquity of the species' nuclear gene diversity throughout the range of the rhesus monkey indicates very high rates of nuclear gene flow (Melnick, 1988). Empirically, we have observed exceptionally high rates of male migration at the local population level in Dunga Gali, resulting in rhesus monkey social groups that are slightly outbred (Melnick, Pearl & Richard, 1984*b*) and genetically very similar to one another (Melnick, Jolly & Kidd, 1984*a*). Nearly identical patterns of migration have been noted in other populations of rhesus (Lindburg, 1969; Sade, 1972) and other macaque species (e.g. Dittus, 1975). Hence, even though migration is essentially limited to one sex, half the genes contributed to any new cohort are migrant genes. This is more than enough to inhibit the development of large nuclear genetic differences between populations (Wright, 1951).

Despite this general picture of genetic homogeneity, the small genetic differences that do exist between the Dunga Gali social groups are statistically significant (Melnick *et al.*, 1984*a*). That is, the collection of genes found in each social group is not simply a random assortment of the genetic material available in the population. There is, however, some reason to believe that the asymmetrical patterns of dispersal will create a complex of random and non-random components in any group's gene pool. Using a multiple randomisation test, we found that the adolescent and adult male portion of any social group was essentially a random genetic sample of the entire population's male gene pool, whereas the adult female and offspring portion was not a random sample of its age–sex specific gene pool (D. J. Melnick, S. J. Goldstein & L. C. Gale, unpublished data). These results parallel what we know of the dispersal of rhesus monkeys, where the adolescent and adult male portion of a social group is primarily a random collection of individuals from the local population in that age–sex class and the adult female and offspring portion represents the descendants of females that have remained in that group generation after generation. Therefore, we find: (1) a general background of broad genetic homogeneity, due largely to high rates of male migration and gene flow; and (2) small statistically significant genetic differences between groups which are primarily the product of differences between each group's non-random genetic component, that contributed by the adult females and their offspring.

While the data we have on macaques other than the rhesus monkey is less detailed, the nuclear population genetic structure of these other species mirrors what we found in *M. mulatta*. In particular, where it was possible to measure we have found that most of the diversity in a local population or a region could be found in its constituent subunits (Table 19.2). This pattern of homogeneity begins to break down as we compare inter-regional differentiation and include species with different geographical distributions. The average proportion of a species diversity that can be found in species with geographically continuous ranges exceeds 90.0%, whether the range is broad (*M. mulatta*) or narrow (*M. sinica*). Conversely, species whose geographical ranges are sub-divided by water barriers (*M. fascicularis, M. nemestrina, M. fuscata*) exhibit much more regional structure to their genetic variation (Table 19.2). Again, this pattern does not appear to be affected by the overall size of the species' range.

The mitochondrial genome

If in the face of overwhelming male gene flow, macaque female sedentism has had a significant effect on nuclear population genetic structure, then we should expect maternally inherited genetic variation to be much more highly differentiated between groups and populations. This is precisely what we find in the rhesus monkey. In a species where nuclear genetic variation is homogeneously spread across a geographically continuous range, we find exceptionally large inter-populational differences (i.e. levels of sequence divergence) in mtDNA. This combination of a continuous geographical distribution of individuals with a geographical discontinuity of mtDNA haplotypes conforms to category II of Avise *et al.*'s (1987) classification of mtDNA population structure. They argued that this pattern, rarely seen among the mammal species studied thus far, is the outcome of recent secondary contact between populations separated for a long time. However, empirical evidence (Melnick *et al.*, 1993) and computer simulations (G. A. Hoelzer, J. Wallman & D. J. Melnick, unpublished data) both suggest that 'large' inter-populational macaque mtDNA distances can develop without geographical barriers. Instead, this divergence reflects that the unusually high degree of female philopatry found in macaques greatly limits the dispersal of any new mtDNA mutation, effectively causing regions to be isolated by distance (Wright, 1943) rather than by geographical barriers. Thus, the differences in population genetic structure between species with continuous and fragmented geographical ranges are much smaller in the mitochondrial genome (Table 19.3), and the general pattern one finds across species is one of intra-populational homogeneity and marked inter-populational divergence.

Computer simulations have also shown that macaque female philopatry, differential reproductive success, and social group fission lead to a rapid loss of matrilines in a local population (Hoelzer *et al.*, unpublished data). This process of lineage sorting (see also Avise *et al.*, 1984), coupled with very low levels of mtDNA gene flow and high mtDNA mutation rates (Brown, George & Wilson, 1979), should result in a highly structured, heterogeneous species with low levels of within-population mtDNA diversity and large between-population differences. Again, this is precisely what we find.

Mating patterns

Male dominance and reproductive success

The relationship between aggression, dominance and reproductive success among primate males has received a great deal of research attention (for reviews of this research, see Bernstein, 1976; Dewsbury, 1982; Fedigan, 1983; Cowlishaw & Dunbar, 1991). The central hypothesis tested in most primate studies (e.g. Carpenter, 1942; Altmann, 1962; Hausfater, 1975; Smith, 1981) has been that dominant males have greater access to fertile females and, consequently, leave more offspring than do their subordinates.

The correlation between aggression and reproductive output in primate males is critical to the understanding of the genetic effects of male migration and the genetic consequences of social group fission. Equating the rate of male migration with gene flow may be somewhat misleading unless we know which of the migrating males father offspring in the groups they finally enter. Similarly, the matrilineal effect of social group fission (Olivier, Ober, & Buettner-Hanusch, 1978) may be significantly altered if paternity is restricted each year to one male, and many individuals in a cohort are more closely related through their patriline than through their matriline (Melnick & Kidd, 1983). Finally, insofar as social dominance affects the frequency of paternity (Dewsbury, 1982; Fedigan, 1983), the sample of genes drawn each year to create a cohort is not random, and this can affect the effective social group size (Nozawa, 1972), effective population size (Crow & Kimura, 1970) and, ultimately, the distribution of genes in a population.

An initial and essential step in the demonstration of a correlation between lifetime male reproductive success and dominance rank is the demonstration of a short-term correlation (i.e. over several mating seasons). However, attempts to establish this relationship in primate populations have obtained contradictory results. Some studies (e.g. Smith, 1981; Witt, Schmidt & Schmitt, 1981; Curie-Cohen *et al.*, 1983) have shown a positive association between dominance rank and reproductive output, while others (e.g. Smuts, 1982; Strum, 1982) have found either a negative correlation or no relationship at all.

Substantively, the lack of a correlation may be due to a number of factors: (a) dominance rank may not be solely determined by aggression (Bernstein, 1976); (b) lifetime reproductive output is complicated by the fact that males change rank position in their lives (Hausfater, 1975); (c) dominance rank may not be the most important determinant of male reproductive output (Saayman, 1971; Packer, 1979; Smuts, 1982; Strum, 1982); and (d) other factors, such as paternal behavior, may also enhance reproductive success (Fedigan, 1983).

Methodologically, several problems may underlie the apparent contradictions between studies. First, different methods have been used to define dominance hierarchies (e.g. Conaway & Koford, 1964 vs Strum, 1982) and to determine the pool of males included in the analysis (e.g. Hausfater, 1975 vs Bercovitch, 1986). It is not clear whether these different measures of agonistic dominance are perfectly correlated with one another (Bernstein, 1976) and, therefore, whether their relationship to reproductive and/or mating success may be different. Of more importance, the inclusion of sub-adult males, when assessing the relationship between dominance rank and mating success, has been seriously questioned (Bercovitch, 1986; MacMillan, 1989). Because young males are socially peripheral, low-ranking, and unlikely to be successful in mating, their inclusion in the analysis may artificially strengthen a correlation between rank and mating success. In some cases, dropping them from the analysis transformed a statistically significant correlation into a non-significant one.

Second, researchers studying wild primate populations have relied on indirect methods of paternity determination that use behavioural or physiological markers, such as copulation frequency (e.g. DeVore, 1965; Drickamer, 1974), consort frequency (e.g. Seyfarth, 1978), or copulations/consorts on the putative day of ovulation (e.g. Hausfater, 1975; Packer, 1979; Smuts, 1982; Strum, 1982). The degree to which these measures reflect paternity, and thus reproductive success is largely unknown. In at least one carefully controlled study (Curie-Cohen *et al.*, 1983) copulation frequency was an extremely poor predictor of paternity. Therefore, in the study of any polygamous primate population, genetic paternity exclusion is the method of choice. However, because animals must be immobilised to obtain samples for genetic analysis, the genetic determination of paternity has, until recently, been confined to studies of captive (Duvall, Bernstein & Gordon, 1976; Smith, 1981; Witt *et al.*, 1981; Curie-Cohen *et al.*, 1983) or provisioned free-ranging groups (Sade, Chepko-Sade & Schneider, 1977). Even now, with the advent of DNA fingerprinting (Jeffreys, Wilson & Thein, 1985) and its application to primates (Martin, Dixson & Wickings, 1992), very few studies of wild primate populations have been completed (de Ruiter *et al.*, 1992; Ménard *et al.*, 1992).

A third related problem concerns the nature of the primate groups studied. Studies using the most accurate means of measuring reproductive output (i.e. genetic paternity exclusion) have been conducted on captive primate groups and have generally yielded a correlation between agonistically defined dominance rank and reproductive success (for an exception, see Duvall *et al.*, 1976). However, studies focusing on wild primate populations have generally used less reliable methods of paternity determination and have not revealed a significant association between dominance and reproduction. The potential for distortion of social behaviour and its reproductive correlates, when animals are confined to relatively small enclosures, has been raised by Curie-Cohen et al. (1983). Hence, a tighter fit between dominance rank and reproductive success in captive primates may be an artifact of captivity itself. For these reasons – confounded measures of dominance, potential inaccuracy of behavioural measures of paternity and the effects of captivity on dominance and reproductive behaviour – many researchers (e.g. Smith, 1981; Dewsbury, 1982; Curie-Cohen et al., 1983; Fedigan, 1983) have recently recognised the need for tests of the association of rank and reproductive success on wild primate populations, using direct genetic means of paternity determination.

The most comprehensive examination of the relationship between dominance and male reproductive success in wild macaques was performed by de Ruiter et al. (1992) on M. fascicularis. In this study, male dominance rank was strongly correlated with short-term reproductive success, though the strength of the correlation was inversely proportional to group size.

Preliminary data from the Dunga Gali populations of M. mulatta (D. J. M., unpublished data) based on polymorphic protein loci, indicate that in the two smallest social groups all but the highest ranking male in each group were eliminated as the father of most infants born between 1977 and 1979. These results, while suggestive of a positive association between male dominance and short-term reproductive output, should be tempered by the fact that the data are limited. Mating is greatly restricted in these groups because they occupy a seasonal, temperate habitat, the groups are small, and the highest ranking male in each group was considerably older than the other adult males (Melnick, 1987).

In a recent review and statistical analysis of 32 primate studies, Cowlishaw & Dunbar (1991) suggested that 'despite conflicting reports, a reliable positive relationship between male dominance rank and mating success amongst animals of the same age class is seen'. The limited results of genetic studies on wild macaque populations do not contradict this conclusion. However, there is no doubt that broad acceptance of this relationship must await more extensive, genetically based research.

Inbreeding and outbreeding among macaques

The social sub-division of a primate population is thought to impede significantly the flow of genes beyond the limits of its basic social units (Bush et al., 1977). If this occurs, one would predict an accumulation of inbreeding in each social group over time (Wright, 1969). Additionally, the potential build-up of inbreeding could be increased by persistent preferential migration between pairs of social groups (Melnick & Pearl, 1987). Indeed, one way for us to measure the effects of the patterns of dispersal and mating in primate populations is to measure the level of inbreeding.

Recent analyses of rhesus macaque, other cercopithecoid primate and other social mammal populations have not produced any genetic evidence of inbreeding (Melnick, 1982; Foltz & Hoogland, 1983; Kawamoto *et al.*, 1984; Melnick *et al.*, 1984*b*). Examining five variable genetic loci, and correcting for bias, such as small sample size (Cannings & Edwards, 1969) and unequal sex-specific parental gene frequencies (Foltz & Hoogland, 1983), there was a consistent excess of heterozygotes in the Dunga Gali social groups. Hence, genotypic estimates of inbreeding (Nei, 1977) were uniformly negative, indicating an avoidance of mating between close relatives in the population.

Because female rhesus in Dunga Gali remain in their natal group throughout their lives, and all males leave their natal group and disperse in the ways described earlier, there is little opportunity for close relatives to mate. Therefore, the lack of consanguineous matings in this population is not necessarily a consequence of behavioural inhibitions against certain mating types (Sade, 1972; Smith, 1982), but may be merely the structural outcome of the dynamics of male migration. The behavioural and demographic patterns that preclude consanguineous matings are also present in other wild rhesus populations (Lindburg, 1969; Southwick *et al.*, 1980), and genetic evidence for a lack of inbreeding can be found in the Cayo Santiago colony, Puerto Rico (C. R. Duggleby, personal communication). Interestingly, in the case of Cayo Santiago, the male offspring of higher ranking females remain in their natal group past sexual maturity (Colvin, 1983), but behavioural mechanisms appear to inhibit mother–son matings (Sade, 1968). These data suggest that other wild rhesus populations are not in-bred. Additionally, the sharp rise in mortality and morbidity after only one or two generations of inbreeding (Curie-Cohen *et al.*, 1981; Ralls & Ballou, 1982) indicates that an avoidance of inbreeding has long characterised the species.

Comparative behavioural and genetic information suggests a similar avoidance of inbreeding in a variety of social mammals. In a survey of 65 mammalian species, most were found to display an asymmetrical pattern of dispersal, where males left their natal group while females remained (Greenwood, 1980). Other behavioural mechanisms have also been reported that would further ensure inbreeding avoidance (McCracken & Bradbury,

1977; Schwartz & Armitage, 1980; Hoogland, 1982). A review of published genetic data (Melnick, 1988) shows that, among rodents, lagomorphs, artiodactyls and primates, average inbreeding coefficients are consistently negative (ranging from 0.0 to -0.1) within social units and entire populations. Therefore, there is little evidence to suggest that the social sub-division of many primate and non-primate mammalian populations (including humans, see Neel & Ward, 1972) results in inbreeding and its genetic effects.

Social group fission

Among rhesus monkeys (Chepko-Sade & Sade, 1979) and other cercopithecoid primates (e.g. Nash, 1976) new social groups are formed by the division or fission of formerly cohesive, larger social groups. There is evidence that this fission occurs along lines of maternal relatedness (Chepko-Sade & Sade, 1979; Furaya, 1968, 1969; Nash, 1976; Dittus, 1988). The genetic consequences of such a division can be significant (Duggleby, 1977; Cheverud, Buettner-Janusch & Sade, 1978; Ober, 1979; Olivier *et al.*, 1981; Buettner-Janusch *et al.*, 1983), resulting in groups genetically more distinct from one another than would be the case if group fission were random (see also Neel, 1967; Smouse, Vitzhum & Neel, 1981). Additionally, as the only effective means of mitochondrial gene flow, the extent to which maternal lineages remain together during group fission will also affect the distribution of mtDNA haplotypes within and between social groups.

The nuclear genome

Empirically, the nuclear genetic consequences of social group fission have been described for only two primate populations: the free-ranging Cayo Santiago colony and the wild Dunga Gali population of rhesus monkeys. The results of these analyses contradict each other and, in so doing, provide information about circumstances under which a matrilineal effect in the fission of primate social groups will be different from the effect of random group division.

On Cayo Santiago, several groups have divided along lines of matrilineal relatedness (Chepko-Sade & Sade, 1979). These fissions have resulted in (1) elevated degrees of relatedness within groups (Chepko-Sade & Olivier, 1979) and (2) levels of genetic differentiation between groups greater than those expected from a random division of parent group members (Cheverud *et al.*, 1978; Buettner-Janusch *et al.*, 1983). In contrast, even though one Dunga Gali rhesus group apparently divided along matrilineal lines (for the methods used to define matrilines, see Pearl, 1982; Pearl & Melnick, 1983), the resulting genetic differences between the two daughter groups were the same as if the group had divided at random (Melnick & Kidd, 1983).

The differing genetic consequences of group fission in these two populations are most likely the outcome of at least two important differences in demographic structure and mating patterns. First, infant mortality, inter-birth intervals and inferred average life spans for female rhesus in Dunga Gali (Melnick, 1981; D. J. M., unpublished data) have resulted in groups composed of many more matrilines containing relatively fewer individuals than comparably defined groups on Cayo Santiago. Thus, the overall average kinship coefficient for a fission group will be very small and the sampling involved in its formation only weakly correlated. This is not the case on Cayo Santiago (Chepko-Sade & Olivier, 1979; MacMillan & Duggleby, 1981), where matrilines are substantially larger and fission group kinship coefficients are relatively higher than those found in the Dunga Gali population. In this case, the assortment of individuals by matrilineal identity can have a major genetic effect.

A second explanation for differences in the genetic consequences of fission lies in patterns of paternity (Melnick & Kidd, 1981, 1983; see above). The Cayo Santiago colony contains an abundance of males (Sade *et al.*, 1985) and partial paternal exclusion tests failed to exclude a large number of adult males from being the fathers of offspring in one group (Sade *et al.*, 1977). If many males actually achieve reproductive success during a given year in a social group, paternal half-sibs would be relatively uncommon. In the Dunga Gali population there is reason to believe that paternity may be much more restricted, particularly in small social groups. If restricted paternity is commonplace, many individuals of an age cohort in a particular social group will be paternal half-sibs (cf. Altmann, 1979). With restricted paternity and low female fecundity over several generations, the average kinship coefficient in Dunga Gali will be determined predominantly by paternal relationships. Despite the fission of groups along matrilines, resulting fission groups would be genetically similar because individuals who share a father may reside in different fission groups.

The genetic consequences of fission in these two very different rhesus populations illustrate the two major effects this process can have on genetic structure. First, group fission accelerates to one degree or another inter-group genetic differentiation. So, for example, in the Dunga Gali population, where the matrilineal effects of group fission were negligible, the fission of a group and the subsequent departure of one of the fission groups still raised the overall level of inter-group differentiation by 13%. The effects of this process on inter-group differentiation in the Cayo Santiago colony, where the matrilineal effects of fission are significant, were even more dramatic (Ober *et al.*, 1984). Hence, a mathematical model that ignores the effects of group fission will always underestimate the actual level of inter-group genetic differentiation.

The second, and perhaps more important effect of fission, is the extent to which it fosters the genetic divergence of founder groups and thus enhances the likelihood of speciation. During a colonising phase, the period most relevant to evolutionary change and diversification (i.e. the emergence of new species), populations are usually growing. Increased survival rates and reduced inter-birth intervals permit large matrilines to develop. In such circumstances, fission groups are probably composed of a few large matrilines with relatively high kinship coefficients. Hence, matrilineal splitting can accentuate the genetic effects of social group fission. This may also be a time when founder groups, formed by this genetically skewed sampling process, become geographically isolated and, therefore, are susceptible to genetic transilience and subsequent speciation. The population dynamics, demographic structure and genetic consequences of group fission in the Cayo Santiago colony represent a good model of this colonising phase.

Alternatively, during phases of stasis, when populations are either stable or growing very slowly, fission groups will consist of many small matrilines with relatively low average kinship coefficients. Under these conditions, genetic effects of matrilineal fission will not be distinguishable from random splitting. The characteristics of the wild Dunga Gali population represent a good model of this phase.

The mitochondrial genome

Social group fission has a more profound effect on the distribution of mtDNA haplotypes. Because individual females generally do not join existing groups, the females and their offspring within a group share the same mtDNA haplotype (see above), unless one of these individuals contains a new mutation. Therefore, when a group splits, both fission groups will exhibit the same haplotype (Hoelzer *et al.*, 1994). As this process occurs over a long time, large areas become occupied by groups related by past fission events. In this way, group fission brings about the observed distribution of mtDNA variation, with boundaries between regions dominated by very different haplotypes (see above; Hoelzer *et al.*, 1994).

The geographical expansion of some mtDNA lineages and the extinction of others (i.e. lineage sorting, Avise *et al.*, 1984) is constrained in macaques because fission groups generally occupy neighbouring territories. Consequently, there is a limit to the distance over which lineage sorting is effective. Boundaries will develop in arbitrary places between effective lineage sorting populations in a widespread species (Hoelzer *et al.*, unpublished data). It is this type of boundary that is observed in toque monkeys at Polonnaruwa, Sri Lanka (Hoelzer *et al.*, 1994).

Summary and conclusions

The primary focus of this review has been the genetic consequences of certain macaque behavioural patterns, with the expressed aim of linking these effects to the distribution of variation in both the nuclear and mitochondrial genomes. Macaque social structure is characterised by social groups in which interaction among males and among females are influenced by dominance hierarchies. To date, the data suggest that a correlation may exist between male dominance rank and reproductive success, which may limit the extent of genetic variability by increasing the degree of polygyny and reducing the effective population size.

The distribution of genetic variation within and between macaque populations is most strongly affected by the gender-biased pattern of dispersal from the natal group. Females are highly philopatric, whereas males disperse before mating. While this can produce small, but significant nuclear genetic differences between social groups, extensive gene flow through male migration generally spreads genetic variants throughout the range of a species and prevents inbreeding in wild populations. Therefore, divergence of nuclear genomes between macaque populations does not develop much in the absence of geographical barriers, such as isolation on oceanic islands. In contrast, female philopatry has strongly constrained gene flow in the mitochondrial genome, which is clonally inherited through the mother. The mechanism of mitochondrial gene flow is the fission of social groups, which generally divides a group while keeping its constituent matrilines intact. Consequently, social groups and local populations exhibit little variation in mtDNA haplotypes, while exceptionally high levels of divergence are seen between populations separated by long distance or oceanic barriers.

The behaviour and genetic structure of macaque populations suggest that nuclear divergence and, therefore, the evolution of new macaque forms (i.e. sub-species and species) usually occurs when geographical barriers prevent male migration between populations. However, because allopatry is not required for the divergence of mtDNA haplotypes, the divergence of 'mtDNA populations' and the branching of mtDNA haplotypes does not necessarily parallel the origins of new macaque species or sub-species (see Melnick *et al.*, 1992; Hoelzer *et al.*, 1994).

Acknowledgements

This chapter has benefited greatly from the contributions of the authors of the articles cited within it. We particularly thank John Avise, Wes Brown, Rodney Honeycutt, Cliff Jolly, Ken Kidd, and, Mary Pearl for discussions of different aspects of this work as it proceeded. D. J. M. also acknowledges the US National Science Foundation and the Henry Frank Guggenheim Foundation for the support of his research in this area over the past 15 years.

References

Altmann, J. (1979). Age cohorts as paternal sibships. *Behavioral Ecology and Sociobiology*, **6**, 161–4.

Altmann, S. A. (1962). A field study of the sociobiology of the rhesus monkey, *Macaca mulatta*. *Annals of the New York Academy of Science*, **102**, 338–435.

Anderson, J. E. & Giblett, E. R. (1975). Interspecific red cell variation in the pig-tailed macaque (*Macaca nemestrina*). *Biochemical Genetics*, **13**, 189–212.

Ashley, M. V., Melnick, D. J. & Western, D. (1989). Conservation genetics of the black rhinoceros (*Diceros bicornis*). I. Evidence from the mitochondrial DNA of three populations. *Conservation Biology*, **4**, 71–7.

Avise, J. C. (1986). Mitochondrial DNA and the evolutionary genetics of higher animals. *Philosophical Transactions of the Royal Society of London*, **312**, 325–42.

Avise, J. C. (1989). A role for molecular genetics in the recognition and conservation of endangered species. *Trends in Ecology and Evolution*, **4**, 279–81.

Avise, J. C., Arnold, J., Ball, R. M., Bermingham, E., Lamb, T., Neigel, J. E., Reeb, C. A. & Saunders, N. C. (1987). Intraspecific phylogeography: the mitochondrial bridge between population genetics and systematics. *Annual Review of Ecology and Systematics*, **18**, 489–522.

Avise, J. C., Neigel, J. E. & Arnold, J. (1984). Demographic influences on mitochondrial DNA lineage survivorship in animal populations. *Journal of Molecular Evolution*, **20**, 99–105.

Berkovitch, F. B. (1986). Male rank and reproductive activity in savanna baboons. *International Journal of Primatology*, **7**, 533–50.

Bernstein, I. S. (1976). Dominance, aggression and reproduction in primate societies. *Journal of Theoretical Biology*, **60**, 459–72.

Brown, W. M. (1983). Evolution of animal mitochondrial DNA. In *Evolution of genes and proteins*, ed. M. Nei & R. K. Koehn, pp. 62–88. Sunderland, MA: Sinauer Press.

Brown, W. M., George, M. & Wilson, A. C. (1979). Rapid evolution of animal mitochondrial DNA. *Proceedings of the National Academy of Sciences USA*, **76**, 1967–71.

Buettner-Janusch, J., Olivier, T. J., Ober, C. L. & Chepko-Sade, B. D. (1983). Models for linear effects in rhesus group fissions. *American Journal of Physical Anthropology*, **61**, 347–53.

Bush, G. L., Case, S. M., Wilson, A. C. & Patton, J. L. (1977). Rapid speciation and chromosomal evolution in mammals. *Proceedings of the National Academy of Sciences USA*, **74**, 3942–6.

Cannings, C. & Edwards, A. W. F. (1969). Expected genotypic frequencies in a small sample: deviation from Hardy–Weinberg equilibrium. *American Journal of Human Genetics*, **21**, 245–7.

Carpenter, C. R. (1942). Sexual behaviour of free-ranging monkeys (*Macaca mulatta*). *Journal of Comparative Psychology*, **33**, 113–42.

Chepko-Sade, B. D. & Olivier, T. J. (1979). Coefficient of genetic relationship and the probability of intra-genealogical fission in *Macaca mulatta*. *Behavioral Ecology and Sociobiology*, **5**, 263–78.

Chepko-Sade, B. D. & Sade, D. S. (1979). Patterns of group splitting within matrilineal kinship groups. *Behavioral Ecology and Sociobiology*, **5**, 67–87.

Cheverud, J. M., Buettner-Janusch, J. & Sade, D. S. (1978). Social group fission and the origin of intergroup genetic differentiation among rhesus monkeys of Cayo Santiago. *American Journal of Physical Anthropology*, **48**, 449–56.

Clutton-Brock, T. H. (1989). Female transfer and inbreeding avoidance in social mammals. *Nature*, **337**, 70–2.

Colvin, J. (1983). Influences of the social situation on male emigration. In *Primate social relationships: an integrated approach*, ed. R. A. Hinde, pp. 160–70. Sunderland, MA: Sinauer Press.

Conaway, C. H. & Koford, C. B. (1964). Estrous cycles and mating behavior in a free ranging band of rhesus monkeys. *Journal of Mammalogy*, **45**, 577–88.

Cowlishaw, G. & Dunbar, R. I. M. (1991). Dominance rank and mating success in male primates. *Animal Behaviour*, **41**, 1045–56.

Crow, J. F. & Kimura, M. (1970). *An introduction to population genetics theory*. New York: Harper & Row.

Curie-Cohen, M., Yoshihara, D., Blystad, C., Luttrell, L., Benforado, K. & Stone, W. H. (1981). Paternity and mating behavior in a captive troop of rhesus monkeys. *American Journal of Primatology*, **1**, 335.

Curie-Cohen, M., Yoshihara, D., Luttrell, L., Benforado, K., MacCluer, J. & Stone, W. (1983). The effects of dominance on mating behavior and paternity in a captive troop of rhesus monkeys (*Macaca mulatta*). *American Journal of Primatology*, **5**, 127–38.

Darga, L. L. (1975). Immunological and electrophoretic investigations of catarrhine evolution. PhD thesis, Wayne State University.

de Ruiter, J. R., Scheffrahn, W., Trommelen, G. J. J. M., Uitterlinden, A. G., Martin, R. D. & van Hooff, J. A. R. A. M. (1992). Male social rank and reproductive success in wild long-tailed macaques: paternity exclusions by blood protein analysis and DNA fingerprinting. In *Paternity in primates: genetic tests and theories*, ed. R. D. Martin, A. F. Dixson & E. J. Wickings, pp. 175–91. Basel: S. Karger.

DeVore, I. (1965). Male dominance and mating behavior in baboons. In *Sex and behavior*, ed. F. A. Beach, pp. 266–89. New York: Krieger Publishing Company.

Dewsbury, D. A. (1982). Dominance rank, copulatory behavior, and differential reproduction. *Quarterly Reviews of Biology*, **57**, 135–59.

Dittus, W. P. J. (1975). Population dynamics of the toque monkey, *Macaca sinica*. In *Socioecology and psychology of primates*, ed. R. H. Tuttle, pp. 125–51. The Hague: Mouton.

Dittus, W. P. J. (1988). Group fission among wild toque macaques as a consequence of female resource competition and environmental stress. *Animal Behaviour*, **36**, 1626–45.

Dowling, T. E., Moritz, C. & Palmer, J. D. (1990). Nucleic acids. II. restriction site analysis. In *Molecular systematics*, ed. D. Hillis & C. Moritz, pp. 250–317. Sunderland, MA: Sinauer Press.

Drickamer, L. C. (1974). Social rank, observability, and sexual behaviour of rhesus monkeys (*Macaca mulatta*). *Journal of Reproduction and Fertility*, **37**, 117–20.

Duggleby, C. R. (1977). Blood group antigens and the population genetics of *Macaca mulatta* on Cayo Santiago. II. Effects of social group division. *Yearbook of Physical Anthropology*, **20**, 263–71.

Duvall, S. W., Bernstein, I. S. & Gordon, T. P. (1976). Paternity and status in a rhesus monkey group. *Journal of Reproduction and Fertility*, **47**, 25–31.

Ehrlich, P. R. & Raven, P. H. (1969). Differentiation of populations. *Science*, **165**, 1228–32.

Ewens, W. J. (1972). The sampling theory of selectively neutral alleles. *Theoretical Population Biology*, **3**, 87–112.

Fedigan, L. M. (1983). Dominance and reproductive success. *Yearbook of Physical Anthropology*, **26**, 91–129.

Foltz, D. & Hoogland, J. (1983). Genetic evidence of outbreeding in the black-tailed prairie dog (*Cynomys Iudovicianus*). *Evolution*, **37**, 273–81.

Fooden, J. (1979). Taxonomy and evolution of the *sinica* group of macaques. 1. Species and subspecies accounts of *Macaca sinica*. *Primates*, **20**, 109–40.

Fooden, J. (1988). Taxonomy and evolution of the *sinica* group of macaques. 6. Interspecific comparisons and synthesis. *Fieldiana Zoology*, **45**, 1–44.

Fooden, J. & Lanyon, S. M. (1989). Blood-protein allele frequencies and phylogenetic relationships in *Macaca*: a review. *American Journal of Primatology*, **17**, 209–41.

Fuerst, P. A., Chakraborty, R. & Nei, M. (1977). Statistical studies on protein polymorphism in natural populations. I. Distribution of single locus heterozygosity. *Genetics*, **86**, 455–83.

Furuya, Y. (1968). On the fission of troops of Japanese monkeys. I. *Primates*, **9**, 323–50.

Furuya, Y. (1969). On the fission of troops of Japanese monkeys. II. *Primates*, **10**, 47–69.

Greenwood, P. J. (1980). Mating systems, philopatry, and dispersal in birds and mammals. *Animal Behaviour*, **28**, 1140–62.

Harihara, S., Saitou, N., Horai, M., Aoto, N., Terao, K., Cho, F., Honjo, S. & Omoto, K. (1988). Differentiation of mitochondrial DNA types in *Macaca fascicularis*. *Primates*, **29**, 117–27.

Hartl, D. L. & Clark, A. G. (1989). *Principles of population genetics.* Sunderland, MA: Sinauer Press.

Hausfater, G. (1975). *Dominance and reproduction in Baboons* (Papio cynocephalus): *a quantitative analysis.* Basel: S. Karger.

Hayasaka, K., Horai, S., Gojobori, T., Shotake, T., Nozawa, K. & Matsunaga, E. (1988). Phylogenetic relationships among Japanese, rhesus, Formosan, and crab-eating monkeys, inferred from restriction enzyme analysis of mitochondrial DNAs. *Molecular Biology and Evolution*, **5**, 270–81.

Hayasaka, K., Horai, S., Shotake, T., Nozawa, K. & Matsunaga, E. (1986). Mitochondrial DNA polymorphism in Japanese monkeys, *Macaca fuscata*. *Japanese Journal of Genetics*, **61**, 345–59.

Hillis, D. M. & Moritz, C. (1990). *Molecular systematics.* Sunderland, MA: Sinauer Press.

Hoelzer, G. A., Dittus, W. P. J., Ashley, M. V. & Melnick, D. J. (1994). The local distribution of highly divergent mitochondrial DNA haplotypes in toque macaques (*Macaca sinica*) at Polonnaruwa, Sri Lanka. *Molecular Ecology*, **3**, 451–8.

Hoelzer, G. A., Hoelzer, M. A. & Melnick, D. J. (1992). The evolutionary history of the *sinica*-group of macaque monkeys as revealed by mtDNA restriction site analysis. *Molecular Phylogenetics and Evolution*, **1**, 215–22.

Honeycutt, R. & Wheeler, W. (1989). Mitochondrial DNA: variation in humans and primates. In *DNA systematics: human and higher primates*, ed. S. K. Dutta & W. Winter, pp. 91–129. Boca Raton, FL: CRC Press.

Hoogland, J. L. (1982). Prairie dogs avoid extreme inbreeding. *Science*, **214**, 1639–41.

Hutchison, C. A., Newbold, C. E., Potter, S. S. & Edgell, M. H. (1974). Maternal inheritance of mammalian mitochondrial DNA. *Nature*, **251**, 536–8.

Jeffreys, A. J., Wilson, V. & Thein, S. L. (1985). Individual-specific 'fingerprints' of human DNA. *Nature*, **316**, 76–9.

Kawamoto, Y., Ischak, T. M. & Supriatna, J. (1984). Genetic variations within and between troops of the crab-eating macaque (*Macaca fascicularis*) on Sumatra, Java, Bali, Lombok, and Sumbawa, Indonesia. *Primates*, **25**, 131–59.

Kawamoto, Y., Takenaka, O., Surgobroto, B. & Brotoisworo, E. (1985). Genetic differentiation of Sulawesi macaques. *Kyoto University Overseas Research Report of Studies on Asian Non-Human Primates*, **4**, 41–61.

Kimura, M. (1968). Evolutionary rate at the molecular level. *Nature*, **217**, 624–6.

Kimura, M. (1983). *The neutral theory of molecular evolution.* Cambridge: Cambridge University Press.

Lindburg, D. G. (1969). Rhesus monkeys: mating season mobility of adult males. *Science*, **166**, 1176–8.

MacMillan, C. A. (1989). Male age, dominance and mating success among rhesus macaques. *American Journal of Physical Anthropology*, **80**, 83–9.

MacMillan, C. & Duggleby, C. R. (1981). Interlineage genetic differentiation among rhesus macaques on Cayo Santiago. *American Journal of Physical Anthropology*, **56**, 305–12.

Martin, R. D., Dixson, A. F. & Wickings, E. J. (1992). *Paternity in primates: genetic tests and theories.* Basel: S. Karger.

McCracken, G. & Bradbury, J. (1977). Paternity and genetic heterogeneity in the polygynous bat, *Phyllostomus hastatus*. *Science*, **198**, 303–6.

Melnick, D. J. (1981). Microevolution in a population of Himalayan rhesus monkeys (*Macaca mulatta*). PhD thesis, Yale University.

Melnick, D. J. (1982). Are social mammals really inbred? *Genetics*, **100**, S46.

Melnick, D. J. (1987). The genetic consequences of primate social organization: a review of macaques, baboons and vervet monkeys. *Genetica*, **73**, 117–35.

Melnick, D. J. (1988). The genetic structure of a primate species: rhesus macaques and other cercopithecine monkeys. *International Journal of Primatology*, **9**, 195–231.

Melnick, D. J. & Hoelzer, G. A. (1992). Differences in male and female macaque dispersal lead to contrasting distributions of nuclear and mitochondrial DNA variation. *International Journal of Primatology*, **13**, 1–15.

Melnick, D. J., Hoelzer, G. A., Absher, R. & Ashley, M. V. (1993). mtDNA diversity in rhesus monkeys reveals overestimates of divergence time and paraphyly with neighboring species. *Molecular Biology and Evolution*, **10**, 282–95.

Melnick, D. J., Hoelzer, G. A. & Honeycutt, R. (1992). The mitochondrial genome: its uses in anthropological research. In *Molecular applications in biological anthropology*, ed. D. Devor, pp. 179–233. Cambridge: Cambridge University Press.

Melnick, D. J., Jolly, C. J. & Kidd, K. K. (1984*a*). The genetics of a wild population of rhesus monkeys (*Macaca mulatta*). I. Genetic variability within and between social groups. *American Journal of Physical Anthropology*, **63**, 341–60.

Melnick, D. J., Jolly, C. J. & Kidd, K. K. (1986). The genetics of a wild population of rhesus monkeys (*Macaca mulatta*). II. The Dunga Gali population in species-wide perspective. *American Journal of Physical Anthropology*, **71**, 129–40.

Melnick, D. J. & Kidd, K. K. (1981). Social group fission and paternal relatedness. *American Journal of Primatology*, **1**, 333–4.

Melnick, D. J. & Kidd, K. K. (1983). The genetic consequences of social group fission in a wild population of rhesus monkeys (*Macaca mulatta*). *Behavioral Ecology and Sociobiology*, **12**, 229–36.

Melnick, D. J. & Kidd, K. K. (1985). Genetic and evolutionary relationships among Asian macaques. *International Journal of Primatology*, **6**, 123–60.

Melnick, D. J. & Pearl, M. C. (1987). Cercopithecines in multimale groups: genetic diversity and population structure. In *Primate societies*, ed. B. B. Smuts, D. L. Cheney, R. M. Seyfarth, R. W. Wrangham & T. T. Struhsaker, pp. 121–34. Chicago: University of Chicago Press.

Melnick, D. J., Pearl, M. C. & Richard, A. F. (1984*b*). Male migration and inbreeding avoidance in wild rhesus monkeys. *American Journal of Primatology*, **7**, 229–43.

Ménard, N., Scheffrahn, W., Vallet, D., Zidane, C. & Reber, C. (1992). Application of blood protein electrophoresis and DNA fingerprinting to the analysis of paternity and social characteristics of wild barbary macaques. In *Paternity in primates: genetic tests and theories*, ed. R. D. Martin, A. F. Dixson & E. J. Wickings, pp. 155–74. Basel: S. Karger.

Moore, J. (1984). Female transfer in primates. *International Journal of Primatology*, **5**, 537–89.

Moore, J. & Ali, R. (1984). Are dispersal and inbreeding related? *Animal Behaviour*, **32**, 94–112.

Napier, J. R. & Napier, P. H. (1967). *A handbook of living primates*. London: Academic Press.

Nash, L. T. (1976). Troop fission in free-ranging baboons in the Gombe Stream National Park, Tanzania. *American Journal of Physical Anthropology*, **44**, 63–78.

Neel, J. V. (1967). The genetic structure of primitive human populations. *Japanese Journal of Human Genetics*, **12**, 1–16.

Neel, J. V. & Ward, R. (1972). The genetic structure of a tribal population, the Yanomamo Indians. VI. Analysis by F-statistics. *Genetics*, **72**, 639–66.

Nei, M. (1973). Analysis of gene diversity in subdivided populations. *Proceedings of the National Academy of Sciences USA*, **70**, 3321–3.

Nei, M. (1977). F-statistics and analysis of gene diversity in subdivided populations. *Annals of Human Genetics*, **41**, 225–33.

Nei, M. (1982). Evolution of human races at the gene level. In *Human genetics*, *part A: the unfolding genome*, ed. B. Bonne-Tamir, P. Cohen & R. N. Goodman, pp. 167–81. New York: Alan R. Liss.

Nei, M. (1987). *Molecular evolutionary genetics*. New York: Columbia University Press.

Nei, M., Fuerst, P. A. & Chakraborty, R. (1976). Testing the neutral mutation hypothesis by distribution of single locus heterozygosity. *Nature*, **262**, 491–3.

Nei, M. & Li, W. H. (1979). Mathematical model for studying genetic variation in terms of restriction endonucleases. *Proceedings of the National Academy of Sciences USA*, **76**, 5269–73.

Nei, M. & Tajima, F. (1983). Maximum likelihood estimation of the number of nucleotide substitutions from restriction sites data. *Genetics*, **105**, 207–17.

Nozawa, K. (1972). Population genetics of Japanese monkeys. I. Estimation of the effective troop size. *Primates*, **13**, 389–93.

Nozawa, K., Shotake, T., Okhura, Y. & Tanabe, Y. (1977). Genetic variations within and between species of Asian macaques. *Japanese Journal of Genetics*, **52**, 15–30.

Nozawa, K., Shotake, T., Okhura, Y. & Tanabe, Y. (1982). Population genetics of Japanese monkeys. II. Blood protein polymorphisms and population structure. *Primates*, **23**, 252–71.

Ober, C. (1979). Demography and microevolution on Cayo Santiago. PhD thesis, Northwestern University.

Ober, C. L., Olivier, T. J., Sade, D. S., Schneider, J. M., Cheverud, J. & Buettner-Janusch, J. (1984). Demographic components of gene frequency change in free-ranging macaques on Cayo Santiago. *American Journal of Physical Anthropology*, **64**, 223–31.

Olivier, T. J., Ober, C. L. & Buettner-Janusch, J. (1978). Genetics of group fissions on Cayo Santiago. *American Journal of Physical Anthropology*, **48**, 424.

Olivier, T. J., Ober, C. L., Buettner-Janusch, J. & Sade, D. (1981). Genetic differentiation among matrilines in social groups of rhesus monkeys. *Behavioral Ecology and Sociobiology*, **8**, 279–85.

Packer, C. (1979). Male dominance and reproductive activity in *Papio anubis*. *Animal Behaviour*, **27**, 37–45.

Pearl, M. C. (1982). *Networks of social relations among Himalayan rhesus monkeys* (Macaca mulatta). PhD thesis, Yale University.

Pearl, M. C. & Melnick, D. J. (1983). Inferring maternity from behavior and genetics of wild rhesus monkeys. *American Journal of Primatology*, **5**, 351.

Pusey, A. E. & Packer, C. (1987). Dispersal and philopatry. In *Primate societies*, ed. B. B. Smuts, D. L. Cheney, R. M. Seyfarth, R. W. Wrangham & T. T. Struhsaker, pp. 250–66. Chicago: University of Chicago Press.

Ralls, K. & Ballou, J. (1982). Inbreeding and infant mortality in primates. *International Journal of Primatology*, **3**, 491–505.

Richard, A. F., Goldstein, S. J. & Dewar, R. E. (1990). Weed macaques: the evolutionary implications of macaque feeding ecology. *International Journal of Primatology*, **10**, 569–94.

Rosenblum, L., Hoelzer, G. A. & Melnick, D. J. (1992). Intraspecific mitochondrial DNA variation in Asian macaques. *American Journal of Physical Anthropology*, Suppl., **14**, 141.

Saayman, G. S. (1971). Behavior of the adult males in a troop of free ranging chacma baboons (*Papio ursinus*). *Folia Primatologica*, **15**, 36–57.

Sade, D. S. (1968). Inhibition of son-mother mating in free-ranging rhesus monkeys. *Scientific Psychoanalysis*, **12**, 18–38.

Sade, D. S. (1972). A longitudinal study of social behavior of rhesus monkeys. In *The functional and evolutionary biology of primates*, ed. R. H. Tuttle, pp. 378–98. Chicago: Aldine.

Sade, D. S., Chepko-Sade, B. D. & Schneider, J. (1977). Paternal exclusions among free-ranging rhesus monkeys on Cayo Santiago. Paper presented at the annual meeting of the Animal Behaviour Society, University Park, PA.

Sade, D. S., Chepko-Sade, B. D., Schneider, J., Roberts, S. S. & Richtsmeier, J. T. (1985). *Basic demographic observation on free-ranging rhesus monkeys*. New Haven, CN: Human Relations Area Files.

Schwartz, O. A. & Armitage, K. B. (1980). Genetic variation in social mammals: the marmot model. *Science*, **207**, 665–7.

Seyfarth, R. M. (1978). Social relationships among adult male and female baboons. I. Behaviour during sexual consortship. *Behaviour*, **64**, 204–26.

Shotake, T. (1979). Serum albumin and erythrocyte adenosine deaminase polymorphisms in Asian macaques with special reference to taxonomic relationships among *M. assamensis*, *M. radiata*, and *M. mulatta*. *Primates*, **20**, 443–51.

Shotake, T., Nozawa, K. & Santiapillai, C. (1991). Genetic variability within and between the troops of toque macaque, *Macaca sinica*, in Sri Lanka. *Primates*, **32**, 283–99.

Shotake, T. & Santiapillai, C. (1982). Blood protein polymorphisms in the troops of the toque macaque, *Macaca sinica*, in Sri Lanka. *Kyoto University Overseas Research Report of Studies on Asian Non-Human Primates*, **2**, 79–95.

Slatkin, M. (1982). Testing neutrality in subdivided populations. *Genetics*, **100**, 533–45.

Smith, D. G. (1981). The association between rank and reproductive success of male rhesus monkeys. *American Journal of Primatology*, **1**, 83–90.

Smith, D. G. (1982). Inbreeding in three captive groups of rhesus monkeys. *American Journal of Physical Anthropology*, **58**, 447–51.

Smith, D. G. & Ahlfors, C. E. (1984). The albumin polymorphism and bilirubin binding in rhesus monkeys (*Macaca mulatta*). *American Journal of Physical Anthropology*, **54**, 37–41.

Smith, D. G. & Ferrel, R. E. (1980). A family study of the hemoglobin polymorphism in *Macaca fascicularis*. *Journal of Human Evolution*, **9**, 557–63.

Smith, D. G. & Rolfs, B. (1984). Segregation distortion and differential fitness at the albumin locus in rhesus monkeys (*Macaca mulatta*). *American Journal of Primatology*, **7**, 285–90.

Smith, D. G. & Small, M. F. (1982). Selection and the transferrin polymorphism in rhesus monkeys (*Macaca mulatta*). *Folia Primatologica*, **37**, 127–36.

Smouse, P., Vitzhum, V. J. & Neel, J. (1981). The impact of random and lineal fission on the genetic divergence of small human groups. *Genetics*, **98**, 179–97.

Smuts, B. (1982). Special relationships between adult male and female olive baboons (*Papio anubis*). PhD thesis, Stanford University.

Southwick, C. H., Richie, T., Taylor, H., Teas, J. & Siddiqi, M. (1980). Rhesus monkey populations in India and Nepal. In *Biosocial mechanisms of population regulation*, ed. M. Cohen, R. Malpass & H. Klein, pp. 151–70. New Haven, CN: Yale University Press.

Strum, S. C. (1982). Agonistic dominance in male baboons: an alternative view. *International Journal of Primatology*, **3**, 175–202.

Watterson, G. A. (1977). Heterosis or neutrality? *Genetics*, **85**, 789–814.

Watterson, G. A. (1978). The homozygosity test of neutrality. *Genetics*, **88**, 405–17.

Weiss, M. L. & Goodman, M. (1972). Frequency and maintenance of genetic variability in natural populations of *Macaca fascicularis*. *Journal of Human Evolution*, **1**, 41–8.

Williams, A. K. (1990). The evolution of mitochondrial DNA in the *silenus-sylvanus* species group of macaques. PhD thesis, Columbia University.

Wilson, A. C., Cann, R. L., Carr, S. M., George, M., Gyllensten, U. B., Helm-Bychowski, K. M., Higuchi, R. F., Palumbi, S. R., Prager, E. M., Sage, R. D. & Stoneking, M. (1985). Mitochondrial DNA and two perspectives on evolutionary genetics. *Biological Journal of the Linnaean Society*, **26**, 375–400.

Witt, R., Schmidt, C. & Schmitt, J. (1981). Social rank and Darwinian fitness in a multimale group of Barbary macaques (*Macaca sylvanus*): dominance reversals and male reproductive success. *Folia Primatologica*, **36**, 201–11.

Wolfheim, J. H. (1983). *Primates of the world*. Seattle: University of Washington Press.

Wright, S. (1943). Isolation by distance. *Genetics*, **28**, 114–38.

Wright, S. (1951). The genetical structure of populations. *Annual Eugenics*, **15**, 323–54.

Wright, S. (1969). *Evolution and the genetics of populations*, vol. 2. Chicago: University of Chicago Press.

20

Variation in social mechanisms by which males attained the alpha rank among Japanese macaques

D. S. SPRAGUE, S. SUZUKI AND T. TSUKAHARA

Introduction

What is the likelihood that a male attains alpha rank during its lifetime in primate species with matrilineal social systems? This question may be answered by a socio-demographic analysis of male rank systems, as some researchers have done with female rank systems (Hausfater, Cairns & Levin, 1987). The major correlates of male rank among macaques are reported to be age and length of tenure in a troop (Drickamer & Vessey, 1973; Vessey & Meikle, 1987; Paul, 1989). However, accurate data on social and demographic processes are prerequisites for building a model. We report here on the social mechanisms by which males attained the alpha rank among Japanese macaques (*Macaca fuscata*), based on our research conducted at the Yakushima study site on Yaku macaques (*M. f. yakui*), and the literature on the history of other selected sites of the main-island sub-species (*M. f. fuscata*).

A linear rank order is a major characteristic of male rank among many primate species (Berenstain & Wade, 1983; Vessey & Meikle, 1987; Walters & Seyfarth, 1987). In such social groups, a single male is dominant to the other males in the group. Researchers refer to these dominant individuals by many terms but they will be called the 'alpha male' in this chapter. Rank can be counted in at least three ways within a linear rank order: absolute rank is counted from the top of the rank order (i.e. 1, 2, 3 etc.); relative rank refers to the dominance relationship between particular individuals; proportional rank is the individual's proportional position within the rank order (e.g. top 30%). Here, the term 'rank' refers to absolute rank, unless qualified as 'relative rank'. A discussion of proportional rank is beyond the scope of this chapter and is not considered any further.

A long history of research exists on male rank and life history studies of macaques, as well as other primate taxa with matrilineal social structure (Sugiyama, 1976; Berenstain & Wade, 1983; van Noordwijk & van Schaik,

444

1985, 1988; Bercovitch, 1986; Dittus, 1986; Vessey & Meikle, 1987; MacMillan, 1989; Sommer & Rajpurohit, 1989; Sprague, 1992). Several social mechanisms are implicated in changes of male rank. The rank of a male rises within a linear rank order by the death or departure of higher ranking males (Drickamer & Vessey, 1973; Vessey & Meikle, 1983). Relative rank can change to allow some males to surpass other males in the same group (Witt, Schmidt & Schmitt, 1981; Paul, 1989; Hamilton & Bulger, 1990). Troop fission produces more opportunities for males to attain alpha rank (Koyama, 1970; Nash, 1976; Chepko-Sade & Sade, 1979; Yamagiwa, 1985; Paul & Kuester, 1988). Another major factor is that males disperse in most primate species. Alpha rank acquisition is often associated with inter-group transfer, since non-troop males may take over the alpha rank of a troop (Wheatley, 1982; van Noordwijk & van Schaik, 1985, 1988; Newton, 1988; Sprague, 1992).

Four social mechanisms for attaining alpha rank were observed among Japanese macaques. We operationally define them as follows:

1. Succession: a male succeeds to the alpha rank as a result of the death or departure of the previous alpha male.
2. Rank turnover: a change in relative rank results in an alpha male losing his rank to another male of the same troop, usually the second ranking male.
3. Troop fission: a male becomes the alpha male of a fission troop, either by (a) shifting over to the new troop from the original group, or (b) arriving as a non-troop male to join a fission group that is not accompanied by any prior male residents of the original troop.
4. Troop take-over: a non-troop male aggressively takes over the alpha rank of a troop.

Succession was common in large, provisioned troops. Take-overs by non-troops males were common in the Yakushima study site, where the unprovisioned study troops are part of a large natural population. Troop fission often created new opportunities for males to attain alpha rank in both Yakushima and the provisioned populations. However, the new alpha males were more likely to be non-troop males in Yakushima, and prior residents in the large, provisioned troops.

Study sites and data

The research from many study sites provides varying amounts and kinds of data on the rank and life history of male Japanese macaques. In addition to the data from Yakushima, we picked four studies which provided long-term data on a troop or a lineage, and a few studies that provided fragmentary but important information on rare events. Some studies noted the social mechanisms by which males attained alpha rank (e.g. Sugiyama, 1976). For other studies we classified the males on the basis of the description of troop history presented in the papers. The ideal data set for this analysis consists of long-term records of a lineage of troops formed by a series of troop fissions. In the history of a single troop, only the first alpha male has attained alpha rank by troop fission. Even if this troop fissions again, the new alpha male is, strictly speaking, no longer (or never was) a member of that original troop.

The study sites of free-ranging Japanese macaques can be classified into three types: (1) unprovisioned troops in a large population distributed in natural forests, (2) provisioned troops in contact with a larger population, and (3) provisioned and isolated troops. Yakushima is a type (1) study site, where the study groups were habituated without provisioning and lived in an extensive and largely undisturbed natural forest, surrounded by many other troops. At the Yakushima study site, researchers have observed the Ko-troop lineage for about 15 years (Maruhashi, 1982; Furuichi, 1984, 1985; Yamagiwa, 1985; Tsukahara, 1990; Hill, 1991; Sprague, 1991a,b, 1992; Suzuki, 1991). Seventeen males were recorded to have been alpha males in the Ko-troop lineage between 1976 to 1989. The events through which they attained alpha status were observed in 13 cases (81%). Some data is also available from the neighbouring HY troop (Oi, 1988). Dominance rank was determined on the basis of paired agonistic interactions. Age was estimated based on physical appearance (Maruhashi, 1982). Four age groups were recognised: small young, large young, adult, and old (Sprague, 1989). Non-troop males are defined as those males that were not members of a particular study troop at the beginning of the mating season.

The other troops are at provisioned study sites on the main islands of Japan. The Arashiyama and Takasakiyama sites are type (3) study sites, with extensive long-term records on troop histories. The troops of both study sites consist of a single lineage descended from a troop originally provisioned in the 1950s. At the Takasakiyama study site, a total of 27 males have become alpha males in the 35 years between 1952 and 1987 (Itani, 1954; Itani *et al.*, 1963; Mizuhara, 1971; Matsui, 1985; Takasakiyama Park, 1988). At Arashiyama, 13 males have become alpha males in the 34 years between 1955 to 1989 (Koyama, 1967, 1970; Norikoshi & Koyama, 1975; Huffman, 1991). The literature provides detailed histories of these troops. The mechanism by which alpha males attained their status was ascertained for 12 males (92%) at Arashiyama. With additional data

Table 20.1. *Social mechanisms by which males attained alpha status at selected study sites of Japanese macaques*

| Study site or troop | Study site type | Social mechanism (no. of events) | | | | | Total events | Total number of alpha males[a] |
		1. Succession	2. Rank turnover	3a. Fission/ Prior resident	3b. Fission/ Non-troop	4. Takeover		
Yakushima (1976–89)	1	4	0	0	3	6	13	17
Arashiyama (1955–89)	3	8	0	4	0	0	12	13
Takasakiyama (1952–87)	3	16	6	1	1	0	24	27
Ryozen troop (1967–74)	2	7	0	0	0	0	7	8
Shiga A troop (1963–73)	2	5	0	0	0	0	5	6
Total events		40 (65%)	6 (10%)	5 (8%)	4 (7%)	6 (10%)	61 (100%)	71

[a]Number of alpha males recorded during the study periods indicated at each site.

Sources: Yakushima: personal observations; Furuichi, 1985; Yamagiwa, 1985. Arashiyama: Koyama, 1967, 1970; Norikoshi & Koyama, 1975; Huffman, 1991. Takasakiyama: Itani, 1954; Itani et al., 1963; Mizuhara, 1971; Matsui, 1985; Takasakiyama Park, 1988. Ryozen: Sugiyama & Ohsawa, 1975; Sugiyama, 1976. Shiga A troop: Yoshihiro & Tokida, 1976.

provided by T. Matsui (personal communication), the mechanisms were ascertained for 24 alpha males (89%) at Takasakiyama. Although a few non-troop males visit or join the Arashiyama and Takasakiyama populations every year, especially during the mating season (Itani *et al.*, 1963; Wolfe, 1981; Huffman, 1991), no other troop lives close to these populations. Dominance rank and residential status of males are reported in the papers. Ages were known for those males born in the study troops after research started. The Arashiyama A troop after 1972 (which becomes the Arashiyama West troop) is excluded from the analysis.

The Ryozen and Shiga A troops are type (2) study sites, provisioned troops that are in contact with other troops. Data on both are not as extensive as those of type (3) study sites, but the literature provides data on eight alpha males during a seven year period (1967–74) for the Ryozen troop (Sugiyama & Ohsawa, 1975; Sugiyama, 1976), and six alpha males during an 11-year period (1963–73) for the Shiga A troop (Yoshihiro & Tokida, 1976).

Data are summarised separately for the study sites described above, which provide relatively complete data (Table 20.1), and those with only fragmentary data that report unique events from a variety of other troops (Table 20.2). The actual events leading to a new alpha male were often more complex than the classifications imply. The operational classification is based on the timing of events. For example, if after the disappearance of the alpha male, a rank turnover occurs between the former second- and third-ranking males, we count two events, one succession and one turnover. If the timing is reversed, the event is considered a single succession. The social mechanisms could not be ascertained for all alpha males; the mechanism was unknown for at least the first alpha male identified at the start of a study. Some males became alpha male when no human observers were present. Some papers listed the alpha males but did not describe how individuals had attained alpha rank.

Results

Male Japanese macaques attained alpha rank by all four mechanisms (Tables 20.1 and 20.2). In general, succession (mechanism 1, see p. 444) was the most common mechanism. However, considerable variation existed between study sites. Succession was significantly more often the mechanism by which males attained alpha rank in the provisioned type (2) and type (3) study sites on the main islands (Table 20.3). The next most common mechanism of gaining rank was turnover (mechanism 2), followed by troop

Table 23.2. *Five forms of branch shaking*

	Branch/tree shake	Dead log thump	Branch break	Dead standing tree shake	Fir crown shake
Positional behaviour	Horizontal/quadrapedal or vertical clinging	Horizontal/quadrapedal	Vertical/bipedal	Vertical clinging or horizontal/quadrapedal	Vertical clinging
Substrate	Non-specific, all live trees	Specific, dead logs	Specific, dead branches of firs	Specific, dead standing conifers	Specific, crowns of firs
Estimated energy expenditure	Minimum–medium	Medium–maximum	Medium	Medium	Medium
Estimated distance sound travels (m)	<300	>1000	<500	<300	<100
Frequency of occurrence in adult males (%)	39.2	27.6	14.7	11.0	7.4

Table 23.3. *Male dominance rank, estimated age, and degree of peripheralisation. Dominance matrix displays number of losses in rows to winners in columns*

Relative dominance rank	Relative spatial status	Estimated age in years	Total branch shakes	Individual losing	Winners in dyadic agonistic interactions												
					Dia	Gpa	Spl	Why	Ln	Btn	Ch	Old	Hlf	Hd	Gh	Rng	Prg
High	C	10–12	38	Diamond		1											
	C	15+	3	Grandpa‡	1												
	C	12–15	29	Split lip	1	1			1								
	C	12–15	14	Why	8	1	3		1								
	P	8–10	17	Line	2		7	2									
	P	12–15	11	Button	6		3		1								
Medium	P–C	12–15	7	China	5	1	3	1	3	4		1					
	P–C	15+	2	Old male‡	1		2	1	1								
	P–C	15+	2	Halfback	3		2					1					
	P–C	7–8	15	Handsome	2	1	11		13	9	2	2	1				
Low	P	7–8	12	Ghost	4	3	2		6	1			1	2			
	P	6–7	7	Ring			1		10	3			1	5	1		1
	P	6–7	6	Progdot	2			2	5	8			2	1	4	1	

‡Indicates male disappeared during study, presumed dead (also see Mehlman, 1986b, appendix 3). C, centre of group during resting, foraging, or movement; P, peripheral to group, often solo or with other peripheralised males; P–C, peripheral, but limited access to centre of group (e.g. birth season).

Table 20.4. *Prior residents and non-troop males among new alpha males*

Study sites or troops	Non-troop male	Prior resident	Total
Arashiyama, Takasakiyama, Ryozen, and Shiga A	1	47	48
Yakushima	9	4	13
Total	10	51	64

Fisher's exact test: $P < 0.01$ for comparison of non-troop male success at the two sites. Data from Table 20.1.

fission in which a prior resident became the new alpha male of the fission troop (mechanism 3a).

Non-troop males joining a fission troop at the alpha rank (mechanism 3b), or take-overs (mechanism 4), were extremely rare among the main-island study sites (Tables 20.1 and 20.2). By contrast, inter-troop transfer played an important role for the males that became alpha males in Yakushima, a type (1) study site. Non-troop males acquired the alpha rank significantly more often in Yakushima compared with the main-island study sites (Table 20.4).

The Yakushima study site

The main study troops in Yakushima consisted of four troops descended from the original Ko-troop. Thus, three males had attained alpha status as a result of troop fission, and all of them were non-troop males. However, take-over (six cases) was the most common mechanism observed by which males attained alpha rank in the Ko-troop lineage. Four residents attained alpha status through succession. All the new males were classified as 'adults', except for one 'large young' and one 'old' male, who also succeeded to alpha rank. A turnover resulting in a new alpha male has never been observed in Yakushima, although turnovers have been observed among lower ranking males. Infanticide has never been observed. In addition, Oi (1988) reported that the new alpha male of a troop that fissioned from the Yakushima HY troop was a non-troop male (Table 20.2).

Other study sites

Within the Arashiyama lineage, succession and fission were the only mechanisms by which males attained alpha rank (Koyama, 1967, 1970;

Norikoshi & Koyama, 1975; Huffman, 1991). Most males succeeded to the rank (eight cases), often after living in the troop for many years. Fission also played a large role; as of 1989, the Arashiyama troops had experienced four fissions, and all of the new alpha males were prior residents. The average age for new alpha males of known age was 14.6 years (*n* = 5, range 9–22). No male achieved alpha rank through a rank turnover. No non-troop male has ever taken over an Arashiyama troop, or joined as the alpha male of a fissioning Arashiyama troop.

The Takasakiyama lineage showed greater variation (Itani, 1954; Itani *et al.*, 1963; Mizuhara, 1971; Matsui, 1985; Takasakiyama Park, 1988). Although succession (16 cases) was still the most common mechanism, six cases of rank turnover have been observed. Of the two fissions, the new alpha males were a prior resident in one case and a non-troop male in the other. This was the only case involving a non-troop male for the four type (2) and (3) study sites in Table 20.1. As in Arashiyama, no take-over has ever been observed at Takasakiyama. Fewer data were available on the ages of new alpha males, but the youngest male of known age to achieve alpha rank was 14 years old.

The Ryozen and Shiga A troops showed the least variation of the troops reviewed here (Sugiyama & Ohsawa, 1975; Yoshihiro & Tokida, 1976). Succession was the only mechanism observed for the 12 alpha males during the 18 years of study of these two troops. All the alpha males in both studies were classified as 'adult' by the respective researchers. Neither troop fissioned during the periods covered by the data, and no non-troop male ever took over either troop.

Potential variation in social mechanisms

Despite the overall patterns, careful review of the literature revealed cases demonstrating that the potential exists in all populations for a male to utilise any one of the mechanisms to attain alpha rank. Several cases have been observed of non-troop males becoming alpha males, even in the main-island study sites (Table 20.2). As already mentioned, a non-troop male joined a fission troop in the Takasakiyama site, and two further cases have been reported by Furuya (1969) from another study site. Furuya (1969) also described social changes at a study site where troop fissions produced more alpha males than did other mechanisms. One case of what could be a troop take-over at a main-island study site was observed at the Shiga B2 troop, a neighbour of the Shiga A troop described above. In this case, the alpha male of the Shiga A troop suddenly departed, and was identified later as the alpha male of the Shiga B2 troop (Tokida & Wada, 1974).

Discussion

The data reveal three characteristics of male rank order among Japanese macaques. First, the relative ranks of males living in the same troop seem to be very stable among Japanese macaques compared to other species. Second, troop fission is an important factor producing opportunities for males to become alpha males. Third, inter-troop transfer may play an important role in alpha rank acquisition in some populations.

Succession was the most common mechanism by which male Japanese macaques attained alpha rank. This is consistent with the theory that the length of tenure in a troop is the primary determinant of rank among macaques. As Drickamer & Vessey (1973; Vessey & Meikle, 1987, p. 287) stated for rhesus macaques (*Macaca mulatta*), 'the rule basically seems to be that when males join a group, they are at or very near the bottom of the male hierarchy and move up in rank only as higher ranking males leave or die'.

First, one reason for the importance of succession among Japanese (and possibly rhesus) macaques may be the stability of relative ranks among males in the same troop. For example, Norikoshi & Koyama (1975) found only two rank turnovers in six years among the 'central' adult males of the Arashiyama A and B troops. By contrast, the relative ranks of Barbary macaques (*M. sylvanus*) may be much more unstable. Kuester & Paul, (1988) and Paul (1989, p. 464) observed 'several rank changes and at least short periods of non-linearity' among adult males in a troop of Barbary macaques. The rank relations of male baboons (*Papio* spp.) may be even more flexible (Hamilton & Bulger, 1990). Bercovitch (1986, p. 546) stated that 'male baboons maintain an unstable, shifting hierarchical relationship with each other'.

The stability of relative rank may be due partly to what Sapolsky & Ray (1989) called 'styles of dominance', including the tendency to form alliances. Hamilton & Bulger (1990) pointed out that rank changes may occur in either 'passive' or 'active' events among savannah baboons (*P. anubis*), where 'active' rank changes were the result of contested reversals of established rank relationships. Rank changes may also be facilitated by alliances between males, or cooperation among siblings (Witt, Schmidt & Schmitt, 1981; Bercovitch, 1988). Further research is necessary to determine whether male Japanese macaques in some populations have dominance styles that can be considered 'passive' or whether they do not form alliances as often as males of other species.

Second, troop fissions are quite common among species with matrilineal social systems (Nash, 1976; Chepko-Sade & Sade, 1979; Paul & Kuester, 1988). Troop fission necessarily results in a new alpha male as long as at least one male joins the new troop. This is a relatively 'passive' means of acquiring a new rank if the males of the original troop redistribute themselves between the daughter troops without altering their relative ranks (e.g. Koyama, 1970).

Third, inter-troop transfer has the potential to play a large role in rank acquisition among male Japanese macaques. At the Yakushima study site, non-troop males were commonly observed to take over or join fission troops. These were extremely 'active' means of acquiring rank. The new alpha males exhibited a suite of behaviours that researchers at Yakushima called the 'alpha male attitude', such as piloerection, vigorous tree-shaking, displays with barking, and general aggressiveness towards troop residents (Yamagiwa, 1985; Sprague, 1990, 1991*b*, 1992; see also Wolfe, 1981). These males were similar to the 'bluff' immigrants reported for long-tailed macaques (*M. fascicularis*) by van Noordwijk & van Schaik (1985, 1988). In addition, non-troop males often mated at Yakushima (Furuichi, 1985; Yamagiwa, 1985; Sprague, 1991*a,b*, 1992), and Yamagiwa (1985) hypothesised that troop fissions were precipitated by mating between non-troop males and the females.

Finally, it is necessary to explain the striking contrast between the Yakushima macaques (*M. f. yakui*) and the main-island sub-species (*M. f. fuscata*). Prior residents seemed to have a clear advantage over non-troop males in the main-island sub-species, while the reverse seemed to be true for the Yaku macaques. This may be because the main-island troops are generally larger than those at Yakushima, and contain more fully grown adult males (Takasaki & Masui, 1984). The Arashiyama and Takasakiyama populations are enormous. The Arashiyama A and B troops totalled 301 animals in 1972, while the total population of three Takasakiyama troops peaked at 2002 animals in 1979. In addition, prior residents may have an advantage over non-troop males if long-term social alliances with females or adult males are important for males to attain or maintain alpha rank. Fewer non-troop males may come to challenge the residents if a troop is geographically isolated.

At Yakushima, the troops are smaller and inter-troop mobility is extensive among the numerous troops in the large, natural population, especially during the mating season. Newly arrived males frequently challenge the few adult males living in each troop, and mate with the females (Yamagiwa, 1985, Sprague, 1991*a,b*, 1992). The conditions in Yakushima may place greater emphasis on individual fighting ability or immediate attractiveness to the females, rather than long-term social relations (Yamagiwa, 1985; but see Furuichi, 1985; Tsukahara, 1990; Hill, 1991; Suzuki, 1991).

Given the considerable variation in the social mechanisms available to males, a model to assess the likelihood of a male attaining the alpha rank during its life should consider at least the following factors:

1. Survival rates of males: higher survival leads to greater competition between adults and fewer higher ranking males that die to allow others to rise in rank.
2. Migration rates of males: some adults may find opportunities to acquire a new rank by transferring to another troop, and this allows those who remain to rise in rank.
3. Stability of relative rank relations: long-term social relationships and alliances may either increase or decrease the stability of rank relations.
4. Frequency of troop fission: troop fission leads to more troops in a population and provides more opportunities for males to become alpha males. Fission also leads to smaller troops with fewer adult male defenders; this may make it easier for non-troop males to take over.

The data presented here generally support the theory that the basic determinants of non-natal male rank among macaques are age and tenure (Drickamer & Vessey, 1973; Vessey & Meikle, 1987; Paul, 1989). Most new alpha males were adult, and many acquired their status through succession. However, it was also found that inter-troop transfer may reduce or eliminate the association between rank and tenure (but not age) in some populations. Some males took over troops or immediately became the new alpha males of fission troops. These are only the most extreme forms of a more general phenomenon: males may not acquire a rank at the bottom of the male rank order upon joining a troop. For example, Vessey & Meikle (1987, p. 287) noted that, among the Cayo Santiago rhesus monkeys, 'Occasionally an older male moves into a group during the breeding season and comes to rank over some of the younger but more senior males.' This phenomenon deserves further attention. Rank turnovers also have the potential to disrupt the rank-tenure relation, but they probably do not do so within Japanese macaque troops, because rank turnovers occur too infrequently and only among pairs of adjacent ranks.

Acknowledgements

Research in Yakushima has been supported by the Wenner-Gren Foundation, and the Cooperative Research Fund of the Kyoto University Primate

Research Institute. We thank: T. Maruhashi, J. Yamagiwa, Y. Takahata, D. Hill and N. Okayasu for cooperation in the field and use of unpublished data; M. Huffman and Y. Takahata for advice on data concerning Arashiyama; and T. Matsui for providing further information on the history of the Takasakiyama troops. A post-doctoral position at Kyoto University for D. S. S. was made possible by T. Nishida and a joint NSF/JSPS fellowship.

References

Bercovitch, F. B. (1986). Male rank and reproductive activity in savanna baboons. *International Journal of Primatology*, **7**, 533–50.

Bercovitch, F. B. (1988). Coalitions, cooperation and reproductive tactics among adult male baboons. *Animal Behaviour*, **36**, 1198–209.

Berenstain, L. & Wade, T. D. (1983). Intrasexual selection and male mating strategies in baboons and macaques. *International Journal of Primatology*, **4**, 201–35.

Chepko-Sade, B. D. & Sade, D. S. (1979). Patterns of group splitting within matrilineal groups. *Behavioral Ecology and Sociobiology*, **5**, 67–86.

Dittus, W. P. J. (1986). Sex differences in fitness following a group take-over among Toque macaques: testing models of social evolution. *Behavioral Ecology and Sociobiology*, **19**, 257–66.

Drickamer, L. C. & Vessey, S. H. (1973). Group changing in free-ranging male rhesus monkeys. *Primates*, **14**, 359–68.

Furuichi, T. (1984). Symmetrical patterns in non-agonistic social interactions found in unprovisioned Japanese macaques. *Journal of Ethology*, **2**, 109–19.

Furuichi, T. (1985). Inter-male associations in a wild Japanese macaque troop on Yakushima Island, Japan. *Primates*, **26**, 219–37.

Furuya, Y. (1968). On the fission of troops of Japanese monkeys. I. Five fissions and social changes between 1955 and 1966 in the Gagyusan troop. *Primates*, **9**, 323–50.

Furuya, Y. (1969). On the fission of troops of Japanese monkeys. *Primates*, **10**, 47–69.

Hamilton, W. J. & Bulger, J. B. (1990). Natal male baboon rank rises and successful challenges to resident alpha males. *Behavioral Ecology and Sociobiology*, **26**, 357–62.

Hausfater, G., Cairns, S. J. & Levin, R. N. (1987). Variability and stability in the rank relations of nonhuman primate females: analysis by computer simulation. *American Journal of Primatology*, **12**, 55–70.

Hill, D. A. (1991). Patterns of affiliative relationships involving non-natal males in a troop of wild Japanese macaques (*Macaca fuscata yakui*) in Yakushima. In *Primatology today*, ed. A. Ehara, T. Kimura, O. Takenaka & M. Iwamoto, pp. 211–4. Amsterdam: Elsevier.

Huffman, M. A. (1991). A history of the Arashiyama troop of Japanese macaques of Kyoto, Japan. In *The monkeys of Arashiyama: thirty-five years of study in the East and the West*, ed. L. M. Fedigan & J. Asquith, pp. 21–53. New York: State University of New York Press.

Itani, J. (1954). [*The monkeys of Takasakiyama.*] (In Japanese.) Tokyo: Kobunsha.

Itani, J., Tokuda, K., Furuya, Y., Kano, K. & Shin, Y. (1963). The social construction of natural troops of Japanese monkeys in Takasakiyama. *Primates*, **4**, 1–42.

Koyama, N. (1967). On dominance rank and kinship of a wild Japanese monkey troop in Arashiyama. *Primates*, **8**, 189–216.

Koyama, N. (1970). Changes in dominance rank and division of a wild Japanese monkey troop in Arashiyama. *Primates*, **11**, 335–90.

Kuester, J. & Paul, A. (1988). Rank relations of juvenile and subadult natal males of Barbary macaques (*Macaca sylvanus*) at Affenberg Salem. *Folia Primatologica*, **51**, 33–44.

MacMillan, C. A. (1989). Male age, dominance, and mating success among rhesus macaques. *American Journal of Physical Anthropology*, **80**, 83–89.

Maruhashi, T. (1982). An ecological study of troop fissions of Japanese monkeys (*Macaca fuscata yakui*) on Yakushima Island. *Primates*, **23**, 317–37.

Matsui, T. (1985). [*Takasakiyama: the land of monkeys.*] (In Japanese.) Fukuoka: Nishi Nihon Shinbun.

Mizuhara, H. (1971). [*A history of the land of monkeys.*] (In Japanese.) Sogensha, Osaka.

Nash, L. T. (1976). Troop fission in free-ranging baboons in the Gombe Stream National Park, Tanzania. *American Journal of Physical Anthropology*, **44**, 63–78.

Newton, P. N. (1988). The variable social organization of Hanuman langurs (*Presbytis entellus*), infanticide, and the monopolization of females. *International Journal of Primatology*, **9**, 59–77.

Norikoshi, K. & Koyama, N. (1975). Group shifting and social organization among Japanese monkeys. In *Proceedings of the Symposia of the Fifth Congress of the International Primatological Society, Nagoya, Japan 1974*, ed. S. Kondo, M. Kawai & A. Ehara, pp. 43–61. Tokyo: Japan Science Press.

Oi, T. (1988). Sociological study on the troop fission of wild Japanese monkeys (*Macaca fuscata yakui*) on Yakushima Island. *Primates*, **29**, 1–19.

Paul, A. (1989). Determinants of male mating success in a large group of Barbary macaques (*Macaca sylvanus*) at Affenberg Salem. *Primates*, **30**, 461–76.

Paul, A. & Kuester, J. (1988). Life-history patterns of Barbary macaques (*Macaca sylvanus*) at Affenberg Salem. In *Ecology and behavior of food-enhanced primate groups*, ed. J. E. Fa, pp. 199–228. New York: Alan R. Liss.

Sapolsky, R. M. & Ray, J. C. (1989). Styles of dominance and their endocrine correlates among wild olive baboons (*Papio anubis*). *American Journal of Primatology*, **18**, 1–13.

Sommer, V. & Rajpurohit, L. S. (1989). Male reproductive success in harem troops of Hanuman langurs (*Presbytis entellus*). *International Journal of Primatology*, **10**, 293–317.

Sprague, D. S. (1989). Male intertroop mobility during the mating seasons among the Japanese macaques of Yakushima Island, Japan. PhD thesis, Yale University.

Sprague, D. S. (1991a). Influence of mating on the troop choice of non-troop males among the Japanese macaques of Yakushima Island. In *Primatology today*, ed. A. Ehara, T. Kimura, O. Takenaka & M. Iwamoto, pp. 207–10. Amsterdam: Elsevier.

Sprague, D. S. (1991b). Mating by non-troop males among the Japanese macaques of Yakushima Island. *Folia Primatologica*, **57**, 156–8.

Sprague, D. S. (1992). Life history and male intertroop mobility among Japanese macaques (*Macaca fuscata*). *International Journal of Primatology*, **13**, 437–54.

Sugiyama, Y. (1976). Life history of male Japanese monkeys. *Advances in the Study of Behavior*, **7**, 255–84.

Sugiyama, Y. & Ohsawa, H. (1975). Life history of male Japanese macaques at Ryozenyama. In *Contemporary primatology*, ed. S. Kondo, A. Ehara, & M. Kawai, pp. 407–10. Basel: S. Karger.

Suzuki, S. (1991). Dominance relationships of young males in a troop of wild Japanese monkeys in Yakushima, Japan. In *Primatology today*, ed. A. Ehara, T. Kimura, O. Takenaka & M. Iwamoto, pp. 133–6. Amsterdam: Elsevier.

Takasaki, H. & Masui, K. (1984). Troop composition data of wild Japanese macaques reviewed by multivariate methods. *Primates*, **25**, 308–18.

Takasakiyama Natural Animal Park (1988). *The four seasons of Takasakiyama* II. Oita: Oita City Tourism Association.

Tokida, E. & Wada, K. (1974). [Some characters of leaving and joining by males among A-troop and its neighboring troops at Shiga Heights.] (In Japanese.) In *Life history of male Japanese monkeys*, ed. K. Wada, S. Azuma, & Y. Sugiyama, pp. 28–34. Inuyama: Primate Research Institute.

Tsukahara, T. (1990). Solicitation in male-female grooming in a wild Japanese macaque troop in Yakushima Island. *Primates*, **31**, 147–56.

van Noordwijk, M. A. & van Schaik, C. P. (1985). Male migration and rank acquisition in wild long-tailed macaques (*Macaca fascicularis*). *Animal Behaviour*, **33**, 849–61.

van Noordwijk, M. A. & van Schaik, C. P. (1988). Male carers in Sumatran long-tailed macaques (*Macaca fascicularis*). *Behaviour*, **107**, 25–43.

Vessey, S. H. & Meikle, D. B. (1987). Factors affecting social behavior and reproductive success of male rhesus monkeys. *International Journal of Primatology*, **8**, 281–92.

Walters, J. R. & Seyfarth, R. M. (1987). Conflict and cooperation. In *Primate societies*, ed. B. B. Smuts, D. L. Cheney, R. M. Seyfarth, R. W. Wrangham & T. T. Struhsaker, pp. 306–17. Chicago: University of Chicago Press.

Wheatley, B. P. (1982). Adult male replacement in *Macaca fascicularis* of East Kalimantan, Indonesia. *International Journal of Primatology*, **3**, 203–19.

Witt, R., Schmidt, C. & Schmitt, J. (1981). Social rank and Darwinian fitness in a multimale group of Barbary macaques (*Macaca sylvanus*, Linnaeus, 1758): dominance reversals and male reproductive success. *Folia Primatologica*, **36**, 201–11.

Wolfe, L. (1981). Display behavior of three troops of Japanese monkeys (*Macaca fuscata*). *Primates*, **22**, 24–32.

Yamagiwa, J. (1985). Socio-sexual factors of troop fission in wild Japanese monkeys (*Macaca fuscata yakui*) on Yakushima Island, Japan. *Primates*, **26**, 105–20.

Yoshihiro, S. & Tokida, E. (1976). [Japanese monkeys in the Shiga Highlands. I. Male monkeys' leaving and entering a troop in the Yokoyugawa Basin (1).] (In Japanese.) *Nihonzaru*, **2**, 1–50.

21

Determinants of dominance among female macaques: nepotism, demography and danger

D. A. HILL AND N. OKAYASU

Introduction

The social behaviour of macaques has been studied most intensively in two species: the Japanese macaque (*Macaca fuscata*), and the rhesus macaque (*M. mulatta*). In both cases the vast majority of data have come from provisioned populations for which there are long-term records of matrilineal kinship (Rawlins & Kessler, 1986; Koyama *et al.*, 1992). Although research on each species has been conducted quite independently, the findings have revealed many strong similarities in the details of their social behaviour and social structure (*sensu* Hinde, 1976). One of the most striking examples of this is the acquisition of dominance rank by females (Kawamura, 1958, 1965; and references cited under 'Kawamura's principles', below). This process follows a clear and predictable pattern, which has been confirmed by several successive studies. Various attempts have been made to elucidate the proximate and ultimate mechanisms involved in the process, and a variety of models have been proposed (Chapais & Schulman, 1980; Schulman & Chapais, 1980; Hausfater, Saunders & Chapman, 1981; Horrocks & Hunte, 1983a,b; Hausfater, Cairns & Levin, 1987; Thierry, 1990; Datta & Beauchamp, 1991).

It has been suggested that the pattern of female rank acquisition found in Japanese and rhesus macaques is characteristic of the genus (Chapais & Schulman, 1980), or even of all cercopithecines with matrilineal social organisation (Horrocks & Hunte, 1983a), but there is now considerable evidence of variation. In this chapter, we describe the classical pattern of rank acquisition by female macaques, and some of the theories that have been put forward to explain its evolutionary significance. We then summarise evidence of variance from this pattern observed within the genus, and consider attempts that have been made to explain it.

Kawamura's principles

The importance of matrilineal kinship in the determination of female dominance rank was first noted by Kawamura (1958, 1965) in his study of a small, provisioned group of Japanese monkeys at Minoo, near Osaka. Two basic principles can be derived from Kawamura's work.

1. A daughter does not normally rise in rank above her mother.
2. A maturing daughter becomes dominant to all females who are subordinate to her mother. This includes her own older sisters, if she has any, a process that has been termed 'youngest ascendancy' (Datta, 1988).

As a result, dominance rank among mature sisters is inversely correlated with age, with the oldest being least dominant. A few years after Kawamura's discovery, Sade (1967) described exactly the same pattern of dominance rank acquisition for female rhesus macaques in the Cayo Santiago colony in Puerto Rico. Subsequent studies of several groups of both species have confirmed these findings (Koyama, 1967; Missakian, 1972; Sade, 1972; Datta, 1988; Mori, Watanabe & Yamaguchi, 1989; Takahata, 1991).

Kawamura (1958, 1965) proposed that the principles by which the matriarchal rank order developed were a general characteristic of Japanese monkeys, but he also noted that exceptions occurred. Subsequent studies have recorded further exceptions, but most have involved only a minority of females in the group (Chikazawa *et al.*, 1979; Mori *et al.*, 1989; Takahata, 1991). In general, the phenomenon is remarkably widespread and robust in the provisioned groups of Japanese and rhesus macaques, where it has most frequently been studied.

The importance of social context

In social groups of Japanese and rhesus macaques the outcome of any agonistic encounter is likely to be influenced by the social context in which it takes place. Kawai (1958, 1965) demonstrated that the 'basic rank' of an individual may be enhanced by the presence of another, conferring a higher 'dependent rank'. In some cases, this influence may be mediated through direct intervention in disputes (Kaplan, 1977; Datta, 1988), but simply being close to a dominant individual can also result in greater 'dependent rank' (Kitamura, 1975). Kawai (1958, 1965) described several forms of 'dependent rank' most of which involved kin, with the most prevalent being between mother and offspring.

Kawamura (1958, 1965) attributed the phenomenon of youngest

ascendancy to the mother's 'regard for the youngest child' (Kawamura, 1965, p. 107), and several subsequent studies have noted the importance of maternal support in rank acquisition (Watanabe, 1979; Horrocks & Hunte, 1983*b*; Datta, 1983, 1988; Netto & van Hooff, 1986). However, cases have been recorded of youngest ascendancy occurring after the mother's death (Sade, 1972; Mori *et al.*, 1989), indicating that maternal support is not a prerequisite. It may be that the support of other kin is important in such cases. Datta (1988) found that kin were responsible for 92.4% of interventions in sibling disputes among provisioned rhesus, and that almost half of these involved kin other than the mother.

Chapais (1985, 1988*a,b*, 1991) has examined the influence of social context on the acquisition and maintenance of dominance rank experimentally, by producing sub-groups from members of a small captive group of Japanese macaques. The group consisted of three small family units, obtained from the Arashiyama West population (Fedigan, 1991), each of which included a mother and two or three immature daughters. In a series of experiments sub-groups were created, and their composition altered by introducing or removing potential allies.

One experiment demonstrated that a dominant individual may influence the outcome of an agonistic interaction without actively participating in it (Chapais, 1985). A spontaneous rank reversal between a 2.5 year-old female and her mother occurred when the immature female formed alliances with two immature daughters of the alpha female. The alpha female herself took no active part in the reversal. Removal of the two immature allies had no effect on the reversed ranks, but when the alpha female was removed the mother re-established her dominance over the daughter.

Chapais (1988*a*) has also examined the role of more direct maternal support by introducing individual immature females into lower-ranking family groups. In each case, the test female became subordinate to the family group, but when the mother was introduced the test female regained her former rank. Further experiments have demonstrated that all females are dependent on the support of others for the maintenance of their ranks, that intervention is the main mechanism bringing about rank changes, rather than solicited support, and that low rates of intervention are sufficient for the process of rank acquisition and its maintenance (Chapais, 1988*a,b*).

The roots of nepotism

Having established that the support of kin, and particularly that of the mother, is the driving force behind female rank acquisition, another question arises: why does the mother choose to give preferential support to her youngest daughter? Schulman & Chapais (1980) presented a model that attempts to answer this question. The model is based on the assumption that, in disputes between sisters, the mother should preferentially give support to whichever of her daughters has the highest reproductive value. In this case, the peak in reproductive value will coincide with reaching sexual maturity, so the mother always supports her youngest sexually mature daughter and this results in the younger daughter's rise in rank. Horrocks & Hunte (1983a) agreed that the presence of the mother determines the outcome of disputes between sisters, and they presented data for vervet monkeys (*Cercopithecus aethiops*) in Barbados that support this. They argued, however, that many reversals occur well before the younger daughter reaches sexual maturity and, therefore, before the peak of her reproductive value. They also noted that mothers seem to favour younger offspring from birth.

On these grounds Horrocks & Hunte (1983a) rejected the model of Schulman & Chapais, and presented their own alternative explanation. A mother perceives her daughters as present and future competitors and must strive to prevent coalitions among them against herself. By reversing the rank order among her daughters, a mother ensures that the rank of the more dominant (i.e. youngest) daughters is dependent upon her own support. Thus, these daughters would have little to gain, and may actually fall in rank, if they were to join a coalition of daughters that succeeded in outranking the mother.

The 'despot's' daughter

The patterns of female rank acquisition found in Japanese and rhesus macaques were once thought to be a typical characteristic of matrilineal social organisation in Old World monkeys (Horrocks & Hunte, 1983a, p. 417). Subsequent data have shown that, while certain features of rank acquisition still appear to be more or less ubiquitous among cercopithecines, there is great variation in others. For example, long-term studies of both captive bonnet macaques (*M. radiata*; Silk, Samuels & Rodman, 1981) and semi-free-ranging Barbary macaques (*M. sylvanus*; Paul & Kuester, 1987) have found that dominance relations within matrilines did not routinely follow 'Kawamura's principles'. The main departure found in bonnet macaques was that several females rose in rank above their mothers and, in some cases, above females dominant to their mothers (Silk *et al.*, 1981). Horrocks & Hunte (1983a) maintain that these exceptions, and similar cases reported by Chikazawa *et al.* (1979), may indicate the vulnerability of mothers who have no younger daughters who would be dependable allies. However, this cannot explain the pattern of female dominance described by Paul & Kuester (1987) for a provisioned population of Barbary macaques. In this group, rank reversals between mothers and daughters were uncommon, but younger sisters rarely became dominant to older sisters.

Clearly, any attempt to explain the evolutionary significance of the classical patterns of female rank acquisition must be able to explain why such exceptions occur. Why, for example, should it be selectively advantageous for mothers to favour their youngest daughters in rhesus and Japanese macaques, but not in Barbary macaques?

It may be that some species of macaque are more 'despotic' than others (Thierry, 1990). Interspecific differences are difficult to assess because most studies are of just one species. Comparisons between studies tend to be confounded by differences in methodology, composition and history of the group, and conditions under which the animals are kept, any of which could influence the results. Thierry (1985) has conducted comparative research on three species of macaque: rhesus, long-tailed (*M. fascicularis*) and Tonkean (*M. tonkeana*), which he studied in small groups kept under similar conditions in outdoor enclosures. He found that aggression was severe and retaliation was rare in rhesus macaques, whereas Tonkean macaques very rarely showed severe aggression, and frequently had bidirectional aggressive exchanges with appeasement behaviours. In general, agonistic interactions in long-tailed macaques were less aggressive than those of rhesus, and more so than those of Tonkean macaques, and other aspects of their behaviour suggested that they were intermediate between the two.

Partly on the basis of these findings, Thierry (1990) proposed that macaque species can be viewed as varying along a continuum, from 'despotism' to 'egalitarianism'. Japanese and rhesus macaques are at the despotic end of the continuum. Tolerance to the demands of others is low, aggression is frequent and often severe, and retaliation is rare. Kinship exerts a powerful influence over dominance relations, as well as many other aspects of social relationships, and reconciliation behaviours are poorly developed. Tonkean macaques are typical of the egalitarian end of the continuum. They show much more tolerance, aggression is less unidirectional and less severe, and conciliatory behaviours are common and distinctive. Affiliative interactions are more frequent and are not dictated by the relative dominance of the participants. Kinship has much less influence over dominance and other aspects of relationships.

Thierry (1990) suggested that each macaque species could be located on this continuum, given adequate data. Long-tailed and pig-tailed macaques (*M. nemestrina*) would be closer to the despotic end, while bonnet, Barbary and stump-tailed macaques (*M. arctoides*) would be closer to the egalitarian end. As Thierry (1985) noted, the comparison is based on data for only one group of each species, so extrapolations must be made with caution, but the work of other researchers has also revealed interspecific differences in accord with Thierry's classification (de Waal & Lutrell, 1989; Butovskaya, 1993).

Maternal support, and consequently inheritance by daughters of maternal rank, would be characteristic of species at the despotic end of the continuum. So the phenomenon of 'youngest ascendancy' would be widespread among despotic species, less common among species in the intermediate part of the continuum, and rare in egalitarian species.

Modelling the influence of demography

Variation in patterns of female dominance may be influenced by factors other than interspecific differences in the degree of despotism. Several studies have noted that demographic variables appear to play an important role in female rank acquisition (Chikazawa *et al.*, 1979; Hausfater, Altmann & Altmann, 1982; Paul & Kuester, 1987). Two types of computer model have been presented that simulate the influence of demographic variation on patterns of dominance acquisition among females (Hausfater *et al.*, 1981; Hausfater *et al.*, 1987; Datta, 1989; Datta & Beauchamp, 1991). In both cases, the authors of the models concluded that much of the variation that had been attributed to species differences could, in fact, be explained by demographic variation.

The first type of model used simulations of a primate population through 50 generations to assess the degree of nepotism among females, and also the extent to which their ranks were inversely age-graded (Hausfater *et al.*, 1981; Hausfater *et al.*, 1987). Baseline demographic parameters were taken from data for yellow baboons (*Papio cynocephalus*; Altmann, Altmann & Hausfater, 1981). Variables that could be altered in the model included demographic parameters, social 'rules' governing rank acquisition, and the rate of occurrence of spontaneous rank reversals between individuals. The model suggested that stochastic factors alone would produce substantial variation in the occurrence of nepotism and age-graded ranking, and that long-term stability in these two measures would not be expected under any conditions. One clear pattern that emerged was that nepotism was more widespread in a declining population than in an expanding one, while the opposite was true for age-graded ranking. While the model demonstrates ways in which demographic variation could influence female ranking, the results of the simulations are difficult to reconcile with patterns of dominance reported for Japanese and rhesus macaques. The vast majority of data for these species comes from provisioned populations, which are undeniably expanding, but in which nepotism is, nevertheless, widespread. Furthermore, although rank reversals and exceptions to Kawamura's principles do occur, nepotism remains widespread over remarkably long periods.

The simulations of Hausfater and co-workers were based on the assumption that a maturing female would occupy a dominance rank just below that of her mother. The probabilities that she would outrank (a) her mother and (b) other older females were adjustable variables in the model, but the mechanisms by which these rank changes take place were not examined. The second type of model, presented by Datta (1989) and Datta & Beauchamp (1991), considered demographic influences on the proximate mechanisms of rank acquisition.

The model assumes that support from one or more powerful allies, in particular the mother, is an integral part of rank acquisition by maturing females. This assumption is supported by evidence from both non-invasive field studies (Horrocks & Hunte, 1983*b*; Datta, 1988) and experimental manipulation (references under 'The importance of social contact', above). Two types of population were compared by the model. In Population A females reached sexual maturity at 4 years old, and produced an infant every year. In Population B sexual maturity was attained at six years old, and females gave birth every two years. For both populations, the sex of infants was randomly assigned in such a way as to produce equal numbers of daughters and sons. Other demographic parameters were also built into the model, but these need not be considered here. It was assumed that on reaching sexual maturity a female would rise in rank above an older sister, provided that she had at least one surviving ally (mother or other sister) who was dominant to that sister. Simulations under various conditions showed that youngest ascendancy would be much more common in Population A than in Population B. The key factor was the greater availability of potential allies in Population A. Datta & Beauchamp (1991) noted similarities between Population A and expanding groups of provisioned macaques, and between Population B and natural groups of savanna baboons (*Papio anubis*).

Data for a small, natural troop of Japanese macaques in Yakushima provided an opportunity to test the model (Hill & Okayasu, 1995). P Troop included four pairs of sisters, and in each pair the older remained dominant to the younger. Of the two hypothetical populations, P Troop most closely resembled Population B in demographic parameters. Sexual maturity came a year earlier, at 5 years old, the mean interval between births was about two and a half years, and four or more years separated three of the four pairs of sexually mature sisters. Datta's (1989) simulations showed that, under certain conditions, youngest ascendancy would not be expected to occur at all in Population B, because the younger sister lacked any 'powerful support'. For all four pairs of sisters in P Troop, however, the mother was alive when the younger daughter reached sexual maturity. Furthermore, for one pair of sisters, who were only two years apart in age, both the mother and a dominant older sister were alive when the younger reached sexual maturity. Although agonistic interactions were somewhat infrequent, there was no ambiguity in their direction, and no indication that mothers were 'weakly dominant' (Datta, 1989). In all cases, the younger females did have potentially powerful supporters, but youngest ascendancy did not occur (Hill & Okayasu, 1995).

The data for P Troop do not support Datta & Beauchamp's model. This does not mean that demographic variation would not influence patterns of rank acquisition in the way their model predicts, but rather that additional factors must be involved in determining whether or not youngest ascendancy takes place.

Resource distribution and the role of danger

Why should youngest ascendancy fail to occur in P Troop, when Japanese macaques are one of the classically nepotistic species? The only reports of the routine occurrence of youngest ascendancy come from provisioned groups of macaques. This raises the possibility that provisioning somehow enhances the process. A major difference between provisioned and natural groups is in the rate and intensity of aggression. When foraging on natural foods in the forest, Japanese macaques are generally widely dispersed. The rate of aggression is somewhat higher during foraging than during resting and grooming, and is also higher on the ground, where a greater proportion of disputes are over concentrated resources, than in trees (Hill & Okayasu, 1995). Provisioning results in the concentration of the major food resource in space and time, and as a result a marked increase in the frequency and intensity of aggression (Southwick *et al.*, 1976; Mori, 1977;

Clark, 1978; Altmann & Muruthi, 1988; Asquith, 1989; Ihobe, 1989). In Japanese macaques provisioning is associated with rates of aggression that are about 30 times those observed during foraging by non-provisioned groups (Furuichi, 1983, 1986; Hill & Okayasu, 1995). Redirection and escalation of aggression are very common, and under such conditions a female's immature offspring may frequently be exposed to the danger of physical injury. This constant danger, especially in the vicinity of the concentrated resource, may lead a mother to routinely support her youngest, most vulnerable daughter in preference to her older offspring (Kawamura, 1958, 1965). Thus, the danger associated with feeding at a highly concentrated resource may be a major factor facilitating the process of youngest ascendancy.

The fact that clumped food resources are associated with a higher frequency of disputes (Hill & Okayasu, 1995), also means that they provide more opportunities for intervention. Furthermore, the chances of potential allies being close at hand are much greater when animals are gathered together to feed on a concentrated resource. In addition to overt aggressive support from the mother, 'proximity effects' may play an important role in youngest ascendancy (Kitamura, 1975), allowing the younger sister to threaten the older with impunity while close to the mother. These effects would also be more pronounced under conditions where the major food source was highly concentrated, so that the mother was usually nearby.

Data for three different troops of wild Japanese macaques in Yakushima have all found very low rates of intervention in agonistic disputes (Furuichi, 1983; Hill & Okayasu, 1995). The highest rate was found in P Troop, where six cases of intervention in a dispute between siblings were observed in 475 hours of focal samples on females, and none of these involved the intervention of kin in a dispute between sisters. The simplest explanation for the absence of youngest ascendancy in P Troop is the total lack of support from the mother or sisters, and the reason for this is almost certainly that disputes over food are much less frequent, and rarely hold the threat of physical injury (Hill & Okayasu, 1995).

Conclusion

Two aspects of female rank acquisition appear to be common to all macaque species (for which there are adequate data), savanna baboons and vervets. First, that maternal rank is a major determinant of female dominance rank in all species (e.g. Kawamura, 1958, 1965; Sade, 1972; Horrocks & Hunte, 1983*b*; Netto & van Hooff, 1986; Deng & Zhao, 1987;

Paul & Kuester, 1987; Samuels, Silk & Altmann, 1987; Prud'homme & Chapais, 1993), even though some daughters manage to outrank their mothers (Chikazawa *et al.*, 1979; Silk *et al.*, 1981; Hausfater *et al.*, 1982). Second, that the proximate mechanism by which dominance acquisition takes place is intervention in agonistic encounters, primarily by the mother (Watanabe, 1979; Horrocks & Hunte, 1983*b*; Datta, 1983, 1988; Netto & van Hooff, 1986; Chapais, 1988*a*), but also by other close kin (Datta, 1988). Apart from these two characteristics, there appears to be considerable variation in the patterns of female rank acquisition.

Attempts to identify the evolutionary significance of the pattern of rank acquisition defined by 'Kawamura's principles' (Schulman & Chapais, 1980; Horrocks & Hunte, 1983*a*) fail to explain variation in its occurrence. The model presented by Datta & Beauchamp (1991) shows how variation in demographic parameters could give rise to the differences found between expanding groups of provisioned rhesus macaques on the one hand, and natural groups of savanna baboons on the other. However, it cannot explain all of the variation, for example the absence of youngest ascendancy in a small, natural group of Japanese macaques (Hill & Okayasu, 1995), or its rarity in expanding populations of Barbary macaques (Paul & Kuester, 1987; Prud'homme & Chapais, 1993).

The most likely source of this variation lies in differences in the propensity to show severe aggression and to give support in agonistic encounters. There is strong evidence for the existence of inter-specific differences in 'aggressiveness' within the genus *Macaca* that cannot be explained in terms of demographic or environmental conditions (Thierry, 1985, 1990; de Waal & Lutrell, 1989; Butovskaya, 1993). There is also evidence of variation within the same species that appears to be due to environmental conditions (Hill & Okayasu, 1995).

Relatively little attention has been paid to the potential influence of ecological factors on patterns of rank acquisition. This is probably because the majority of data has come from provisioned populations of macaques. If, as we have argued above, the concentration of food resources enhances the process of youngest ascendancy, its occurrence in natural groups of the more 'despotic' (*sensu* Thierry, 1990) macaque species should correspond to the distribution of major resources. Unfortunately, there are still very few long-term data sets for natural populations of macaques, so this idea cannot be tested. The only records of youngest ascendancy in macaques come from provisioned populations, and the possibility that the phenomenon is largely a product of these artificial conditions cannot be dismissed. This illustrates the potential dangers of constructing evolutionary explanations for phenomena observed only under artificial conditions, and demonstrates the importance of confirming findings from studies of provisioned macaques with data from natural populations.

Acknowledgements

We thank colleagues in the Yakushima Research Group, the Center for African Area Studies and Laboratory for Human Evolution Studies, Kyoto University, and Professor J. Itani for their support. Thanks also to Peter Lucas for encouragement at a vital moment, and to Anne Main for comments, patience and kindness.

During the preparation of the manuscript D. A. H. was funded by an RGC Grant to Dr Peter Lucas, Department of Anatomy, University of Hong Kong. Fieldwork was funded by fellowships from the Royal Society (D. A. H.), and the Japan Society for the Promotion of Science (D. A. H. and N. O.).

References

Altmann, J., Altmann, S. A. & Hausfater, G. (1981). Physical maturation and age estimates of yellow baboons, *Papio cynocephalus*, in Amboseli National Park, Kenya. *American Journal of Primatology*, **1**, 389–99.

Altmann, J. & Muruthi, P. (1988). Differences in daily life between semiprovisioned and wild-feeding baboons. *American Journal of Primatology*, **15**, 213–21.

Asquith, P. J. (1989). Provisioning and the study of free-ranging primates: history, effects and prospects. *Yearbook of Physical Anthropology*, **32**, 129–58.

Butovskaya, M. (1993). Kinship and different dominance styles in groups of three species of the genus *Macaca* (*M. arctoides*, *M. mulatta*, *M. fascicularis*). *Folia Primatologica*, **60**, 210–24.

Chapais, B. (1985). An experimental analysis of a mother-daughter rank reversal in Japanese macaques (*Macaca fuscata*). *Primates*, **26**, 407–23.

Chapais, B. (1988a). Experimental matrilineal inheritance of rank in female Japanese macaques. *Animal Behaviour*, **36**, 1025–37.

Chapais, B. (1988b). Rank maintenance in female Japanese macaques: experimental evidence for social dependency. *Behaviour*, **104**, 41–59.

Chapais, B. (1991). Matrilineal dominance in Japanese macaques: the contribution of an experimental approach. In *The monkeys of Arashiyama*, eds. L. M. Fedigan & P. J. Asquith, pp. 251–73. New York: SUNY Press.

Chapais, B. & Schulman, S. R. (1980). An evolutionary model of female dominance relations in primates. *Journal of Theoretical Biology*, **82**, 47–89.

Chikazawa, D., Gordon, T. P., Bean, C. A. & Bernstein, I. S. (1979). Mother–daughter dominance reversals in rhesus monkeys (*Macaca mulatta*). *Primates*, **20**, 301–5.

Clark, T. W. (1978). Agonistic behavior in a transplanted troop of Japanese macaques: Arashiyama West. *Primates*, **19**, 141–51.

Datta, S. B. (1983). Relative power and the acquisition of rank. In *Primate social relationships*, ed. R. A. Hinde, pp. 93–103. Oxford: Blackwell Scientific Publications.

Datta, S. B. (1988). The acquisition of dominance among free-ranging rhesus monkey siblings. *Animal Behaviour*, **36**, 754–72.

Datta, S. B. (1989). Demographic influences on dominance structure among female primates. In *Comparative socioecology*, ed. V. Standen & R. Foley, pp. 265–84. Oxford: Blackwell Scientific Publications.

Datta, S. B. & Beauchamp, G. (1991). Effects of group demography on dominance relationships among female primates. I. Mother–daughter and sister–sister relations. *American Naturalist*, **138**, 201–26.

de Waal, F. B. M. & Lutrell, L. M. (1989). Towards a comparative socioecology of the genus *Macaca*: different dominance styles in rhesus and stumptail monkeys. *American Journal of Primatology*, **19**, 83–109.

Deng, Z. & Zhao, Q. (1987). Social structure in a wild group of *Macaca thibetana* at Mount Emei, China. *Folia Primatologica*, **49**, 1–10.

Fedigan, L. M. (1991). History of the Arashiyama West Japanese macaques in Texas. In *The monkeys of Arashiyama*, ed. L. M. Fedigan & P. J. Asquith, pp. 54–73. New York: SUNY Press.

Furuichi, T. (1983). Interindividual distance and influence of dominance on feeding in a natural Japanese macaque troop. *Primates*, **24**, 445–55.

Furuichi, T. (1986). [Dominance relations and social co-existence in a troop of wild Japanese macaques.] (In Japanese.) In *Yakushima-no yasei nihonzaru*, ed. T. Maruhashi, J. Yamagiwa & T. Furuichi, pp. 126–85. Tokyo: Tokai University Press.

Hausfater, G., Altmann, J. & Altmann, S. A. (1982). Long-term consistency of dominance relations among female baboons (*Papio cynocephalus*). *Science*, **217**, 752–5.

Hausfater, G., Cairns, S. J. & Levin, R. N. (1987). Variability and stability in the rank relations of nonhuman primate females: analysis by computer simulation. *American Journal of Primatology*, **12**, 55–70.

Hausfater, G., Saunders, C. D. & Chapman, M. (1981). Some applications of computer models to the study of primate mating and social systems. In *Natural selection and social behavior*, eds. R. D. Alexander & D. W. Tinkle, pp. 345–60. New York: Chiron Press.

Hill, D. A. & Okayasu, N. (1995). Absence of 'youngest ascendancy' in the dominance relations of sisters in wild Japanese macaques (*Macaca fuscata yakui*). *Behaviour*, **132**, 267–79.

Hinde, R. A. (1976). Interactions, relationships and social structure. *Man*, **11**, 1–17.

Horrocks, J. A. & Hunte, W. (1983a). Rank relations in vervet sisters: a critique of the role of reproductive value. *American Naturalist*, **122**, 417–21.

Horrocks, J. A. & Hunte, W. (1983b). Maternal rank and offspring rank in vervet monkeys: an appraisal of the mechanisms of rank acquisition. *Animal Behaviour*, **31**, 772–82.

Ihobe, H. (1989). How social relationships influence a monkey's choice of feeding sites in a troop of Japanese macaques (*Macaca fuscata fuscata*) on Koshima Islet. *Primates*, **30**, 17–26.

Kaplan, J. R. (1977). Patterns of fight interference in free-ranging rhesus monkeys.

American Journal of Physical Anthropology, **47**, 279–87.

Kawai, M. (1958). [On the system of social ranks in a natural troop of Japanese monkeys.] (In Japanese.) *Primates*, **1**, 111–48.

Kawai, M. (1965). On the system of social ranks in a natural troop of Japanese monkeys. I. Basic rank and dependent rank. In *Japanese monkeys*, ed. K. Imanishi & S. A. Altmann, pp. 66–86. Chicago: Altmann.

Kawamura, S. (1958). [Matriarchal social ranks in the Minoo-B troop: a study of the rank system of the Japanese monkeys.] (In Japanese.) *Primates*, **1**, 149–56.

Kawamura, S. (1965). Matriarchal social ranks in the Minoo-B troop: a study of the rank system of the Japanese monkeys. In *Japanese monkeys*, ed. K. Imanishi & S. A. Altmann, pp. 105–12. Chicago: Altmann.

Kitamura, K. (1975). [Special proximity relations between individuals in Japanese monkeys.] (In Japanese.) *Kikan Jinruigaku* [*Quarterly Anthropology*], **8**, 3–39.

Koyama, N. (1967). On dominance rank and kinship of a wild Japanese monkey troop in Arashiyama. *Primates*, **8**, 189–216.

Koyama, N., Takahata, Y., Huffman, M. A., Norikoshi, K. & Suzuki, H. (1992). Reproductive parameters of female Japanese macaques: thirty years' data from the Arashiyama troops, Japan. *Primates*, **33**, 33–47.

Missakian, E. A. (1972). Genealogical and cross-genealogical dominance relations in a group of free-ranging rhesus (*Macaca mulatta*) on Cayo Santiago. *Primates*, **13**, 169–80.

Mori, A. (1977). Intra-troop spacing mechanism of the wild Japanese monkeys of the Koshima troop. *Primates*, **18**, 331–57.

Mori, A., Watanabe, K. & Yamaguchi, N. (1989). Longitudinal changes of dominance rank among females of the Koshima group of Japanese monkeys. *Primates*, **30**, 147–73.

Netto, W. J. & van Hooff, J. A. R. A. M. (1986). Conflict interference and the development of dominance relationships in immature *Macaca fascicularis*. In *Primate ontogeny, social cognition and social behaviour*, ed. J. G. Else & P. C. Lee, pp. 291–300. Cambridge: Cambridge University Press.

Paul, A. & Kuester, J. (1987). Dominance, kinship and reproductive value in female Barbary macaques (*Macaca sylvanus*) at Affenberg Salem. *Behavioral Ecology and Sociobiology*, **21**, 323–31.

Prud'homme, J. & Chapais, B. (1993). Rank relations among sisters in semi-free-ranging Barbary macaques (*Macaca sylvanus*). *International Journal of Primatology*, **14**, 405–20.

Rawlins, R. G. & Kessler, M. J. (1986). *The Cayo Santiago macaques: history, behavior and biology*. New York: SUNY Press.

Sade, D. S. (1967). Determinants of dominance in a group of free-ranging rhesus monkeys. In *Social communication in primates*, ed. S. A. Altmann, pp. 99–114. Chicago: University of Chicago Press.

Sade, D. S. (1972). A longitudinal study of social behavior of rhesus monkeys. In *The functional and evolutionary biology of primates*, ed. R. Tuttle, pp. 378–98. New York: Aldine-Atherton.

Samuels, A., Silk, J. B. & Altmann, J. (1987). Continuity and change in dominance relations among female baboons. *Animal Behaviour*, **35**, 785–93.

Schulman, S. & Chapais, B. (1980). Reproductive value and rank relations among macaque sisters. *American Naturalist*, **115**, 580–93.

Silk, J. B., Samuels, A. & Rodman, P. S. (1981). Hierarchical organization of female *Macaca radiata* in captivity. *Primates*, **22**, 84–95.

Southwick, C. H., Siddiqi, M. F., Farooqi, M. Y. & Pal, B. C. (1976). Effects of artificial feeding on aggressive behaviour of rhesus monkeys in India. *Animal Behaviour*, **24**, 11–15.

Takahata, Y. (1991). Diachronic changes in the dominance relations of adult female Japanese monkeys in the Arashiyama B group. In *The monkeys of Arashiyama*, ed. L. M. Fedigan & P. J. Asquith, pp. 123–39. New York: SUNY Press.

Thierry, B. (1985). Patterns of agonistic interactions in three species of macaque (*Macaca mulatta, M. fascicularis, M. tonkeana*). *Aggressive Behavior*, **11**, 223–33.

Thierry, B. (1990). Feedback loop between kinship and dominance: the macaque model. *Journal of Theoretical Biology*, **145**, 511–22.

Tsukahara, T. (1990). Initiation and solicitation in male-female grooming in a wild Japanese macaque troop in Yakushima island. *Primates*, **31**, 147–56.

Watanabe, K. (1979). Alliance formation in a free-ranging troop of Japanese macaques. *Primates*, **20**, 459–74.

22

A twenty-one-year history of a dominant stump-tail matriline

R. J. RHINE AND A. MARYANSKI

Introduction

One of the reasons why social dominance among cercopithecines has generated a lively interest among researchers is its role in the enhancement of survival and reproductive success (Fedigan, 1983; Samuels, Silk & Altmann, 1987; Rhine, Wasser & Norton, 1988). Among cercopithecine species, high-ranking matrifocal networks commonly maintain their relative dominance positions over other matrilines for long periods of time (see e.g. Bernstein, 1969; Sade, 1972; Estrada, 1978; Cheney, Lee & Seyfarth, 1981; Hausfater, Altmann & Altmann, 1982; Samuels et al., 1987; Chapais, 1988; Rhine, Cox & Costello, 1989). Greater access to social and material resources by females in high-ranking matrilines should benefit both them and their progeny. Thus, the longer a cercopithecine female is able to retain a high status position, the better are her chances of achieving above-average reproductive success (Rhine et al., 1989).

This chapter describes the 21-year dominance history of three members of a dominant matriline living in a colony of stump-tailed macaques (*Macaca arctoides*). The matriline consisted of a matriarch, Maria, her first-born daughter, Emma, and her son, Paul (another daughter, Zaria, was not an adult group member long enough to be included). Maria was one of the founding adult females when the study began in 1968, while Paul and Emma, respectively, were over 19 and 20 years old when the study ended in 1989. Because of their importance in macaque social structure, dominance relations were monitored regularly, with particular interest in determining the alpha male and female. Consequently, considerable attention was paid to Maria and her children during their entire lives in the colony.

Such long-term observations of dominance and other social behaviors

473

can reveal subtleties of social interchange that would be expected if higher intelligence arose from improved reproductive success for those who were best able to cope with a complex social world (Small, 1990). For this reason, the chronology of Maria's family bears upon the question of whether advanced intelligence evolved as an adaptation to a complex social environment (Humphrey, 1976; Byrne & Whiten, 1988).

It also bears upon a second question: in achieving, maintaining, and changing dominance relations, what are the relative weights of personality, a stable core of attributes *within* individuals, and social structure, the patterns of relationships *among* individuals? Social structure, in the form of matrilineal blood ties, is known to be an important determining factor in cercopithecine dominance relations; personality, which is less well understood, may also be significant in the dominance status of some individuals (Small, 1990; Rhine, 1992). The role of personality may be magnified when social structure is disrupted or changed. In studies where the membership of *M. fascicularis* groups was periodically reorganized, it was found that a given individual tended consistently to have the same rank in different groups (Welker & Luehrmann, 1982; Kaplan, Manuck & Gatsonis, 1990; Shively & Kaplan, 1991), presumably due to personality attributes.

Stability of personality traits

Allport (1937) and others (Phares, 1991) have conceived of human personality in terms of a constellation of core traits that characterize individuals in a variety of circumstances over substantial periods of life history. Many human traits, such as confidence, sociability, and aggressiveness, are applicable also to non-human primates, especially advanced ones like macaques, for whom significant individual differences have been described by trait constellations (Kellerman & Plutchik, 1968; Chamove, Eysenck & Harlow, 1972; Stevenson-Hinde & Zunz, 1978; Stevenson-Hinde, Stillwell-Barnes & Zunz, 1980; Cox, 1989; Santillan-Doherty *et al.*, 1991).

Trait theories of personality have been questioned by those who saw, at best, weak evidence for trait stability because a given individual often behaved differently in different situations (Mischel, 1968). Proponents of trait theories countered this criticism with the argument that trait consistency cannot be validly determined from brief observations of individuals in a few limited social settings (Epstein & O'Brien, 1985; Kendrick & Funder, 1988). What is needed to investigate adequately the question of trait stability are long-term longitudinal studies of the spontaneous behavior of known individuals interacting within a wide array of different social situations. Such in-depth studies of humans are difficult, because humans have a very long life span and cannot be kept in managed groups for regular, convenient observation. Instead, long-term and intensive scrutiny of spontaneously interacting, non-human primates can usefully cast light on the issue of trait stability.

Nomothetic and ideographic methods

Bem & Allen (1974) argued that cross-situational stability is more likely to be documented if the nomothetic method is supplemented by ideographic research. The nomothetic method yields statistical analyses of groups of subjects and allows generalizations about the non-existent average individual, while sacrificing a sense of the full richness of real, individual lives. Most analytical studies in leading primate journals are largely nomothetic analyses of the effects of defined variables upon groups of animals. The ideographic method focuses upon case histories of individuals, preferably over extended periods, with attention to individual variation and complexity. An example is de Waal's (1982) well-known work with the Arnhem chimpanzees. Because both methods have their unique strengths, Allport (1937) recommends an artful blending of the two as the optimal approach to understanding what makes individuals tick behaviorally. The present chapter supplements nomothetic data with an ideographic analysis of Maria and her kin.

Dominance style

The relative importance of personality versus social structure in determining dominance relations may depend upon a species' typical dominance style. Following fights, stump-tails compared to rhesus macaques were more conciliatory and used a richer repertoire of reassurance gestures (de Waal & Ren, 1988). Although the frequency of aggressive interactions was greater among stump-tails than among rhesus (Bernstein, Williams & Ramsey, 1983; de Waal & Luttrell, 1989), stump-tails were considerably less violent, more tolerant of unrelated adults, and more often engaged in positive social activities (de Waal & Luttrell, 1989). Adult stump-tails in agonistic conflicts tended to support victims more than attackers and to protect unrelated sub-adults (Butovskaya & Ladygina, 1989). While stump-tails are easily aroused, they are ready thereafter to reconcile, which allows non-kin to take chances in a relatively tolerant social system.

The stump-tail social system has strong group cohesion because powerful kinship ties are supplemented by important, weaker ties with non-relatives (friendship and acquaintance bonds) that provide the structural network needed for overall group cohesion (Granovetter, 1973; Wellman, 1983; Maryanski, 1987, 1988; Maryanski & Turner, 1990). Thierry (1990) suggested that macaque species can be placed on a social continuum ranging from relatively despotic to relatively egalitarian. He placed stump-tails on the egalitarian side, even though their formal dominance structure is as rigidly linear as that of rhesus. Because species with like formal structures can differ considerably in the style by which these structures are maintained (de Waal, 1989), a double-layered macaque dominance structure has been proposed: a formal layer and a style layer (de Waal & Luttrell, 1989). The notion of a social network consisting of strong kinship ties supplemented by weaker but important non-kin ties, complements the views of de Waal and Thierry and provides a useful framework for the evaluation and understanding of the dominance history of Maria's family.

Social and physical environments

Maria was among a group of wild-born, young adults (Harvey & Rhine, 1983) that arrived at Riverside, California, from Thailand in mid-1968. These adults and some of their offspring were observed periodically until either they died or the study ended in 1989. Animals that were adults during all or part of this period are listed in Table 22.1. In July and August 1968, the new arrivals from Thailand were placed in two separate indoor groups (groups 1 and 2, Table 22.2) whose composition was later modified (groups 1.1 and 2.1) for research purposes (Rhine, 1972, 1973; Rhine & Kronenwetter, 1972). In April 1969, the two groups were moved outside to the newly constructed quarters described by Bunyak et al. (1982) and shown in Fig. 22.1, and in February 1972 a tunnel connecting the two outdoor cages was left open to form a single group that was studied until it was disbanded in November 1989.

Establishing long-term dominance

Long-term dominance

Long-term dominance rank is the time-weighted average of standardized, shorter-term ranks (Rhine et al., 1989). Long-term rank was estimated by the mean of dominance ranks sampled periodically during the 17 years of

Table 22.1. *Animals which were adults (10 male and 13 female) during all or part of the study (from Rhine* et al., *1989)*

| Wild-born, arrived as adults | Born into colony[a] | |
	Male (birthdate: day-month-year)	Female (birthdate)
Males		
Abe		
Karl[b]		
Ute		
Yuri[b]		
Females		
Carol[b]		
Doris[b]		
Gail	Sam (6-8-69)	Fran (29-9-70)
Heather	Ivan (27-11-70)	
Joan	Ned (4-8-69)	Queen (12-8-70)
Maria	Paul (21-2-70)	Emma (27-5-69)
		Zaria (7-9-82)
Thelma	Vic (31-10-70)	Terry[c] (7-11-79)
Lois	Russ (18-5-70)	

[a]Offspring are shown in the same row as their mother. [b]Adults removed from the colony or dead prior to the formation of a single group in February 1972. [c]Low-ranking female who died in April 1986.

the single group, using methods described by Rhine *et al.* (1989). Changes in dominance rank, which can occur for a variety of reasons, are virtually inevitable in nature due to death, maturation, and transfer between troops. In the semi-undisturbed setting of the Riverside colony, all adults except one occupied different rank positions for various lengths of time. Early short-term ranks are provided by the order in which animals are listed in Table 22.2; alpha males and females during the life of the colony are identified in Fig. 22.2.

Long-term ranks of adults in the single group are given by matriline in Table 22.3, which shows that variation in dominance rank was quite limited. Maria and all members of her family had higher long-term ranks than all other adults except Sam, the son of the second-ranked matriarch. The road by which Maria rose to be alpha female is not reflected in the data of Table 22.3 because it was traversed before the single group was established. It was a road with some rocky patches.

Table 22.2. *Early adult groups in dominance order, showing the month and year when the groups were first constituted[a] (from Rhine et al., 1989)*

Indoor groups			
Group 1 (8-68)	Group 1.1 (1-69)	Group 2 (7-68)	Group 2.1 (1-69)
Abe	Maria	*Yuri*	*Abe*
Doris	Heather	*Karl*	*Yuri*
Heather	Doris	Gail	Gail
Maria	Carol	Joan	Joan
Carol	Lois	Thelma	Thelma

Outdoor groups					
Group 1.2 (4-69)	Group 1.3 (4-70)	Group 1.4 (6-70)	Group 2.2 (4-69)	Group 2.3 (4-70)	Group 2.4[b] (7-70)
Abe	*Yuri*	*Abe*	*Yuri*	*Karl*	*Karl*
Ute	Gail	*Ute*	*Karl*	Joan	*Yuri*
Maria	*Abe*	Gail	Gail	Thelma	Joan
Doris	Maria	Maria	Joan	*Ute*	Thelma
Heather		Heather	Thelma	Heather	Doris
Carol			Lois	Doris	Lois
				Carol	

[a]Names of adult males are in italics. Dominance of some mid-ranking individuals was sometimes quite close or difficult to specify confidently. This applies especially to Doris and Heather in groups 1.2 and 2.3 and to Ute in all outdoor groups. [b]Karl, Yuri, Joan and Thelma were placed together in June 1970, and Doris and Lois were added one month later.

Establishment of dominance by Maria

As the story of Maria's rise to be alpha matriarch was told previously (Rhine, 1992), it need only be summarized here. The key events of her rise are the first seven items listed in Table 22.4.

Early in colony life, Maria was the third-ranked of four females (group 1, Table 22.2), but the dominance difference between her and the two females ranked above her, Doris and Heather, seemed relatively small (Rhine & Kronenwetter, 1972). The unequivocally top-ranked animal in group 1 was its only male, Abe, for whose affiliation Doris, Heather, and Maria seemed to compete. (Boyd & Silk (1983) incorrectly placed Doris above Abe due to a limitation of their method and a failure to take all available information into account; Rhine *et al.*, 1989.) Grooming and other affiliative behavior

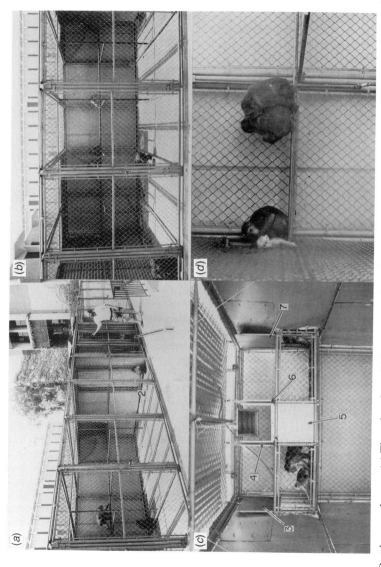

Fig. 22.1. Outdoor enclosure. (a) There is a door (1) into a central work area and a door (2) from that area into one of the two connected cages. (b) One of the two cages with an animal on the rear sitting rails. (c) The central work area with the five guillotine doors of the connecting tunnel indicated by arrows numbered (3) to (7). (d) The rear sitting rails at the entrance into the tunnel (from Bunyak et al., 1982).

Table 22.3. *Mean rank from 1972 to 1988 by matriline (based upon*
Rhine et al., 1989)

Animal	N^a	Mean	SD	Median	Range
Maria	16	3.61	1.46	3.44	1.00–6.00
Paul	14	1.67	0.99	1.00	1.00–4.00
Zaria	2	4.41	0.96	4.41	3.73–5.09
Emma	17	5.31	1.66	4.33	3.14–9.18
Gail	13	8.85	2.50	8.50	5.29–12.25
Sam	14	3.58	2.21	4.50	1.00–6.36
Frankie	1	8.00		8.00	
Heather	15	9.36	2.18	9.00	7.00–13.27
Ivan	14	7.93	1.24	8.50	5.50–9.33
Joan	16	12.03	2.02	11.98	9.57–14.64
Ned	14	11.22	2.47	11.61	7.82–14.50
Queen	14	12.45	1.37	12.80	10.55–14.50
Thelma	7	12.89	1.23	12.79	11.00–14.13
Vic	5	12.27	1.48	12.79	10.00–13.86
Lois	16	15.80	0.42	16.00	14.93–16.00
Russ	5	16.00	0.00	16.00	16.00–16.00

[a]Number of dominance hierarchies upon which statistics are based.
Males in italics.

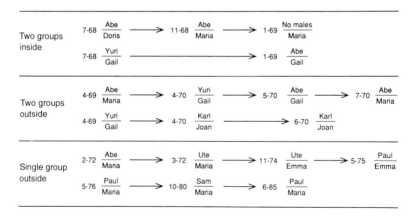

Fig. 22.2. Dates of changes in dominance of alpha males and females from July 1968
(7–68) to June 1985 (6–85). Males are above the short lines and females below. Paul
and Maria were still dominant in November 1989. The upper panel shows
dominance for two separate indoor groups, the middle panel for two separate
outdoor groups, and the lower panel for a single outdoor group.

Table 22.4. *Major dominance events (based upon Table II of Rhine, 1991)*

Date (month-year)	Event
8-68	Maria living with adult alpha male Abe, and adult females Doris, Heather, and Carol. Doris was dominant female
11-68	Maria rose to dominant female
1-69	Abe removed and replaced with adult female Lois. Maria was alpha in this all-female group
4-69	Abe returned and Lois removed. Maria still top female
4-70	Inadvertent mixture of the two groups. Maria ended up in a group with adult alpha male Yuri, Abe, and adult dominant female Gail. Maria lowest ranking
5-70	Yuri removed. Adults of group were alpha male Abe, dominant female Gail, adult male Ute, Heather, and Maria
7-70	Maria rose to dominant female
2-72	Two groups combined, with Abe the alpha male and Maria the dominant female of the combined groups
3-72	Ute became alpha male. Maria remained dominant female
11-74	Maria and her daughter, Emma, exchanged dominance positions
5-76	Maria again became top female when her son, Paul, became alpha
10-80	Sam replaced Paul as alpha male. Maria still top female
6-85	Maria retained top female rank as Paul again became alpha. Paul and Maria still top-ranked in 11-89 when the colony was disbanded

revealed a developing closeness between Abe and Maria. Although Doris groomed him most, he preferred to groom Maria. Meanwhile, Maria groomed all other females more than they were groomed by any other, and more than they groomed her. The relationships in group 1 were summarized by Rhine & Kronenwetter (1972, pp. 31–2) as follows:

Abe himself favored *Maria*. She, in turn, established close relationships with him, and also attended to the other females. Dominance reactions to *Maria* by *Doris* and *Heather* were greater than for any other dominance dyad . . . and *Maria* almost never reversed the relationship. These dominance reactions probably reflected some recognition of *Maria* by *Doris* and *Heather* as a threat to their status. Similarly, grooming of *Doris* and *Heather* by *Maria* and her lack of aggressive reprisal may have reflected a kind of primitive 'politics' in which the competition was kept from becoming too intimidated while *Maria* was placed near the center of power.

By November 1968, Maria had probably risen to alpha female, though the status differences between her, Doris, and Heather, as indicated by amounts of chasing, threatening, charging and fleeing, were still slight

(Rhine, 1972). Affiliative behaviors were more revealing. Grooming relationships had reversed, with all other females now grooming Maria much more than she groomed them. Now Maria instead of Doris groomed Abe the most, and he still preferred to groom Maria. While Maria was monopolizing much of Abe's social time, Doris and Heather were treating her with increasing caution.

In January 1969, an all-female group was formed to determine whether Maria's tentative hold on the alpha position would be loosened without Abe (group 1.1, Table 22.2). While agonistic indicators were still somewhat equivocal, with Maria being dominant by some measures and Heather by others, affiliative behaviors indicated that Maria had indeed replaced Abe, albeit tentatively, at the top of the hierarchy. As indicated by the following quote, the difference between Maria's and Heather's affiliative attention to Doris probably reflects a personality difference that was crucial to Maria's standing (Rhine, 1992):

Interestingly, Maria groomed third-ranking Doris over twice as often as she groomed second-ranking Heather. Similarly in the other affiliative behaviors, Maria was more often with Doris than with Heather. It is possible that Maria was pursuing a strategy either of alliance (ingratiation?) with Doris or of inducing social distance between Doris and Heather, for it seems highly unlikely that Maria could have withstood a determined coalition of these two females. Maria and Heather, respectively, threatened, charged or chased Doris six and 17 times. The slight dominance difference between Maria and Heather might have been reversed if Heather had been less aggressive toward Doris and more positively affiliative.

From this point forward, the strength of Maria's dominance position gradually increased. In April 1969, she and three other adult females were placed in an outdoor group with Abe and a second-ranked male, Ute (group 1.2, Table 22.2). She re-established her close relationship with Abe, and by the time she delivered her first offspring, Emma, she was by all measures unequivocally the dominant female of her group. Emma, and later her brother, Paul, benefited from their mother's close relationship with Abe, the certain father of Emma and the probable father of Paul. Abe paid more attention to Maria's children than to any others (Hendy-Neely & Rhine, 1977; Rhine & Hendy-Neely, 1978).

Cage-mixture incident and its aftermath

A feisty female called Emma, living under Abe's and Maria's protection, was in for a shock shortly before the end of her first year, when the doors between the two group cages were inexplicably opened in April 1970. A violent fight ensued after which the doors were locked on whichever animals happened to be together. Abe was beaten and severely wounded by Yuri, the alpha male of the other group. When the cage doors were shut, Abe, Maria, Yuri and Gail, the alpha female of the Yuri's group, were the adults in group 1.3 (Table 22.2). As a result of the cage mixture, Maria became a frightened, bottom-ranking shadow of her former self, abused on occasion even by Abe, who quickly made his peace with Yuri and began to associate with him and Gail. As a harassed loner, Maria ignored or threatened off her confused infant daughter, Emma, who had reached an age when adult tolerance is far from guaranteed, had also to deal with Gail and her infant, and, as a result, was suddenly being threatened and chased more than ever before in her short life. After a week, Maria seemed adjusted to her new subordinate status. She began spending time with the others, and grooming Abe. She was less frightened about taking food, and was again receiving Emma maternally.

Yuri was removed from group 1.3 a month after the start of the cage-mixture incident. As soon as Yuri was removed, Abe asserted himself over Gail and within a few minutes was obviously the dominant animal of the group, demonstrating that his former standing beneath Gail was strongly dependent upon her relationship with Yuri. Group 1.4 (Table 22.2) was formed shortly thereafter by returning Ute and Heather once again to the group with Abe and Maria.

Now all adults in Maria's group were old acquaintances, except Gail whose position as alpha female was not immediately challenged. The change to come was to be far more subtle. It was doubtful at the onset of Abe's renewed dominance that Maria could have successfully enlisted him as an ally against Gail, who was now the female with whom Abe was interacting most frequently. Instead, Maria began grooming and affiliating more and more with Gail and to some extent with Abe. Her confidence noticeably increased as she gradually turned the Abe–Gail duet into a trio, not with aggression, but with persistent affiliation and appeasement.

When the end of Gail's reign came, it was quick and violent. Maria defeated her in a series of bloody battles occurring on a single day about two months after group 1.4 was constituted. The blood was almost entirely Gail's, who became less capable of defending herself after each fight. The origin of these decisive battles is uncertain, but it probably revolved around Emma and Gail's infant son, Sam. While Gail was still dominant, she sometimes intervened aggressively against Emma on the side of Sam. On these occasions, Maria, disturbed and vocalizing, did not dare to protect Emma, but occasionally during these incidents Abe chased and bit Gail.

Probably either Maria took advantage of one of these chases to join in and injure Gail, or Abe first injured Gail after which Maria progressively weakened her in a subsequent series of fights. Whatever the case, from this point forward, either Maria or Emma was alpha female.

Long-term stability

As indicated in Fig. 22.2, Maria was alpha female for all but 1.5 years from 1970 until the study ended in 1989, and during the 1.5 years she and Emma swapped alpha and beta positions. Similarly, Paul was the alpha or beta male for most of his adult life in the colony. At different periods, there were four alpha males of the single group (Fig. 22.2). A male losing alpha status was never forced low in the male hierarchy; instead, he was tolerated in the second position, which is consistent with the stump-tail dominance style described above. The variances and ranges of long-term ranks (Table 22.3) attest to the fact that the other colony members also tended to occupy a limited range of dominance positions, especially the lowest-ranked animals. Correlations among 18 hierarchies, spread across the years of the single group's existence, are high and positive (Rhine *et al.*, 1989). Although short-term fluctuations occurred, dominance was remarkably stable over the long term for every adult in the study group.

Despite this strong stability over the long term, considerable variability will appear to occur in descriptions given below of the single group. What is more, these descriptions, together with the cage-mixture incident, will seem to portray an almost endless sequence of intense aggression. It is important to bear in mind that this chapter covers only a tiny sample of the events occurring over 21 years, and that the topic of the chapter dictates the selection of key events associated with dominance changes. The dominance events described here are among the most intensely violent that occurred during the life of the colony, but the total amount of time taken by these events is a very small proportion of 21 years. During most of the time, life in the colony was more peaceful.

Combining the two groups in 1972

In February 1972, in an attempt to reduce wounding, it was decided to leave open the tunnel between the two cages to allow more escape space during conflict and to provide an opportunity for more distance between individuals at other times. To avoid a repetition of the bloodshed between adult males during the earlier cage-mixture incident, the two adult males of the second group, Yuri and Karl (Table 22.2), were permanently removed from the study site a few weeks before the groups were combined.

Shortly after Yuri and Karl were removed, Gail was placed in her former group (group 2.2, Table 22.2), where she had been the alpha female (Fig. 22.2) until the cage-mixture incident had trapped her in Maria's group. Twenty-two months had passed before Gail was returned to her former group, now devoid of adult males. She was received ambivalently by the three adult females, Joan, Thelma and Lois, and by their immatures. At first Gail was ignored, but the other adult females soon began directing at her a mixture of mild threats and grimacing/teeth-chattering. Gradually, they became bolder as Gail stayed away and exhibited nervousness. Periodically Gail submitted to the hold-bottom ritual (de Waal & Ren, 1988; Rhine, 1992). As the others grew more confident, Gail was chased and bitten, especially by Thelma, but Gail asserted herself enough to deter an attacker if bitten too hard. Even Lois, the habitually bottom-ranked matriarch, had her brief moment of glory. Of particular interest, Joan, who was now alpha female and who had been Gail's closest female affiliate 22 months earlier, began protecting Gail from the others if they bothered her too much. Within a few hours of her entrance into the group, Gail's confidence had noticeably risen. She no longer cringe-crouched in a corner, and she began to threaten away immatures and take food.

Gail was an accepted low-ranking group member on 25 February 1972 when the doors to the tunnel between her group and Maria's were permanently opened. The animals in both groups oriented toward the sounds of the opening doors, and for a few minutes there was an eerie silence. Joan, Thelma, Gail, and company were in the left-hand cage and Abe, Maria, Heather, and the others were to the right (Fig. 22.1). Almost all of the action that followed took place in the left cage into which first Abe, and then Maria and Heather entered.

The silence was suddenly broken by a furious mixture of screaming, fighting females into which Abe jumped on the side of his cage-mates. For several hours, thereafter, Joan and Thelma, often shoulder to shoulder, confronted Maria and Heather, also shoulder to shoulder, in periodic, intense bouts of threatening, interspersed with chases and fights. From time to time Abe backed up Maria and Heather, but as the day wore on he increasingly spent time breaking up fights. Gail, who had lived with both sets of opponents, mostly stayed clear of the main conflict. She was treated by Maria and company as a member of their group, and was rarely bothered. It was hard to believe that only a few hours earlier Gail had been a low-ranking member of Joan's group.

By the end of the day, as the dust settled, it was clear that Maria would be the alpha matriarch of the combined group, and this was repeatedly confirmed during the next few days and weeks as the females of Joan's group, except Gail, settled into the roles of middle- to low-ranking matriarchs. In part, Maria owed this outcome to the support she solicited and received from Abe, and probably also to her benign attitude toward Gail, who thereafter became second ranked matriarch and never again received the kind of beating she had previously endured. It certainly would not have been politic for Maria to do anything during the formation of the single group that would have added Gail to the alliance of Joan and Thelma.

Key dominance events of the single group
Ute becomes dominant male

In early March 1972, Ute and Abe reversed dominance positions, an outcome not unexpected from signs of growing confidence by Ute before, during and after the single group was constituted. Ute spent considerable time reminding the others of his new-found alpha status by 'branch shaking' (shaking cage wire), stiff-legged bouncing on sitting rails, chasing, biting, and mounting. When Ute took up his new position, Maria, like other group members, behaved quite cautiously, even fearfully, around him; Gail seemed more relaxed with Ute, and they spent much time together. Abe, Maria's old ally, was clearly submissive to Ute, but, as in his relationship with Yuri, the two males associated peacefully together.

Gail now clearly dominated all adult females except Maria and Emma. At one point, unseen by observers, Gail's foot was injured causing her to limp. A week later, the following comment was made in *ad libitum* notes: 'So far, Maria seems to have maintained her position, but it is not certain.' In early April 1972, a series of food tests – throwing a bit of favored food between two animals to see who gets it – established that Maria and Emma both dominated Gail.

Maria's dominance never seemed thoroughly secure during Ute's time. For example, in August 1974, when Thelma was probably in estrus, she was protected by Ute from Maria and Emma. Of the mid- to low-ranking adult females, Thelma was the one who seemed least satisfied with her status after the single group was established. There was tension between her and Maria's family as she sought the company of the dominant males, especially Ute. When Maria and Emma attempted to gang up on Thelma, Ute seemed deliberately to place himself between the opponents. More telling, at this time, Maria and Emma had numerous minor cuts and bruises on their bodies, probably inflicted over several days by Ute. If they persisted in hostility to Thelma, Ute sometimes chased and bit them. Nevertheless, they kept threatening and chasing Thelma, who often managed to avoid them by staying near Ute. During this period Thelma and Ute were frequent mutual groomers and affiliates.

Maria deposed

Although Maria remained the alpha female during much of Ute's reign, her relationship with him was never as close as her earlier relationship with Abe. Yet she managed to stay at the top of the female hierarchy for over 18 of the months during which Ute was the alpha male. In November 1974, her relationship with Ute broke down and she was deposed, but not by Thelma.

An incident involving Emma's yearling son precipitated Maria's fall from grace, though subtle events over a longer time may have set the stage. The yearling interrupted and annoyed Emma while she was grooming Ute. Emma hit and chased her child. He ran for protection to Maria who then threatened Emma. Emma went after Maria, who first fled and then fought when trapped. Ute joined in aggressively against Maria. He dragged her across the floor and bit her, drawing blood. Paul tried unsuccessfully to distract or stop Emma, who joined Ute in giving Maria a thorough going over. Maria was attacked several times during about half an hour until, bloodied and hurt, she gave up and simply tried to escape. Emma took the occasion of the conflict to threaten and chase several of the other colony members.

During the next few days, it was clear that Emma was the new alpha female, allied with Ute and her brother, Paul, now nearing 5 years of age. Despite her alliance with Paul, she regularly chased him away from Ute, and she resumed peaceful relations with her mother, who was still ranked above her son.

Like Emma, Paul took the opportunity of heightened group arousal to threaten and chase other group members, especially the several other males who were near his age. A notable exception, in view of future events, was Sam, Gail's son, who was several months older than Paul (Table 22.1). At one point, Maria redirected aggression to Gail, who in turn threatened Heather. Paul chased Gail and wounded her on the buttocks. When Maria tried to join in, it was she and not Paul who Ute attacked and bit severely. Paul was beginning to make his presence felt. The security of Ute's position was probably better served at this time by tolerance of Paul than by intolerance.

Paul and Maria move up

In March 1975, Ute was observed with a torn ear and a wound on the back of his head, suggesting that it was inflicted while attempting to escape an attacker. Several other group members had minor cuts, including Paul, but not including Sam who was unscathed. During the next several weeks many animals, including Sam, displayed minor injuries, which were a result of spasms of general arousal in the group.

Reports from regular observers indicated that Ute was coming under pressure from Paul and Sam individually (not together). Paul was taking more liberties with Ute and by late April 1975 may have surpassed him. Yet the two were still allied. In Sam's presence, Ute backed into Paul's grasp, with both displaying excited grimacing and teeth-chattering. They directed their attention to Sam and faced him together. Sam's redirects were a main reason for many of the cuts and bruises received by other colony members. All the males had been recently weighed. Sam was the heaviest and probably the best fighter one-on-one.

In early May 1975, dominance was reassessed. Paul was now dominant, followed by Ute, Emma and Maria. After her son became alpha male, Maria proceeded to rise again to the top of the female heap seemingly without a serious struggle. When Paul first became dominant, he was seen several times biting his mother and mounting her sexually, which she tried to resist. Sam was an ominous third among the males. Ute had no family members in the group. By submitting to Paul, while still maintaining an alliance with him and his family, Ute had the backing to remain, at least temporarily, higher placed than Sam.

Inhibition of reproduction in Maria and Emma

Harvey's (1983) intensive, year-long study of reproductive behavior of the Riverside stump-tail colony was done after Paul became the dominant male. Harvey almost never observed a male copulating successfully with a female to whom he was subordinate. Paul was seen copulating with his female kin, but only during times when he was the alpha male.

Harvey's findings create an unusual opportunity for an analysis of inhibition of inbreeding. Studies of incest avoidance typically examine the frequency of copulations between kin as compared to non-kin. The

Table 22.5. *Conceptions of Maria's family and of other females during times when Paul or others were dominant male*[a]

Dominant male at time of conception	Conceptions		
	Maria's family	Other females	Total
Paul	0	24	24
Abe, Ute, Sam	13	45	58
Total	13	69	82

[a] $\chi^2 = 4.79$, $p < 0.05$.

presumed reason for incest avoidance is inhibition of inbreeding. In a spontaneously interacting, mixed-sex group, inbreeding inhibition can not be confidently assessed from copulations alone if any copulations occur with female kin who bear offspring. Here it is possible to go beyond copulations by examining the offspring production of Maria and Emma during times when Paul was alpha or not alpha.

Inhibition of inbreeding in Maria's family was incomplete as measured by copulations, which occurred between Paul and his kin, but was 100% effective as measured by reproductive output. The average of eight timed gestations in the colony was 179.6 days (Harvey & Rhine, 1983), so times of conception could be estimated for other offspring by counting back 180 days from their dates of birth. Table 22.5 is a two-by-two table showing the number of offspring sired by Paul or other males that were born to Paul's kin or other females. These frequencies yield a significant chi-square ($\chi^2 = 4.79$, $p < 0.05$) due primarily to a frequency of zero out of 24 of Paul's offspring being conceived by his female kin during the time when he was dominant (Fig. 22.2). In contrast, at times when Paul ranked beneath his female kin, they bore 13 offspring to an alpha male who was the only male ranked above them (at different times, Abe, Ute or Sam).

Two consequences of Maria's resistance to Paul's sexual advances are (1) all of his observed copulations with her during Harvey's (1983) study were incomplete (no ejaculate seen) and (2) his resulting sexual frustration was re-directed to Gail, the highest-ranking female after Paul's kin. On one occasion, Paul became visibly aroused after he investigated his mother during her mid-cycle. Then he left her, went to Gail and forced her to present for sexual investigation. Seemingly dissatisfied with her unattractive reproductive condition – she was in the luteal phase of her cycle – he bit, pushed, and hit her before returning to his mother, where the whole sequence started once again. Only after switching back and forth several times did Paul become sufficiently aroused to ejaculate with Gail.

Although he dominated Maria and Emma, Ute was unable to inseminate them during Paul's ascendancy. Ute's dominance over Maria and Emma was quite tenuous (Harvey, 1981), probably because Paul frequently supported his female kin in agonistic encounters with Ute. Of the seven adult males in the colony during Harvey's study, Paul and Ute did 98% of the mating. Paul performed 75% of the observed copulations and Ute 23%. Only Paul and Ute attempted to copulate with Maria and Emma, but Ute's mating interests were severely inhibited by Paul's interference. This interference was a main factor in the occurrence of incomplete copulations during mid-cycle, when Paul was seen attacking Ute and pulling him off females. Maria and Emma remained barren until Paul was defeated by Sam.

Sam takes over

Sam's rise to the alpha position was far less surprising than his subsequent decline. Several signs preceded the rise, including an incident occurring in late October 1978. At this time, the adult females were being vaginally swabbed for a study of reproductive behavior (Harvey, 1983). The swabbing technique (Bunyak et al., 1982) required herding the females to one side of the living space (Fig. 22.1) and closing the males off briefly on the other side. During one of the swabbing sessions, Paul and Sam were seen in a serious fight with biting and wrestling; both received wounds. What is fascinating about this fight is that it was, to quote the log, 'absolutely silent,' as if there were no supporters to solicit. The other adult males were unrelated to the combatants, and all were lower-ranking (Ute had been removed permanently a year earlier). They sat and watched from a distance; Sam, by repeated pushes, backed Paul into a corner and bit him. Then he turned and walked away, only to be attacked from the rear. In continued fighting, the larger, stronger Sam threw Paul down and bit his neck. They finally parted, tired, avoiding each other. Similar fighting and pausing was repeated. In the end, as the females were re-entering his side, Sam teeth-chattered and grimaced to Paul, who walked away an uneasy victor. Observers who witnessed the incident reported that Sam had the better of Paul for about two-thirds of the time. Paul ended several encounters by backing away from Sam's attacks, but Sam did not pursue.

A week later, during swabbing, Sam and Paul went at it again. Both males were lightly wounded. They had separated when the females re-entered. Paul went straight to Gail, Sam's mother, and threatened her.

Then he sat next to Maria, put his arms around her, and laid his head on her back. Meanwhile, Sam sat and watched the group. A lower-ranking adult male approached him and presented.

It was not until almost two years later, in September and early October 1980, that the gathering storm of change finally struck. During August and September, especially the last two weeks of September, the colony was easily aroused and the number of fights escalated, leading to the wounding of several colony members. In mid-September a limited amount of a favorite food was thrown into the colony, and it was taken by Sam as Paul stood by. By October there could be no doubt that Sam had ascended to the alpha position. This is made clear by paraphrased excerpts from the log:

> The change in Sam's rank was noticed a few minutes after arrival when he was seen contacting and threatening Maria to present for sex investigation. Females do not respond to a male's attempt to induce a sex present unless the male is dominant over the female [Harvey, 1983]. Previously, Maria and Emma had been dominant over Sam. There was a cut on Maria's lower back plus several bite wounds around her neck and shoulders, characteristic of an aggressive sexual attack. When Sam and Paul were a few feet apart, bits of favored food were thrown between them several times. Each time Sam took the food; Paul got nothing.

With Gail's son in unequivocal command, the time was ripe for her and her $2\frac{1}{2}$-year old daughter to join him in replacing the Maria matriline at the top of the hierarchy. The immature daughter did try to take advantage of her brother's position and protection, and thereby caused a great deal of arousal and trouble in the group. Gail, however, never replaced Maria's family, all three of whom continued to be ranked just below Sam and just ahead of Gail.

Abdication or peaceful coup?

Sam was the dominant male for nearly five years. In July 1985, when no systematic observations were in progress, Paul was noticed to be acting again as the dominant male. When signs of a possible dominance change emerged, it was research policy to step up *ad libitum* recording, but in this case no unusual changes or signs had appeared. Food testing was done, and clearly Paul was now dominant. Once the change was realized, it was repeatedly verified by further observations. As far as anyone knows, either Sam abdicated or he was replaced by a peaceful coup too subtle for experienced observers to detect during *ad libitum* observations. Maria and Emma were already the most dominant females and, with the surprise of Paul's new standing, Maria's clan once again held sway, just as they had for so many years previously. They were still dominant when the colony was disbanded in November 1989.

Personality profiles

Cox study

Personality profiles of ten colony adults were worked out by Cox (1989) from data collected during five of the years between 1980 and 1988, a period during which first Sam, and then Paul, was dominant. The ten subjects included Maria, Emma and Paul, whose profiles will be contrasted with those of Gail and Sam. Unless otherwise indicated, all findings on personality traits described herein are from Cox (1989), whose starting point was a list of defined adjectives determined by Stevenson-Hinde & Zunz (1978) to describe personalities of rhesus macaques.

A total of 25 experienced observers – four or more for each year over a period of five years – independently rated the degree to which 21 reliable trait descriptions applied to each subject. Ratings were factor-analyzed and, with only one exception, the traits grouped into the same three-component clusters previously reported for rhesus (Stevenson-Hinde & Zunz, 1978; Stevenson-Hinde *et al.*, 1980). These three broad dimensions of monkey personality were called *Confident-Fearful* (sometimes just *Confident*), *Active-Slow* (*Excitable*), and *Sociable-Solitary* (*Sociable*). *Confident-Fearful* is a cluster of traits including aggressive, confident, strong, and, with reverse scoring, fearful, insecure and subordinate. Traits clustering under *Sociable-Solitary* are *Sociable, Opportunistic, Protective, Playful*, and (reverse scored) *Solitary* and *Eccentric*. The third component, *Active-Slow*, is a cluster of active, curious, equable, excitable, and (reverse scored) slow.

Overall findings

There were two main findings for the group as a whole. First, long-term dominance was strongly correlated with the *Confidence-Fearful* cluster for all five years of data, and with *Sociable-Solitary* for four years. Highly ranked animals were confident and sociable. Dominance, however, did not correlate significantly with *Active-Slow* for which low correlations were negative for three years and positive for two. Activity seems correlated with age regardless of dominance; all adult offspring, except one, were more active than their mothers, who were quite old. A relationship between activity and age was also found in a study showing that the old females of the Riverside colony tended to rest more than younger ones and to locomote less (Hauser & Tyrrell, 1984). In a four-month stump-tail study, Santillan-Doherty *et al.* (1991) found a strong relationship between age and activity, but also that dominant animals were less active than subordinates, which would be expected if the subordinates tended to be younger (subjects included immatures).

The second main finding was the stability of ratings from 1980 to 1988. Despite ratings by several observers each year, and mostly different observers from one year to the next, the three components of macaque personality were well correlated across years. This was especially so for Confident-Fearful and least so for Active-Slow. Of course, none of the ten correlations per component was perfect. Personality is neither 100% rigid nor forever etched in stone at maturity. It is virtually certain to be associated with behavioral variation due to life-span development, including physical aging, and to events such as the reversal in male dominance rank between Paul and Sam. Nevertheless, personality does not start from scratch on the first day of every year; a basic core develops that, for the most part, is modified or embellished only gradually. The across-year correlations impressively reflect a stable core. Twenty-nine of these 30 correlations are positive and 18 are statistically significant, mostly at beyond the 0.01 level.

Individual profiles

The overall relationship between dominance and components of personality overlooks the uniqueness of individuals, which is more fully portrayed by individual profiles of traits within components, supplemented by information obtained from *ad libitum* and systematic observations over the years. Although there were ups and downs over these years, an attempt is made in the five brief summary profiles that follow to portray the basic core that repeatedly shines through. In these personality profiles, which contrast Maria's and Gail's matrilines, traits from Cox's study are italicized.

Maria

Maria was the most *Confident* of the females and one of the most *Social*, which included being *Protective* and *Opportunistic*. As a mother, she was attentive and supportive. When two frightened yearlings were introduced into the colony, she tried to protect them; in contrast, she was *Aggressive* toward introduced adult females (Rhine & Cox, 1989). She often came to the aid of her adult offspring. She was *Effective* in inter-personal relations and good at money 'politics'. She started as a middle-ranking female without relatives to support her. Her later high status was due, not to social structure, but to personality, including an ability to manipulate alliances. She used affiliation, friendship and appeasement when they paid off and toughness when she had the stronger hand. She did not press lower-ranking individuals too hard as long as they did not present a challenge or were allied with her.

Emma

Maria's daughter was the next most *Confident* of the females and was as *Sociable* as her mother in the sense of being near to and interacting with others, but these others were mainly her kin. She was brought up and spent her adult life mostly in the company of dominant companions. As a juvenile, she was described as a spoiled brat who made trouble for other animals that attempted to avoid her. Emma seems not to have acquired the more subtle social skills that characterized her mother and, therefore, was neither as socially *Effective* nor as *Popular* as Maria. She was the second most *Excitable* of the females and could be *Strong* and *Aggressive* with subordinates. As she aged, she mellowed.

Gail

Gail's personality was quite different from Maria's. Gail was the second-ranked matriarch and third-ranked female, but her long-term rank among all adults (8.85) was closer to the next lower-ranked matriarch (9.36) than to Maria (3.61) and Emma (5.31) (Table 22.3). Though somewhat aloof (*Eccentric?*), she was one of the most sexually attractive of the females. Interestingly, Gail's aloofness, noticed in 1968 and 1969 (Rhine & Kronenwetter, 1972), was reflected in the 1980s in the second-lowest *Sociable* rating received in Cox's study. During that time, Gail was one of the least *Excitable* animals in the group. As a young adult, in groups 2, 2.1, and 2.2 (Table 22.2), she was *Strong* and *Confident*. Yet, in later life her confidence score was one of the lowest, probably due to the *Insecurity* engendered by painful treatment at Maria's hands over a long period of time, before settling somewhat *Apprehensively* into a non-threatening, mid-ranking role. For this reason, her personality may have changed over the 21-year study more than that of any other group member.

Paul

Paul, who was the alpha animal of the entire group longer than any other individual, was rated highest in *Confidence*, *Sociability*, and *Excitability*. He seemed to have acquired some of his mother's social skill, illustrated by his alliance with Ute against Sam. On the one hand, he was attractive (*Popular*) to all group members who competed to groom him, and on the other hand he could become quite *Aggressive* when aroused. He was tolerant with infants and other immatures who looked to him for *protection*. Paul was always at or near the top of the heap, and it showed as clearly in his *Confident* behavior as did *Tenseness* and *Apprehensiveness* in the behavior of low-ranked individuals.

Sam

Just as Gail contrasted with Maria, Sam made an interesting contrast with Paul. Sam was the largest and, probably, the *Strongest* animal in the group. Like Paul, he was high in *Confidence*, but unlike Paul, he was middle-ranked in *sociable*, somewhat *eccentric*, and below average in *Excitable*. He became over-weight and *Slow*-moving, giving the impression of not wanting to expend any more energy than was absolutely necessary. This may explain why he gave up the position of alpha male without a fight; it takes considerable energy to sustain the alpha role.

Summary and conclusions

Trait constellations of stump-tailed macaques exhibited a high degree of long-term stability over a wide array of different social situations. For example, Maria, the dominant matriarch, repeatedly displayed characteristic traits for dominance, and other related personality attributes, despite dramatic changes in her social environment over 21 years. This finding is consistent with expectations from trait theories of personality.

The data are also consistent with the hypothesis that sociality and the evolution of intelligence are causally linked in cercopithecine societies, where complex, learned social skills probably contribute to reproductive success. The social skills of Maria's family, particularly Maria's political adeptness, illustrate the importance of intelligence in manipulating the social environment to personal advantage (Byrne & Whiten, 1988; Small, 1990).

Birth data support the hypothesis that selection has operated to inhibit incestuous reproduction. In a colony setting with limited space and without the possibility of male transfer, stump-tail females of the dominant matriline were in danger of attenuated reproductive success due to inhibition of breeding. When the alpha male was from the dominant matriline, blood-line females did not reproduce; otherwise, they did. In nature, dispersal of macaque males may serve to enhance both gene flow and the reproductive success of dominant females, who can then breed successfully with unrelated, prime males.

Personality plays an important role in the establishment and maintenance of dominance. Personality attributes have received less attention than structural variables in studies of cercopithecine societies, where ascribed statuses and ranked matrilineages are known to play a major role in group cohesion and integration. Yet, without support from blood relatives, Maria rose from a middle rank to become the long-term dominant female. When strangers were placed together to form a colony, it was the personality traits of group members that laid the building blocks of social structure: Maria's personality facilitated the establishment of ties with resident members, which created a structural basis for group stability and cohesion.

Acknowledgements

This research was supported briefly by a small grant from NIH and mainly by intra-mural grants from the University of California, Riverside, and was made possible by the diligent work during 21 years of a large number of first-rate undergraduate and graduate students. The manuscript benefited from the critical and editorial comments of Roberta Cox, Nancy Harvey and Doris Rhine.

References

Allport, G. (1937). *Personality*. New York: Henry Holt.
Bem, D. J. & Allen, A. (1974). On predicting some of the people some of the time: the search for cross-situational consistencies in behavior. *Psychological Review*, **81**, 506–20.
Bernstein, I. S. (1969). The stability of the status hierarchy in a pigtail monkey group (*Macaca nemestrina*). *Animal Behaviour*, **17**, 452–8.
Bernstein, I., Williams, L. & Ramsey, M. (1983). The expression of aggression in Old World monkeys. *International Journal of Primatology*, **4**, 113–25.
Boyd, R. & Silk, J. B. A. (1983). A method for assigning cardinal dominance ranks. *Animal Behaviour*, **31**, 45–58.
Bunyak, S. C., Harvey, N. C., Rhine, R. J. & Wilson, M. I. (1982). Venipuncture and vaginal swabbing in an enclosure occupied by a mixed-sex group of stump-tailed macaques (*Macaca arctoides*). *American Journal of Primatology*, **2**, 201–4.
Butovskaya, M. L. & Ladygina, O. N. (1989). Support and cooperation in agonistic encounters of stumptail macaques (*Macaca arctoides*). *Anthropologie*, **27**, 73–81.
Byrne, R. W. & Whiten, A. (1988). *Machiavellian intelligence: social expertise and the evolution of intellect in monkeys, apes and humans*. Oxford: Clarendon Press.

Chamove, A. S., Eysenck, H. J. & Harlow, H. F. (1972). Personality in monkeys: factor analysis of rhesus social behavior. *Quarterly Journal of Experimental Psychology*, **24**, 496–504.

Chapais, B. (1988). Rank maintenance in female Japanese macaques: experimental evidence for social dependency. *Behaviour*, **104**, 41–59.

Cheney, D. L., Lee, P. C. & Seyfarth, R. M. (1981). Behavioral correlates of nonrandom mortality among free-ranging female vervet monkeys. *Behavioral Ecology and Sociobiology*, **9**, 153–61.

Cox, R. L. (1989). Personality profiles of mother and offspring stumptailed macaques (*Macaca arctoides*): behavior, personality, and dominance rank. PhD thesis, University of California, Riverside.

de Waal, F. (1982). *Chimpanzee politics*. New York: Harper & Row.

de Waal, F. B. M. (1989). Dominance 'style' and primates social organization. In *Comparative socioecology*, V, ed. V. Standen & R. Foley, pp. 243–63. Oxford: Blackwell Scientific Publications.

de Waal, F. B. M. & Luttrell, L. M. (1989). Toward a comparative socioecology of the genus *Macaca*: different dominance styles in rhesus and stumptail monkeys. *American Journal of Primatology*, **19**, 83–109.

de Waal, F. B. M. & Ren, R. (1988). Comparison of the reconciliation behavior of stumptail and rhesus macaques. *Ethology*, **78**, 129–42.

Epstein, S. & O'Brien, E. J. (1985). The person-situation debate in historical and current perspective. *Psychological Bulletin*, **98**, 513–37.

Estrada, A. (1978). A study of the social relationships in a free-ranging troop of stumptail macaques (*Macaca arctoides*). *Boletin de Estudios Medicos y Biologicos*, **23**, 313–94.

Fedigan, L. M. (1983). Dominance rank and reproductive success in primates. *Yearbook of Physical Anthropology*, **26**, 91–129.

Granovetter, M. S. (1973). The strength of weak ties. *American Journal of Sociology*, **78**, 1360–80.

Harvey, N. C. (1981). Social and sexual behavior during the menstrual cycle in colony living stumptail macaques (*Macaca arctoides*). PhD thesis, Rutgers University.

Harvey, N. C. (1983). Social and sexual behavior during the menstrual cycle of a colony of stumptail macaques (*Macaca arctoides*). In *Hormones, drugs, & social behavior in primates*, vol. II, ed. D. Steklis & A. S. Kling, pp. 141–74. New York: SP Medical and Scientific Books.

Harvey, N. C. & Rhine, R. J. (1983). Some reproductive parameters of stumptailed macaques (*Macaca arctoides*). *Primates*, **24**, 530–6.

Hauser, M. D. & Tyrrell, G. (1984). Old age and its behavioral manifestations: a study of two species of macaque. *Folia Primatologica*, **43**, 24–35.

Hausfater, G., Altmann, J. & Altmann, S. (1982). Long-term consistency of dominance relationships among female baboons (*Papio cynocephalus*). *Science*, **217**, 752–5.

Hendy-Neely, H. & Rhine, R. J. (1977). Social development of stumptail macaques (*Macaca arctoides*): momentary touching and other interactions with adult males during the infants' first 60 days of life. *Primates*, **18**, 589–600.

Humphrey, N. K. (1976). The social function of intellect. In *Growing points in ethology*, ed. P. P. G. Bateson & R. S. Hinde, pp. 303–17. Cambridge: Cambridge University Press.

Kaplan, J. R., Manuck, S. B. & Gatsonis, C. (1990). Heart rate and social status among male cynomolgus monkeys (*Macaca fascicularis*) housed in disrupted social groupings. *American Journal of Primatology*, **21**, 175–87.

Kellerman, H. & Plutchik, R. (1968). Emotion-trait interrelations and the measurement of personality. *Psychological Report*, **23**, 1107–14.

Kendrick, D. T. & Funder, D. C. (1988). Profiting from controversy: lessons from the person-situation debate. *American Psychologist*, **43**, 23–34.

Maryanski, A. R. (1987). African ape social structure: is there strength in weak ties? *Social Networks*, **9**, 191–215.

Maryanski, A. R. (1988). The supportive networks of monkeys and apes: an overview. *Connections*, **9**, 29–35.

Maryanski, A. R. & Turner, J. H. (1990). In *Structure of sociological theory*, 5th edn, ed. J. H. Turner, pp. 540–72. California: Wadsworth.

Mischel, W. (1968). *Personality and assessment*, New York: John Wiley & Sons.

Phares, E. J. (1991). *Introduction to personality*. New York: Harper Collins.

Rhine, R. J. (1972). Changes in the social structure of two groups of stumptail macaques (*Macaca arctoides*). *Primates*, **13**, 181–94.

Rhine, R. J. (1973). Variation and consistency in the social behavior of two groups of stumptail macaques (*Macaca arctoides*). *Primates*, **14**, 21–35.

Rhine, R. J. (1992). Dominance life history of the alpha matriarch of a colony of stumptail macaques (*Macaca arctoides*): establishment of long-term dominance. *Primate Report*, **32**, 107–18.

Rhine, R. J. & Cox, R. L. (1989). How not to enlarge a stable group of stumptailed macaques (*Macaca arctoides*). In *Housing, care, and psychological wellbeing of captive and laboratory primates*, ed. E. F. Segal, pp. 255–69. Park Ridge, NJ: Noyes.

Rhine, R. J., Cox, R. L. & Costello M. B. (1989). A twenty-year study of long-term and temporary dominance relations among stumptailed macaques (*Macaca arctoides*). *American Journal of Primatology*, **19**, 69–82.

Rhine, R. J. & Hendy-Neely, H. (1978). Social development of stumptail macaques (*Macaca arctoides*): synchrony of changes in mother-infant interactions and individual behaviors during the first 60 days of life. *Primates*, **19**, 681–92.

Rhine, R. J. & Kronenwetter, C. (1972). Interaction patterns of two newly formed groups of stumptail macaques (*Macaca arctoides*). *Primates*, **13**, 19–33.

Rhine, R. J., Wasser, S. K. & Norton, G. W. (1988). Eight-year study of social and ecological correlates of mortality among immature baboons of Mikumi National Park, Tanzania. *American Journal of Primatology*, **16**, 199–212.

Sade, D. S. (1972). A longitudinal study of social behavior of rhesus monkeys. In *The functional and evolutionary biology of primates*, ed. R. Tuttle, pp. 378–98. Chicago: Aldine.

Samuels, A., Silk, J. B. & Altmann, J. (1987). Continuity in dominance relationships among female baboons. *Animal Behaviour*, **35**, 785–93.

Santillan-Doherty, A., Chiappa, P., Arenas-Frias, V. & Mondragon-Ceballos, R. (1991). The relation between 'personality' traits and rank, sex and age in stumptail macaques. *American Journal of Primatology*, **24**, 133.

Shively, C. A. & Kaplan, J. R. (1991). Stability of status ranking of female cynomolgus monkeys, of varying reproductive condition, in different social groups. *American Journal of Primatology*, **23**, 239–45.

Silk, J. B. (1987). Social behaviour in evolutionary perspective. In *Primate societies*, ed. B. B. Smuts, D. L. Cheney, R. M. Seyfarth, R. W. Wrangham, & T. T. Struhsaker, pp. 77–100. Chicago: University of Chicago Press.

Small, M. F. (1990). Social climber: independent rise in rank by a female Barbary macaque (*Macaca sylvanus*). *Folia Primatologica*, **55**, 85–91.

Stevenson-Hinde, J., Stillwell-Barnes, R. & Zunz, M. (1980). Subjective assessment of rhesus monkeys over four successive years. *Primates*, **21**, 66–82.

Stevenson-Hinde, J. & Zunz, M. (1978). Subjective assessment of individual rhesus monkeys. *Primates*, **19**, 473–82.

Thierry, B. (1990). Feedback between kinship and dominance: the macaque model. *Journal of Theoretical Biology*, **145**, 511–21.

Welker, C. & Luehrmann, B. (1982). Social behavior of the crab-eating monkey *Macaca fascicularis*. I. The rank as an attribute of the individual. *Zoologischer Anzeiger*, **208**, 175–91.

Wellman, B. (1983). Network analysis: some basic principles. In *Sociological theory*, ed. R. Collins, pp. 155–200. San Francisco: Jossey-Bass.

IV
Communication

23

Branch shaking and related displays in wild Barbary macaques

P. T. MEHLMAN

Branch shaking behaviour, common for many species of the Anthropoidea (Modahl & Eaton, 1977), is a rapid, repeated bouncing in place, while grasping a flexible substrate. It is characterised by stereotyped locomotor components and the production of audible sound by shaking of the substrate. Most authors denote these behaviours as 'displays', implicitly presuming that natural and/or sexual selection has shaped them for purposes of communication (Andrew, 1972; Modahl & Eaton, 1977; Wolfe, 1981).

For species of the genus *Macaca*, researchers have qualitatively noted that branch shaking displays (BSDs) are primarily performed by adult males and that they appear to serve a variety of communicatory functions, such as inter- and intra-group aggressive threat, attraction of females in oestrus, and more non-specific contexts of 'excitement' (Imanishi, 1957; Altmann, 1962; Bernstein, 1967; Lindburg, 1971; Hausfater, 1972; Deag, 1973; Taub, 1978; reviewed in Modahl & Eaton, 1977). Only two studies have quantitatively examined BSD in macaques, and both focused on rate increases of these displays by adult male Japanese macaques (*M. fuscata*) in the mating season (Modahl & Eaton, 1977; Wolfe, 1981). Both studies reached similar conclusions: BSDs served as 'broadcast stimuli' (Modahl & Eaton, 1977) by adult males to 'advertise' (Wolfe, 1981) their location to sexually receptive females.

Among some macaque species, there also seems to be a high degree of intra-specific variation in the contextual use of BSD. In Japanese macaques, BSD forms, and their contextual uses within the general category of courtship, have been described as being quite variable between groups (Stephenson, 1973; Wolfe, 1981). In Barbary macaques (*M. sylvanus*), two studies of the same contiguous population reached different conclusions as to the principle communicative function of BSD: Deag (1974) reported them as functioning primarily as inter-group agonistic displays, while Taub (1978) suggested they functioned as intra-group sexual displays.

In the present study, BSDs by adult males in a wild group of Barbary macaques are categorised by three variables: (1) form; (2) social context; and (3) male dominance rank. BSDs are then quantitatively analysed in a three-way analysis to determine whether different display forms are significantly associated with different communicatory contexts, and whether adult male dominance rank affects the frequency and expression of BSD. An hypothesis is proposed that BSDs in this Barbary macaque group often functioned in a context of 'intra-group locational signalling', disassociated from sexual and agonistic contexts.

Methods

Study site, study group and data collection

The results presented in this chapter are derived from a two-year study (September 1981 to August 1983) of wild Barbary macaques in the Ghomaran fir forest region of the Moroccan Rif mountains (Mehlman 1986a,b, 1988, 1989; Mehlman & Parkhill, 1988). The present research focuses on an habituated study group that contained between 54 and 63 individuals, and between 11 and 13 adult males at any given time during the study. Beginning in October 1981, the study group was observed for 746 h. Because adult male and female heterosexual behaviour in Barbary macaques is highly seasonal (Mehlman, 1986a,b), yearly cycles were divided into: (1) the mating season (October to January, 357 observation h); and (2) months exclusive of the height of the mating season (February to September, 389 observation h).

All occurrences of branch shaking were observed *ad libitum* with other observational protocols. Since BSDs were highly visible and audible, these data are assumed to approximate an unbiased sample. The analyses herein are restricted to adult males, since they performed 75% of branch shaking (Table 23.1), and their total sample size was relatively large ($n = 163$).

Forms of branch shaking

All BSDs comprised a vigorous, rhythmic shaking of the substrate produced by rapid hindlimb and forelimb extension. Typically, most displays lasted about 8–10 s, consisted of 4–10 bounces, and were not

Table 23.1. *Occurrences of BSD by age–sex class: adult males aged 6 years or more; sub-adult males aged 4–5 years; adult females aged 5 years or more; immature females aged 4 years or less; immature males aged 3 years or less*

	Adult males	Sub-adult males	Adult females	Immatures	Unknown	Total
n	163	18	9	28	19	237
%	68.8%	7.6%	3.8%	11.8%	8.0%	

accompanied by audible vocalisations. Five different forms were identified, based on positional behaviour, type of substrate used, relative energy expenditure, and the acoustic properties of the sound produced (Table 23.2).

Branch/tree shakes were performed on large branches of mature conifers, one to several metres out from the main trunk and with the body in a quadrapedal position; the subject would bounce up and down, relying on the flexibility of branches to create audible sound. In smaller trees, subjects performed this display horizontally (as above) or vertically, by clinging to the main trunk and repeatedly pushing off with the rear limbs.

Dead log thumping was easily differentiated from all other BSD forms by the production of a series of sounds that could often be heard at distances exceeding 1000 m. For this form, subjects would use large dead conifer logs that typically extended out over cliff edges. These logs were loose and positioned a few centimetres above limestone bedrock at the cliff face; the subject would move to the very end of the log and bounce it rhythmically, striking it against the limestone and producing a series of hollow thumps or knocking sounds (Fig. 23.1).

Branch breaking typically took place in mature conifers at or below half the height of the tree. The subject would bipedally raise himself on a lower, weaker branch, and then reach up to a stronger branch. Hanging on to the branch above, the subject would bounce the weak branch and forcefully thrust his hindlimbs downward, until it broke and fell through the foliage.

Dead standing tree shakes occurred in large conifers (firs and cedars) that had died and lost their foliage. Subjects would cling vertically to the trunk near the crown, repeatedly pushing off with the rear limbs and shake the entire tree.

In *Fir crown shaking*, the individual climbed to the crown of a 15–30 m fir and, while vertically clinging to the trunk, would shake it as described for dead standing tree shakes above.

Table 23.2. *Five forms of branch shaking*

	Branch/tree shake	Dead log thump	Branch break	Dead standing tree shake	Fir crown shake
Positional behaviour	Horizontal/ quadrapedal or vertical clinging	Horizontal/ quadrapedal	Vertical/bipedal	Vertical clinging or horizontal/ quadrapedal	Vertical clinging
Substrate	Non-specific, all live trees	Specific, dead logs	Specific, dead branches of firs	Specific, dead standing conifers	Specific, crowns of firs
Estimated energy expenditure	Minimum–medium	Medium–maximum	Medium	Medium	Medium
Estimated distance sound travels (m)	<300	>1000	<500	<300	<100
Frequency of occurrence in adult males (%)	39.2	27.6	14.7	11.0	7.4

Contexts of branch shaking

Throughout the study, 85% of branch shaking episodes by adult males ($n = 139$) were classified into one of six mutually exclusive contexts: *Intergroup contexts* were scored whenever two groups were in auditory or visual contact, regardless of the apparent context of a particular BSD episode. This method of scoring was chosen to clearly distinguish inter-group contexts from the five types of intra-group contexts described below.

Sexual contexts were scored whenever an adult male directed any form of BSD towards a sexually receptive female either: (1) immediately preceding a sexual interaction; or (2) when she was already in close proximity during a sexual association. Males would often perform BSDs as an apparent invitation to females in oestrus passing below them; the female would then join him in a sexual association. On other occasions, a male already with a female in oestrus would leave her after one to two partial mounts, perform BSDs nearby, and then rejoin the female (or the female would move to him) to continue the copulatory sequence. In two cases, these were scored as unidentified contexts, since other adult males were nearby and the display may have contained threat components directed to a nearby male.

Agonistic contexts were scored whenever any form of BSD was embedded in a sequence of agonistic interactions between two adult males. On most occasions, it appeared to be used as an elaborate threat gesture, since it was embedded in a sequence of threat gestures directed from the subject to another adult male. Sometimes, it appeared to function as a 'protest' because one male, without giving submissive gestures, performed BSDs as he retreated from another that had threatened him. BSD was also used to displace other individuals as for example, when adult males used BSD to displace others from fruiting trees.

Polyadic agonistic contexts were observed only six times during the study. In these instances, aggressive chases developed into large polyadic agonistic interactions and, as the males involved ran through rather large areas, this elicited BSDs from other adult males located in arboreal positions above the chase sequence.

Two other contexts were disassociated from contexts of sex and aggression, and were differentiated from each other on the basis of whether or not the group was involved in group movement throughout its range. *Neutral, group movement contexts* were scored whenever the group was moving, and in the absence of all identified contexts above. *Neutral, group stationary contexts* were scored whenever the group was not engaged in group movement, and in the absence of all other identified contexts.

Fig. 23.1. *Dead log thumping*. An adult male climbs out on a dead fir log overhanging a 30 m cliff. He shakes the log, bouncing it against the limestone cliff face producing a hollow, knocking sound that carries for more than 1000 m. This log, positioned along a well travelled route, was used by many of the dominant males. To prevent a deadly fall, this male and others must have 'pre-tested' the log for its safety and, perhaps, its sound quality.

Adult male dominance rank, age and degree of peripheralisation

Throughout the study, all agonistic behaviours between males were recorded *ad libitum*. To determine the dominance rank relations of adult males, losers in all dyadic agonistic conflicts were tabulated in a standard matrix; cases involving polyadic interactions or ambiguous outcomes were omitted (Sade, 1967; Barrette & Vandal, 1986). Behavioural categories for threat and submission follow those established for other macaque species (Altmann, 1962; van Hooff, 1967; de Waal, van Hooff & Netto, 1976) as well as for the Barbary macaque (Deag, 1974; Taub, 1978; White & Hosey, 1981). A loser was defined by his avoidance behaviour, flight, bared-teeth face, teeth-chattering, lip-smacking, or bark-screaming subsequent to, or simultaneous with, one or more aggressive threats from another individual. Interactions in which one male avoided another or directed submissive behaviours at another male, without previously receiving an aggressive threat (stare, open mouth, head-bob, ground slap, and/or chase and physical assault, etc.), were omitted from the matrix. These methods were employed because many gestures of the Barbary macaque, which might be scored as submissive in other macaque species, often occurred in dyadic, affiliative male interactions such as grooming, homosexual mounting, and male–male interactions that mimic triadic male–male–infant interactions (Taub, 1980), but did not include an infant.

Two other variables – age and degree of peripheralisation – were also analysed (Table 23.3). Age was estimated (Mehlman, 1986*a,b*) by grouping males in categories according to their physical characteristics and comparing these traits to a large sample of accurately aged males described in Fa (1984, appendix 1). Male peripheralisation was assessed qualitatively by examining each male's focal and *ad libitum* samples of recorded positions in the group relative to a central core of high-ranking females (Mehlman, 1986*b*). Central and peripheral group tendencies have also been reported in other wild Barbary macaques (Deag, 1974).

Data analyses

Branch shaking data were analysed in a three-way frequency table (rank × form × context) using log-linear models (*G*-tests of independence), evaluated by critical values ($P < 0.05$) of the chi-square distribution (Sokal & Rohlf, 1981; Norusis, 1986). Unidentified contexts, inter-group contexts, and polyadic agonistic contexts were omitted due to small sample sizes and/or ambiguous contexts. *G*-values lying between $0.05 > P > 0.001$ were adjusted with Williams correction for low sample sizes. Although this method is considerably more robust than the chi-square test, there is a lack of consensus as to what constitutes the lower limits for observed and expected cell frequencies (reviewed in Sokal & Rohlf, 1981). Therefore, in the three-factor analysis, adult males were pooled into three dominance classes (low, medium, high) to increase expected cell frequencies. Given the unknown effect of low expected cell frequencies in the three-factor analysis (and to verify and confirm certain results from the latter), more cell pooling was performed, and a secondary, more conservative, two-factor analysis (form × context) was tested with the *G*-statistic using the simultaneous test procedure (Sokal & Rohlf, 1981). To locate cells that contributed to significant heterogeneity in all sub-analyses, individual cell parameter estimates were divided by their standard errors to obtain *z*-values; if any *z*-value exceeded $+1.96$, that particular cell was considered as significantly 'boosting' heterogeneity (Norusis, 1986). Methods for Spearman's rank correlation coefficient and Fisher's exact test (two-tailed) follow those of Siegel (1956).

Table 23.3. *Male dominance rank, estimated age, and degree of peripheralisation. Dominance matrix displays number of losses in rows to winners in columns*

Relative dominance rank	Relative spatial status	Estimated age in years	Total branch shakes	Individual losing	Winners in dyadic agonistic interactions												
					Dia	Gpa	Spl	Why	Ln	Btn	Ch	Old	Hlf	Hd	Gh	Rng	Prg
High	C	10–12	38	Diamond		1											
	C	15+	3	Grandpa#	1												
	C	12–15	29	Split lip	1	1			1								
	C	12–15	14	Why	8	1	3		1								
	P	8–10	17	Line	2		7	2									
	P	12–15	11	Button	6		3		1			1					
Medium	P–C	12–15	7	China	5	1	3	1	3	4							
	P–C	15+	2	Old male#	1		2	1	1		1						
	P–C	15+	2	Halfback	3	2	2	1				1					
	P–C	7–8	15	Handsome	2	1	11		13	9	2	2	1				
Low	P	7–8	12	Ghost	4	3	2		6	1		2	1	2			
	P	6–7	7	Ring			1		10	3			1	5	1		1
	P	6–7	6	Progdot	2			2	5	8			2	1	4	1	1

#Indicates male disappeared during study, presumed dead (also see Mehlman, 1986b, appendix 3). C, centre of group during resting, foraging, or movement; P, peripheral to group, often solo or with other peripheralised males; P–C, peripheral, but limited access to centre of group (e.g. birth season).

Results

Male dominance rank and rate of branch shaking

A directional matrix of dyadic agonistic losses was arranged to minimise reversals above the diagonal (Table 23.3), revealing the presence of a linearly ranked adult male dominance hierarchy (also reported in other wild Barbary macaque groups: Deag, 1977; Taub, 1980). For adult males, there was a clear association between dominance rank and total number of BSDs (Table 23.3). Three aged males, however, were rarely observed to perform BSDs. Two were arthritic and were assumed to have died during the study (Mehlman, 1986*a* & 1986*b*, appendix 3). Presumably, the physical condition of these males impeded their ability to perform BSDs. If these two aged males were removed from the analysis, there was a significant correlation between dominance and total number of BSDs ($r_s = 7.97$, $n = 10$, $P < 0.01$).

Branch shaking in inter-group contexts

The only BSDs observed to occur in inter-group contexts were performed by adult and sub-adult males, and accounted for only 4.2% (10/237) of the total BSDs given by the group, or 6.1% (10/163) of all BSDs by adult males (Table 23.4). The overall rate during inter-group encounters was 0.48/h (10/21 observation hours); during intra-group contexts, the total rate was 0.21/h (153/725 observation hours). Note, however, that the inter-group BSD rate ranged from 0.00/h (9/11 encounters) to a rate of 2.15/h during one encounter (Mehlman & Parkhill, 1988). These results indicate that BSDs occurred primarily in intra-group contexts, although on one occasion the presence of another group probably increased the rate of branch shaking.

Branch shaking in intra-group contexts

Branch shaking episodes, as entered in the three-way table (male rank × form × context, Table 23.4), were tested against a hierarchy of interactive models by the use of the *G*-test (Table 23.5). Adult males were pooled into three classes based on their dominance ranks of high, medium, and low (three males older than 15 years were dropped from the analysis because of their small sample sizes, Table 23.4). BSDs were then partitioned by all forms of BSD against the four contexts with the highest frequency of occurrence (columns 1–4, Table 23.4A).

Two initial steps in the test procedure revealed the presence of significant

Table 23.4A. *Adult male branch shaking in the focal group by rank, form and context (n = 163)*

	Neutral, group movement	Agonistic	Sexual	Neutral, group stationary	Totals	Other[a]
High-ranking males						
Branch/tree shake	9	1	2	2	14	8
Dead log thump	18	5	4	0	27	5
Branch break	5	1	2	1	9	3
Dead standing shake	3	0	2	3	8	0
Fir crown shake	0	3	0	2	5	2
Sub-totals	35	10	10	8	63	18
Middle-ranking males						
Branch/tree shake	0	4	10	0	14	3
Dead log thump	3	1	0	0	4	0
Branch break	1	2	0	1	4	2
Dead standing shake	0	1	3	1	5	0
Fir crown shake	0	2	0	0	2	1
Sub-totals	4	10	13	2	29	6
Low-ranking males						
Branch/tree shake	2	6	1	2	11	8
Dead log thump	1	2	1	0	4	4
Branch break	1	0	1	2	4	2
Dead standing shake	1	0	1	2	4	1
Fir crown shake	0	0	0	1	1	1
Sub-totals	5	8	4	7	24	16
Oldest males						
Branch/tree shake	2	2	0	2	6	0
Dead log thump	1	0	0	0	1	0
Total all males	47	30	27	19	123	40

[a]Other contexts are explained in Table 23.4B.

Table 23.4B. *Other contexts for adult male branch shaking*

	High rank	Middle rank	Low rank	Total
Other				
Unidentified contexts	10	6	8	24
Polyadic agonistic contexts	3	0	3	6
Inter-group contexts	5	0	5	10

heterogeneity in the overall sample, and the absence of any three-way interactions (Table 23.5, model: male rank × form × context). Three further steps, testing for two-way interactions, revealed that significant associations were present between male rank and context, and between context and BSD form (Table 23.5, models: two-way effects and partial associations). An examination of cell parameters revealed that the significant heterogeneity of this sample was due to: (1) *middle-ranking males* performing *Branch/tree shakes* significantly more often in *sexual contexts*; (2) *high-ranking males* performing significantly more in *neutral, group movement contexts*; and (3) *all males*, with the effects of rank removed, performing *Dead log thumping* significantly more often in *neutral, group movement contexts*.

These results can be summarised as follows. Males classed by rank did not differ significantly in their use of BSD forms, but for all males, the *branch/tree shake* was used more frequently than other forms (for middle-ranking males, this resulted from a significantly higher proportion of sexual contexts). High-ranking males performed more overall BSDs than did other males, and this resulted from their significantly higher proportion of BSDs used in neutral, group movement contexts.

Given the low observed and expected cell frequencies in the above tests, a more conservative test was employed to verify the association between *Dead log thumping* and *neutral, group movement contexts*. All BSDs for males were pooled, and the *Fir crown shake* data was omitted (small sample size). This produced a 4 × 4 table (Table 23.6) that could be tested with a simultaneous *G*-test procedure with an adjusted type I error rate (Sokal & Rohlf, 1981) to assess all ten possible groupings of forms by their contextual distributions. These tests revealed that : (1) *Branch/tree shaking, Branch breaking*, and *Dead upright tree shaking* did not differ from each other with respect to their contextual distributions; and (2) the contextual distribution of *Dead log thumping* differed from the contextual distributions of all other forms with the exception of branch breaking (Table 23.6). To confirm these

Table 23.5. *Results of log-linear modelling for a three-way contingency table: branch shaking classified by adult male rank (a = 3), form (b = 5), and context (c = 4). For context, only the four highest frequencies of occurrence were used: neutral, group movement; dyadic agonistic; sexual; and neutral, group stationary (n = 116)*

Rank × Form × Context test for three-way and higher order effects = 0

K	d.f.	G (Williams)	Probability	Iterations		
3	18	25.23	>0.05	3 →	Model three-way effects (below)	
2	50	113.06	<0.001	2 →	Model three-way effects (A, B, C, below)	
1	59	178.81	<0.001	0 →	H_0 = Heterogeneity in sample (cannot reject)	

	d.f.	G	G (Williams)	Probability	Iterations
Model three-way effects **(Rank × Form) (Rank × Context) (Form × Context)** H_0 = three-way independence (cannot reject)	18	29.58	25.23	>0.05	3
Model two-way effects (A) **(Rank × Context) (Form × Context)** H_0 = given the level of Context, Rank and Form are independent (cannot reject)	26	40.17	35.30	>0.05	2
Model two-way effects (B) **(Context × Form) (Rank × Form)** H_0 = given the level of Form, Rank and Context are independent (reject)	24	53.61	na	<0.0001	2
Model two-way effects (C) **(Form × Rank) (Context × Rank)** H_0 = given the level of Rank, Form and Context are independent (reject)	36	71.34	na	<0.0001	2

Partial associations and location of significant heterogeneity

Effects	d.f.	G (partial)	Probability	Significant cell divisions	z-values
Rank × Form	8	(10.59)	0.2262	None	
Rank × Context	6	(24.03)	0.0005	High rank & neutral, group movement	2.06 ($P < 0.05$)
Form × Context	12	(41.76)	0.0000	*Dead log thump* & neutral, group movement	2.01 ($P < 0.05$)
Rank	2	(21.93)	0.0000	High rank	2.42 ($P < 0.05$)
Form	4	(31.12)	0.0000	*Branch tree/shake*	2.13 ($P < 0.05$)
Context	3	(12.70)	0.0530	None	
Rank × Form × Context	59	178.81	0.0000	Middle rank & *branch/tree shake & sexual*	2.37 ($P < 0.05$)

na represents Williams correction factor not applicable.

Table 23.6. *Statistical tests of association between branch shake forms based on contextual frequency distributions. The F × C table used for the first step of the analysis and an overall G-test for heterogeneity is displayed above. A simultaneous test procedure with ten intended comparisons (experimentwise error rate, a = 0.05) is displayed below*

	Neutral group movement	Agonistic	Sexual	Neutral group stationary	Totals
Branch/tree shake	13	13	13	6	45
Branch break	7	3	3	4	17
Dead standing shake	4	1	6	6	17
Dead log thump	23	8	5	0	36
Totals	47	25	27	16	115

G statistic (Williams correction) = 26.08, significant heterogeneity ($P < 0.05$). Critical χ^2 value (d.f. 9, $a = 0.05$) = 16.92.

Tests for heterogeneity	G (Williams)
Branch break & branch/tree shake	2.45 *ns*
Branch break & dead standing shake	3.03 ns
Branch/tree shake & dead upright shake	6.39 ns
Branch break & branch/tree & dead standing	8.23 ns
Branch break & dead log thump	9.52 ns
Branch shake & dead log thump	14.32*
Branch shake & branch break & dead log thump	17.51**
Dead standing & dead log thump	20.81*
Branch break & dead standing & dead log thump	22.41*
Branch shake & dead standing & dead log thump	26.36*

ns, non-significant; *significant heterogeneity $P < 0.05$; **$P < 0.07$.

results for *Fir crown shaking* (low sample size), two Fisher's exact tests, pooling all contexts against the neutral, group movement context, revealed that the pooled contextual distribution of the *Fir crown shake* differed significantly from *Dead log thumping*, yet did not differ from all the other BSD forms combined (Table 23.7). These results indicate that *Dead log thumping* differed significantly from other forms by the higher frequency of occurrence in neutral, group movement contexts (63.9%), confirming an important component of the three-factor analysis above.

Table 23.7. *Two Fisher's exact tests (two-tailed), testing the similarity of the fir crown shake to other forms*

	Neutral, group movement contexts	Agonistic + sexual + neutral, group stationary contexts	Fisher's exact test, P
Fir crown shakes	0	8	
Dead log thumps	23	13	0.0011
Fir crown shakes	0	8	
Branch/tree shakes + branch breaks +			
dead standing shakes	24	55	0.1000

The effect of the mating season on branch shaking

During the non-mating season, adult males performed 38 BSDs (0.098/h). During the mating season, they performed 101 BSDs (0.283/h), nearly three times the frequency of that in the non-mating season months. This rate increase was not even, however, with regard to rank, form and context. Because *Dead log thumping* differed with respect to its significant association with neutral, group movement contexts, BSD forms were classified as *Dead log thumping* or all other forms combined, and then plotted by rate of contextual occurrence for the mating and non-mating seasons (Fig. 23.2). This revealed that the rate of neutral, group movement contexts was less affected by the mating season than all other contexts, regardless of the form used.

Discussion

Intra-group locational signalling hypothesis

Approximately half (52.5%) of the BSD contexts observed in the present study were associated with sex, intra-group agonism, and inter-group interactions. Although these results do not differ qualitatively from contexts reported for BSD among several macaque species, the degree of contextual heterogeneity in this study group appears higher than that reported for any one macaque species (Imanishi, 1957; Altmann, 1962; Bernstein, 1967; Lindburg, 1971; Hausfater, 1972; Deag, 1973; Modahl & Eaton, 1977; Taub, 1978). Even more interesting, however, is the remaining

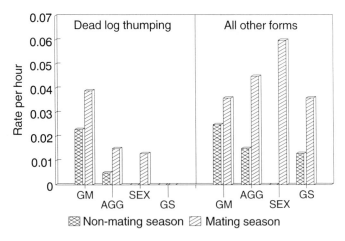

Fig. 23.2. Rates of *Dead log thumping* versus all other forms (combined) relative to context and the mating season. GM, neutral, group movement contexts; AGG, dyadic agonistic contexts; SEX, sexual contexts; GS, neutral, group stationary contexts.

47.5% of BSDs that appear to be disassociated from contexts of sex and agonism. Similar non-agonistic and non-sexual displays have gone unreported in other studies or have been treated anecdotally.

One hypothesis to account for these non-sexual and non-agonistic BSDs is that intra-group locational signalling, previously described as being only a component of sexual displays in Japanese macaques (Modahl & Eaton, 1977; Wolfe, 1981), may be a functional context in itself, disassociated from agonism and sex. Therefore, these BSDs in neutral contexts might serve as 'broadcast stimuli' (Modahl & Eaton, 1977), functioning not only to communicate the actor's position and identity to other group members ('identifier', Marler, 1965), but, perhaps more importantly, functioning to elicit responses from other group members, such that the subject is able to locate the position of individuals (or the group as a whole). During periods of coordinated group movement, these displays may further function to communicate to other group members the intended direction of travel of the actor, as well as to attract other members in the actor's direction.

Many qualitative observations appear to support the locational hypothesis. For example, neutral displays by peripheralised males generally produced responses in other group members, if not other BSD by males, followed by movement of others into more visible arboreal and terrestrial positions, or, on occasion, a specialised intra-group locational vocalisation, the 'booming roar' (Mehlman, 1984, 1986*b*). From these responses, the subject could easily derive locational information. For example, on two occasions, single adult males were observed to apparently lose visual and auditory contact with the rest of the group. Each performed BSDs, remained in the tree until group members responded, and then descended and moved toward the rest of the group.

In 34.0% of neutral, group movement contexts, there was also a subsequent movement of individuals and/or sub-groups of females and immatures in the direction of the actor. These situations occurred when subjects and others were separated by relatively small distances (30–100 m), and there was an obvious cause and effect relationship between BSDs and other macaques' movement toward the subject. At longer distances, this may also have occurred, but cause and effect were impossible to ascertain. Neutral BSDs were also performed as the group descended from their sleeping trees in the morning and initiated movement. On five occasions, it appeared that two or more males employing BSDs were competing to influence the direction of the group toward their intended travel paths for the day (similar to a system described for *Papio hamadryas*, Kummer, 1971).

The quantitative results of this study also support the locational hypothesis. First, the rate of neutral, group movement contexts, independent of BSD form, was less affected by the mating season than the rate of all other contexts (Fig. 23.2). Note also that high-ranking males performed 74% (35/47) of the BSDs in neutral, group movement contexts. These males, by virtue of their high dominance rank and central status, were more involved in group leadership throughout the year. Second, the occurrence of a specific form, *Dead log thumping*, was significantly associated with neutral, group movement contexts (Table 23.4), and when it occurred in this context, its rate change from non-mating to mating season was relatively small (Fig. 23.2). The additional fact that *Dead log thumping* could be differentiated from all other forms by the production of a hollow knocking sound, an acoustic property that enabled it to be clearly located and heard for long distances, further suggests that *Dead log thumping* functioned to transmit locational information.

Locational branch shaking as a locale-specific communicatory adaptation

In two previous long-term studies of the Barbary macaque in cedar forest habitat of the Moyen Atlas mountains of Morocco, BSD was reported to be either an inter-group agonistic display (Deag, 1973) or an intra-group sexual display (Taub, 1978). Neither study, however, suggested that BSD functioned in purely locational contexts. The Ghomaran fir forest habitat differs ecologically from the Moyen Atlas cedar forest in three principal ways: (1) the conifer forest is less dense; (2) the topography is more complex; and (3) the vegetation stand types are more patchily distributed (Mehlman, 1986b). This results in a poorer and more patchily distributed food base and, as a consequence, the Ghomaran Barbary macaques display increased home range size (Mehlman, 1989) and greater intra-group dispersion (Mehlman, 1986b). For example, Deag (1974) in a study of ten groups (12–36 individuals) reported that they rarely exceeded a maximum intra-group 'spread' of 1.5 ha. In contrast, in the Ghomara, groups used areas of 2–3.5 ha during dispersed foraging, and during group movement this area often increased to 5–7 ha (Mehlman, 1986b).

Despite the high intra-group dispersion in the Ghomara, the study group maintained its integrity quite well during day ranging; individuals rarely became lost, and only two cases of sub-division were verified (Mehlman, 1986b). This integrity was maintained by several behavioural characteristics, such as the use of traditional pathways and resting points, and an intra-group locational vocalisation, the *Booming roar* (Mehlman, 1984, 1986b; E. Merz, personal communication). The present study suggests that another important way in which individual members may have maintained group integrity was by the modification of the adult male BSD repertory. This shift in function to contexts of location and intended travel direction was further enhanced by the development of a new form, *Dead log thumping*, unreported in previous studies of Barbary macaques. Locational branch shaking was employed principally by high-ranking, non-aged adult males in what appeared to be their leadership roles, and was used in tandem with many aspects of behaviour that enabled members to disperse widely and forage in a poor quality habitat, yet retain group integrity.

Branch shaking in the genus Macaca

In the present study, as in other studies of macaques, BSD was primarily an adult male activity (Lindburg, 1971; Hausfater, 1972; Modahl & Eaton, 1977; Wolfe, 1981). For adult males estimated to be less than 15 years of age, there was a significant correlation between dominance rank and overall BSDs. In the two quantitative studies of Japanese macaques, the relationship between male rank and BSDs was equivocal (no correlation, Modahl & Eaton, 1977; significant correlation, Wolfe, 1981), but in Japanese macaques BSD is thought to function primarily as a sexual display of adult males. In the present study, there was no significant association between dominance rank and BSD performed in sexual contexts.

In some studies of macaques, adult male BSD is qualitatively associated with inter-group contexts (*M. fuscata*, Imanishi, 1957; *M. mulatta*, Altmann, 1962; Hausfater, 1972). Although Deag (1973, 1974) concluded that BSD in his study group of Barbary macaques was primarily an inter-group display, the difference between his conclusions and the present results may have involved methodological constraints. Deag reached his conclusion by comparing inter- and intra-group rates outside the mating season and his sample size was low (*n* = 17).

In the present study, the rate increase in BSD in sexual contexts during the mating season agrees well with other quantitative studies of the Japanese macaque (Modahl & Eaton, 1977; Wolfe, 1981), as well as Taub's (1978) observations of a wild Barbary macaque group. Taub observed that the lowest-ranking adult males in his group, seeking sexual associations with females in oestrus, used a 'peripheralise and attract' strategy (1978, p. 86) that included BSDs. To some degree, the present results corroborate Taub's findings. Middle-ranking males (also peripheralised) differed significantly from other males by their use of the *Branch/tree shake* in sexual contexts. In contrast to Taub's findings, however, the low-ranking, peripheralised males of the present study had the lowest frequency of BSD in sexual contexts. This difference probably results from the high number of adult males in the present study group (*n* = 11–13) relative to Taub's (1978) group (*n* = 7) and the fact that in the Ghomara, male competition for females in oestrus was intense. In general, low-ranking males in the Ghomara appeared inhibited from even attempting to attract females. Note also that in the Ghomara, the large rate increase in agonistic BSD for high- and middle-ranking males during the mating season was in large part due to the increased inter-male competition and aggression over access to females in oestrus.

The above discussion, and the results of the present study, suggest that the only conclusions that can be reached currently with respect to adult male branch shaking in species of *Macaca* is that it can function to: (1) attract females in oestrus in sexual contexts; (2) repulse other conspecifics in intra- and inter-group agonistic contexts; and (3) attract other group members in locational contexts, and/or elicit other BSD responses or vocalisations in other group members for locational purposes. Given that BSDs occur in agonistic, sexual and locational contexts, and can produce responses in conspecifics as diverse as attraction and repulsion, explanations that these behaviours are affective, emotional displays, facilitated by 'incompatible tendencies' to move in two different directions (Andrew, 1972; Hausfater, 1972), or that they are sexual displays canalised by female choice (Modahl & Eaton, 1977; Wolfe, 1981) may be too narrow in perspective. Much as we have begun to view vocal communication in cercopithecines as being information-rich, influenced by learning, referential to various degrees, modifiable, voluntary, and presumably under some degree of cortical control (Struhsaker, 1967; Marler, 1973; Sutton *et al.*, 1973; Green, 1975; Seyfarth, Cheney & Marler, 1980; Dittus, 1984; Gouzoules, Gouzoules & Marler, 1984; Seyfarth, 1986; Seyfarth & Cheney, 1986), the results of the present study suggest that branch shaking might be viewed as a similar communicatory phenomenon.

Therefore, taking this view, branch shaking may be precipitated by a certain level of arousal or excitement, but the arousal may not necessarily be associated with sex and/or aggression; moreover, an individual would have the ability to control and channel an 'excited' display into a diverse array of communicatory functions contingent upon the challenges of a particular environment. For example, in the present study it was clear that *Dead log thumping* was not an 'emotional rush' onto a precariously positioned log overhanging a 30–80 m cliff face. Rather, these logs had to have been 'pre-tested' for their safety and, perhaps, even for their sound quality. Similarly, *Branch breaking* also necessitated some deliberate choice as males moved into trees to begin an episode, as only certain, limited, branch combinations were suitable for this display.

Clearly, answers to these questions should be sought in the ontogenetic learning processes associated with the expression of 'power' (e.g. the correlation between BSD frequency and dominance), while carefully controlling for environmental contingencies that may influence the functional use of these displays. Therefore, while one macaque male growing up in a captive corral may learn to shape his 'power display' either to intimidate males or to attract females in oestrus (cuing on social responses), another male in a wild habitat may 'discover' that his 'power display' can also be used to orient and lead other group members.

For these reasons, it would be premature at this time to conclude that locational branch shaking does not occur in other macaque species. The only other two quantitative studies of BSD in macaques took place either in a 0.81 ha corral (Modahl & Eaton, 1977) or in semi-free and free ranging, provisioned groups, in which data were collected in 1 ha areas near a food storage area and feeding station (Wolfe, 1981, p. 25). These are not environmental conditions conducive to modifying BSDs to longer-distance locational contexts during group movement.

In conclusion, branch shaking in macaques may contain much more communicative heterogeneity than was previously believed, and under appropriate conditions, these behaviours may be *adaptively modified* to suit communicatory needs shaped by local environmental conditions. Further investigations into the ontogeny and contextual variability of branch shaking in wild and captive groups may reveal to what degree these displays are voluntary, and the extent to which learning and experience shape the expression, modification, and communicatory adaptiveness of these displays.

Summary

The majority of branch shaking displays (BSDs) in a wild group of Barbary macaques were adult male behaviours (74.8%) that occurred only rarely in inter-group contexts (6.1%), and primarily in five intra-group contexts: sexual (19.4%); dyadic agonistic (21.6%); polyadic agonistic (4.3%); neutral, group stationary (13.7%); and neutral, group movement (33.8%). BSDs in this one study group were at least as contextually heterogeneous as has been previously reported for the entire genus. Males classified by high, middle, and low dominance rank did not differ in their frequencies of the five *forms* of BSD, but high-ranking males performed more overall BSDs than did other males (49.7%), because they indulged in a high proportion of BSDs in neutral, group movement contexts. Middle-ranking males performed significantly more BSDs in sexual contexts. Independent of male rank, there was a significant association between the neutral, group movement context and *Dead log thumping* (63.9% of *Dead log thumping* was performed in this context), a specialised BSD form that produced a hollow knocking sound that could be heard at distances of more than 1 km. BSDs in neutral contexts (non-sexual and non-agonistic) are hypothesised to function as locational broadcast stimuli, signalling the position of the subject, enabling him to locate other group members by their responses, and sometimes signalling the subject's intended direction of travel, attracting others in his direction. Locational displays were used primarily by high-ranking, non-aged adult males, and their frequency of occurrence was less affected by the mating season than other BSDs occurring in other contexts. Locational BSDs are suggested to be an adaptive communicatory modification, enabling group members to disperse widely during foraging in poor quality habitat while maintaining group integrity.

Acknowledgements

I thank the Division des Eaux et Forêts du Maroc for permission to study in Morocco. I am also grateful to Shane Parkhill and Mohammad Halifa for their assistance in the field and to John Fa for an introduction to the study site. I thank Bernard Chapais, David Taub, Becky Sigmon, Shane Parkhill, and Bob Martin for their comments on the manuscript, as well as the late Melissa Knauer, who through her stimulating discussions first helped to shape the initial analyses. A special thanks go to Xavier Carot for his technical assistance with the statistics. This research was supported by a grant from the L. S. B. Leakey Foundation, University of Toronto Doctoral Fellowships, and postdoctoral grants from the Université de Montreal.

References

Altmann, S. A. (1962). A field study of the sociobiology of rhesus monkeys, *Macaca mulatta*. *Annals of the New York Academy of Science*, **102**, 338–435.

Andrew, R. J. (1972). The information potentially available in mammal displays. In *Nonverbal communication*, ed. R. J. Hinde, pp. 179–206. Cambridge: Cambridge University Press.

Barrette, C. & Vandal, D. (1986). Social rank, dominance, antler size and access to food in snow-bound wild woodland caribou. *Behaviour*, **97**, 118–46.

Bernstein, I. S. (1967). A field study of the pigtail macaque. *Primates*, **8**, 217–28.

Deag, J. M. (1973). Intergroup encounters in the wild Barbary macaque, *Macaca sylvanus*. In *Comparative ecology and behavior of primates*, ed. R. P. Michael & J. H. Crook, pp. 315–73. London: Academic Press.

Deag, J. M. (1974). A study of the social behaviour and ecology of the wild Barbary macaque *Macaca sylvanus*. PhD thesis, University of Bristol.

Deag, J. M. (1977). Aggression and submission in monkey societies. *Animal Behaviour*, **25**, 465–74.

de Waal, F. B., van Hooff, J. A. R. A. M. & Netto, W. J. (1976). An ethological analysis of types of agonistic interaction in a captive group of Java monkeys (*Macaca fascicularis*). *Primates*, **17**, 257–90.

Dittus, W. P. J. (1984). Toque macaque food calls: semantic communication concerning food distribution in the environment. *Animal Behaviour*, **32**, 470–7.

Fa, J. E. (1984). Definitions of age–sex classes for the Barbary macaque (Appendix I). In *The Barbary macaque – a case study in conservation*, ed. J. E. Fa, pp. 335–46. New York & London: Plenum Press.

Gouzoules, S., Gouzoules, H. & Marler, P. (1984). Rhesus monkey (*Macaca mulatta*) screams: representational signaling in the recruitment of agonistic aid. *Animal Behaviour*, **32**, 182–93.

Green, S. (1975). Variation of vocal pattern with social situation in the Japanese monkey (*Macaca fuscata*): a field study. In *Primate behavior*, vol. 4, ed. L. A. Rosenblum, pp. 1–102. New York: Academic Press.

Hausfater, G. (1972). Intergroup behavior of free-ranging rhesus monkeys *Macaca mulatta*. *Folia Primatologica*, **18**, 78–107.

Imanishi, K. (1957). Social behavior in Japanese monkeys, *Macaca fuscata*. In *Primate social behavior*, ed. C. H. Southwick, pp. 68–81. Princeton: Van Nostrand.

Kummer, H. (1971). *Primate societies*, Chicago: Aldine.

Lindburg, D. G. (1971). The rhesus monkey in North India: an ecological and behavioral study. In *Primate behavior: developments in field and laboratory research*, vol. 2, ed. L. A. Rosenblum, pp. 1–106. New York: Academic Press.

Marler, P. (1965). Communication in monkeys and apes. In *Primate behavior: field studies of monkeys and apes*, ed. I. DeVore, pp. 544–85. New York: Holt, Rinehart & Winston.

Marler, P. (1973). A comparison of vocalizations of red tailed monkeys and blue monkeys, *Cercopithecus ascanius* and *C. mitis*, in Uganda. *Zietschrift für Tierpsychologie*, **33**, 223–47.

Marler, P. (1977). Primate vocalization: affective or symbolic? In *Progress in ape research*, ed. G. H. Bourne, pp. 85–97. New York: Academic Press.

Mehlman, P. T. (1984). Aspects of the ecology and conservation of the Barbary macaque in fir forest habitat of the Moroccan Rif mountains. In *The Barbary macaque – a case study in conservation*, ed. J. E. Fa, pp. 165–99. New York & London: Plenum Press.

Mehlman, P. T. (1986a). Male intergroup mobility in a wild population of the Barbary macaque (*Macaca sylvanus*), Ghomaran Rif mountains. *American Journal of Primatology*, **10**, 67–81.

Mehlman, P. T. (1986b). Population ecology of the Barbary macaque (*Macaca sylvanus*) in the fir forests of the Ghomara, Moroccan Rif mountains. PhD thesis, University of Toronto.

Mehlman, P. T. (1988). Food resources of the wild Barbary macaque (*Macaca sylvanus*) in high altitude fir forest, Ghomaran Rif, Morocco. *Journal of Zoology, London*, **214**, 469–90.

Mehlman, P. T. (1989). Comparative density, demography, and ranging behavior of Barbary macaques (*Macaca sylvanus*) in marginal and prime conifer habitats. *International Journal of Primatology*, **10**, 269–92.

Mehlman, P. T. & Parkhill, R. S. (1988). Intergroup interactions in wild Barbary macaques (*Macaca sylvanus*), Ghomaran Rif mountains, Morocco. *American Journal of Primatology*, **15**, 31–44.

Modahl, K. B. & Eaton, G. G. (1977). Display behaviour in a confined troop of Japanese macaques (*Macaca fuscata*). *Animal Behaviour*, **25**, 525–35.

Norusis, M. J. (1986). *Advanced statistics: SPSS-PC +*. Chicago: SPSS Inc.

Sade, D. S. (1967). Determinants of dominance in a group of free-ranging rhesus monkeys. In *Social communication among primates*, ed. S. A. Altmann, pp. 99–114. Chicago: University of Chicago Press.

Seyfarth, R. M. (1986). Vocal communication and its relation to language. In *Primate societies*, ed. B. B. Smuts, D. L. Cheney, R. M. Seyfarth, R. W. Wrangham & T. T. Struhsaker, pp. 440–51. Chicago: University of Chicago Press.

Seyfarth, R. M. & Cheney, D. L. (1986). Vocal development in vervet monkeys. *Animal Behaviour*, **34**, 1450–68.

Seyfarth, R., Cheney, D. L. & Marler, P. (1980). Vervet monkey alarm calls: semantic communication in a free-ranging primate. *Animal Behaviour*, **28**, 1070–94.

Siegel, S. (1956). *Nonparametric statistics for the behavioral sciences*. New York: McGraw-Hill.

Sokal, R. R. & Rohlf, F. J. (1981). *Biometry*. New York: W. H. Freeman.

Stephenson, G. R. (1973). Testing for group specific communication patterns in Japanese macaques. In *Precultural primate behavior*, Proceedings of the Fourth International Congress of Primatology, vol. 1, ed. E. W. Menzel, pp. 51–70. Basal: S. Karger.

Struhsaker, T. T. (1967). Auditory communication among vervet monkeys (*Cercopithecus aethiops*). In *Social communication among primates*, ed. S. A. Altmann, pp. 281–325. Chicago: University of Chicago Press.

Sutton, D., Larson, C., Taylor, E. M. & Lindeman, R. C. (1973). Vocalization in rhesus monkeys: conditionability. *Brain Research*, **52**, 225–31.

Taub, D. M. (1978). Aspects of the biology of the wild Barbary macaque (Primates: Cercopithecinae, *Macaca sylvanus* L. 1758): biogeography, the mating system and male-infant interactions. PhD thesis, University of California, Davis.

Taub, D. M. (1980). Testing the 'agonistic buffering' hypothesis. I. The dynamics of participation in the triadic interaction. *Behavioral Ecology and Sociobiology*, **6**, 187–97.

van Hooff, J. A. R. A. M. (1967). The facial displays of the catarrhine monkeys and apes. In *Primate ethology*, ed. D. Morris, pp. 7–68. London: Weidenfield & Nicholson.

White, D. & Hosey, G. R. (1981). Social organization in captive Barbary macaques (*Macaca sylvana*). *Primates*, **22**, 487–93.

Williams, D. A. (1976). Improved likelihood ratio tests for complete contingency tables. *Biometrika*, **63**, 33–7.

Wolfe, L. D. (1981). Display behavior of three troops of Japanese macaques (*Macaca fuscata*). *Primates*, **22**, 24–32.

24

The inter-play of kinship organisation and facial communication in the macaques

A. ZELLER

Introduction

Macaques are an extremely adaptable genus of monkeys found in many habitats spread over a wide area of the Old World. Part of their success comes from their labile social organisation and ability to subsist on a wide variety of foods. The three species of macaques under particular consideration in this chapter, Barbary macaque (*Macaca sylvanus*), Japanese macaque (*Macaca fuscata*) and long-tailed macaque (*Macaca fascicularis*), live in habitats ranging from the forest-covered mountains of Morocco and Japan to the lowland swamp forests of South-East Asia. Nonetheless they show marked similarities in social organisation. They tend to live in multi-male–multifemale social groups with a fairly constant female membership, although some Barbary macaques live in single male groups. Migration of males into and out of the groups provides some gene flow and the potential for males to negotiate their political position in a new group where they may be relatively unknown. Females tend to stay in their natal groups and rely on family bonds and interactions with known animals to establish their place in the social system. All of these negotiations require the use of communicative skills in both self-presentation and in responding appropriately to messages being sent. The development of these complex skills is due not only to the multiple functions that communication serves but also to the fact that the various codes developed are based on anatomical substrates that evolved for other purposes. For example, the musculature of the face, which is so important in the production of complex signals, was originally designed for efficient eating and observations of the environment. Macaques may have a larger range of fine facial movements used in communication because they tend to lack the array of colours around the face, which serves to enhance the message in many *Cercopithecus* species (Figs. 24.1 and 24.2).

Fig. 24.1. Male *Macaca fascicularis*: *Open mouth threat* showing incisors and canines, with ears back, eyebrows lowered, piloerection and eyelid flicker.

Over past decades of research it has become clear that communication is not a single process, but rather the outcome of a concatenation of events based on the animal's anatomy, the ecological situation, the message transmitted and the coding pattern utilised. In the early 1960s communication was often studied in captivity by putting two animals in one cage and observing their interaction. These animals were often strangers to one another, and sometimes belonged to different genera (van Hooff, 1962, 1967; Andrew, 1963*a*,*b*,*c*). Experiments were undertaken to assess 'innate' communicative abilities (e.g. Sackett, 1966), and the development of facial gestures was also studied (e.g. Chevalier-Skolnikoff, 1974). Group interactions were observed with a view to searching for patterns in communicative behaviours (Rowell & Hinde, 1962; Hinde, 1964). Much of this early work focused on visual or auditory vocal signals and most of it was conducted in captivity. Some field research was occurring during the same time period, and it was in the late 1960s that Struhsaker (1967) first published on the possibility that vervet monkeys were utilising indicators in their call system to refer to particular classes of predators. This was the beginning of a new approach to the complexity of communication systems. In addition to the transmission of simple emotionally based responses to particular stimuli,

Fig. 24.2. Female *Macaca fascicularis*: *Open mouth threat* with mouth stretched, ears back, eyes staring, eyebrows raised, and nostrils flared.

further research revealed multiple functions for communication. Primates can indicate individual identity (e.g. Waser, 1977; Epple, Alveario & Katz, 1982; Smith, Newman & Symes, 1982; Haimoff & Tilson, 1985), and differential responses to vocalisations that, to human observers, seem very similar to each other (Green, 1975; Seyfarth, Cheney & Marler, 1980; Cheney & Seyfarth, 1982*a,b*, 1990). Screams by out of sight immature rhesus monkeys can elicit different responses depending on whether the antagonist is higher- or lower-ranking and thus poses a greater or lesser threat of physical danger to the youngster. Non-maternal female kin also respond to such screams, even in the absence of the young animal's mother, which reinforces patterns of kin solidarity (Gouzoules, Gouzoules & Marler, 1984). Other current work has focused on the uses of communication in establishing ranges (Deputte, 1982; Epple *et. al.*, 1982; Raemakers & Raemakers, 1984), individual spacing (Robinson, 1982; Waser, 1982), localising individuals (Brown, 1982) and responding to strangers (Waser, 1982).

Much of this work has focused on primate vocalisations (e.g. Waser, 1977, 1982; Cheney & Seyfarth, 1980, 1981, 1982*a,b*, 1990; Brown, 1982; Deputte, 1982; Robinson, 1982; Hertzog & Hohmann, 1984; Seyfarth &

Cheney, 1984, 1986), although a few researchers have studied olfactory cues (e.g. Schilling, 1974, 1979; Shorey 1976; Epple *et al.*, 1982) and facial gestures of communication (e.g. Chevalier-Skolnikoff, 1982). These aspects of primate communication, the visual gesture, vocal expression, and olfactory emission, are three of the four major channels primates use to transmit information. The fourth or tactile channel, is frequently used in close contact situations, and can vary from the friendly groom to the punitive slap or bite. In many cases several channels are used together to provide refinement, specificity, and occasionally redundancy to the message. However, in most cases they are studied separately due to the complexities of attempting to gather data in more than one channel at a time.

Because communication is primarily a social activity it must be studied in a social situation. The episodic dyadic encounters that formed the foundation for much early communication research are inadequate for testing more than the bare elicitation of gesture. Even in artificially constructed groups the social basis is in flux and, according to Rowell (1972), cannot really be considered stable until several generations have passed and kin groups have been established. Moreover, the animals require enough space to interact and to refuse to interact, in order for researchers to assess adequately the social ramifications of particular communicative gestures.

Variation in macaque facial gestures

One of the major subjects under current investigation is variability in communicative gestures. Several researchers have worked on this problem, mainly in the area of vocal signals (e.g. Green, 1975; Newman & Symes, 1982; Smith *et al.*, 1982; Gouzoules *et al.*, 1984; Seyfarth & Cheney, 1984). Vocal expressions are tape recorded and spectrographic analysis of sounds can be done. My research is focused on variation in the visual channel particularly in the facial gestures used to send threats. Due to the potential complexity of interactions and the fact that the data were recorded on film, I concentrated on the initial sender of the threat. Data were gathered over a 10-year period, 1975–85, from three species of macaques: *M. sylvanus* from Gibraltar, *M. fuscata* from Arashiyama West in Texas, and *M. fascicularis* from Monkey Jungle in Florida.

All of these are well-established intact social groups of long-standing, living in semi-free-ranging conditions in areas of from 4 to 20 ha. I concentrated on gathering data from adult animals of both sexes, although some sub-adults were included in the *M. sylvanus* and *M. fascicularis* samples because of the small population sizes. The study populations

Table 24.1. *Number of macaques observed*

	sylvanus	fuscata	fascicularis
No. of animals	11	18	12
Males	3	5	8
Females	8	13	4
Old adults	3	13	6
Young adults	5	5	4
Sub-adults	3		2
Frames	10 191	6217	11 632
Units	67	69	86

consisted of 11 Barbary macaques, 18 Japanese macaques and 12 long-tailed macaques (Table 24.1).

The data were all gathered by using Super 8 film to record interactions between known animals; this was then analysed in order to describe facial expressions used during these episodes. Super 8 film rather than video was used because it permits clear viewing of each frame of film and thus a tight level of time control since all the film was taken at 1/24 s.

When the film records were examined frame by frame, starting with *M. sylvanus*, 34 components of the face and head were noted. Of these, 14 were components that had not been described previously in macaque communication gestures. One of these has since been described for squirrel monkeys (Marriot & Salzen, 1978). These components were defined as aspects of the facial anatomy coupled with movement in a particular direction, such that 'eyebrows raised' was distinguished from 'eyebrows lowered'. The initial procedure involved filming a number of episodes of gesture transmission (here called units) and examining them frame by frame. Check sheets were compiled of how many and which components were present in each individual's expression of threat.

Two types of data are derived from the records of component use in macaque threat. One is frequency data, which is based on the number of units in which a component occurs expressed as a percentage of the total number of units between various classes such as *sylvanus* vs *fuscata* vs *fascicularis*, male vs female, and old vs young animals. The other type of information is a summation of all the frames of film in which a particular component was used, expressed as a percentage of the total number of frames analysed for each species. These figures express the rate at which particular components are used in the overall group patterns of a threat. An example of the difference is the calculation that, although the *M. fuscata* threat *Eyelid flicker* occurred in 78.2% of the units, it was only present in 10.2% of the frames. Therefore, it was a brief, but presumably informative, component in a large proportion of the units expressing threat. This difference between frequency and rate allows us to discriminate between high levels of use by one or two animals, and consistent use by the entire group.

In order to study the level of similarity or difference in the threat patterns of the three species under consideration, a number of questions must be formulated. They can be summarised as:

1. Do these three macaque species use the same pattern of threat?
2. Does differential use of a constant component characterise classes based on species, age or sex?
3. Is the variability seen in these facial gestures random?

My analysis of the data on component use indicates that all three macaque species use basically the same components in quite different patterns. Various types of analysis reveal organisation of component use that differs by species, and within species by age and sex. The only cross-species grouping that seems to have a pattern is the rate of component use by sub-adults. Therefore, all three species have individual patterns of threat gestures. In addition, analysis by kin group for *M. sylvanus* reveals a clear discrimination of kin groups, suggesting that this category is also important. Discrimination of the social categories age, sex and kin group by component use in macaques may well reveal some aspects of how they order their social systems.

Data

Component use

It was discovered that none of the animals used all 34 of the components possible in a threat gesture (Tables 24.2, 24.3, 24.4). This was very informative because it meant that there is a pool of components on which the animals can draw to construct their threats. The range in number of components over all species was from 7 to 30 of the 34 available. However, the overlap between species was almost complete.

Barbary and long-tailed macaques did not use *Upper lip back* or *Lower jaw displaced left* in threat, while Japanese macaques did. Japanese macaques did not use *Mandrill mouth*, the enigmatic infinity smile, which was seen at a low level in *sylvanus*, though not in *fascicularis*. However, the

Table 24.2. *List of components in descending order of use for*
M. sylvanus – *Threat (from Zeller, 1994)*

Components[a]	1[b]	2	3
1. *Eyes staring*	65	97.01	51.9
2. *Eyebrows raised*	64	95.52	46.4
3. *Piloerection*	62	92.53	80.0
4. *Eyebrows forward*	61	91.04	37.4
5. *Ears back*	58	86.56	52.9
6. *Mouth open threat*	56	83.58	20.2
7.[a] *Mouth stretch*	55	82.09	14.4
8. *Nostrils flared*	52	77.61	22.9
9. *Head raised*	52	77.61	27.8
10. *Eyelid flicker*	50	74.63	20.5
11. *Head forward*	50	74.63	40.0
12.[a] *Lower jaw protrude*	44	65.67	10.8
13. *Ears forward*	43	64.18	24.9
14. *Mouth closed*	40	59.70	6.9
15.[a] *Mouth forward*	30	44.78	10.7
16.[a] *Mouth protrude*	26	38.80	7.9
17.[a] *Upper jaw protrude*	25	37.31	9.5
18. *Mouth open grimace*	17	25.37	2.1
19.[a] *Skin around eyes back*	11	16.41	1.5
20. *Incisors show*	8	11.94	1.5
21.[a] *Upper lip puffy*	7	10.45	0.7
22.[a] *Mouth puffy*	6	8.96	0.5
23. *Tongue protrude*	6	8.96	0.5
24. *Lower canine*	5	7.46	1.0
25. *Tongue show*	5	7.46	0.7
26.[a] *Rear crouch*	3	4.48	0.5
27.[a] *Mouth open wider right*	3	4.48	0.2
28.[a] *Mandrill mouth*	2	2.99	0.1
29.[a] *Mouth open wider left*	2	2.99	0.1
30. *Upper canine*	2	2.99	0.3
31.[a] *Lower jaw displaced right*	1	1.49	0.1
32.[a] *Body crouch*	1	1.49	0.6
33.[a] *Upper lip back*	0	0	0
34.[a] *Lower jaw displaced left*	0	0	0

[a]The components indicated are the 16 components discovered by detailed observation of the data. Breaks in the column emphasise major differences in the frequency of use. Note differences between species.
[b]1, Number of units (out of 67) in which component occurred; 2, Frequency of units in which component occurred (%); 3, Frequency of frames in which component occurred (%).

Table 24.3. *List of components in descending order of use for*
M. fuscata – *Threat (from Zeller, 1994)*

Components[a]	1[b]	2	3
1. *Eyes staring*	69	100	81.3
2. *Piloerection*	67	97.1	81.3
3. *Mouth closed*	67	97.1	54.1
4. *Nostrils flared*	62	89.8	55.0
5.[a] *Mouth stretch*	61	88.4	82.1
6. *Mouth open threat*	57	82.6	42.1
7. *Eyelid flicker*	54	78.2	10.2
8.[a] *Mouth forward*	54	78.2	28.9
9. *Eyebrows raised*	51	73.9	50.1
10. *Eyebrows forward*	51	73.9	33.0
11. *Ears forward*	50	72.4	39.7
12. *Ears back*	48	69.5	36.2
13. *Head forward*	48	69.5	48.5
14.[a] *Lower jaw protrude*	33	47.8	23.5
15.[a] *Upper jaw protrude*	30	43.4	10.1
16.[a] *Mouth protrude*	29	40.2	10.7
17.[a] *Lower jaw displaced right*	21	30.4	4.5
18. *Head raised*	17	24.6	5.6
19.[a] *Mouth puffy*	17	24.6	8.6
20.[a] *Skin around eyes back*	17	24.6	6.1
21.[a] *Upper lip puffy*	15	21.7	2.3
22.[a] *Mouth open wider left*	14	20.2	1.5
23.[a] *Lower jaw displaced left*	12	17.3	1.5
24. *Incisors show*	11	15.9	4.7
25.[a] *Mouth open wider right*	9	13.0	1.9
26. *Mouth open grimace*	7	10.1	1.2
27. *Lower canine*	5	7.2	0.5
28.[a] *Body crouch*	5	7.2	1.3
29. *Upper canine*	4	5.7	0.5
30.[a] *Rear crouch*	3	4.3	1.5
31.[a] *Upper lip back*	3	4.3	0.2
32. *Tongue show*	2	2.8	0.2
33. *Tongue protrude*	2	2.8	0.04
34.[a] *Mandrill mouth*	0	0	0

[a]The components indicated are the 16 components discovered by detailed observation of the data. Breaks in the column emphasise major differences in the frequency of use. Note differences between species.
[b]1, as for Table 24.2 but out of 69; 2 and 3 as for Table 24.2.

Table 24.4. *List of components in descending order of use for*
M. fascicularis – *Threat (from Zeller, 1994)*

Components[a]	1	2	3
1. *Mouth open threat*	80	93.0	51.2
2.[a] *Mouth stretch*	78	90.6	76.3
3. *Mouth closed*	78	90.6	37.5
4. *Eyebrows raised*	76	88.3	68.0
5. *Nostrils flared*	74	86.0	71.5
6. *Piloerection*	73	84.8	82.6
7. *Eyelid flicker*	72	83.7	16.1
8. *Eyes staring*	71	82.5	75.2
9. *Ears back*	67	77.9	64.0
10. *Head raised*	58	67.4	36.1
11. *Head forward*	56	65.1	42.9
12. *Eyebrows forward*	44	51.1	29.4
13. *Ears forward*	40	46.5	30.1
14.[a] *Upper jaw protrude*	40	46.5	25.1
15. *Mouth open grimace*	33	38.3	7.0
16.[a] *Skin around eyes back*	26	30.2	17.0
17.[a] *Mouth protrude*	25	29.0	14.2
18. *Tongue show*	23	26.7	2.2
19. *Incisors show*	18	20.9	4.1
20. *Tongue protrude*	16	18.9	1.7
21.[a] *Upper lip puffy*	13	15.1	3.2
22.[a] *Mouth forward*	11	12.7	7.2
23. *Upper canine*	9	10.4	1.9
24. *Lower canine*	9	10.4	1.6
25.[a] *Mouth open wider left*	6	6.9	1.5
26.[a] *Lower jaw protrude*	6	6.9	3.1
27.[a] *Mouth puffy*	4	4.6	2.7
28.[a] *Rear crouch*	3	3.4	1.5
29.[a] *Lower jaw displaced right*	3	3.4	0.45
30.[a] *Mouth open wider right*	2	2.3	0.18
31.[a] *Body crouch*	2	2.3	0.96
32.[a] *Mandrill mouth*	0	0	0
33.[a] *Upper lip back*	0	0	0
34.[a] *Lower jaw displaced left*	0	0	0

[a]The components indicated are the 16 components discovered by detailed observation of the data. Breaks in the column emphasise major differences in the frequency of use. Note differences between species.
[b]1, As for Table 24.2 but out of 86 (note this is not as published in Zeller, 1994); 2 and 3 as for Table 24.2.

Table 24.5. *Relative use of components by species during units*

	sylvanus	fuscata	fascicularis
Mouth open grimace		−	+
Mouth closed	−	+	+
Tongue show		−	+
Eyebrows raised	+	−	
Eyebrows forward	+	+	−
Head raised	+	−	+
Lower jaw protrude	+	+	−
Mouth forward	+	+	−
Mouth puffy		+	−
Lower jaw displaced right	−	+	−

Plus sign indicates a component used by a species in at least 20% more units compared to cases marked with a minus sign.

other 31 components recognised as contributing to threats were seen in all three species, although at different frequencies and rates.

Some 14 of the total list of 34 components were established as contributing to threat by this research (Zeller, 1978, 1980, 1986, 1987, 1994).

Component use during units

Component use analysis in units (Table 24.5) revealed that there were three components used in at least 20% more units by *M. sylvanus* than by *M. fascicularis* and two used in at least 20% more units than by *M. fuscata*. The component *Lower jaw protrude* showed the most marked difference between *sylvanus* and any other species, occurring in 58.8% more units for Barbary than for long-tailed macaques. This was closely followed by *Head raised* used in 53% more units by Barbary than by Japanese monkeys. *Macaca fuscata* used two components more frequently than did *M. sylvanus* and five more frequently than did *M. fascicularis*. The greatest divergence between the species occurred in *Mouth forward*, which Japanese macaques used in 65.5% more units than did *M. fascicularis*. *Macaca fascicularis* used one component, *Mouth closed*, at over 30% more than did *M. sylvanus*, and three components including *Head raised* at a higher level than did *M. fuscata*. If low is taken to represent a use level at least 30% less than both other species, the pattern here suggests a low use of *Mouth closed* by *M. sylvanus*, a low use of *Head raised* by *M. fuscata*, and a low use of *Lower jaw protrude* by *M. fascicularis*. The exceptionally frequent (at least 30% higher) components included *Mouth forward* and *Lower jaw displaced*

Table 24.6. *List distinguishing mouth from non-mouth components*

Mouth components	Non-mouth components
Mouth open threat	*Nostrils flared*
Mouth open grimace	*Eyelid flicker*
Mouth closed	*Eyebrows forward*
Incisors show	*Eyebrows raised*
Upper canine	*Eyes staring*
Lower canine	*Ears back*
Tongue show	*Ears forward*
Tongue protrude	*Head forward*
Mouth stretch	*Head raised*
Lower jaw protrude	*Piloerection*
Upper jaw protrude	*Rear crouch*
Mouth forward	*Body crouch*
Mouth protrude	*Skin around eyes back*
Upper lip puffy	
Mouth puffy	
Mandrill mouth	
Upper lip back	
Lower jaw displaced right	
Mouth open wider right	
Lower jaw displaced left	
Mouth open wider left	

right by Japanese macaques. In addition, *Lower jaw displaced left* is used in 17.3% of units by Japanese macaques and not at all by the other two species. The data suggest that mouth components (Table 24.6) are more frequent than non-mouth components as indicators distinguishing between the threat patterns of the three species at a ratio of 2:1. Previous research on Barbary macaques has revealed that mouth components show up significantly more frequently as the variable elements of a threat gesture pattern within that species (Zeller, 1986).

Component use in frames

Repeating this pattern of analysis utilising the number of frames in which each component appeared (Table 24.7) reveals nine components that show more than a 20% difference in occurrence rate. Of these, *M. sylvanus* demonstrated more frequent use of one component, *Head raised*, and 20% less frequent use of six components. *Macaca fuscata* used five components more frequently and two less frequently than did one or both of the others, while *M. fascicularis* used eight more frequently and one less frequently

Table 24.7. *Relative use of components by species, comparing frames*

Component	sylvanus	fuscata	fascicularis
Mouth open threat	−		+
Mouth closed	−	+	+
Nostrils flared	−	+	+
Eyes staring	−	+	+
Eyebrows raised	−		+
Ears back		−	+
Head raised	+	−	+
Mouth stretch	−	+	+
Mouth forward		+	−

Plus sign indicates a component used by a species in at least 20% more frames compared to cases marked by a minus sign.

(Table 24.7). The largest level of difference was the 67.7% separating the incidence of *Mouth stretch* in *M. sylvanus* and *M. fuscata*.

The patterns of overall difference between the three species do not reveal any component used more frequently (over 30% more) than the other two. However, *M. sylvanus* used three components at least 30% less than did the other species. These components were *Mouth closed*, *Nostrils flared* and *Mouth stretch*, which is used 61.9% less by *M. sylvanus* than by the next nearest species. The pattern of differential component use to express variability between species is maintained by the analysis of frame frequency. At a 20% difference level higher use rates for components occurred 14 times and lower rates 9 times in between-species comparisons (Table 24.7). Therefore, both component use in units and frame counting should reveal substantial levels of difference in the component use patterns of the three species.

Use of constant components

Due to the common perception of threat as utilising a basic core of components, analysis of those components used in every unit of threat by different species, ages and sex will allow comparisons between groups. However, when this idea of a core of components is examined, only one species, *M. fuscata*, uses one component, *Eyes staring*, in all threat expressions. Constant components were, therefore, defined as those used in all or all but one unit by every individual in the group. However, there were a small number of components used regularly by all members of each species, although they were by no means the same components for each species (Table 24.8). In fact, overlap occurred in only two of the six

Table 24.8. *Constant components used by each species*
(from Zeller, 1994)

Component	*sylvanus*	*fuscata*	*fascicularis*
Eyes staring	+	+	
Eyebrows raised	+		
Eyebrows forward	+		
Piloerection	+	+	
Mouth closed		+	
Mouth open threat			+
Mouth stretch			+

components used regularly. Another feature of this analysis was the realisation that long-tailed macaques use only mouth components as constants, while six of the seven used by the other species were the non-mouth type. This understanding of the tremendous variability in precisely which components are used regularly by a species helps to emphasise the level of difference distinguishing the codes utilised by members of the same genus.

The contrast between species in rates and frequencies of use uncovered by the species component analysis above is confirmed by examination of the most consistent aspects of threat by sex (Table 24.9). For *M. sylvanus*, only one of the six constant components used by males was also used constantly by females. The females had only one that they used constantly and the males did not. Japanese macaque males also used a higher number of constant components than did conspecific females and three of these distinguished the sexes. However, there were more constant components exhibited by *M. fascicularis* females than by the males, and four out of five of these distinguished their patterns of threat expression. In other words, *M. sylvanus* and *M. fuscata* males showed more consistency in their threat patterns than did the females, while the reverse was true for the long-tailed macaques. Also, the level of similarity between male and female constant component use in *M. sylvanus* suggested a highly similar threat pattern. In this analysis by sex it is also interesting to note that use of the non-mouth constant components occured 16 times compared to only 6 uses of mouth components, reinforcing the suggestion that there may be a difference in function of movements occurring in different regions of the face.

Table 24.10, which details the pattern of constant component use differentiated by age also suggests similar trends. This material is less clearly demarcated due to two factors. One is the fact that the average life span of *sylvanus* in Gibraltar seemed to be much shorter than the 20 to 25

Table 24.9. Constant components used differentiated by species and sex

	Male			Female		
	sylvanus	*fuscata*	*fascicularis*	*sylvanus*	*fuscata*	*fascicularis*
Eyes staring	+	+	+			
Eyebrows raised	+				+	+
Eyebrows forward	+			+		
Head raised	+					
Ears back	+					+
Eyelids flicker	+					+
Piloerection		+		+	+	
Mouth open threat		+	+			+
Mouth closed		+				+
Mouth stretch		+				

Table 24.10. *Consistent component use differentiated by age class across species*

	Old			Young			Sub-adult	
	s	fu	fa	s	fu	fa	s	fa
Eyes staring	+	+	+	+	+		+	+
Mouth open threat	+		+		+	+	+	+
Piloerection	+	+		+	+			
Head raised	+						+	
Ears back					+	+		
Head forward					+			
Eyebrows raised	+					+		
Eyelids flicker						+		
Eyebrows forward							+	+
Mouth stretch			+		+			+
Mouth closed		+			+			+

s, *Macaca sylvanus*; fu, *M. fuscata*; fa, *M. fascicularis*.

years regularly achieved by *M. fuscata* and *M. fascicularis*. Barbary macaques in Gibraltar rarely lived past 15 years. Therefore, I have grouped their ages as 3–4 years sub-adults, 5–8 years young adults and more than 9 years as older animals. In *M. fuscata* and *M. fascicularis* I placed the boundary between young adults and older animals at 11 years of age, and classed as sub-adults those *M. fascicularis* who were 3–4 years old. Some of the *M. fascicularis* were over 20 years old by estimate, but no actual records of their ages had been kept. This lack of records is the other factor that may have caused confusion in the *M. fascicularis* data.

In Table 24.10, the almost universal occurrence of *Eyes stare* and *Mouth open threat* reduces their utility in discriminating categories. Of the components remaining, old *sylvanus* constantly used *Eyebrows raised*, which differed from the young adult and sub-adult categories. The other components overlapped, one with young adults and one with sub-adult animals. Old *fuscata* did not use any components that distinguished them from young adult ones, but the young adult category exhibited regular use of four components that distinguished them from old animals. These were *Mouth open threat*, *Ears back*, *Head forward* and *Mouth stretch* on a regular basis, which served to discriminate their category. Therefore six of the nine consistent components that distinguish age classes are non-mouth, which supports the argument of different functions for components in different face regions, with non-mouth ones forming the predictable basis of an expression. In answer to the second question above (p. 532), each species

Table 24.11. *Separation by kin group on the basis of overall use (prevalent components), in* Macaca sylvanus *(from Zeller, 1994)*

Threat

| | \multicolumn{4}{c|}{Kin group} | Non-kin |
	W	B	M1	O	A1
Nostrils flared	+				
Head raised	+				
Lower jaw protrude	+				
Eyelid flicker	+				
Ears forward		+			
Upper jaw protrude		+			
Mouth stretch		+			
Mouth protrude			+		
Mouth forward			+		
Mouth closed				+	
Eyes staring				+	
Mouth open grimace					+
Incisors show					+
Rear crouch					+
Mouth open threat					+

Friendly

| | \multicolumn{4}{c|}{Kin group} | Non-kin |
	W	B	O	M1	A1
Mouth stretch	+				
Eyes staring	+				
Incisors show	+				
Upper canine	+				
Nostrils flared	+				
Ears forward	+				
Eyebrows raised		+			
Upper jaw protrude		+			
Lower jaw protrude			+		
Low canine				+	
Eyelid flicker				+	
Ears back					+
Head forward					+
Mouth closed					+

Table 24.11. (*cont*).

	Fear		
	Kin group		
	W	B	O
Nostrils flared	+		
Eyebrows forward	+		
Head forward	+		
Head raised	+		
Piloerection	+		
Lower jaw protrude	+		
Ears back		+	
Incisors show			+
Upper canine			+
Lower canine			+
Tongue show			+
Eyelid flicker			+
Upper jaw protrude			+

exhibits some components that distinguish between the ages and sexes of its members owing to the consistency with which they are used.

Another feature of research on *M. sylvanus* is the analysis of kin group, which I was able to undertake on this species. Due to more extensive data collection this analysis could be carried out for threatening, friendly, and fear expressions. The group had four kin lines and one non-kin individual. As Table 24.11 indicates, kin group could be distinguished by overall use pattern in all three types of expressions analysed. This type of analysis was called *prevalent use* analysis in which the percentage of frames in which an individual used a particular component was compared with the group mean for that component. The component was considered prevalent for a particular kin group if all members of the group used it at least 5% more than the mean group, and it was not used at that level by other kin groups. There were four kin lines and one non-kin individual present in this group. The non-kin individual, A1, was clearly distinguished from the others. In analysis of the friendly expressions only one member of kin group M was recorded; therefore, they are referred to as M1. There is a smaller number of kin groups indicated in the fear expression category because some animals did not express enough fear to be included in the analysis. This discrimination of kin groups has not been demonstrated in the other species due to lack of data, but it may well exist.

Pattern and functions of variability in facial gesture

Comparison of species

The initial discovery of this research is the high level of variability present in facial expressions of threat exhibited between these species of macaque. Although they all use a very similar selection of components, they do so in very different ways. When the number of frames in which various components occur are compared between species, it becomes clear that they all use piloerection at a very high rate. No other component is used by all three species at levels of over 80%. Japanese macaques and long-tails both use *Eyes staring* and *Mouth stretch* threat frequently, but *M. sylvanus* rates are much lower. In Tables 24.2, 24.3 and 24.4 a marked break is evident in all three species between those components occurring in units at rates of well over 50% and those occurring in about 50% and under. The frequently used ones range from 11 to 14 components out of 34 available for each species. Of the possible total of 14 components used at these high rates, 10 are used by all three species, which would certainly provide them with a common basis for inter-specific communication. Low rates of use for certain components probably distinguishes a species more clearly than high use rates. *Macaca sylvanus* shows low use rates of many components when compared with the other two species. In fact, Japanese and long-tailed macaques seem to show more similarity with each other in higher rates of component use than either shows with *M. sylvanus*. In the pattern of constant component use, Barbary macaques show more within-group consistency in their regular use of four components. Japanese macaques also use *Eyes staring* as a constant like *M. sylvanus*, but unlike them use a mouth component, *Mouth closed*, as their other regular expressive feature. In this they more closely resemble *M. fascicularis*, who use two mouth components as their discriminating consistent components.

The fact that the three species can be discriminated by use of particular constant components suggests a regularity of pattern difference between the species. The presence of an initial level of species codification, as a regular aspect of the expressions of all adult members, forms a foundation for the overlay of patterns due to other social factors.

Comparison of males and females

Examination of sex differences within the species by constant component use reveals discrimination by sex in the use of particular components.

However, this does not result in a particular 'male' or 'female' pattern of component use in macaques' threats.

If a 'male' or 'female' pattern across the three species had been revealed, it would have suggested either a very stable evolutionary basis underlying a long continued pattern arising from a common past, or a very tight functional relationship between the components required by each sex to send its particular version of the message. Since, under most circumstances, animals of various species are not communicating with one another, this lack of pattern universality does not interfere with information transfer. In some situations where mixed species groups exist, younger animals may experience some difficulty in learning appropriate codes (F. D. Burton, personal communication). The only type of pattern consistency revealed by sex discrimination is the use of more constant components by male *M. fuscata* and *M. sylvanus* than by females. Given the different life strategies of males and females, a base level of pattern consistency on the part of males, who may require a somewhat simplified pattern when moving into a new group, may reduce ambiguity. This suggestion is not supported by the work on *M. fascicularis*; however, no new macaques have moved into that particular captive population for the last 50 years, and all the males are group-born.

Comparison of age classes

Constant component use did not distinguish between the threat patterns of old, young and sub-adults considered across all three species. Two components were used in at least six of the eight possible age categories and several of the other components were constants in more than one age class. However, when each species was compared with itself differences in age-class markers were evident. The sub-adult category had only a few animals but exhibited quite a distinct difference to the threat pattern of adults. Their class distinguishing component was *Eyebrows forward*, which sub-adults of both species use in contrast to the other age classes.

Noise or pattern?

One interpretation of the variability present in facial communication is the idea that expressions are constructed from a basic core of elements accompanied by random inclusions of components. These random extras would then merely be 'noise' in the system. Since, as demonstrated above, there is no irreducible component core for each type of message in any of the

Table 24.12. *Chi square analysis of constant and variable component use by attribution as mouth and non-mouth location, in* M. sylvanus *(from Zeller, 1994)*

	Constant component	Variable component
Threat		
Mouth	6	98
Non-mouth	39	58
d.f. = 1	$\chi^2 = 34.17$	$P < 0.001$
Friendly		
Mouth	17	51
Non-mouth	45	45
d.f. = 1	$\chi^2 = 9.55$	$P < 0.001$
Fear		
Mouth	35	26
Non-mouth	13	30
d.f. = 1	$\chi^2 = 7.48$	$P < 0.01$

three species, this interpretation is probably not correct. If each individual's set of components functioned in this way, it would imply that components that are vital to one animal's expressions are 'noise' in another's. This would lead to the conclusion that each animal has its own code system, which would be a very complex and extravagant way for a communication system to operate.

As an alternative, I would like to propose that the use of components is a regularised phenomenon. The choice of components is not random, as the previous interpretation suggests, but structured according to facial location. In particular, those components that tend to characterise a particular social category tend to be mouth components. This can be seen in the pattern of constant component use referred to above. In addition, the analysis of friendly and fearful interactions, which has only been undertaken for *M. sylvanus*, also reveals non-mouth and mouth association with constant and variable components, respectively (variable components are those only used in some units by an individual, Table 24.12). These preliminary results from analysing different types of messages suggest that this pattern of meaning associated with facial location is not unique to threat expressions.

There are certainly differences among the macaque species in how closely they adhere to this pattern. The long-tailed macaques do use mouth components as their group constant ones, but these are at a much lower level of consistency than is found in the other two species. The importance of these discoveries is two-fold. First, it means that facial gestures occur over a substrate of anatomical localisation for different aspects of a message. In other words, the animals can pay attention to the non-mouth regions of the face in order to acquire the basic 'information' of the message and to the mouth region for the markers indicating age, sex and perhaps kin group of the sender. This ability to discriminate social category would be very valuable to an animal, such as a male, moving into the group from the outside. It would also assist youngsters growing up in a large group to learn the social identities of other animals. A second asset of this system is that it would permit animals such as sub-adults to indicate unambiguously their status. This may be very useful to them because if they can clearly express themselves as non-adults they may be less likely to provoke a punitive response from an adult. Since the borderline between a sub-adult and an adult male can be one of self-perception and self-presentation rather than anatomical maturity, this may allow less confident animals to remain unmolested in their natal groups a little longer, before they move out to find a new social unit.

The ubiquitous nature of the mouth/non-mouth discrimination, and the presence of age and sex markers for each species clearly supports an argument that the complex levels of variation seen in macaque facial gestures are not merely random additions of meaningless components. This realisation parallels the discovery that vocal communication systems exhibit patterned variability that can change the meaning of very similar sounding calls or can be used as individual markers (Haimoff & Tilson 1985). In some cases the researchers are not aware of the nature of the differences in the calls but find that the animals respond to them in such a way that attribution to specific individuals can be deduced (Seyfarth & Cheney, 1988). Some individual vocal markers are being discovered but the system underlying them is not clear. This facial gesture research does not focus on individual markers, but instead reveals discriminators based on social class that can operate to permit animals to learn a system and then to classify individuals in terms of it. The evidence suggesting that this is true for a number of macaque species supports an argument that differential component use may be a widespread feature of facial gesture systems. The complexity of discrimination by age, sex, and kin group maybe more intense for macaques because of their relatively large, complex and long-lasting groups, compared with pairs or single male groups. Also, as mentioned above, the lack of differential colouring in the face may be a factor affecting the intricacy of gesture patterns. For these reasons macaques may demonstrate one of the higher levels of complexity in monkey facial gesture systems.

Acknowledgements

I acknowledge the generosity of several people who have helped make my research possible. These include Frances Burton, for helping me establish contact with the Gibraltar macaques, Linda Fedigan for inviting me to visit the *M. fuscata* at Arashiyama West, and the owners and staff at Monkey Jungle and the DuMond Conservancy for their support during my study of *M. fascicularis* at their facility.

References

Andrew, R. J. (1963*a*). Evolution of facial expression. *Science*, **142**, 1034–41.

Andrew, R. J. (1963*b*). Trends apparent in the evolution of vocalization in the Old World monkeys and apes. *Symposia of the Zoological Society of London*, **10**, 89–101.

Andrew, R. J. (1963*c*). Origin and evolution of the calls and facial expressions of the primates. *Behaviour*, **20**, 1–109.

Brown, J. G. (1982). Auditory localization and primate vocal behaviour. In *Primate communication*, ed. C. T. Snowdon, C. H. Brown & M. R. Petersen, pp. 144–64. Cambridge: Cambridge University Press.

Cheney, D. L., & Seyfarth, R. M. (1980). Vocal recognition in free ranging vervet monkeys. *Animal Behaviour*, **28**, 362–7.

Cheney, D. L. & Seyfarth, R. M. (1981). Selective forces affecting the predator alarm calls of vervet monkeys. *Behaviour*, **76**, 25–61.

Cheney, D. L. & Seyfarth, R. M. (1982*a*). How vervet monkeys perceive their grunts. Field playback experiments. *Animal Behaviour*, **30**, 739–51.

Cheney, D. L. & Seyfarth, R. M. (1982*b*). Recognition of individuals within and between groups of free ranging vervet monkeys. *American Zoologist*, **22**, 519–29.

Cheney, D. L. & Seyfarth, R. M. (1990). *How monkeys see the world*. Chicago: University of Chicago Press.

Chevalier-Skolnikoff, S. (1974). *The ontogeny of communication in the stump-tail macaque* (Macaca arctoides). Basel: S. Karger.

Chevalier-Skolnikoff, S. (1982). A cognitive analysis of facial behaviour in Old World monkeys, apes and human beings. In *Primate communication*, ed. C. T. Snowdon, C. H. Brown & M. R. Petersen, pp. 303–62. Cambridge: Cambridge University Press.

Deputte, B. L. (1982). Dueting in male and female songs of the white-cheeked gibbon (*Hylobates concolor leucogenys*). In *Primate communication*, ed. C. T. Snowdon, C. H. Brown & M. R. Petersen, pp. 67–93. Cambridge: Cambridge University Press.

Epple, G., Alveario, M. C. & Katz, Y. (1982). The role of chemical communication in aggressive behaviour and its gonadal control in the tamarin (*Saguinis fusicollis*). In *Primate communication*, ed. C. T. Snowdon, C. H. Brown & M. R. Petersen, pp. 279–302. Cambridge: Cambridge University Press.

Gouzoules, S., Gouzoules, H. & Marler, P. (1984). Rhesus monkey (*Macaca mulatta*) screams. Representational signalling in the recruitment of agonistic aid. *Animal Behavior*, **32**, 182–93.

Green, S. (1975). Communication by a graded vocal system in Japanese monkeys. In *Primate behavior*, vol. 4, ed. L. A. Rosenblum, pp. 1–102. New York: Academic Press.

Haimoff, E. H. & Tilson, R. L. (1985). Individuality in the female songs of wild Koss's gibbon *Hylobates klossi*, Siberut Island, Indonesia. *Folia Primatologica*, **44**, 129–37.

Hertzog, M. O. & Hohmann, G. M. (1984). Male loud calls in *Macaca silenus* and *Prebbytis johnii*: a comparison. *Folia Primatologica*, **43**, 189–97.

Hinde, R. (1964). Ritualization and social communication in rhesus monkeys. *Philosophical Translations of the Royal Society of London*, **251**, 285–94.

Hohmann, G. (1991). Comparative analyses of age- and sex-specific patterns of vocal behaviours in four species of Old World monkeys. *Folia Primatologica*, **56**, 133–56.

Marriott, B. M. & Salzen, E. A. (1978). Facial expressions in captive squirrel monkeys (*Saimiri sciereus*). *Folia Primatologica*, **29**, 1–18.

Newman, J. D. & Symes, D. (1982). Inheritance and experience in the acquisition of primate acoustic behavior. In *Primate communication*, ed. C. T. Snowdon, C. H. Brown & M. R. Petersen, pp. 259–78. Cambridge: Cambridge University Press.

Raemakers, J. J. & Raemakers, P. M. (1984). The Oooa duet of the gibbon (*Hylobates lar*). A group call which triggers other groups to respond in kind. *Folia Primatologica*, **42**, 209–15.

Robinson, J. G. (1982). Vocal systems regulating within-group spacing. In *Primate communication*, ed. C. T. Snowdon, C. H. Brown & M. R. Petersen, pp. 94–116. Cambridge: Cambridge University Press.

Rowell, T. E. (1972). *Social behaviour of monkeys*. Harmondsworth: Penguin.

Rowell, T. E. & Hinde, R. A. (1962). Vocal communication by the rhesus monkeys (*Macaca mulatta*). *Proceedings of the Zoological Society of London*, **138**, 279–94.

Sackett, G. P. (1966). Monkeys reared in isolation with pictures as visual input: evidence for an innate releasing mechanism. *Science*, **154**, 1470–3.

Schilling, A. (1974). A study of marking behaviour in *Lemur catta*. In *Prosimian biology*, ed. R. D. Martin, G. A. Doyle & A. C. Walker, pp. 347–62. London: Duckworth Press.

Schilling, A. (1979). Olfactory communication in primates. In *The study of prosimian behaviour*, ed. G. A. Doyle & R. D. Martin, pp. 461–542. New York: Academic Press.

Seyfarth, R. M. & Cheney, D. L. (1980). The ontogeny of vervet monkey alarm calling behaviour. A preliminary report. *Zeitschrift für Tierpsychologie*, **54**, 37–56.

Seyfarth, R. M. & Cheney, D. L. (1984). The acoustic features of vervet monkey grunts. *Journal of the Acoustic Society of America*, **75**, 1623–8.

Seyfarth, R. M. & Cheney, D. L. (1986). Complexities in the study of vervet monkey grunts. In *Current perspectives in primate social dynamics*, ed. D. M.

Taub & F. A. King, pp. 378–88. New York: Van Nostrand Reinhold.

Seyfarth, R. M. & Cheney, D. L. (1988). Do monkeys understand their relations. In *Machiavellian intelligence: social expertise and the evolution of intellect in monkeys, apes and humans*, ed. R. W. Byrne & A. Whiten, pp. 69–93. Oxford: Oxford University Press.

Seyfarth, R. M., Cheney, D. L. & Marler, P. (1980). Vervet monkey alarm calls. Semantic communication in a free ranging primate. *Animal Behaviour*, **28**, 1070–94.

Shorey, H. H. (1976). *Animal communication by pheromones*. London: Academic Press.

Smith, H. J., Newman, J. D. & Symes, D. (1982). Vocal concomitants of affiliative behaviour in squirrel monkeys. In *Primate communication*, ed. C. T. Snowdon, C. H. Brown & M. R. Petersen, pp. 30–49. Cambridge: Cambridge University Press.

Struhsaker, T. (1967). Auditory communication among vervet monkeys (*Cercopithecus aethiops*). In *Social communication among primates*, ed. S. A. Altmann, pp. 281–324. Chicago: University of Chicago Press.

van Hooff, J. A. R. A. M. (1962). Facial expressions in higher primates. *Symposia of the Zoological Society of London*, **8**, 97–125.

van Hooff, J. A. R. A. M. (1967). The facial displays of the catarrhine monkeys and apes. In *Primate ethology*, ed. D. Morris, pp. 7–68. London: Weidenfield & Nicholson.

Waser, P. M. (1977). Individual recognition, intra group cohesion and spacing. Evidence from sound playback to forest monkeys. *Behaviour*, **60**, 28–74.

Waser, P. M. (1982). The evolution of male loud calls among mangabeys and baboons. In *Primate communication*, ed. C. T. Snowdon, C. H. Brown & M. R. Petersen, pp. 117–43. Cambridge: Cambridge University Press.

Zeller, A. (1978). Film analysis as a field technique. A study of communication in *Macaca sylvanus* of Gibraltar. PhD thesis, University of Toronto.

Zeller, A. (1980). Primate facial gestures. A study of communication. *International Journal of Human Communication*, **13**, 565–606.

Zeller, A. (1986). Comparison of component patterns in threatening and friendly gestures in *Macaca sylvanus* of Gibraltar. In *Current perspectives in primate social dynamics*, ed. D. M. Taub & F. A. King, pp. 487–504. New York: Van Nostrand Reinhold.

Zeller, A. (1987). Communication by sight and smell. In *Primate societies*, ed. B. B. Smuts, D. L. Cheney, R. M. Seyfarth, R. W. Wrangham & T. T. Struhsaker, pp. 433–9. Chicago: University of Chicago Press.

Zeller, A. (1994). Evidence of structure in macaque communications. In *The ethological roots of culture*, ed. R. A. Gardner, B. Gardner, B. Chiarelli & F. Plooij, pp. 15–39. Dordrecht: Kluwer Academic Publishers.

25

Vocal communication in macaques: causes of variation

M. D. HAUSER

Introduction

Like so many issues in primatology, studies of vocal communication have been motivated by the 'human uniqueness challenge'. Because human language is, perhaps, the last stronghold for those who wish to defend the uniqueness of *Homo sapiens* in the animal kingdom, it is not surprising that intellectual exchanges have been somewhat heated. Within the past 15 years, however, the explosion of research on non-human primate vocal communication (reviewed by Cheney & Seyfarth, 1990; Snowdon, 1990; Hauser, 1995) has placed primatologists in a substantially stronger position with regard to critically evaluating which aspects of human language are evolutionarily revolutionary, and which are part of the expected continuity based on descent from a common ancestor. Although studies have been conducted on a variety of species, ranging from ring-tailed lemurs (*Lemur catta*) to chimpanzees (*Pan troglodytes*), the macaques (*Macaca*) represent the most comprehensively investigated taxonomic group, both in terms of the number of species and the breadth of topics covered. The diversity of topics is particularly important because it provides the opportunity to assess, for the first time, how ontogeny, phylogeny, proximate causation and evolutionary functions have shaped the structure of communication in the order Primates. It is, therefore, the goal of this chapter to provide a comprehensive review of the factors influencing repertoire variation, both within and between species of macaques. To achieve this goal, I review published material and, where possible, combine results from different species to assess how and why the patterns of communication have diverged and converged.

Ontogeny

Changes in call morphology

A characteristic feature of all human languages is that infants go through a transition from babbling to clearly articulated speech (Oller & Eilers, 1992; Locke, 1993). More specifically, the acoustic morphology of the child's utterances changes from approximations of the adult form to the adult form. Guiding the transition from babbling to speech are changes in neuromotor function, anatomy of the peripheral organs involved in sound production and perception, and experience with the linguistic environment. Over the past 20 years, primatologists have searched for evidence that experience guides the ontogeny of sound production. Unfortunately, as I describe below, there is only limited evidence that experience influences the observed changes in call morphology, and this is the case for macaques as well as other non-human primates (reviewed by Symmes & Biben, 1992; Hauser & Marler, 1992). This finding is paradoxical given the apparent importance of learning in other aspects of primate social development.

One approach used by avian biologists to assess the importance of auditory experience on vocal development is to isolate individuals from species-specific vocal signals (e.g. placing a bird in an acoustic isolation chamber with white noise). Unfortunately, this technique is not well-suited for non-human primates because it is difficult to disentangle social deficits from acoustic deficits. Nonetheless, early work by Newman & Symmes (1974) showed that acoustically isolated rhesus monkeys (*Macaca mulatta*) produced 'coo' vocalisations that differed significantly from those of socially reared individuals.

In the second approach, if learning plays a role in vocal modification, then conditioning techniques should be successful in shifting the acoustic structure of a given call type. Research by Sutton *et al.* (1974) provided limited evidence that rhesus monkey calls are subject to conditioning. Specifically, they showed that in response to a conditioned stimulus, three rhesus monkeys were capable of calling at a higher rate and produced calls of significantly longer duration than during the pre-training period. Unfortunately, no other study has been able to replicate these findings.

A third approach is to search for developmental changes in acoustic morphology that are unlikely to correlate with maturational changes. Gouzoules & Gouzoules (1989) have taken this approach furthest, and focused their research on the recruitment screams of captive pig-tailed macaques (*M. nemestrina*). Specifically, they showed, using a discriminant function analysis, that the calls of infants and juveniles were often misclassified to the inappropriate context. Although there was clear evidence that the pitch of a pig-tail's voice dropped with age – as would be predicted from the relationship between changes in mass of the vocal folds and changes in pitch (for relevant data on Japanese macaques, *M. fuscata*, see Inoue, 1988) – there were also changes in parameters such as bandwidth and frequency modulation that could not be accounted for by simple maturational changes. Therefore, data on pig-tails suggest that learning may play a role in vocal modification.

A fourth approach focuses on geographical variation in the acoustic morphology of a clearly defined call type, or what linguists and avian biologists call 'dialects'. As defined, dialects represent variations that have been learned from experience. For non-human primates, Green (1981) provided some suggestive evidence of dialects within *fuscata* populations, and Masataka (1987) provided evidence of dialects in red-bellied marmoset populations living in Brazil. Two recent studies on macaques, however, provide stronger evidence of dialects. Hauser's (1991) analyses of the rhesus monkey's coo vocalisation revealed that the primary factor influencing acoustic variation was individual identity. Moreover, results from a discriminant function analysis indicated that there was greater similarity between the coos of closely related kin than between those of distantly related individuals. A follow-up study (Hauser, 1992*a*) showed that, in one matriline, all adults consistently produced a perceptually nasal version of the coo; related males, and unrelated males and females, rarely if ever produced nasal coos. What human observers perceived as nasality was associated with significant changes in the spectral properties of the call. Specifically, whereas non-nasal coos were characterised by a series of energy bands or harmonics separated by silence, nasal coos were characterised by energy bands in between the primary harmonics and a significant diminution in overall amplitude between 1 and 2 kHz.

Although we cannot rule out the possibility that perceptually nasal coos are produced as a result of some genetically inherited morphological defect (e.g. cleft palate), four observations support the hypothesis that nasal coos are derived from a learned mode of production. First, other vocalisations in the repertoire were not produced nasally by members of the target matriline, even though the acoustic structure of such vocalisations was not radically different from the structure of coos. Second, related females, but not related males, produced nasal coos. However, it is unlikely that the production of nasal coos is the result of a sex-linked trait. Third, female infants born to members of the nasal matriline produced non-nasal coos. And finally, two females who consistently produced nasal coos have recently moved into a group with no adult female relatives and they no longer produce nasal coos. These observations suggest, therefore, that the production of nasal coos is a learned mode of production – a peculiar dialect of one rhesus matriline.

More recently, Masataka (1992) has carried out a series of analyses on the acoustic structure of coos produced by Japanese macaques living in geographically isolated populations. Acoustic analyses revealed that the fundamental frequency of coos produced by the Yakushima populations was significantly greater than for the Ohirayama population. In addition to the fundamental frequency differences, Masataka also found populational differences in the duration of time between consecutive productions of the coo. Therefore, Masataka's results suggest that there are dialects within Japanese macaque coos and, consequently, that learning plays at least some role in the modification of call structure.

The most powerful technique for looking at experiential influences on call structure comes from cross-fostering studies. To date, two cross-fostering studies have been conducted, both using Japanese macaques and rhesus macaques. Unfortunately, the results are conflicting. Masataka & Fujita (1989) showed that for both cross-fostered rhesus and Japanese macaques, the fundamental frequency of food coos was more like the cross-fostered species than its own species. In contrast, Owren *et al.* (1992) showed that in terms of the spectral properties of food coos, the cross-fostered infants' call was more like that of their own species than like those of their cross-fostered species. Therefore, Masataka & Fujita claimed that learning influences developmental changes in call morphology, whereas Owren and colleagues claimed that learning plays a limited role.

In summary, current research on macaques provides weak support for the role of experience in shaping call structure. The same conclusion is reached for other non-human primate species (Hauser & Marler, 1992), although recent data on callitrichids are somewhat more promising with regard to experiential effects on vocal development (Snowdon, 1990). Future research focusing on geographical variation and, perhaps, different cross-fostered species may shed new light on this paradoxical finding.

Changes in usage and comprehension

In humans, infants must not only learn to produce the sounds of speech, they must also learn the meaning of these sounds, and thus their correct usage. For most human children, comprehension of sound meaning precedes correct usage. Although there is considerable evidence that learning plays a role in the development of correct usage and comprehension in cotton-top tamarins *Saguinus oedipus oedipus,* and vervet monkeys, *Cercopithecus aethiops* (reviewed by Snowdon, 1990; Hauser & Marler, 1992; Symmes & Biben, 1992), data on macaques are much more limited and restricted to the observations of pig-tailed macaque recruitment screams (Gouzoules & Gouzoules, 1989, 1990*a,b*). Specifically, focusing on screams given in four different contexts, a discriminant function analysis revealed that the calls of individuals under the age of 3 years were successfully classified at a rate of only 30%. This classification accuracy stands in striking contrast to the rate of 75% achieved with individuals older than 3 years. Intriguingly, immature females produced screams in the appropriate context more often than did immature males. Together, these results demonstrate the importance of experience in learning to produce agonistic screams in the appropriate context. Playback experiments have not yet been conducted to determine at what age comprehension starts and whether its developmental timing precedes or follows correct usage.

Phylogeny
Similarities between repertoires

At present, it is difficult to assess repertoire size in macaques. There are two reasons for this difficulty. First, all of the existing studies have either provided a fairly qualitative description of the repertoire, including limited acoustic and behavioural analyses (e.g. Green, 1975; Hohmann & Herzog, 1985; Hohmann, 1989; Palombit, 1992*b*), or have provided an in-depth set of analyses on a small sample of calls from the repertoire (e.g. Dittus, 1988; Bauers & de Waal, 1991; Hauser, 1991, 1992*a*; Hauser & Marler, 1993*a*). Second, even in those cases where detailed acoustic analyses have been provided, it is unclear whether the acoustic classifications made by the researcher are perceptually salient to the focal species. As discussed below, there is only limited work on the features used by macaques, and other non-human primates, in classifying calls within the repertoire. In what follows, therefore, I use existing data to provide a rough assessment of some of the commonalities between macaque repertoires.

Table 25.1 is a summary of some of the major contexts in which macaques vocalise. As this table illustrates, there is a great deal of overlap in terms of socio-ecological contexts eliciting vocal signals. Thus, all macaques use vocalisations to signal submission and aggression, although in some species, there appears to be extensive acoustic variation that maps

Table 25.1. *Preliminary assessment of the socio-ecological contexts associated with call production in macaques (Macaca)*[a]

Species	Contact/ Affiliative	'Coo' contact	Group move	Food discovery	Sexual	Submissive	Agonistic recruitment	Dominant aggression	Inter-group aggression	Alert/ predator
M. arctoides[b]	Yes[1]	Yes[1]	?Yes	No	Yes	Yes[1]	Yes	Yes	?	Yes[1]
M. fascicularis[c]	Yes	Yes	Yes[1]	No	Yes	Yes[1]	Yes	Yes	Yes[1]	Yes[1]
M. fuscata[d]	Yes[1]	Yes[1]	Yes	No	Yes	Yes[1]	Yes	Yes[1]	Yes	Yes
M. mulatta[e]	Yes	Yes	Yes	Yes[1]	Yes	Yes[1]	Yes[1]	Yes[1]	Yes	Yes
M. nemestrina[f]	Yes	Yes[1]	?	?	?	Yes	Yes[1]	Yes[1]	?	Yes
M. radiata[g]	Yes	Yes	Yes	Yes	Yes	Yes[1]	Yes	Yes[1]	Yes[1]	Yes
M. silenus[h]	Yes	Yes	Yes	Yes	Yes	Yes[1]	Yes	Yes[1]	Yes[1]	Yes
M. sinica[i]	Yes[1]	Yes[1]	Yes	Yes	?	Yes	?	Yes	Yes	Yes[1]
M. sylvanus[j]	Yes	No	?	No	Yes	Yes	?	Yes	?	Yes

Yes, call is present in the repertoire; No, call is absent from the repertoire; ?, no relevant data on presence/absence of call in the repertoire; [1], sub-contextual variation in call structure.

[a]Data presented were obtained from published material, manuscripts in preparation and personal communications; [b], Bauers, 1989, personal communication; Bauers & de Waal, 1991; [c], Palombit, 1992a; [d], Green, 1975, 1981; [e], Gouzoules et al., 1984; Hauser, 1991, 1992a; Hauser & Marler, 1993a,b; [f], Grimm, 1967; Gouzoules & Gouzoules, 1989; H. Gouzoules & S. Gouzoules, personal communication; [g], Hohmann, 1989; [h], Hohmann & Herzog, 1985; [i], Dittus, 1984, 1988, personal communication, personal observation; [j], Hammerschmidt, Ansorge & Fischer, 1992, personal communication; D. Todt, personal communication; J.-P. Gautier & A. Gautier-Hion, personal communication.

onto sub-contextual variation. Lack of sub-contextual variation may be due to the lack of quantification on the part of the investigator or to the fact that the social organisation is such that contexts are lumped rather than split.

The most noticeable differences can be seen with regard to the 'coo' vocalisation and calls given during the discovery of food. Specifically, only *M. sylvanus* appears to lack a coo, although it does produce structurally different calls in the same context (i.e. general contact). In addition, in some species (e.g. *M. fuscata*, *M. sinica*) there is extensive acoustic variation within the coo class and such variation maps onto changes in the socio-ecological environment. In other species, there is either no variation within the coo class, or existing variation covaries with differences in caller identity rather than in call context. Regarding food-associated calls, only *M. mulatta* appears to have a highly differentiated suite of vocalisations. In other species, individuals apparently do not call when they discover or consume food or the call given is similar to that given in other contexts, thereby suggesting that it conveys information of a much more general sort, such as affiliative contact.

The relationship between frequency, motivational state and body weight

On the basis of the early intuitions of Darwin (1872), Morton (1977, 1982) proposed that the structure of avian and mammalian vocalisations can be explained by the relationship between motivational state and two acoustic parameters, pitch and tonality. Specifically, Morton predicted that in the context of aggression, animals will produce atonal (i.e. noisy) and low pitched vocalisations, whereas in the context of submission or non-aggression animals will produce tonal and high pitched vocalisations. A qualitative test, using published material on birds and mammals, provided support for the relationship between motivational state and acoustic structure, or what Morton called the 'motivational–structural rules'.

Recently, Hauser (1993) has provided a quantitative test of Morton's (1977, 1982) motivational–structural rules with data on non-human primates. The first step in the analysis involved an assessment of the relationship between body weight and pitch – or in acoustic terms, the fundamental frequency. This analysis is important, because large individuals produce relatively lower frequency calls than do small individuals, and because large individuals tend to be dominant over smaller ones, low frequency calls will be produced in the context of aggression and high frequency calls in the context of submission. Using data from 36 species, representing 23 genera, Hauser's analyses provided overall support for the

M.D. Hauser

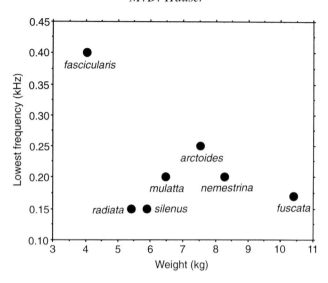

Fig. 25.1. The relationship between body weight (kg) and the lowest frequency (kHz) of a call in the repertoire for seven species of macaque.

relationship between body weight and pitch. As revealed in Figs. 25.1 and 25.2, however, the macaques provide an important exception to the general pattern. Specifically, if one looks at both the lowest and highest frequency calls in the repertoire of each species, body weight explains only a small fraction of the variation. For example (Fig. 25.2), although *M. radiata* has the highest frequency call and is one of the lightest species, *M. fascicularis* and *M. fuscata* have calls that are similar in frequency and yet their average weights differ by 6 kg. Although comparative anatomical work has not yet been conducted on the different macaque species, it is unlikely that these differences are due to micro-anatomical differences in vocal tract morphology. As Hauser (1993) suggests, one factor that may contribute to the differences in call frequency is the species-typical habitat. For example, *M. silenus* lives in a relatively dense habitat and its repertoire consists of low frequency calls. In contrast, *M. radiata* lives in relatively open habitat and produces higher frequency calls. These differences make sense in light of current views on environmental acoustics (e.g. Morton, 1975; Brown & Waser, 1988) which suggest that in dense habitats, low frequency calls are subject to significantly less degradation than high frequency calls.

 To test the relationship between motivational state and acoustic morphology, Hauser (1993) restricted the data set to calls given in two unambiguously described contexts: aggression and fear. This data set,

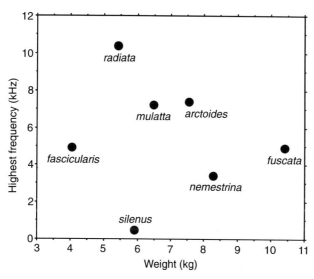

Fig. 25.2. The relationship between body weight (kg) and the highest frequency (kHz) of a call in the repertoire for seven species of macaque.

therefore, excluded putatively affiliative contexts and situations where mixed motivational states were likely. Considering primates in general, there was support for the predicted relationship between motivational state and frequency, but not tonality. When the data set is restricted to the macaques, however, the picture is somewhat less clear (Fig. 25.3). To repeat, Morton's prediction is that high frequency calls should be produced by fearful individuals, whereas low frequency calls should be produced by aggressive individuals. Although all of the macaques except *M. radiata*, produce a majority of high frequency fear vocalisations, only *M. mulatta* and *M. nemestrina* produce low frequency aggressive calls. Therefore, other factors must play a role in explaining the acoustic structure of macaque vocalisations. One factor, discussed in greater detail below, is that the acoustic morphology of the call is related to external objects (e.g. predators, food) and events (e.g. agonistic interactions) in the environment.

Mechanisms of sound production and perception

Neural control of vocal production and perception

In humans, both the production and perception of speech is dependent upon structures in the neocortex. On the basis of behavioural and

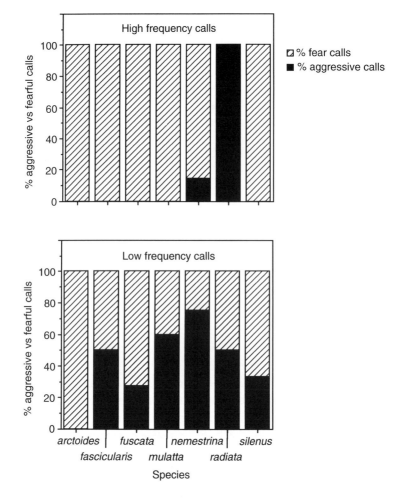

Fig. 25.3. The proportion of high and low frequency calls in the repertoire that are produced in the context of fear (striped bars) as opposed to aggression (black bars).

neuroanatomical observations, the left hemisphere appears to play a more dominant role than the right hemisphere in both language production and comprehension.

Although early work on human speech was guided by the view that there is a language 'organ', and language 'centres', such as Broca's and Wernicke's areas, more recent work has demonstrated that speech production and comprehension are influenced by a set of interconnected cortical and subcortical structures. Many of these structures are also involved in non-speech functions (e.g. reviewed by Deacon, 1991; Greenfield, 1991;

Damasio & Damasio, 1992). Nonetheless, research on the neural circuitry underlying non-human primate vocal communication has been heavily influenced by studies of human speech and, therefore, by attempts to find homologues to Broca's and Wernicke's areas. With the exception of the squirrel monkey (Jürgens, 1990), the macaques are by far the most carefully studied non-human primate group. Because the literature on neural control of macaque vocal communication is extensive (reviewed by Sutton & Jürgens, 1988; Deacon, 1991), I provide only a brief review.

Our understanding of the neuroanatomical correlates of macaque vocal production and perception is based on three techniques: electrical stimulation, axonal tracers, and lesions. Results from electrical stimulation have indicated that sub-cortical, rather than neocortical, areas are responsible for call production (Robinson, 1967a,b; Aitken, 1981; Larson, Ortega & DeRosier, 1988; Deacon, 1991). Specifically, stimulation of the periaqueductal grey area and the anterior cingulate gyrus has led to the production of a wide range of species-typical vocalisations. Emphasizing the importance of these results are lesion studies that have demonstrated significant deficits in call production, including the rate of delivery and call structure. There are two problems, however, with current neuroanatomical and neurophysiological studies on macaque vocal production. First, although electrical stimulation elicits an acoustically variable array of signals, it is unclear from the reports whether these are in fact species-specific vocalisations used during natural social interactions. Moreover, from my own work on rhesus communication (see references cited within this chapter) it is clear that only a small fraction of the vocal repertoire has been elicited by means of electrical stimulation. Second, in the absence of perceptual experiments, it is difficult to assess whether the reported structural differences emerging from the lesion studies are biologically meaningful.

Studies involving the cortical structures underlying sound production have reported conflicting results. Early work by Green & Walker (1938) showed that lesioning of the motor face area in rhesus led to a mildly weakened voice. Franzen & Myers (1973) administered bilateral lesions of the frontal and temporal lobes to rhesus and observed a significant decrease in spontaneous vocalisations. Sutton, Larson & Lindeman (1974) found that lesions of the anterior cingulate gyrus caused a significant decrement in the ability of rhesus to acquire a classically conditioned vocal task. Lastly, Aitken (1981) lesioned the homologue of Broca's and Wernicke's areas in rhesus and found no significant effect on vocalisation. As with the reported studies on sub-cortical structure, it is difficult to draw any firm conclusions regarding the cortical structures presumed to underlly vocal production in macaques. Specifically, some of the lesioned structures that were assumed to be linked to vocal communication alone, also resulted in significant changes in social behaviour. Thus, for example, changes in the rate of spontaneous vocalisation may not be the result of changes in the operation of vocal control pathways but, rather, of a change in the motivation to call or engage in social interaction mediated by vocal signals. In addition, although some of the lesion studies reported no effect on call structure, the basis for this claim is weakened by the fact that no quantitative acoustical measurements were collected.

The role of the peripheral organs in vocal production

There are striking differences between human and non-human primates with regard to the anatomy of their supralaryngeal vocal tracts (Negus, 1929, 1949; Crelin, 1987). In humans, the oropharynx connects to the larynx at a 90° angle. In non-human primates, the laryngeal cavity is short and positioned high in the neck. Consequently, there is virtually no bend in the vocal tract as one descends from the lips to the larynx. Researchers have long claimed that these anatomical differences are critical for understanding why non-human primates cannot produce human speech (reviewed by Lieberman, 1984). Although it can be argued that the acoustic properties of human speech are only tangentially relevant to the discussions of the special nature of human language (Hauser, 1992c), it is important to address the claim that the non-human primate vocal tract lacks flexibility and, consequently, lacks the ability to generate a broad array of meaningful utterance.

To date, extremely little research has been conducted to explore the role of the peripheral organs in structuring call morphology. Using anaesthetised rhesus macaques, Sapir, Campbell & Larson (1981) induced vocalisations by stimulating the periaqueductal grey area of the brain. At the same time, the cricothyroid, geniohyoid and sternothyroid were also electrically stimulated to assess whether changes in laryngeal muscle activity have an effect on the fundamental frequency of the uttered call. Results showed that each of the muscles contributed to an increase in the fundamental frequency but that the cricothyroid provided the most substantial contribution. In addition, there was suggestive evidence that changes in the relative activity of the geniohyoid and sternothyroid produced changes in the call's 'hoarseness' or voice quality. Although these results demonstrate a role for laryngeal activity in non-human primate phonation, they are limited. Specifically, only one call type was elicited through brain stimulation (i.e.

what appears spectrographically to be a coo) and because animals were anaesthetised during call production, it is difficult to interpret to what extent laryngeal activity differs for non-anaesthetised animals and for other calls in the repertoire.

Under semi-natural conditions, Hauser (Hauser, Evans & Marler, 1993; Hauser & Schön Ybarra, 1994) has explored the effects of changes in mandibular position and lip configuration on the structure of rhesus monkey vocalisations. In one study (Hauser *et al.*, 1993), audio–video recordings of free-ranging individuals producing coos were obtained. The video footage was used to quantify, frame-by-frame, changes in mandibular position. The audio recordings were used to obtain measurements of the fundamental frequency (i.e. the acoustic parameter generated by vibrations of the vocal folds) and resonance frequencies (i.e. the component of the call derived from the filtering properties of the supralaryngeal tract). Results showed that during the production of coos, the mandible gradually drops and then returns to a resting position. Regression analyses revealed that the changes in mandibular position were highly correlated with changes in the first resonance frequency, but not with changes in the fundamental frequency. This suggests that for at least one call produced by rhesus monkeys, the vocal source (i.e. the larynx) is primarily independent of the filtering effects of the supralaryngeal tract. These results contradict earlier claims (reviewed by Lieberman, 1984) that Fant's (1960) source–filter theory is only applicable to human speech production.

In a second study (Hauser & Schön Ybarra, 1994), xylocaine was used to block temporarily control over lip configuration during call production. The primary aim of this study was to assess whether, in the absence of control over lip configuration, individuals were capable of using an alternative articulatory routine to produce a call with the appropriate acoustic morphology. The study involved two captive, but unrestrained rhesus monkeys who, following injections of xylocaine into the upper and lower lips, were allowed to interact with their social group. One individual produced a relatively large sample of coos, whereas the other individual produced a majority of 'noisy screams' (Gouzoules, Gouzoules & Marler, 1984). During the production of coos the lips are protruded, thereby extending the length of the vocal tract. In contrast, the lips are retracted during the production of noisy screams, thereby shortening the length of the vocal tract. The primary effect of the xylocaine treatment, therefore, is to cause individuals to produce coos with a relatively shorter vocal tract (i.e. relative to the length of the vocal tract *typically* used and involving lip protrusion) and to produce noisy screams with a relatively longer vocal tract.

On the basis of acoustic theory (Fant, 1960), one would predict that the length of the vocal tract would have minimal effects on the fundamental frequency and duration of the call, but would have significant effects on the resonance frequencies of the call. Specifically, a relatively longer vocal tract should result in lowered resonance frequencies. Acoustic analyses of coos revealed that xylocaine had no effect on the fundamental frequency or call duration. The xylocaine treatment was, however, associated with a significant increase in the first and second resonance frequencies as would be predicted on the basis of the relationship between tract length and the location of the resonance frequencies. Therefore, the focal subject did not appear to compensate for the experimentally shortened vocal tract. One problem with these results, however, is that perceptual experiments have not yet been conducted to assess whether the acoustic changes observed are salient to the animals themselves – that is, whether the statistically significant differences in resonance frequencies are biologically meaningful. Nonetheless, the results suggest that for coos articulatory reorganisation may not be possible, or may require time and learning to achieve the intended acoustic target (for a more detailed discussion see Hauser & Schön Ybarra, 1994).

In contrast to coos, there were no detectable difference in call morphology between noisy screams produced with and without xylocaine. There are two possible explanations for these results. First, lip retraction during screams causes only minor changes in vocal tract length, so xylocaine injection does not impose a significant constraint on the articulatory trajectory. Second, lip retraction during screams does represent a significant change in vocal tract length and when xylocaine is injected, a new articulatory routine (e.g. a shift in the height of the larynx) is implemented so as to compensate for the longer tract. At present, it is not possible to distinguish between these alternatives. Future research, using cineradio-graphy, will provide the opportunity to assess, quantitatively, whether alternative articulatory routines are being called up or whether the observed changes in vocal tract length are inconsequential for configuring call morphology.

Perception of conspecific stimuli

During the late 1950s, linguists interested in the unique features of language argued that, unlike other auditory signals or stimuli in other sensory modalities, speech was processed categorically – what is known as the phenomenon of categorical perception (reviewed by Harnad, 1987). Tests of this hypothesis came from three directions, experiments comparing speech with non-speech, studies of human infants who lacked experience with speech, and psychophysical experiments on non-human animals, including chinchilla, macaques and quail. The non-human animal data have shown, quite convincingly, that subjects with non-linguistic experience process speech categorically. Although this demonstrates that non-human animals, and especially non-human primates, can extract some of the salient features of human speech, it tells us little about the perceptual processes involved in the perception of conspecific stimuli.

In the mid-1970s, ethologists working at Rockefeller University and psychophysicists working at the University of Michigan initiated a collaborative project to investigate the factors guiding the perception of species-typical vocalisations in Japanese macaques (reviewed by Moody, Stebbins & May, 1990; Stebbins & Sommers, 1992). The primary motivation for this work was the observation made by Green (1975) that, under natural conditions, Japanese macaques produce subtle variations in the spectral properties of the coo vocalisation that map onto variations in the social context. Thus, for example, coos with a smooth transition to a late frequency peak (SLH-coos) are produced by females in oestrus attempting to attract males, whereas coos with an early frequency peak followed by a smooth frequency drop (SEH-coos) are produced when young individuals are sitting alone or when the group is dispersed. Using these two coo types, Stebbins and his colleagues initiated a series of studies to assess whether differences in the peak position of the frequency maximum was a salient feature for both Japanese macaques and other closely related species, whether the acoustic continuum from one call type to the other was perceived continuously or categorically, and whether the right and left hemispheres played similar or different roles in processing the calls.

To address the various perceptual issues, Japanese macaques were brought into an acoustic isolation chamber and were trained with operant conditioning techniques to make a discrimination between SLH-coos and SEH-coos. Results showed that Japanese macaques learned the discrimination more rapidly than the other species (e.g. *M. nemestrina*, *M. radiata*). When the discrimination task was changed to absolute frequency rather than frequency peak location, results indicated that the other species performed better than the Japanese macaques. These results suggest that in Japanese macaques there is perceptual specialisation for processing species-typical vocalisations.

During these initial experiments, call presentation varied randomly from the right to the left ear. Interestingly, the Japanese macaques performed at a higher level in the discrimination task when stimuli were presented to the right ear (i.e. left hemisphere bias), than when they were presented to the left ear. Moreover, when the left temporal lobe is lesioned, the perceptual deficit is significantly greater than if the right temporal lobe is lesioned (Heffner & Heffner, 1986, 1990). Therefore, and as also demonstrated for studies of human speech perception, Japanese monkeys show evidence of hemispheric specialisation when processing species-typical vocal signals.

Because of the acoustic continuum between SEH- and SLH-coos, it is possible to ask whether Japanese macaques perceive a continuum of variation or whether the variation is divided up into discrete categories. Using synthetic coos to gain precise control over the frequency peak position, May, Moody & Stebbins (1989) showed that the SEH to SLH continuum was perceived categorically. In other words, most of the calls were classified as belonging to either the SEH or SLH category, with little perceptual differentiation within each of these categories. Although this study provides support for categorical perception within an animal's natural vocal communication system (for other examples, see review by Ehret, 1987), a more recent study (Hopp *et al.*, 1992), using the same set of stimuli has failed to confirm the results reported by May and colleagues. In discussing the conflicting data, Hopp *et al.* point out two important methodological differences: first, the procedures used by May and colleagues may have been more demanding and second, the subjects used to test for categorical perception had been used in earlier experiments involving coos (May *et al.*, 1988) – evidence from human studies suggest that prior training may actually create a categorical processing effect. Therefore, we must remain agnostic with regard to the existence of categorical perception in macaques.

Function and adaptive significance

Referential signals: recruitment screams and food-associated calls

Until the pioneering work of Seyfarth, Cheney & Marler (1980) on the alarm calls of vervet monkeys *Cercopithecus aethiops*, it was commonly believed that non-human primates, like other non-human animals, produced vocalisations that simply reflected internal state. However, research on the vervet monkey's alarm call system revealed that predators with different hunting strategies, and possibly different levels of threat and danger, elicit acoustically distinctive calls (Struhsaker, 1967). More importantly, based on playback experiments, Seyfarth and colleagues (Seyfarth *et al.*, 1980) demonstrated that the acoustic properties of these calls were sufficient to elicit behaviourally adaptive responses. Consequently, one can argue that the vervet monkeys's alarm call system is 'functionally referential' (reviewed by Marler, Evans & Hauser, 1992), providing information about both external objects and events.

Following the work on vervet monkeys, a series of papers on referential signalling in macaques appeared. The first of these was a study conducted by Gouzoules, Gouzoules & Marler (1984) on the recruitment screams of rhesus monkeys living on the island of Cayo Santiago, Puerto Rico. Natural observations and acoustic analyses revealed that in the context of agonistic interactions, individuals produced one of five acoustically distinct vocalisations, apparently in an attempt to recruit aid. The acoustic features of the call varied non-randomly and depended upon the rank, kinship and probability of receiving physical contact. For example, 'tonal screams' are given to relatives or higher ranking individuals in the absence of physical contact. In contrast, 'noisy screams' are given to unrelated high-ranking individuals in the presence of physical contact. Playback experiments of these different screams elicited different responses, thereby supporting the claim that rhesus monkey recruitment screams are functionally referential. Interestingly, more recent work by Gouzoules & Gouzoules (1989) has revealed that pig-tailed macaques also have five different recruitment screams given in similar contexts to those described for rhesus. Although both species call in similar contexts, the acoustic structure of the calls used in the same context differs between pig-tails and rhesus.

The discovery of food represents a second context where functionally referential vocalisations have been described. Natural observations of toque macaques in Sri Lanka revealed that when individuals found high quality food items, they gave an acoustically distinct vocalisation (Dittus, 1984, 1988). One consequence of producing such calls is that the signaller recruits other group members to the resource and, thereby, increases feeding competition. With a small number of exceptions, these calls were never heard outside of the feeding context. Because animals who were out of sight of the caller responded by rapidly approaching, Dittus argued that the acoustic properties of the call alone were sufficient to designate the context – the discovery of food.

Just like toque macaques, rhesus monkeys living on the island of Cayo Santiago, Puerto Rico, also produce vocalisations when they discover food (Hauser, 1992*d*; Hauser & Marler, 1993*a,b*). In contrast to toque macaques, however, rhesus produce five acoustically distinct food-associated calls (Fig. 25.4). Three of these calls ('warbles', 'harmonic arches', 'chirps') are

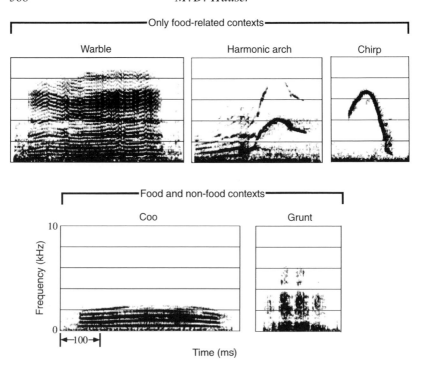

Fig. 25.4. Spectrograms of five call types produced by rhesus monkeys on the island of Cayo Santiago, Puerto Rico, during the discovery and/or consumption of food. The top three calls, the 'warble', 'harmonic arch' and 'chirp' are restricted to the context of food, whereas the 'coo' and 'grunt' are given in food and non-food contexts. The x-axis indicates time in milliseconds (ms) and the y-axis indicates frequency in kilohertz (kHz).

restricted to the context of food and are typically produced during the discovery or consumption of high quality foods. The other two vocalisations, 'coos' and 'grunts' are given in both food and non-food contexts. To quantify the relative contributions of affective and referential components of the situation in configuring call structure, Hauser conducted a series of field experiments. Specifically, individuals were presented with either chow or coconut (relatively low vs high quality food items), during the morning or afternoon (relatively hungry vs satiated individuals). Results showed that in 18 of the 40 trials, food-associated calls were produced, with females calling more often than males. Animals called at a higher rate in the morning than in the afternoon, and at a higher rate with coconut than with chow. Regarding call type, coconut was the only food item to elicit warbles, harmonic arches or chirps, whereas chow elicited coos and grunts. These data, therefore, suggest that call structure conveys information about food quality, whereas call rate provides information about the discoverer's hunger level.

The most surprising observation to emerge from the rhesus experiments was the fact that discoverers who failed to call, but were detected by other group members, received significantly more aggression than discoverers who called. Differences in aggression received by silent and vocal discoverers were observed even when dominance rank was taken into account (i.e. high-ranking males and females were targets of aggression if they were silent). One important consequence of such targeted aggression was that for females, silent discoverers consumed less food than vocal discoverers. A likely interpretation of these results (Hauser, 1992*d*; Hauser & Marler, 1993*b*) is that the production of food-associated calls is treated by other group members as a sign of ownership (see Kummer & Cords, 1991). Consequently, if an animal calls it can maintain access to the resource without being challenged. In the absence of calling, priority of access to the resource has not been established and thus competition ensues.

Conclusions and future directions

Relative to work on avian and anuran species, research on primate vocal communication is in its infancy. An increasing body of the primatological literature, however, suggests that there are a number of promising avenues for future research. The macaques provide, perhaps, the most promising taxonomic group for exploring such avenues because of the existing comparative data base and the diversity of topics addressed. In concluding this chapter, I discuss two areas where additional research is desperately needed – the role of articulation in configuring call morphology and the perceptual salience of different call features.

Generally speaking, there are two ways to assess how the peripheral organs influence sound structure: (1) using the spectrally varying features of the sound to derive the relative contributions of the larynx and supralaryngeal cavity and, (2) direct observation and manipulation of the vocal tract. As we know from studies of human speech, the first approach is problematic without at least some understanding of how the different articulators contribute to sound structure. Given some basic understanding, it is then possible to build acoustic models that predict sound morphology. At present, the only attempt to reconstruct articulation from sound structure is Owren's research on the vervet monkey's alarm call system (see e.g. Owren & Bernacki; 1988). Using a technique known as linear predictive coding (LPC), Owren was able to show that the source of sound production (i.e. the larynx) was essentially independent of the filtering effects of the supralaryngeal tract. Although the potential of LPC to separate source from filter is powerful, it does depend on the researcher's ability to input a set of critical parameters concerning vocal tract shape and length. In general, quantitative data on vocal tract shape, both during resting position and during articulation, are unavailable. Consequently, the validity of the LPC technique for non-human primate calls is questionable, although checking the output of the LPC with the raw spectrum, as Owren has done, may be a reasonable compromise.

Direct observation and manipulation of the operative vocal tract is limited to two studies, one on chimpanzees (Bauer, 1987) and one on rhesus macaques (Hauser et al., 1993; Hauser & Schön Ybarra, 1994). Although both studies suggest that changes in articulation play an important role in configuring call morphology, both studies are also limited to a small sample of subjects, a small sample of call types within the repertoire, and the results presented are correlational, rather than causal, thereby permitting only weak inferences about the degree to which changes in one articulator (e.g. the lips) can account for changes in particular acoustic features of the call. The macaques are, however, ideally suited for further exploration of these issues because several investigators have provided rich acoustic descriptions of both within and between call type variation (see Table 25.1). As a preliminary, and relatively inexpensive step in this direction, it should therefore be possible for most investigators to document the changes in lip configuration and mandibular position by simply videotaping individuals during naturally occurring vocal interactions. Given current video technology, one can obtain extremely high quality footage of calling animals and use computer graphics to quantify changes in the vocal gesture (Hauser et al., 1993). Such measurements, hooked up with existing sound analysis tools (e.g. SIGNAL, Beeman, 1995), provide a relatively straightforward approach to assessing the relationship between articulation and sound structure.

The methodological suggestion made above is limited to changes in articulation that are visible to the human eye, such as changes in lip configuration and mandibular position. Studies of human speech, however, make it clear that a good deal of the 'action' during vocal production occurs within the vocal tract and, in particular, between the velum and the teeth. Perhaps the most important articulator in human speech is the tongue because of its fundamental contribution to changes in the cross-sectional area of the vocal tract. To assess tongue position during vocal production, speech scientists have used techniques such as cineradiography and palatography. Such techniques could certainly be used with macaques, but would require chair-restrained subjects. If we hope to improve our understanding of the role of articulation in macaque communication, such invasive techniques will be necessary.

Several times during this chapter, I have alluded to the fact that we know relatively little about the perceptual salience of different acoustic features. The reason why this issue has surfaced a number of times is because it is absolutely central to any substantive understanding of a species' communication system. Documenting the sources of acoustic variation – be they physiological, ontogenetic or socio-ecological – is an important task, but in the absence of perceptual experiments to confirm the biological importance of such variation, we may be describing insignificant noise in the system rather than information that is both detectable and functionally meaningful. In this sense, the work conducted by Stebbins and his colleagues (reviewed by Stebbins & Sommers, 1992) on captive Japanese macaques, and the field research by Cheney & Seyfarth (1990) on vervet monkeys, represent important steps. However, they are only the first steps.

In my opinion, a two-pronged approach is needed. Scientists working on captive macaques need to use some of the available psychophysical techniques to dissect the acoustic morphology of a call into its basic acoustic components, and then ask the monkeys both what they can discriminate and which of the suite of features is the most salient. I want to stress looking at a suite of features because, in general, most studies have focused on the fundamental frequency of the call and yet, for many non-human primates, a good deal of information is carried by other spectral and temporal properties of the calls examined (e.g. Hauser, 1992a). Owren's (1990a,b) research with vervet monkey alarm calls represents the only set of perceptual experiments where spectral features other than the fundamental frequency have been examined in detail.

The psychophysical experiments will, of course, need to be guided by careful studies of wild macaques, where the full range of social and ecological contexts eliciting vocal signals can be observed. Studies on wild macaques can, however, make advances on their own. In this respect, research on vervet monkeys is perhaps the most exemplary to date, having tapped into a set of extremely powerful methodological tools developed within developmental psycholinguistics (reviewed by Cheney & Seyfarth, 1990). Specifically, to assess whether acoustic similarity or semantic similarity underlies classification of call exemplars into categories, Cheney & Seyfarth employed a habituation–dishabituation paradigm. Briefly, the paradigm taps into the fact that sensory systems tend to habituate to repeated exposure to the *same* stimulus but if exposed to a *different* stimulus, dishabituation ensues. The key, therefore, is to find out what the animal considers to be a *different* stimulus. In one experiment, Cheney & Seyfarth demonstrated that although 'wrrs' and 'chutters' are acoustically different, they are given in the same context (i.e. inter-group encounters). Experiments revealed that following habituation to the wrr – indicated by the low level of response observed – subjects were also unresponsive to the chutter (i.e. no evidence of dishabituation). Therefore, for the purpose of responding under natural conditions, wrrs and chutters are classified together. Although this technique has been used to test for the referential or semantic properties of vervet monkey calls, it is an 'all-purpose' experimental tool that can be readily adopted to a wide variety of perceptual problems. Because scientists working with captive subjects can also use the habituation–dishabituation paradigm, it is possible that in the future, data collected in the field will be better suited for comparison with data collected in the laboratory.

Returning to the theme set out in the introduction to this chapter, 'macaque-ologists' are in an ideal position to tackle some of the claims made by linguists regarding the uniqueness of human language. By combining detailed studies of animals living in their natural habitats with current techniques in neurobiology, psychophysics and bioacoustics, it will be possible to assess, more rigorously than before, how neuroanatomical constraints limit what macaques express vocally and the extent to which their expressions are functionally similar to human expressions.

Acknowledgements

For comments on an earlier draft of the manuscript, I thank Peter Marler. Support during the writing of this chapter was provided by an NSF Young Investigator Award.

References

Aitken, P. G. (1981). Cortical control of conditioned and spontaneous vocal behavior in rhesus monkeys. *Brain and Language*, **13**, 636–42.

Bauer, H. (1987). Frequency code: orofacial correlates of fundamental frequency. *Phonetica*, **44**, 173–91.

Bauers, K. A. (1989). The role of vocal communication in the intra-group social dynamics of stumptailed macaques (*Macaca arctoides*). PhD thesis, University of Wisconsin.

Bauers, K. A. & de Waal, F. B. M. (1991). 'Coo' vocalizations in stumptailed macaques: a controlled functional analysis. *Behaviour*, **119**, 143–60.

Beeman, K. (1995). *SIGNAL user's guide*. Belmont, MA: Engineering Design.

Brown, C. & Waser, P. (1988). Environmental influences on the structure of primate vocalizations. In *Primate vocal communication*, ed. D. Todt, P. Goedeking & D. Symmes, pp. 51–68. Berlin: Springer-Verlag.

Cheney, D. L. & Seyfarth, R. M. (1990). *How monkeys see the world*. Chicago: Chicago University Press.

Crelin, E. (1987). *The human vocal tract*. New York: Vantage Press.

Damasio, A. R. & Damasio, H. (1992). Brain and language. *Scientific American*, **267**, 88–109.

Darwin, C. (1872). *The expression of the emotions in man and animals*. London: John Murray.

Deacon, T. W. (1991). The neural circuitry underlying primate calls and human language. In *Language origins: a multidisciplinary approach*, ed. J. Wind, pp. 131–72. Dordrecht: Kluwer Academic Publishers.

Dittus, W. P. G. (1984). Toque macaque food calls: semantic communication concerning food distribution in the environment. *Animal Behaviour*, **32**, 470–7.

Dittus, W. G. (1988). An analysis of toque macaque cohesion calls from an ecological perspective. In *Primate vocal communication*, ed. D. Todt, P. Goedeking & D. Symmes, pp. 31–50. Berlin: Springer-Verlag.

Ehret, G. (1987). Categorical perception of sound signals: facts and hypotheses from animal studies. In *Categorical perception*, ed. S. Harnad, pp. 301–31. Cambridge: Cambridge University Press.

Fant, G. (1960). *Acoustic theory of speech production*. The Hague: Mouton.

Franzen, E. A. & Myers, R. E. (1973). Neural control of social behavior: prefrontal and anterior temporal cortex. *Neuropsychologia*, **11**, 141–57.

Gouzoules, H. & Gouzoules, S. (1989). Design features and developmental modification in pigtail macaque (*Macaca nemestrina*) agonistic screams. *Animal Behaviour*, **37**, 383–401.

Gouzoules, H. & Gouzoules, S. (1990a). Matrilineal signatures in the recruitment screams of pigtail macaques (*Macaca nemestrina*. *Behaviour*, **115**, 327–47.

Gouzoules, H. & Gouzoules, S. (1990b). Body size effects on the acoustic structure of pigtail macaque (*Macaca nemestrina*) screams. *Ethology*, **85**, 324–34.

Gouzoules, S., Gouzoules, H. & Marler, P. (1984). Rhesus monkey (*Macaca mulatta*) screams: representational signalling in the recruitment of agonistic aid. *Animal Behaviour*, **32**, 182–93.

Green, H. D. & Walker, A. E. (1938). The effects of ablation of the cortical face area in monkeys. *Journal of Neurophysiology*, **1**, 262–80.

Green, S. (1975). Variation of vocal pattern with social situation in the Japanese monkey (*Macaca fuscata*): a field study. In *Primate behavior*, vol. 4, ed. L. A. Rosenblum, pp. 1–102. New York: Academic Press.

Green, S. (1981). Sex differences and age gradations in vocalizations of Japanese and lion-tailed macaques. *American Zoologist*, **21**, 165–83.

Greenfield, P. M. (1991). Language, tools, and brain: the ontogeny and phylogeny of hierarchically organized sequential behavior. *Behavioral and Brain Sciences*, **14**, 531–95.

Grimm, R. J. (1967). Catalogue of sounds of the pigtailed macaque (*Macaca nemestrina*). *Journal of Zoology, London*, **152**, 361–73.

Hammerschmidt, K., Ansorge, V. & Fischer, J. (1992). The vocal repertoire of barbary macaques. *Abstracts of the XIVth Congress of the International Primatological Society*, Strasbourg. p. 224.

Harnad, S. (ed.) (1987). *Categorical perception: the groundwork of cognition*. Cambridge: Cambridge University Press.

Hauser, M. D. (1991). Sources of acoustic variation in rhesus macaque vocalizations. *Ethology*, **89**, 29–46.

Hauser, M. D. (1992a). Articulatory and social factors influence the acoustic structure of rhesus monkey vocalizations: a learned mode of production? *Journal of Acoustical Society of America*, **91**, 2175–9.

Hauser, M. D. (1992b). A mechanism guiding conversational turn-taking in vervet monkeys and rhesus macaques. In *Topics in primatology*, vol. 1, *Human origins*, ed. T. Nishida, W. C. McGrew, P. Marler, M. Pickford & F. de Waal, pp. 235–48. Tokyo: Tokyo University Press.

Hauser, M. D. (1992c). Review of 'Uniquely human: the evolution of speech, thought, and selfless behavior' by P. Lieberman. *Applied Psycholinguistics*, **13**, 237–43.

Hauser, M. D. (1992d). Costs of deception: cheaters are punished in rhesus monkeys. *Proceedings of the National Academy of Sciences, USA*, **89**, 12 137–9.

Hauser, M. D. (1993). The evolution of nonhuman primate vocalizations: effects of phylogeny, body weight and motivational state. *American Naturalist*, **142**, 528–42.

Hauser, M. D. (1996). Nonhuman primate vocal communication. In *Handbook of acoustics*, ed. M. Cochran. New York: John Wiley & Sons, in press.

Hauser, M. D., Evans, C. S. & Marler, P. (1993). The role of articulation in the production of rhesus monkey (*Macaca mulatta*) vocalizations. *Animal Behaviour*, **45**, 423–33.

Hauser, M. D. & Fowler, C. (1991). Declination in fundamental frequency is not unique to human speech: evidence from nonhuman primates. *Journal of the Acoustical Society of America*, **91**, 363–9.

Hauser, M. D. & Marler, P. (1992). How do and should studies of animal communication affect interpretations of child phonological development? In *Phonological development*, ed. C. Ferguson, L. Menn & C. Stoel-Gammon, pp. 663–80. Maryland: York Press.

Hauser, M. D. & Marler, P. (1993a). Food-associated calls in rhesus macaques (*Macaca mulatta*). I. Socioecological factors influencing call production. *Behavioral Ecology*, **4**, 194–205.

Hauser, M. D. & Marler, P. (1993b). Food-associated calls in rhesus macaques (*Macaca mulatta*). II. Costs and benefits of call production and suppression. *Behavioral Ecology*, **4**, 206–12.

Hauser, M. D. & Schön Ybarra, M. (1994). The role of lip configuration in monkey vocalizations: experiments using xylocaine as a nerve block. *Brain and Language*, **46**, 423–33.

Heffner, H. E. & Heffner, R. S. (1986). Effect of unilateral and bilateral auditory cortex lesions on discrimination of vocalizations by Japanese macaques. *Journal of Neurophysiology*, **56**, 683–701.

Heffner, H. E. & Heffner, R. S. (1990). Role of primate auditory cortex in hearing. In *Comparative perception*, vol. 2, *Complex signals*, ed. W. C. Stebbins & M. A. Berkley, pp. 143–68. New York: John Wiley & Sons.

Hohmann, G. (1989). Vocal communication of wild bonnet macaques (*Macaca radiata*). *Primates*, **30**, 325–45.

Hohmann, G. & Herzog, M. O. (1985). Vocal communication in lion-tailed macaques. *Folia Primatologica*, **45**, 148–78.

Hopp, S. L., Sinnott, J. M., Owren, M. J. & Petersen, M. R. (1992). Differential sensitivity of Japanese macaques (*Macaca fuscata*) and humans (*Homo sapiens*) to peak position along a synthetic coo call continuum. *Journal of*

Comparative Psychology, **106**, 128–36.

Inoue, M. (1988). Age gradations in vocalization and body weight in Japanese monkeys (*Macaca fuscata*). *Folia Primatologica*, **51**, 76–86.

Jürgens, U. (1979). Vocalizations as an emotional indicator. A neuroethological study in the squirrel monkey. *Behaviour*, **69**, 88–117.

Jürgens, U. (1990). Vocal communication in primates. In *Neurobiology of comparative cognition*, ed. R. P. Kesner & D. S. Olton, pp. 51–76. Hillsdale, NJ: Lawrence Erlbaum.

Kummer, H. & Cords, M. (1991). Cues of ownership in long-tailed macaques, *Macaca fascicularis*. *Animal Behaviour*, **42**, 529–49.

Larson, C. R., Ortega, J. D. & DeRosier, E. A. (1988). Studies on the relation of the midbrain periaqueductal gray, the larynx and vocalization in the awake monkey. In *The physiological control of mammalian vocalizations*, ed. J. D. Newman, pp. 43–65. New Jersey: Plenum Press.

Lieberman, P. (1984). *The biology and evolution of language*. Cambridge, MA: Harvard University Press.

Lieberman, P. (1985). The physiology of cry and speech in relation to linguistic behavior. In *Infant crying*, ed. B. M. Lester & C. F. Z. Boukydis, pp. 29–57. New York: Plenum Press.

Lieberman, P., Klatt, D. H. & Wilson, W. H. (1969). Vocal tract limitations on the vowel repertoires of rhesus monkeys and other nonhuman primates. *Science*, **164**, 1185–7.

Locke, J. (1993). *The path to spoken language*. Cambridge, MA: Harvard University Press.

Marler, P., Evans, C. S. & Hauser, M. D. (1992). Animal vocal signals: reference, motivation, or both? In *Nonverbal vocal communication: comparative and developmental approaches*, ed. H. Papoušek, U. Jürgens & M. Papoušek, pp. 66–86. New York: Cambridge University Press.

Masataka, N. (1987). The response of red-chested moustached tamarins to long calls from their natal and alien populations. *Animal Behaviour*, **36**, 55–61.

Masataka, N. (1992). Flexibility in vocal behavior of coo calls in Japanese macaques. *Abstracts of the XIVth Congress of the International Primatological Society*, Strasbourg, p. 225.

Masataka, N. & Fujita, K. (1989). Vocal learning of Japanese and rhesus monkeys. *Behaviour*, **109**, 191–9.

May, B., Moody, D. B. & Stebbins, W. C. (1988). The significant features of Japanese monkey coo sounds: a psychophysical study. *Animal Behaviour*, **36**, 1432–44.

May, B., Moody, D. B. & Stebbins, W. C. (1989). Categorical perception of conspecific communication sounds by Japanese macaques, *Macaca fuscata*. *Journal of the Acoustical Society of America*, **85**, 837–47.

Moody, D. B., Stebbins, W. C. & May, B. J. (1990). Auditory perception of communication signals by Japanese monkeys. In *Comparative perception: complex perception*, vol. 2, *Complex signals*, ed. W. C. Stebbins & M. A. Berkley, pp. 311–44. New York: John Wiley & Sons.

Morton, E. S. (1975). Ecological sources of selection on avian sounds. *American Naturalist*, **109**, 17–34.

Morton, E. S. (1977). On the occurrence and significance of motivation–structural rules in some bird and mammal sounds. *American Naturalist*, **111**, 855–69.

Morton, E. S. (1982). Grading, discreteness, redundancy and motivation–structural rules. In *Acoustic communication in birds*, vol. 1, ed. D. Kroodsma & E.

Miller, pp. 183–212. New York: Academic Press.

Negus, V. (1929). *The mechanism of the larynx.* St Louis, MO: C. V. Mosby Company.

Negus, V. E. (1949). *The comparative anatomy and physiology of the larynx.* New York: Hafner Publishing Co.

Newman, J. D. & Symmes, D. (1974). Vocal pathology in socially deprived monkeys. *Developmental Psychobiology,* **7**, 351–8.

Oller, D. K. & Eilers, R. E. (1992). Development of vocal signaling in human infants: toward a methodology for cross-species comparisons. In *Nonverbal communication: comparative and developmental approaches,* ed. H. Papousek, U. Jürgens & M. Papousek, pp. 174–91. New York: Cambridge University Press.

Owren, M. J. (1990*a*). Acoustic classification of alarm calls by vervet monkeys (*Cercopithecus aethiops*) and humans. I. Natural calls. *Journal of Comparative Psychology,* **104**, 20–8.

Owren, M. J. (1990*b*). Acoustic classification of alarm calls by vervet monkeys (*Cercopithecus aethiops*) and humans. II. Synthetic calls. *Journal of Comparative Psychology,* **104**, 29–40.

Owren, M. J. & Bernacki, R. (1988). The acoustic features of vervet monkey (*Cercopithecus aethiops*) alarm calls. *Journal of the Acoustical Society of America,* **83**, 1927–35.

Owren, M. J., Dieter, J. A., Seyfarth, R. M. & Cheney, D. L. (1992). Evidence of limited modification in the vocalizations of cross-fostered rhesus (*Macaca mulatta*) and Japanese (*M. fuscata*) macaques. In *Topics in primatology,* vol. 1, *Human origins,* ed. T. Nishida, W. C. McGrew, P. Marler, M. Pickford & F. de Waal, pp. 235–48. Tokyo: Tokyo University Press.

Palombit, R. (1992*a*). A preliminary study of vocal communication in wild long-tailed macaques (*Macaca fascicularis*). I. Vocal repertoire and call emission. *International Journal of Primatology,* **13**, 143–82.

Palombit, R. (1992*b*). A preliminary study of vocal communication in wild long-tailed macaques (*Macaca fascicularis*). II. Potential of calls to regulate intragroup spacing. *International Journal of Primatology,* **13**, 183–207.

Robinson, B. W. (1967*a*). Neurological aspects of evoked vocalizations. In *Social communication among primates,* ed. S. A. Altmann, pp. 135–47. Chicago: Chicago University Press.

Robinson, B. W. (1967*b*). Vocalization evoked from the forebrain in *Macaca mulatta. Physiology of Behavior,* **2**, 345–54.

Rowell, T. E. & Hinde, R. A. (1962). Vocal communication by the rhesus monkey (*Macaca mulatta*). *Symposia of the Zoological Society of London,* **8**, 91–6.

Sapir, S., Campbell, C. & Larson, C. (1981). Effect of geniohyoid, cricothyroid and sternothyroid muscle stimulation on voice fundamental frequency of electrically elicited phonation in rhesus macaque. *Laryngoscope,* **91**, 457–68.

Seyfarth, R. M., Cheney, D. L. & Marler, P. (1980). Monkey responses to three different alarm calls: evidence of predator classification and semantic communication. *Science,* **210**, 801–3.

Snowdon, C. T. (1979). Response of nonhuman animals to speech and to species-specific sounds. *Brain, Behavior and Evolution,* **16**, 409–29.

Snowdon, C. T. (1987). A naturalistic view of categorical perception. In *Categorical perception,* ed. S. Harnad, pp. 332–54. Cambridge: Cambridge University Press.

Snowdon, C. T. (1990). Language capacities of nonhuman animals. *Yearbook of*

Physical Anthropology, **33**, 215–43.

Stebbins, W. C. & Sommers, M. S. (1992). Evolution, perception, and the comparative method. In *The evolutionary biology of hearing*, ed. D. B. Webster, R. F. Fay & A. N. Popper, pp. 211–28. New York: Springer-Verlag.

Struhsaker, T. T. (1967). Auditory communication among vervet monkeys (*Cercopithecus aethiops*). In *Social communication among primates*, ed. S. A. Altmann, pp. 281–324. Chicago: Chicago University Press.

Sutton, D. & Jürgens, U. (1988). Neural control of vocalization. In *Comparative primate biology*, vol. 4, *Neurosciences*, ed. J. Erwin, pp. 625–47. New Jersey: Alan R. Liss.

Sutton, D., Larson, C. R. & Lindeman, R. C. (1974). Neocortical and limbic lesion effects on primate phonation. *Brain Research*, **71**, 61–75.

Sutton, D., Larson, C., Taylor, E. M. & Lindeman, R. C. (1974). Vocalization in rhesus monkeys: conditionability. *Brain Research*, **52**, 225–31.

Symmes, D. & Biben, M. (1992). Vocal development in nonhuman primates. In *Nonverbal vocal communication: comparative and developmental approaches*, ed. H. Papoušek, U. Jürgens & M. Papoušek, pp. 123–42. New York: Cambridge University Press.

Index

M. = *Macaca* : m. = macaque.

Underlined figures refer to tables: *underlined italics* to illustrations.